Dear Ed,

A little bed-time reading for you: see pages 115–133.

Jon

AF777265

Progress in Probability
Volume 47

Series Editors
Thomas Liggett
Charles Newman
Loren Pitt

High Dimensional Probability II

Evarist Giné
David M. Mason
Jon A. Wellner
Editors

Birkhäuser
Boston • Basel • Berlin

Evarist Giné
Departments of Mathematics & Statistics
University of Connecticut
Storrs, CT 06269
USA

David M. Mason
Department of Food and Resource Econ.
University of Delaware
Newark, DE 19717
USA

Jon A. Wellner
Department of Statistics
University of Washington
Seattle, WA 98195
USA

Library of Congress Cataloging-in-Publication Data

High dimensional probability II/Evarist Giné, David M. Mason, Jon A. Wellner, editors.
p. cm. – (progress in probability ; 47)
Includes bibliographical references.
ISBN 0-8176-4160-2 (alk. paper) – ISBN 3-7643-4160-2 (alk. paper)
1. Probabilities–Congresses. 2. Linear topological spaces–Congresses. 3. Gaussian processes–Congresses. I. Giné, Evarist, 1944- II. Mason, David M., 1946- III. Wellner, Jon A., 1945- IV. International Conference on High Dimensional Probability (2nd : 1999 : University of Washington) V. Series

QA273.A1 H54 2000
519.2–dc21 00-058525

AMS Subject Classifications: 60B11, 60B12, 60E05, 60E07, 60E15, 60F05, 60F10, 60F15, 60F17, 60G15, 60G17, 60G40, 60G44, 60G50, 60H25, 60J25, 60J30, 60J60, 60J65, 62E17, 62E20, 62G07, 63G09, 62G20

Printed on acid-free paper.
© 2000 Birkhäuser Boston

Birkhäuser ®

All rights reserved. This work may not be translated or copied in whole or in part without the written permission of the publisher (Birkhäuser Boston, c/o Springer-Verlag New York, Inc., 175 Fifth Avenue, New York, NY 10010, USA), except for brief excerpts in connection with reviews or scholarly analysis. Use in connection with any form of information storage and retrieval, electronic adaptation, computer software, or by similar or dissimilar methodology now known or hereafter developed is forbidden.
The use of general descriptive names, trade names, trademarks, etc., in this publication, even if the former are not especially identified, is not to be taken as a sign that such names, as understood by the Trade Marks and Merchandise Marks Act, may accordingly be used freely by anyone.

ISBN 0-8176-4160-2 SPIN 10749208
ISBN 3-7643-4160-2

Reformatted from editors' files in LaTeX by TeXniques, Inc., Cambridge, MA.
Printed and bound by Hamilton Printing Company, Rensselaer, NY.
Printed in the United States of America.

9 8 7 6 5 4 3 2 1

Contents

Preface

High dimensional probability, in the sense that encompasses the topics represented in this volume, began about thirty years ago with research in two related areas: limit theorems for sums of independent Banach space valued random vectors and general Gaussian processes. An important feature in these past research studies has been the fact that they highlighted the essential probabilistic nature of the problems considered. In part, this was because, by working on a general Banach space, one had to discard the extra, and often extraneous, structure imposed by random variables taking values in a Euclidean space, or by processes being indexed by sets in R or R^d. Doing this led to striking advances, particularly in Gaussian process theory. It also led to the creation or introduction of powerful new tools, such as randomization, decoupling, moment and exponential inequalities, chaining, isoperimetry and concentration of measure, which apply to areas well beyond those for which they were created. The general theory of empirical processes, with its vast applications in statistics, the study of local times of Markov processes, certain problems in harmonic analysis, and the general theory of stochastic processes are just several of the broad areas in which Gaussian process techniques and techniques from probability in Banach spaces have made a substantial impact.

Parallel to this work on probability in Banach spaces, classical probability and empirical process theory were enriched by the development of powerful results in strong approximations. They now constitute additional and irreplaceable tools in the study of high dimensional probability problems.

The papers in this volume display the breadth of the subjects that the notion of high dimensional probability has grown to include. Topics covered include the development of new tools, in particular, probabilistic inequalities and results on strong approximations as well as further results on classical and more modern topics, such as Gaussian processes, empirical processes, and sums of independent and weakly dependent random vectors. Also included are applications to nonparametric function estimation, goodness of fit testing and the bootstrap.

This volume is an outgrowth of the *Second International Conference on High Dimensional Probability*, held at the University of Washington, August 1–6, 1999. Many of the papers were presented there, and all of them are based on topics covered at this conference. The first two conferences on high dimensional probability and their proceedings continue a long tradition of conferences on this subject that began at Oberwolfach, July 20–26, 1975, under the title of *Probability in Banach Spaces*. Nine volumes of the proceedings of conferences on *Probability in Banach Spaces* have appeared since then. Actually, this number is ten, if one counts a special 1974 volume of the *Annales de l'Institut Fourier* devoted to a conference at the Université de Strasbourg on "processus gaussiens et distributions aléatoires." In spite of the name changes all these conferences are united by several recurring themes.

The participants of this conference are grateful for the support of the National Science Foundation and the National Security Agency.

Evarist Giné
David M. Mason
Jon A. Wellner

May, 2000

High Dimensional Probability II

PART 1

Inequalities

Moment Bounds for Self-Normalized Martingales*

Victor H. de la Peña, Michael J. Klass and Tze Leung Lai

ABSTRACT Let τ be any stopping time of Brownian motion w_t. Recently, Graversen and Peskir (1998) derived an upper bound for the mean of the supremum of the self-normalized process $\left(\frac{|w_t|}{\sqrt{1+t}}\right)$ over $0 \le t \le \tau$. At the cost of adding a universal constant, we extend their result by applying p-th powers and exponential functions to it, as well as by considering more general processes, including martingales.

1 Introduction

Graversen and Peskir (1998) (see also Yang (1999) for related bounds) recently proved the following interesting result for any stopping time τ of Brownian motion w_t: There exists a universal constant C such that

$$E\left(\sup_{0\le t\le\tau}\left(\frac{|w_t|}{\sqrt{1+t}}\right)\right) \le CE\sqrt{\log(1+\log(1+\tau))}\ . \tag{1.1}$$

They also showed that the above inequality is sharp, as there is a corresponding lower bound. It is natural to ask if w_t can be replaced by a discrete-time martingale, or other process of interest.

Note that for Brownian motion, $\exp\{\lambda w_{t\wedge\tau} - \frac{1}{2}\lambda^2(t\wedge\tau)\}$, $t \ge 0$, is a martingale for all $\lambda \in \mathbf{R}$ and any stopping time τ. In place of $w_{t\wedge\tau}$ and $t\wedge\tau$, we consider more general processes A_t and B_t^2 such that $A_0 = 0$, $B_0 = 0$ and

$$\exp\{\lambda A_t - (\lambda^2 B_t^2/2)\}\ ,\quad t\ge 0,\quad \text{is a supermartingale for all } \lambda\in\mathbf{R}. \tag{1.2}$$

Our methods and results are applicable to both discrete-time and continuous-time processes. In the continuous-time case we shall also assume that the supermartingale in (1.2) is right-continuous. Clearly (1.2) implies that for any $t \ge 0$,

$$E\exp\{\lambda A_t - \lambda^2\left(B_t^2/2\right)\} \le 1 \quad \text{for all } \lambda\in\mathbf{R}\ . \tag{1.3}$$

*Partially supported by National Science Foundation grants DMS–96–26175, 99 –72237, 96–26236, 97–04324, 99–72417.

We first consider random variables A $(= A_t)$ and B $(= B_t)$ satisfying (1.3) with $B > 0$. Note that if we were allowed to maximize over λ inside the expectation, then the maximizing value $\lambda = \left(\frac{A}{B^2}\right)$ would give us $E\exp(A^2/2B^2) \leq 1$, which in turn would imply that $P(|\frac{A}{B}| \geq x) \leq e^{-x^2/2}$, yielding bounds on the moments and moment generating function of $\left|\frac{A}{B}\right|$. Although we cannot interchange the order of $\max_\lambda$ and E, we can integrate over λ with respect to a probability measure F and interchange the order of integration with respect to P and F. In this paper we restrict attention to probability measures F which concentrate on $(0, \infty)$ and random variables $B \geq 0$. Thereby, we obtain information on the tail behavior of A^+/B based on (1.2) holding for $\lambda > 0$. This approach was used in de la Peña, Klass and Lai (1999) to prove exponential inequalities and moment bounds for random variables satisfying (1.3) and to derive limit theorems for self-normalized processes satisfying (1.2). Here we use this approach to prove several extensions of (1.1).

Let $L_0(x) = 2(\log xe^e)(\log\log xe^e)^2 I(x \geq 1)$ (where $\log(x)$ stands for the logarithm to the base e) and observe that for $x \geq 1$, $\log(2\log xe^e) \leq \log(L_0(x)) \leq 2\log(2\log xe^e)$ and $\lim_{y\to\infty} \log L_0(y)/\log\log(y) = 1$. Our first extension applies to random variables satisfying (1.3) with B $(= B_t) > 0$. We show that

$$E\left|\frac{A^+}{B}\right|^p \leq C_{1,p} + C_{2,p}\, E\left(\log^+ L_0\left(B \vee \frac{1}{B}\right)\right)^{p/2} \tag{1.4}$$

for all $p > 0$ and some universal constants $C_{1,p}$ and $C_{2,p}$. This is done in the same spirit as (1.1) with the more general L_p-norm replacing the L_1-norm, at the expense of introducing an additional constant term. In fact, we consider the more general problem of bounding $Eh\left(\frac{A}{B}\right) I(A > 0)$ for non-decreasing $h : [0,\infty) \to [0,\infty)$. We also show that

$$Eh\left(\frac{A^+}{B\sqrt{1 \vee \log^+ L_0\left(B \vee \frac{1}{B}\right)}}\right) \leq C_h \tag{1.5}$$

for some universal constant C_h depending only on h. We next consider processes A_t and B_t satisfying (1.2) and such that B_t is positive and monotone in t. In this case we derive analogous bounds for

$$Eh\left(\sup_{s\leq\tau} \frac{A_s^+}{\sqrt{B_s^2+1}}\right), \quad Eh\left(\sup_{s\leq\tau} \frac{A_s^+}{B_s\sqrt{1 \vee \log^+ L_0(\sqrt{B_s^2+1})}}\right), \tag{1.6}$$

for any stopping time τ, thereby extending (1.1) subject to the addition of a universal constant. The main results are presented in Section 2, and Section 3 provides some examples.

2 Main results

Let $L(\cdot)$ be a non-negative, non-decreasing continuous function such that

$$L(cy) \leq 3cL(y) \quad \text{for all} \quad c \geq 1 \quad \text{and} \quad y \geq 1 , \tag{2.1}$$

$$L(y^2) \leq 3L(y) \quad \text{for all} \quad y \geq 1 , \tag{2.2}$$

$$\lim_{y \to \infty} \frac{\log(L(y))}{\log(\log y)} = 1 , \tag{2.3}$$

$$\int_1^\infty \frac{dx}{xL(x)} = \tfrac{1}{2} . \tag{2.4}$$

An example is the function $L_0(x) = 2(\log xe^e)(\log\log xe^e)^2 1(x \geq 1)$, introduced earlier. Let

$$g(x) = \exp\{x^2/2\}\, 1(x \geq 1) , \quad g_r(x) = \exp(rx^2/2)1(x \geq 1) . \tag{2.5}$$

By a change of variables,

$$\int_0^1 \left(\lambda L\left(\frac{1}{\lambda}\right)\right)^{-1} d\lambda = \int_1^\infty (\lambda L(\lambda))^{-1} d\lambda ,$$

so condition (2.3) ensures that

$$f(\lambda) = \frac{1}{\lambda L\left(\max\left(\lambda, \frac{1}{\lambda}\right)\right)} , \qquad \lambda > 0 , \tag{2.6}$$

is a probability density on $(0, \infty)$. Therefore (1.3) implies that

$$1 \geq E \int_0^\infty \exp\left(Ax - \left(B^2x^2/2\right)\right) I(A > 0) f(x) dx . \tag{2.7}$$

The following four lemmas will be used to prove the main results. Lemma 1 is taken from Shao (1998) for discrete-time nonnegative supermartingales. A standard discretization argument and Fatou's lemma give the corresponding result for right-continuous nonnegative supermartingales. Lemmas 2 and 3 are taken from de la Peña, Klass and Lai (1999). The proof of Lemma 4 is straightforward and is omitted.

Lemma 1. *Let T_k, $k \geq 1$, be a nonnegative supermartingale. Then, for all $0 < \alpha < 1$,*

$$E\left(\sup_{k \geq 1} T_k\right)^\alpha \leq \frac{(ET_1)^\alpha}{1 - \alpha} .$$

Lemma 2. *Take any* $A \geq B > 0$ *and any* $-\frac{A}{B} < x \leq 0$. *Then*

$$\left(x+\frac{A}{B}\right)L\left(\frac{x+\frac{A}{B}}{B}\vee\frac{B}{x+\frac{A}{B}}\right)\leq 3\frac{A}{B}\left\{L\left(\frac{A}{B}\right)\vee L\left(B\vee\frac{1}{B}\right)\right\}. \tag{2.8}$$

Lemma 3. *Let* $A, B > 0$ *be two random variables such that* $E\exp\{\lambda A-\lambda^2B^2/2\}I(A>0)\leq 1$ *for all* $\lambda > 0$. *Then*

$$E\frac{g\left(\frac{A}{B}\right)}{\frac{A}{B}\left(L\left(\frac{A}{B}\right)\vee L\left(B\vee\frac{1}{B}\right)\right)}\leq\frac{3}{\int_0^1\exp\{-x^2/2\}dx}. \tag{2.9}$$

Lemma 4. *Take any* $0 < \delta < 1$. *Let* $g_r(x)$ *be defined as in (2.5). If*

$$g_r^{1-\delta}(x)\leq L\left(B\vee\frac{1}{B}\right), \tag{2.10}$$

then

$$x\leq\sqrt{\frac{2}{r(1-\delta)}\log^+ L\left(B\vee\frac{1}{B}\right)}. \tag{2.11}$$

Theorem 1. *Let* $A, B > 0$ *be two random variables such that* $E\exp\{\lambda A-(\lambda^2B^2/2)\}I(A>0)\leq 1$ *for all* $\lambda>0$. *Fix any* $0<\delta<1$. *Take any non-decreasing function* $h:[0,\infty)\to[0,\infty)$ *such that* $\limsup_{y\to\infty} yh(y)/g^\delta(y)<\infty$, *and any* $L(\cdot)$ *satisfying* (2.1)-(2.4). *Then*

$$Eh\left(\frac{A^+}{B}\right)\leq 4\sup_{y\geq 1}\frac{y(L(y)\vee g^{1-\delta}(y))h(y)}{g(y)} + Eh\left(1\vee\sqrt{\frac{2}{1-\delta}\log^+ L\left(B\vee\frac{1}{B}\right)}\right), \tag{2.12}$$

and

$$Eh\left(\frac{A^+/B}{\sqrt{1\vee\log^+ L(B\vee\frac{1}{B})}}\right)\leq h\left(\sqrt{\frac{2}{1-\delta}}\right) + \sup_{y\geq 1}\frac{y(L(y)\vee g^{1-\delta}(y))h(y)}{g(y)}. \tag{2.13}$$

Proof. Let

$$Q_\delta=\left\{g^{1-\delta}\left(\frac{A}{B}\right)\leq L\left(B\vee\frac{1}{B}\right)\right\}. \tag{2.14}$$

Then, by (2.11), with $r = 1$,

$$
\begin{aligned}
Eh\left(\frac{A^+}{B}\right) &\leq Eh(1)I\left(\frac{A}{B} \leq 1\right) \\
&+ Eh\left(\sqrt{\frac{1}{1-\delta}\log^+ L\left(B \vee \frac{1}{B}\right)}\right) I\left(\frac{A}{B} > 1, Q_\delta\right) \\
&+ E\frac{g\left(\frac{A}{B}\right)}{4\frac{A}{B}\left(L\left(\frac{A}{B}\right) \vee L\left(B \vee \frac{1}{B}\right)\right)} \\
&\times \frac{4(\frac{A}{B})\left(L\left(\frac{A}{B}\right) \vee g^{1-\delta}\left(\frac{A}{B}\right)\right) h\left(\frac{A}{B}\right)}{g\left(\frac{A}{B}\right)} I\left(\frac{A}{B} > 1, Q_\delta^c\right) \\
&\equiv I + II + III .
\end{aligned}
$$

Dropping the event Q_δ from II, it is easy to see that

$$
I + II \leq Eh\left(1 \vee \sqrt{\frac{2}{1-\delta}\log^+ L\left(B \vee \frac{1}{B}\right)}\right) .
$$

By Lemma 3,

$$
III \leq \sup_{y \geq 1} \frac{4y(L(y) \vee g^{1-\delta}(y))h(y)}{g(y)} ,
$$

which establishes (2.12). To establish (2.13) we employ the analogous three-term decomposition, using a cruder upper-bound for the third term:

$$
\begin{aligned}
&Eh\left(\frac{A^+/B}{\sqrt{1 \vee \log^+ L(B \vee \frac{1}{B})}}\right) \\
&\quad\leq Eh\left(\frac{1}{\sqrt{1 \vee \log^+ L(B \vee \frac{1}{B})}}\right) I(A^+/B \leq 1) \\
&\quad+ Eh\left(\frac{\sqrt{\frac{2}{1-\delta}\log^+ L(B \vee \frac{1}{B})}}{\sqrt{1 \vee \log^+ L(B \vee \frac{1}{B})}}\right) I\left(\frac{A^+}{B} > 1, Q_\delta\right) \\
&\quad+ E\frac{g(A/B)}{4\frac{A}{B}\left(L\left(\frac{A}{B}\right) \vee L\left(B \vee \frac{1}{B}\right)\right)} \\
&\quad\times \frac{4\frac{A}{B}\left(L\left(\frac{A}{B}\right) \vee g^{1-\delta}\left(\frac{A}{B}\right)\right) h\left(\frac{A}{B}\right)}{g\left(\frac{A}{B}\right)} I\left(\frac{A}{B} \geq 1, Q_\delta^c\right) \\
&\quad\leq h\left(\sqrt{\frac{2}{1-\delta}}\right) + 4\sup_{y \geq 1} \frac{y(L(y) \vee g^{1-\delta}(y))}{g(y)} h(y) .
\end{aligned}
$$

□

Theorem 2. *Let* $T = \{0, 1, 2, ...\}$ *or* $T = [0, \infty)$. *Assume that for each fixed* $\lambda > 0$, $y_t(\lambda) = \exp\{\lambda A_t - \lambda^2 B_t^2/2\}$ *is a supermartingale for* $t \in T$ *such that* B_t *is positive and non-decreasing in* $t > 0$ *and* $A_0 = B_0 = 0$. *In the case* $T = [0, \infty)$, A_t *and* B_t *are assumed right-continuous. Fix any* $0 < \delta$, $p < 1$ *and any non-decreasing function* $h(\cdot) : [0, \infty) \to [0, \infty)$. *Let* $g_p(y) = (\exp(py^2/2))I(y \geq 1)$. *Assume that* $\limsup_{y\to\infty} (yh(y)/g_p^\delta(y)) < \infty$. *Then for any stopping time* τ *and any positive non-decreasing* $L(\cdot)$ *on* $(0, \infty)$ *satisfying* (2.1)-(2.4),

$$(2.15) \quad Eh\left(\sup_{s\leq\tau} \frac{A_s}{\sqrt{B_s^2+1}}\right) \leq \sup_{y\geq 1} \frac{4y(L(y) \vee g_p^{1-\delta}(y))h(y)}{g_p(y)} + Eh\left(1 \vee \left(\sqrt{\frac{2}{p(1-\delta)}\log^+ L(\sqrt{B_\tau^2+1})}\right)\right),$$

and

$$(2.16) \quad Eh\left(\sup_{s\leq\tau} \frac{A_s}{(\sqrt{B_s^2+1})\sqrt{1\vee\log^+ L(\sqrt{B_s^2+1})}}\right) \leq \sup_{y\geq 1} \frac{4y(L(y) \vee g_p^{1-\delta}(y))h(y)}{g_p(y)} + h\left(\sqrt{\frac{2}{p(1-\delta)}}\right).$$

Proof. Let τ be any stopping time. Once the theorem is proved for bounded stopping times it applies to $\tau \wedge t$. Letting $t \to \infty$, the results extend to arbitrary τ by the monotone convergence theorem on one or both sides of the inequalities. Hence we may assume τ is bounded. There exists a sequence of nonnegative random times $t_n \leq \tau$ such that

$$\lim_{n\to\infty} \frac{A_{t_n}^+}{\sqrt{B_{t_n}^2+1}} = \sup_{0\leq s\leq\tau} \frac{A_s^+}{\sqrt{B_s^2+1}} = \sup_{0\leq s\leq\tau} \frac{A_s}{\sqrt{B_s^2+1}},$$

since $A_0 = 0$. Take any $0<p<1$. As a consequence of Lemma 1,

$$\begin{aligned} E\left(\exp\{\lambda A_{t_n} - \frac{\lambda^2}{2}\{B_{t_n}^2+1\}\right)^p &\leq E\left(\exp\{\lambda A_{t_n} - \frac{\lambda^2}{2}B_{t_n}^2\}\right)^p \\ &\leq E\left(\sup_{0\leq s\leq\tau} \exp\{\lambda A_s - \frac{\lambda^2}{2}B_s^2\}\right)^p \\ &= E\left(\sup_{s\geq 0} \exp\{\lambda A_{\tau\wedge s} - \frac{\lambda^2}{2}B_{\tau\wedge s}^2\}\right)^p \\ &\leq \frac{1}{1-p}(E\exp\{\lambda A_0 - \frac{\lambda^2}{2}B_0^2\})^p = \frac{1}{1-p}. \end{aligned}$$

For the moment, let $A = A_{t_n}$ and $B = B_{t_n}$. Multiplying by the $f(\lambda)$ defined in (2.6) and integrating, $\frac{1}{1-p}$ is bounded from below by

$$\begin{aligned}
&\int_0^\infty E\exp\{p(\lambda A - \frac{\lambda^2}{2}\{B^2+1\})\}\frac{1}{\lambda L(\lambda \vee \frac{1}{\lambda})}\,d\lambda \\
&= E\int_0^\infty \frac{\exp\{p\left(\frac{Ay}{\sqrt{B^2+1}} - \frac{y^2}{2}\right)\}}{yL\left(\frac{y}{\sqrt{B^2+1}} \vee \frac{\sqrt{B^2+1}}{y}\right)}\,dy \quad \text{(by Fubini, letting } y = \lambda\sqrt{B^2+1}) \\
&= E\left\{\exp\left(\frac{pA^2}{2(B^2+1)}\right)\int_{-\frac{A}{\sqrt{B^2+1}}}^\infty \frac{\exp(-px^2/2)dx}{(x+\frac{A}{\sqrt{B^2+1}})L\left(\frac{x+\frac{A}{\sqrt{B^2+1}}}{\sqrt{B^2+1}} \vee \frac{\sqrt{B^2+1}}{x+\frac{A}{\sqrt{B^2+1}}}\right)}\right\} \\
&\geq E\left\{\frac{I\left(\frac{A}{\sqrt{B^2+1}} \geq 1\right)\exp\left(\frac{pA^2}{2(B^2+1)}\right)\int_{-1}^0 \exp(-px^2/2)dx}{3\frac{A}{\sqrt{B^2+1}}(L\frac{A}{\sqrt{B^2+1}}) \vee L(\sqrt{B^2+1} \vee \frac{1}{\sqrt{B^2+1}}))}\right\} \quad \text{by (2.8)} \\
&\geq E\frac{g_p(\frac{A}{\sqrt{B^2+1}})}{4\frac{A}{\sqrt{B^2+1}}(L(\frac{A}{\sqrt{B^2+1}}) \vee L(\sqrt{B^2+1}))},
\end{aligned}$$

where the second equality is obtained via the change of variables $x = y - \frac{A}{\sqrt{B^2+1}}$ and the last one uses the fact that $\sqrt{B^2+1} \geq 1/\sqrt{B^2+1}$. Replacing $g(x)$ by $g_p(x)$ and arguing as in Theorem 1,

$$\begin{aligned}
Eh\left(\sup_{0\leq s\leq\tau}\frac{A_s^+}{\sqrt{B_s^2+1}}\right) &= \lim_{n\to\infty} Eh\left(\frac{A_{t_n}^+}{\sqrt{B_{t_n}^2+1}}\right) \\
&\leq \sup_{y\geq 1}\frac{4y(L(y)\vee g_p^{1-\delta}(y))}{(1-p)g_p(y)}\,h(y) \\
&\quad + \lim_{n\to\infty} Eh\left(1 \vee \sqrt{\frac{2}{p(1-\delta)}\log^+ L(\sqrt{B_{t_n}^2+1})}\right) \\
&\leq \sup_{y\geq 1}\frac{4y(L(y)\vee g_p^{1-\delta}(y))}{(1-p)g_p(y)}\,h(y) \\
&\quad + Eh\left(1 \vee \sqrt{\frac{2}{p(1-\delta)}\log^+ L(\sqrt{B_\tau^2+1})}\right),
\end{aligned}$$

with the last inequality following because $t_n \leq \tau$ and B_t is increasing in t, giving (2.15). The proof of (2.16) is similar and follows by replacing $\sqrt{B_t^2+1}$ with $(\sqrt{B_s^2+1})\sqrt{1\vee\log^+ L(\sqrt{B_s^2+1})}$ in the denominator of the argument of $h(\;\cdot\;)$; see de la Peña, Klass and Lai (1999) for the details used to derive a similar result. □

3 Applications

The application of Theorem 1 (or Theorem 2) with $h(x) = x^p$ $(x \geq 0,\ p > 0)$ and

$$L(x) = L_0(x) = 2(\log xe^e)(\log\log xe^e)^2 I(x \geq 1)$$

yields (1.4) and (1.5) (or similar bounds for (1.6)). In particular, letting $B_t = \sqrt{t},\ \ A_t = w_t$ (and also $A_t = -w_t$), Theorem 2 yields the following pth moment extension of (1.1).

$$E\left(\sup_{0\leq t\leq \tau}\left(\frac{|w_t|}{\sqrt{1+t}}\right)^p\right) \leq C_{1,p} + C_{2,p}E\left(\log(1+\log(1+\tau))\right)^{p/2}.$$

Let $\{Y_i\}$ be a sequence of i.i.d. random variables with $P(Y_i = 1) = P(Y_i = -1) = \frac{1}{2}$ and $T = \{\inf n \geq e^e : \sum_{j=1}^n Y_j \geq \sqrt{2n\log\log n}\}$ with $T = \infty$ if no such n exists. By a result of Erdös (1942), $P(T < \infty) = 1$. Let $X_{n,j} = Y_j I(T \geq j)$ for $1 \leq j \leq n$ with $X_{n,j} = 0$ when $j > n$. Then

$$\frac{X_{n,1} + \dots + X_{n,n}}{\sqrt{X_{n,1}^2 + \dots + X_{n,n}^2}} = \frac{\sum_{i=1}^{T\wedge n} Y_i}{\sqrt{T\wedge n}} \to \sqrt{2\log\log T}.$$

Letting $A = \sum_{i=1}^{T\wedge n} Y_i$ and $B = \sqrt{T\wedge n}$ shows that (1.4) is sharp.

Condition (1.2) is satisfied by several types of processes including continuous local martingales A_t, with $B_t^2 = \langle M\rangle_t$, assuming that $A_0 = 0$. More generally, let $\{M_t,\ t \geq 0\}$ be a locally square-integrable martingale with $M_0 = 0$. Let $\{V_t,\ t \geq 0\}$ be an increasing process, which is adapted, purely discontinuous and locally integrable, and let $V^{(p)}$ be its dual predictable projection. Let $X_t = M_t + V_t$,

$$C_t = \sum_{s\leq t}((\Delta X_s)^+)^2, \quad D_t = \{\sum_{s\leq t}((\Delta X_s)^-)^2\}_t^{(p)}, \quad H_t = \langle M^c\rangle_t + C_t + D_t.$$

Then as shown by Barlow, Jacka and Yor (1986), condition (1.2) holds with $A_t = X_t - V_t^{(p)}$ and $B_t^2 = H_t$.

An example without integrability conditions which automatically satisfies (1.2) is discussed in de la Peña, Klass, and Lai (1999). It can be obtained as follows: Let $\{d_i, \mathcal{F}_i\}$ be a sequence of conditionally symmetric random variables ($\mathcal{L}(d_i|\mathcal{F}_{i-1}) = \mathcal{L}(-d_i|\mathcal{F}_{i-1})$) and take $A = \sum_{i=1}^n d_i$, $B = \sqrt{\sum_{i=1}^n d_i^2}$.

4 Acknowledgment

We thank the referee for his thorough and exacting job. His comments helped us refine this paper.

References

1. Barlow, M.T., Jacka, S.D., and Yor, M. (1986). Inequalities for a pair of processes stopped at a random time. *Proc. London Math. Soc. Ser. III*, **53**, 152–172.

2. de la Peña, V. H., Klass, M.J. and Lai, T.L. (1999). Self-normalized processes: Exponential inequalities, moments and limit theorems. Technical Report, Dept. of Statistics, Columbia University.

3. Erdös, P. (1942). On the law of the iterated logarithm. *Ann. Math.* **43**, 419-436.

4. Graversen, S.E. and Peskir, G. (1998). Maximal inequalities for the Ornstein-Uhlenbeck process. Technical report, Dept. of Theoretical Statistics, Aarhus University.

5. Shao, Q.M. (1998). A comparison theorem on maximal inequalities between negatively associated and independent random variables. To appear in *J. Theoret. Probab.*

6. Yang, M. (1999). Occupation times and beyond. To appear in *Stoch. Process. Appl.* (Included in Yang (1999). *On the order of the γth-moment of the reflecting level crossing time and some general results.* PhD Thesis, Columbia University, Dept. of IEOR).

Victor H. de la Peña
Department of Statistics
Columbia University
New York, NY 10027
vp@stat.columbia.edu

Tze Leung Lai
Department of Statistics
Stanford University
Stanford, CA 94306–4065
lait@stat.stanford.edu

Michael J. Klass
Department of Mathematics and Statistics
University of California
Berkeley, CA 94720–3860

Exponential and Moment Inequalities for U-Statistics

Evarist Giné,* Rafał Latała,** and Joel Zinn

ABSTRACT A Bernstein-type exponential inequality for (generalized) canonical U-statistics of order 2 is obtained, and the Rosenthal and Hoffmann-Jørgensen inequalities for sums of independent random variables are extended to (generalized) U-statistics of any order whose kernels are either nonnegative or canonical.

1 Introduction

Exponential inequalities, such as Bernstein's and Prohorov's, and moment inequalities, such as Rosenthal's and Hoffmann-Jørgensen's, are among the most basic tools for the analysis of sums of independent random variables. Our object here consists in developing analogues of such inequalities for generalized U-statistics, in particular, for U-statistics and for multilinear forms in independent random variables.

Hoffmann-Jørgensen type moment inequalities for canonical (that is, completely degenerate) U-statistics of any order m were first considered by Giné and Zinn (1992), and their version for U-statistics with nonnegative kernels turned out to be useful for obtaining the best possible necessary integrability conditions in limit theorems for U-statistics. (By Khinchin's inequality it is irrelevant whether one considers canonical or nonnegative kernels in moment inequalities, at least if multiplicative constants are not at issue). Klass and Nowicki (1997) also obtained moment inequalities for nonnegative generalized U-statistics, but only for order $m = 2$, and their decomposition of the moments is more complete than that in Giné and Zinn (1992). Ibragimov and Sharakhmetov (1998, 1999) recently obtained analogues of Rosenthal's inequality for nonnegative and for canonical U-statistics. The moment inequalities we present in the first part of this article, valid for canonical and for nonnegative generalized U-statistics of any order m, when specialized to $m = 2$, represent the same level of moment decomposition as the Klass-Nowicki inequalities, coincide with theirs for powers $p > 1$ (except for constants) and are expressed in terms of different, simpler quantities for powers $p < 1$. Proposition 2.1 below, which

*Research partially supported by NSF Grant No. DMS-96–25457.

**Research partially supported by Polish Grant KBN 2 PO3A 043 15.

constitutes the first step towards more elaborate bounds such as those in Theorem 2.3 below, has also been obtained, up to constants, by Ibragimov and Sharakhmetov. Our proofs consist of simple iterations of the classical moment inequalities for sums of independent random variables.

The moment inequalities in the first part of this article do imply exponential bounds for canonical U-statistics of any order and with bounded kernels which are sharper than those in Arcones and Giné (1993); however, they are not of the best kind as they do not exhibit Gaussian behavior for part of the tail, which they should in view of the tail behavior of Gaussian chaos.

In the second part of this article we improve the moment inequalities from the first part in the case of generalized canonical U-statistics of order 2, and for moments of order $p \geq 2$ (Theorem 3.2). The bounds not only involve moments but also the L_2 operator norm of the matrix of kernels. Then we show how these improved moment inequalities imply what we believe is the correct analogue (up to constants) of Bernstein's exponential inequality for generalized canonical U-statistics of order 2 (Theorem 3.3). This exponential inequality, which does exhibit Gaussian behavior for small values of t, is strong enough to imply the law of the iterated logarithm for canonical U-statistics under conditions which are also necessary. The main new ingredient in this part of the paper is Talagrand's (1996) exponential bound for empirical processes, which gives a Rosenthal-Pinelis type inequality for moments of empirical processes (Proposition 3.1), basic for the derivation of the moment inequality for U-statistics of order 2.

Because of the decoupling results of de la Peña and Montgomery-Smith (1995), we can work with decoupled U-statistics, and this allows us to proceed by conditioning and iteration.

2 Moment inequalities

We consider estimation of moments of *generalized decoupled U-statistics*, defined as

$$\sum_{1\leq i_1,\ldots,i_m\leq n} h_{i_1,\ldots,i_m}(X_{i_1}^{(1)},\ldots,X_{i_m}^{(m)}), \tag{2.1}$$

where the random variables $X_i^{(j)} : 1 \leq i \leq n, 1 \leq j \leq m$, $m \leq n$, are independent (not necessarily with the same distribution) and take values in a measurable space $(S, \mathcal{S})$, and $h_{i_1,\ldots,i_m}$ are real valued measurable functions on S^m. For short, this sum is denoted by $\sum_{\mathbf{i}} h_{\mathbf{i}}$.

Given $J \subseteq \{1,\ldots,m\}$ ($J = \emptyset$ is not excluded), and $\mathbf{i} = (i_1,\ldots,i_m) \in \{1,\ldots,n\}^m$, we set $\mathbf{i}_J$ to be the point of $\{1,\ldots,n\}^{|J|}$ obtained from $\mathbf{i}$ by deleting the coordinates in the places not in J (e.g., if $\mathbf{i} = (3,4,2,1)$, then $\mathbf{i}_{\{1,3\}} = (3,2)$). Also, $\sum_{\mathbf{i}_J}$ indicates the sum over $1 \leq i_j \leq n$, $j \in J$ (for instance, if $m = 4$ and $J = \{1,3\}$, then

$$\sum_{\mathbf{i}_J} h_{\mathbf{i}} = \sum_{\mathbf{i}_{\{1,3\}}} h_{i_1,i_2,i_3,i_4} = \sum_{1\le i_1,i_3\le n} h_{i_1,i_2,i_3,i_4}(X_{i_1}^{(1)},\dots,X_{i_4}^{(4)}).)$$

By convention, $\sum_{\mathbf{i}_\emptyset} a = a$.

Likewise, while E will denote expected value with respect to all the variables, E_J will denote expected value only with respect to the variables $X_i^{(j)}$ with $j \in J$ and $i \in \{1,\dots,n\}$. By convention, $E_\emptyset a = a$.

Rosenthal's inequality is easiest to extend to U-statistics because it involves only moments of sums (as opposed to moments of maxima and quantiles for Hoffmann-Jørgensen's inequality). So, we will first obtain analogues of Rosenthal's inequality, and then we will transform these inequalities into analogues of Hoffmann-Jørgensen's by first showing that some moments of sums can be replaced by moments of maxima, and then, that the lowest moment can in fact be replaced by a quantile. We will illustrate this three-steps procedure first in the case of nonnegative kernels and moments of order $p \ge 1$. Then we will see that this also solves, via Khinchin's inequality, the case of canonical kernels and moments of order $p \ge 2$. Finally we will consider the cases of moments of order $p < 1$ for positive kernels and moments of order $p > 2$ for canonical kernels, for which the inequalities are less neat, but still useful. We will pay some attention to the behavior of the constants as $p \to \infty$ in these inequalities, since such behavior translates into (exponential) integrability properties.

2.1. Nonnegative kernels, moments of order $p \ge 1$. For nonnegative independent random variables ξ_i, we have the following two improvements of Rosenthal's inequalities, valid for $p \ge 1$:

1) Latała's, (1997):

$$(R_1) \qquad E\Big(\sum \xi_i\Big)^p \le (2e)^p \max\Big[\frac{e}{p} p^p \sum E\xi_i^p,\ e^p\Big(\sum E\xi_i\Big)^p\Big], \quad p > 1,$$

(see Pinelis (1994) for the corresponding inequality when the random variables are centered);

2) Johnson, Schechtman and Zinn's, (1985):

$$(R_2) \qquad E\Big(\sum \xi_i\Big)^p \le K^p\Big(\frac{p}{\log p}\Big)^p \max\Big[\sum E\xi_i^p,\ \Big(\sum E\xi_i\Big)^p\Big], \quad p > 1,$$

where K is a universal constant. See Utev (1985) and Figiel, Hitczenko, Johnson, Schechtman and Zinn (1997) for more precise inequalities of the same type.

And for general $p > 0$, we have the following improved Hoffmann-Jørgensen inequality that follows from Kwapień and Woyczyński (1992) and which can be obtained as in the proof of Theorem 1.2.3 in de la Peña and Giné (1999):

3)

$$(H)\quad E\Big\|\sum \xi_i\Big\|^p \le 2^{p-2}\cdot 2^{(p-1)\vee 0}\cdot (p+1)^{p+1}\Big[t_0^p + E\max\|\xi_i\|^p\Big],\quad p>0,$$

where

$$t_0 := \inf\Big[t>0 : \Pr\Big\{\Big\|\sum \xi_i\Big\| > t\Big\} \le \frac{1}{2}\Big],$$

and where we write norm for absolute value in order to include not only independent nonnegative real random variables, but also independent nonnegative random functions ξ_i taking values in certain 'rearrangement invariant spaces' such as $L_s(\Omega,\Sigma,\mu)$, $0<s<\infty$, with $\|\xi\| := \big(\int |\xi|^s d\mu\big)^{1/(s\vee 1)}$, or $\ell_\infty(L_s)$. Note that, by Markov,

$$t_0 \le 2^{1/r}\Big(E\Big\|\sum \xi_i\Big\|^r\Big)^{1/r},$$

so that (H) becomes:
4) for $0<r<p<\infty$,

$$E\Big\|\sum \xi_i\Big\|^p \le 2^{p-2}\cdot 2^{(p-1)\vee 0}\cdot (p+1)^{p+1}\Big[2^{p/r}\Big(E\Big\|\sum \xi_i\Big\|^r\Big)^{p/r} + E\max\|\xi_i\|^p\Big]. \tag{H_r}$$

Inequalities (H) and (H_r) hold for spaces of functions which are quasinormed measurable linear spaces whose quasinorm $\|\cdot\|$ has the property that $\|x\|\le\|y\|$ whenever $0\le x\le y$.

In the following proposition we extend inequalities (R_1) and (R_2) by means of an easy induction.

Proposition 2.1. *Let $m\in\mathbf{N}$, $p>1$, and, for all $\mathbf{i}\in\{1,\dots,n\}^m$, let $h_{\mathbf{i}}$ be a nonnegative function of m variables whose p-th power is integrable for the law of $\mathbf{X}_{\mathbf{i}} = (X_{i_1}^{(1)},\dots,X_{i_m}^{(m)})$. Then,*

$$\begin{aligned}\max_{J\subseteq\{1,\dots,m\}}\Big[\sum_{\mathbf{i}_J} E_J\Big(\sum_{\mathbf{i}_{J^c}} E_{J^c} h_{\mathbf{i}}\Big)^p\Big] &\le E\Big(\sum_{\mathbf{i}} h_{\mathbf{i}}\Big)^p\\ &\le (2e^2)^{mp}\sum_{J\subseteq\{1,\dots,m\}}\Big[p^{|J|p}\sum_{\mathbf{i}_J} E_J\Big(\sum_{\mathbf{i}_{J^c}} E_{J^c} h_{\mathbf{i}}\Big)^p\Big],\end{aligned}\tag{2.2}$$

and also, there exists a universal constant $K<\infty$ such that

$$E\Big(\sum_{\mathbf{i}} h_{\mathbf{i}}\Big)^p \le K^{mp}\Big(\frac{p}{\log p}\Big)^{mp}\max_{J\subseteq\{1,\dots,m\}}\Big[\sum_{\mathbf{i}_J} E_J\Big(\sum_{\mathbf{i}_{J^c}} E_{J^c} h_{\mathbf{i}}\Big)^p\Big].\tag{2.2$'$}$$

Proof. The proof of (2.2′), with a sum over the subsets J instead of a maximum, differs from that of (2.2) only in the starting point ((R_2) instead of (R_1)); then, replacing sum by maximum simply increases the constant by a factor of 2^m. The left side inequality in (2.2) follows by Hölder since $p \geq 1$. Consider the right-hand side inequality. For $m = 1$ this is just inequality (R_1) and we can proceed by induction. Suppose the result holds for $m-1$. By applying the induction hypothesis to

$$E\Big(\sum_{\mathbf{i}} h_{\mathbf{i}}\Big)^p = E_m E_{\{1,\ldots,m-1\}}\Big[\sum_{\mathbf{i}_{\{1,\ldots,m-1\}}}\Big(\sum_{i_m} h_{\mathbf{i}}\Big)\Big]^p,$$

we only have to consider the generic term in the decomposition (2.2) for the new kernels $\left(\sum_{i_m} h_{\mathbf{i}}\right)$ with the $X_i^{(m)}$ variables fixed. In other words, letting J_{m-1} be any subset of $\{1,\ldots,m-1\}$ and J_{m-1}^c its complement with respect to $\{1,\ldots,m-1\}$, we must estimate

$$\begin{aligned} E_m \sum_{\mathbf{i}_{J_{m-1}}} E_{J_{m-1}}\Big(\sum_{\mathbf{i}_{J_{m-1}^c}} E_{J_{m-1}^c}\Big(\sum_{i_m} h_{\mathbf{i}}\Big)\Big)^p \\ = \sum_{\mathbf{i}_{J_{m-1}}} E_{J_{m-1}} E_m\Big(\sum_{i_m}\Big(E_{J_{m-1}^c}\sum_{\mathbf{i}_{J_{m-1}^c}} h_{\mathbf{i}}\Big)\Big)^p. \end{aligned}$$

Rosenthal's inequality (R_1) applied to the kernels $E_{J_{m-1}^c}\sum_{\mathbf{i}_{J_{m-1}^c}} h_{\mathbf{i}}$ with the variables in J_{m-1} fixed, gives

$$\begin{aligned} E_m\Big(\sum_{i_m}\Big(E_{J_{m-1}^c}\sum_{i_{J_{m-1}^c}} h_{\mathbf{i}}\Big)\Big)^p \leq (2e^2)^p\Big[\Big(\sum_{i_m, i_{J_{m-1}^c}} E_m E_{J_{m-1}^c} h_{\mathbf{i}}\Big)^p \\ + p^p \sum_{i_m} E_m\Big(E_{J_{m-1}^c}\sum_{\mathbf{i}_{J_{m-1}^c}} h_{\mathbf{i}}\Big)^p\Big]. \end{aligned}$$

Upon integrating each term with respect to $E_{J_{m-1}}$ and summing over $\mathbf{i}_{J_{m-1}}$, we obtain

$$\begin{aligned} E_m \sum_{\mathbf{i}_{J_{m-1}}} E_{J_{m-1}}\Big(\sum_{\mathbf{i}_{J_{m-1}^c}} E_{J_{m-1}^c}\Big(\sum_{i_m} h_{\mathbf{i}}\Big)\Big)^p \\ \leq (2e^2)^p\Big[\sum_{\mathbf{i}_{J_{m-1}}} E_{J_{m-1}}\Big(\sum_{\mathbf{i}_{J_{m-1}^c\cup\{m\}}} E_{J_{m-1}^c\cup\{m\}} h_{\mathbf{i}}\Big)^p \\ + p^p \sum_{\mathbf{i}_{J_{m-1}\cup\{m\}}} E_{J_{m-1}\cup\{m\}}\Big(E_{J_{m-1}^c}\sum_{\mathbf{i}_{J_{m-1}^c}} h_{\mathbf{i}}\Big)^p\Big]. \end{aligned}$$

Multiplying by $(2e^2)^{(m-1)p}p^{|J_{m-1}|}$, this is the sum of two terms of the form $(2e)^{mp}p^{|J|p}\sum_{\mathbf{i}_J} E_J\Big(\sum_{\mathbf{i}_{J^c}} E_{J^c}h_{\mathbf{i}}\Big)^p$ (for $J = J_{m-1}$ and for $J = J_{m-1}\cup\{m\}$), proving the proposition. □

This proposition solves the problem of estimating, up to constants, the moments of a decoupled U-statistic by 'computable' expressions. For instance, if the functions $h_{\mathbf{i}}$ are all equal and if the variables $X_i^{(j)}$ are i.i.d., then the typical term at the right of (2.1) just becomes $n^{|J|+p|J^c|}E_J(E_{J^c}h)^p$, a 'mixed moment' of h. For $m = 2$ the right-hand side of inequality (2.2) is just

$$E\Big(\sum_{i,j} h_{i,j}(X_i^{(1)}, X_j^{(2)})\Big)^p \le (2e^2)^{2p}\Big[\Big(\sum_{i,j} Eh_{i,j}(X_i^{(1)}, X_j^{(2)})\Big)^p + p^p\sum_i E_1\Big(\sum_j E_2h_{i,j}(X_i^{(1)}, X_j^{(2)})\Big)^p + p^p\sum_j E_2\Big(\sum_i E_1h_{i,j}(X_i^{(1)}, X_j^{(2)})\Big)^p + p^{2p}\textstyle\sum_{i,j} Eh_{i,j}^p(X_i^{(1)}, X_j^{(2)})\Big]. \tag{2.2''}$$

We have been careful with the dependence on p of the constants because it is of some interest to obtain constants of the best order, as $p \to \infty$. In fact, (2.2′) exhibits constants of the best order as can be seen by taking the product of two independent copies of the example in Johnson, Schechtman and Zinn (1985), Proposition 2.9.

Next we replace the external sums of expected values at the right side of the above inequalities by expectations of maxima, without significantly altering the order of the multiplicative constants. If ξ_i are independent nonnegative random variables, then,

$$\frac{1}{2}\Big[\delta_0^p \vee \sum E\xi_i^p I_{\xi_i>\delta_0}\Big] \le E\max\xi_i^p \le \delta_0^p + \sum E\xi_i^p I_{\xi_i>\delta_0}, \quad 0<p<\infty, \tag{2.3}$$

where

$$\delta_0 = \inf\Big[t>0 : \sum \Pr\{\xi_i > t\} \le 1\Big] \tag{2.4}$$

(Giné and Zinn (1983); see also de la Peña and Giné (1999), page 22). The left-hand side of (2.3) gives that, for $0 < r < p$ and ξ_i independent,

$$\sum E|\xi_i|^p \le 2E\max|\xi_i|^p + 2\Big(\sum E|\xi_i|^r\Big)\Big(E\max|\xi_i|^p\Big)^{(p-r)/p} \tag{2.5}$$

(e.g., de la Peña and Giné (1999), page 48). This inequality, applied with $r = 1 < p$, yields

$$p^{\alpha p}\sum E|\xi_i|^p \le 2(1+p^\alpha)\max\Big[p^{\alpha p}E\max|\xi_i|^p,\ \Big(\sum E|\xi_i|\Big)^p\Big] \tag{2.6}$$

for all $\alpha \ge 0$. There are similar inequalities for other values of r; $r = 1$ is adequate for $\xi_i \ge 0$, but $r = 2$ is better for centered variables. If we use inequality (2.6) in (2.2″), iteratively for the last term, we obtain that, for a universal constant K (easy but cumbersome to compute), $h_{i,j} \ge 0$, $p > 1$,

$$\begin{aligned} E\Big(\sum_{i,j} h_{i,j}\Big)^p \ \le\ & K^p(2e^2)^p p^4\Big[\Big(\sum_{i,j} Eh_{i,j}\Big)^p + p^p E_1\max_i\Big(\sum_j E_2 h_{i,j}\Big)^p \\ & + p^p E_2\max_j\Big(\sum_i E_1 h_{i,j}\Big)^p + p^{2p}E\max_{i,j} h_{i,j}^p\Big]. \end{aligned} \tag{2.7}$$

Inequality (2.7) was obtained, up to constants, by Klass and Nowicki (1997) (it is their inequality (4.14)). Our proof is different, and it is contained in the proof of the next corollary, which extends inequality (2.7) to any m.

Corollary 2.2. *Under the same hypotheses as in Proposition 2.1, there exist universal constants K_m such that*

$$\begin{aligned} \max_{J\subseteq\{1,\dots,m\}}\Big[E_J\max_{\mathbf{i}_J}\Big(\sum_{\mathbf{i}_{J^c}} E_{J^c}h_{\mathbf{i}}\Big)^p\Big] &\le E\Big(\sum_{\mathbf{i}} h_{\mathbf{i}}\Big)^p \\ &\le K_m^p \sum_{J\subseteq\{1,\dots,m\}}\Big[p^{|J|p}E_J\max_{\mathbf{i}_J}\Big(\sum_{\mathbf{i}_{J^c}} E_{J^c}h_{\mathbf{i}}\Big)^p\Big], \end{aligned} \tag{2.8}$$

and

$$E\Big(\sum_{\mathbf{i}} h_{\mathbf{i}}\Big)^p \le K_m^p\Big(\frac{p}{\log p}\Big)^{mp}\max_{J\subseteq\{1,\dots,m\}}\Big[E_J\max_{\mathbf{i}_J}\Big(\sum_{\mathbf{i}_{J^c}} E_{J^c}h_{\mathbf{i}}\Big)^p\Big]. \tag{2.8′}$$

Proof. The left side of (2.8) follows by Hölder. Inequality (2.8′) has a proof similar to that of the right-hand side of (2.8), and therefore we only prove the latter. We will prove it by induction over m simultaneously with the inequality

$$p^{mp}\sum_{\mathbf{i}} Eh_{\mathbf{i}}^p \le \tilde{K}_m^p \sum_{J\subset\{1,\dots,m\}}\Big[p^{|J|p}E_J\max_{\mathbf{i}_J}\Big(\sum_{\mathbf{i}_{J^c}} E_{J^c}h_{\mathbf{i}}\Big)^p\Big]. \tag{2.9}$$

Let us first note that the inequalities (2.9) for $1,\dots,m-1$ together with (2.2) imply (2.8). It is therefore enough to show that if (2.8) and (2.9) hold for $1,\dots,m-1$ then (2.9) is satisfied for m. We will follow the notation of the proof of Proposition 2.1. Inequality (2.9) for $m = 1$ is just (2.6), and

(2.8) for $m = 1$ is (H_1) (which also follows from (R_1) and (2.6)). By the induction assumptions we have

$$p^{mp} \sum_{\mathbf{i}} E h_{\mathbf{i}}^p = p^p \sum_{i_m} E_m p^{(m-1)p} \sum_{\mathbf{i}_{\{1,\dots,m-1\}}} E_{\{1,\dots,m-1\}} h_{\mathbf{i}}^p \tag{2.10}$$

$$\le \tilde{K}_{m-1}^p$$

$$\times \sum_{J_{m-1} \subset \{1,\dots,m-1\}} \Big[p^{(|J_{m-1}|+1)p} E_{J_{m-1}} E_m \sum_{i_m} \max_{\mathbf{i}_{J_{m-1}}} \Big(\sum_{\mathbf{i}_{J_{m-1}^c}} E_{J_{m-1}^c} h_{\mathbf{i}} \Big)^p \Big].$$

Now, by (2.6), for any $J_{m-1} \subset \{1, \dots, m-1\}$ we have

$$p^{(|J_{m-1}|+1)p} E_{J_{m-1}} E_m \sum_{i_m} \max_{\mathbf{i}_{J_{m-1}}} \Big(\sum_{\mathbf{i}_{J_{m-1}^c}} E_{J_{m-1}^c} h_{\mathbf{i}} \Big)^p$$

$$\le 2(1+p) \Big[p^{(|J_{m-1}|+1)p} E_{J_{m-1} \cup \{m\}} \max_{\mathbf{i}_{J_{m-1} \cup \{m\}}} \Big(\sum_{\mathbf{i}_{J_{m-1}^c}} E_{J_{m-1}^c} h_{\mathbf{i}} \Big)^p$$

$$+ p^{|J_{m-1}|p} E_{J_{m-1}} \Big(\sum_{i_m} E_m \max_{\mathbf{i}_{J_{m-1}}} \sum_{\mathbf{i}_{J_{m-1}^c}} E_{J_{m-1}^c} h_{\mathbf{i}} \Big)^p \Big]. \tag{2.11}$$

To estimate the last term we note that

$$p^{|J_{m-1}|p} E_{J_{m-1}} \Big(\sum_{i_m} E_m \max_{\mathbf{i}_{J_{m-1}}} \sum_{\mathbf{i}_{J_{m-1}^c}} E_{J_{m-1}^c} h_{\mathbf{i}} \Big)^p$$

$$\le p^{|J_{m-1}|p} E_{J_{m-1}} \Big(\sum_{\mathbf{i}} E_{J_{m-1}^c \cup \{m\}} h_{\mathbf{i}} \Big)^p \tag{2.12}$$

$$\le \tilde{K}_{|J_{m-1}|}^p \sum_{J \subset J_{m-1}} p^{|J|p} E_J \max_{\mathbf{i}_J} \Big(\sum_{\mathbf{i}_{(J_{m-1} \setminus J) \cup J_{m-1}^c \cup \{m\}}} E_{(J_{m-1} \setminus J) \cup J_{m-1}^c \cup \{m\}} h_{\mathbf{i}} \Big)^p,$$

where in the last line we use the induction assumption (2.8) for $|J_{m-1}| < m$. Finally (2.10), (2.11) and (2.12) imply (2.9) and complete the proof. □

Remark. The proof of Proposition 2.6 below will use a version of Corollary 2.2 for nonnegative random functions taking values in L_r. The inequality is as follows: for $p > 1$ there exists $K_{m,p,r} < \infty$ such that

$$E \Big\| \sum_{\mathbf{i}} h_{\mathbf{i}} \Big\|^p \le K_{m,p,r} \max_{J \subseteq \{1,\dots,m\}} \Big[E_J \max_{\mathbf{i}_J} \Big(E_{J^c} \Big\| \sum_{\mathbf{i}_{J^c}} h_{\mathbf{i}} \Big\| \Big)^p \Big]. \tag{2.8''}$$

The proof is similar to the previous ones and is omitted: one takes (H_p) as the starting point of the induction.

Finally we come to the third step, which will extend Hoffmann-Jørgensen's inequality (H) for $p \geq 1$. If we want to use the inequalities from Corollary 2.2 to obtain boundedness of moments from stochastic boundedness of a sequence of U-statistics, we need to replace the term corresponding to $J = \emptyset$ by the p-th power of a quantile of $\sum_{\mathbf{i}} h_{\mathbf{i}}$. For this we use Paley-Zygmund's inequality (e.g., Kahane (1968) or de la Peña and Giné (1999)): if A is a nonnegative random variable and $0 < r < p < \infty$, then, for all $0 < \lambda < 1$,

$$\Pr\Big\{A > \lambda \|A\|_r\Big\} \geq \left[(1-\lambda^r)\frac{\|A\|_r}{\|A\|_p}\right]^{p/(p-r)}, \tag{2.13}$$

where $\|A\|_r = \big(E|A|^r\big)^{1/r}$ for $0 < r < \infty$. Consider for instance inequality (2.8). It has the form

$$EA^p \leq B + K_m^p (EA)^p, \quad p > 1,$$

with $A = \sum_{\mathbf{i}} h_{\mathbf{i}}$. Then, either $B \geq K_m^p (EA)^p$, in which case we have $EA^p \leq 2B$, or $B < K_m^p (EA)^p$, in which case we have $EA^p \leq 2K_m^p (EA)^p$ and we can apply Paley-Zygmund's (2.13) with $\lambda = 1/2$ and $r = 1$. It gives

$$\Pr\Big\{A > \frac{1}{2} EA\Big\} \geq \frac{1}{2^{(p+1)/(p-1)} K_m^{p/(p-1)}}.$$

Hence, if we define

$$t_0 = \inf\left[t \geq 0 : \Pr\{A > t\} \leq \frac{1}{2^{(p+1)/(p-1)} K_m^{p/(p-1)}}\right], \tag{2.14}$$

we obtain $EA \leq 2t_0$. So, in either case,

$$EA^p \leq 2B + 2^{1+p} K_m^p t_0^p.$$

Also, by Markov's inequality,

$$\frac{1}{2^{(p+1)/(p-1)} K_m^{p/(p-1)}} t_0^p \leq EA^p.$$

We then have:

Theorem 2.3. *Under the hypotheses of Proposition 2.1, there exist a universal constant $K_m < \infty$ such that, if t_0 is as defined by (2.14) for*

$A = \sum_{\mathbf{i}} h_{\mathbf{i}}$, *then*

$$\frac{1}{(4K_m)^{p/(p-1)}} t_0^p \vee \max\Big[\max_{\substack{J \subseteq \{1,\dots,m\} \\ J \neq \emptyset}} \Big[E_J \max_{\mathbf{i}_J} \Big(\sum_{\mathbf{i}_{J^c}} E_{J^c} h_{\mathbf{i}} \Big)^p \Big]$$

$$\leq E\Big(\sum_{\mathbf{i}} h_{\mathbf{i}} \Big)^p$$

$$\leq (4K_m)^p \Big\{ 2^{1+p} t_0^p + \sum_{\substack{J \subseteq \{1,\dots,m\} \\ J \neq \emptyset}} \Big[p^{|J|p} E_J \max_{\mathbf{i}_J} \Big(\sum_{\mathbf{i}_{J^c}} E_{J^c} h_{\mathbf{i}} \Big)^p \Big] \Big\}. \tag{2.15}$$

A similar inequality with different constants can be obtained from (2.8′). This is the most elaborate form we will give to our bounds for $h \geq 0$ and $p > 1$.

The right-hand side of (2.15) for $m = 2$ becomes, disregarding constants,

$$E\Big(\sum_{i,j} h_{i,j} \Big)^p \leq C \max\Big[E_1 \max_i \Big(\sum_j E_2 h_{i,j} \Big)^p,\ E_2 \max_j \Big(\sum_i E_1 h_{i,j} \Big)^p,\ E \max_{i,j} h_{i,j}^p,\ t_0^p \Big]. \tag{2.15′}$$

So, we get the p-th moment of the double sum controlled by moments of partial maxima of conditional expectations plus a quantile. The Giné-Zinn (1992) inequality (for $m = 2$),

$$E\Big(\sum_{i,j} h_{i,j} \Big)^p \leq C \max\Big[E \max_i \Big(\sum_j h_{i,j} \Big)^p, t_0^p \Big], \quad p \geq 1,$$

is slightly weaker in appearance than (2.15′) (actually, we only published the result for canonical U-statistics, but we applied it as well to nonnegative variables, for which the proof is the same: see, e.g., Giné and Zhang (1996)). For applications of this inequality in the asymptotic theory of U-statistics see Giné and Zhang (1996), Giné, Kwapień, Latała and Zinn (1999) and de la Peña and Giné (1999).

Remark. The constants in the definition of t_0 in (2.15) depend on p, hence, so does t_0. This is not the case when $m = 1$ (as a consequence of the improved Hoffmann-Jørgensen's inequality of Kwapień and Woyczyński —see, de la Peña and Giné (1999) p. 11). But in most applications it does not matter whether the definition of the quantile depends on p.

2.2. Canonical kernels, moments or order $p \geq 2$. If ξ_i are centered and independent and $p \geq 2$, then, by convexity and the Khinchin-Bonami inequality (e.g., de la Peña and Giné, 1999, p. 113), we have

$$2^{-p}E\Big(\sum \xi_i^2\Big)^{p/2} \le 2^{-p}E\Big|\sum \varepsilon_i\xi_i\Big|^p \le E\Big|\sum \xi_i\Big|^p$$

$$\le 2^pE\Big|\sum \varepsilon_i\xi_i\Big|^p \le 2^p(p-1)^{p/2}E\Big(\sum \xi_i^2\Big)^{p/2}, \tag{2.16}$$

where ε_i are independent identically distributed Rademacher random variables, independent from $\{\xi_i\}$. Suppose $h_{\mathbf{i}}$ is canonical for the variables $\{X_i^{(j)}\}$ given in the previous subsection, that is, suppose

(2.17)
$$E_j h(X_{i_1}^{(1)}, \dots, X_{i_m}^{(m)}) = 0 \text{ a.s. for all } j = 1, \dots, m, \quad 1 \le i_1, \dots, i_m \le n.$$

Let $\varepsilon_i^{(j)}$ be an independent Rademacher array independent of $\{X_i^{(j)}\}$, and set

$$\varepsilon_{\mathbf{i}} := \varepsilon_{i_1}^{(1)} \cdots \varepsilon_{i_m}^{(m)}.$$

Then, recursive application of inequality (2.16) gives

$$2^{-mp}E\Big(\sum_{\mathbf{i}} h_{\mathbf{i}}^2\Big)^{p/2} \le 2^{-mp}E\Big|\sum_{\mathbf{i}} \varepsilon_{\mathbf{i}}h_{\mathbf{i}}\Big|^p \le E\Big|\sum_{\mathbf{i}} h_{\mathbf{i}}\Big|^p$$

$$\le 2^{mp}E\Big|\sum_{\mathbf{i}} \varepsilon_{\mathbf{i}}h_{\mathbf{i}}\Big|^p \le 2^{mp}(p-1)^{mp/2}E\Big(\sum_{\mathbf{i}} h_{\mathbf{i}}^2\Big)^{p/2}. \tag{2.18}$$

This inequality reduces estimation of moments of canonical U-statistics to estimation of moments of nonnegative ones (and conversely), at least if constants are not an issue. Combined with Proposition 2.1, it gives the analogue of Rosenthal's inequality for centered variables and $p > 2$, and if we apply it in conjunction with Corollary 2.2, we obtain the following inequality:

Proposition 2.4. *If, for $p > 2$ and all $\mathbf{i} \in \{1, \dots, n\}^m$, $h_{\mathbf{i}}(X_{i_1}^{(1)}, \dots, X_{i_m}^{(m)})$ is p-integrable and $E_j h_{\mathbf{i}}(X_{i_1}^{(1)}, \dots, X_{i_m}^{(m)}) = 0$ a.s. for all $j = 1, \dots, m$, then*

$$2^{-mp} \max_{J\subseteq\{1,\dots,m\}} \Big[E_J \max_{\mathbf{i}_J}\Big(\sum_{\mathbf{i}_{J^c}} E_{J^c}h_{\mathbf{i}}^2\Big)^{p/2}\Big] \le E\Big|\sum_{\mathbf{i}} h_{\mathbf{i}}\Big|^p$$

$$\le K_m^p \sum_{J\subseteq\{1,\dots,m\}} \Big[p^{(m+|J|)p/2}E_J \max_{\mathbf{i}_J}\Big(\sum_{\mathbf{i}_{J^c}} E_{J^c}h_{\mathbf{i}}^2\Big)^{p/2}\Big] \tag{2.19}$$

for universal constant $K_m < \infty$.

And, applying Paley-Zygmund with $r = 2$, we finally have:

Theorem 2.5. *Let $h_{\mathbf{i}}$ be as in Proposition 2.4, and let $p > 2$. Then, there exist universal constants $K_m < \infty$ such that, if t_0 is defined as*

$$t_0 = \inf\Big[t \geq 0 : \Pr\Big\{\Big|\sum_{\mathbf{i}} h_{\mathbf{i}}\Big| > t\Big\} \leq \Big(\frac{3}{4}\Big)^{p/(p-2)} \frac{1}{\big(2K_m^p p^{mp/2}\big)^{1/(p-2)}}\Big],$$

then

$$\frac{1}{(4K_m p^{m/2})^{p/(p-2)}} t_0^p \vee \Big[2^{-mp} \max_{\substack{J\subseteq\{1,\dots,m\}\\ J\neq\emptyset}} \Big[E_J \max_{\mathbf{i}_J}\Big(\sum_{\mathbf{i}_{J^c}} E_{J^c} h_{\mathbf{i}}^2\Big)^{p/2}\Big]\Big]$$

$$\leq E\Big|\sum_{\mathbf{i}} h_{\mathbf{i}}\Big|^p \tag{2.20}$$

$$\leq 2K_m^p \Big\{(2p^{m/2})^p t_0^p + \sum_{\substack{J\subseteq\{1,\dots,m\}\\ J\neq\emptyset}} \Big[p^{(m+|J|)p/2} E_J \max_{\mathbf{i}_J}\Big(\sum_{\mathbf{i}_{J^c}} E_{J^c} h_{\mathbf{i}}^2\Big)^{p/2}\Big]\Big\}.$$

If, instead of inequality (2.2), we wish to obtain an analogue of inequality (2.2′), that is, if we want to replace the constants at the right-hand side of (2.19) by $(Kp/\log p)^{mp}$, then we cannot use Khinchin's inequality and must proceed directly with an induction as in Proposition 2.1 with the following change: we must consider the variables $\sum_{\mathbf{i}_{J_{m-1}^c}} h_{\mathbf{i}}$ as taking values in $L_2(J_{m-1}^c)$ and apply inequality (1.5) in Kwapień and Szulga (1991), which gives Rosenthal's inequality with best constants for centered independent random variables in Banach spaces. We skip the details.

2.3. Nonnegative kernels, moments of order $p \leq 1$. It seems impossible to obtain inequalities as simple as in the previous section for this case. However, one can still obtain inequalities that may become useful when combined with Paley-Zygmund. Here is an analogue of Corollary 2.2 for $h \geq 0$ and $p \leq 1$. The method of proof is inefficient regarding constants as Hoffmann-Jørgensen is applied twice at each step. Hence, constants will not be specified.

Proposition 2.6. *Let $0 < r < p \leq 1$, $m < \infty$ and assume that the kernels $h_{\mathbf{i}} \geq 0$ have integrable p-th powers. Then*

$$\max_{J\subseteq\{1,\dots,m\}} \Big[E_J \max_{\mathbf{i}_J}\Big(E_{J^c}\Big(\sum_{\mathbf{i}_{J^c}} h_{\mathbf{i}}\Big)^r\Big)^{p/r}\Big] \leq E\Big(\sum_{\mathbf{i}} h_{\mathbf{i}}\Big)^p$$

$$\leq K_{r,p,m} \max_{J\subseteq\{1,\dots,m\}} \Big[E_J \max_{\mathbf{i}_J}\Big(E_{J^c}\Big(\sum_{\mathbf{i}_{J^c}} h_{\mathbf{i}}\Big)^r\Big)^{p/r}\Big], \tag{2.21}$$

where $K_{r,p,m}$ depends only on the parameters r, p, m.

Note that all the terms in this bound represent a reduction in the number of sums *except* for the term corresponding to $J = \emptyset$, which consists of a power of the r-th moment of a U-statistic of order m. We will deal later with this term by means of the Paley-Zygmund argument.

Proof. The inequality at the left side of (2.21) follows from Hölder. Inequality (H_r) is just the right-hand side of inequality (2.21) for $m = 1$ and we can proceed by induction. We still use the notation from Proposition 2.1. By the induction hypothesis we have

$$E\Big(\sum_{\mathbf{i}} h_{\mathbf{i}}\Big)^p = E_m E_{\{1,\dots,m-1\}}\Big(\sum_{\mathbf{i}_{\{1,\dots,m-1\}}} \sum_{i_m} h_{\mathbf{i}}\Big)^p \tag{2.22}$$

$$\leq K_{r,p,m-1} \sum_{J_{m-1}\subset\{1,\dots,m-1\}} E_{J_{m-1}} E_m \max_{\mathbf{i}_{J_{m-1}}} \Big[E_{J_{m-1}^c}\Big(\sum_{\mathbf{i}_{J_{m-1}^c}} \sum_{i_m} h_{\mathbf{i}}\Big)^r\Big]^{p/r}.$$

Let us fix $J_{m-1} \subset \{1,\dots,m-1\}$ and note that, for fixed $(X_j^{(i)})_{i\in J_{m-1}}$, we have

$$\max_{\mathbf{i}_{J_{m-1}}} E_{J_{m-1}^c}\Big(\sum_{\mathbf{i}_{J_{m-1}^c}} \sum_{i_m} h_{\mathbf{i}}\Big)^r := \Big\|\sum_{i_m} \tilde{h}_{i_m}\Big\|$$

for suitably chosen independent r.v.'s $\tilde{h}_{i_m}$ in $l^\infty(L^r)$. Therefore by (H_r), which still holds in this space (as the norm, restricted to nonnegative vectors, is monotone increasing), we have

$$E_{J_{m-1}} E_m \max_{\mathbf{i}_{J_{m-1}}} \Big[E_{J_{m-1}^c}\Big(\sum_{\mathbf{i}_{J_{m-1}^c}} \sum_{i_m} h_{\mathbf{i}}\Big)^r\Big]^{p/r} = E_{J_{m-1}} E_m \Big\|\sum_{i_m} \tilde{h}_{i_m}\Big\|^{p/r}$$

$$\leq C_{p,r} E_{J_{m-1}}\Big[E_m \max_{i_m} \|\tilde{h}_{i_m}\|^{p/r} + \Big(E_m\Big\|\sum_{i_m} \tilde{h}_{i_m}\Big\|\Big)^{p/r}\Big]$$

$$= C_{p,r}\Big[E_{J_{m-1}\cup\{m\}} \max_{\mathbf{i}_{J_{m-1}\cup\{m\}}} \Big(E_{J_{m-1}^c}\Big(\sum_{\mathbf{i}_{J_{m-1}^c}} h_{\mathbf{i}}\Big)^r\Big)^{p/r}$$

$$+ E_{J_{m-1}}\Big(E_m \max_{\mathbf{i}_{J_{m-1}}} E_{J_{m-1}^c}\Big(\sum_{\mathbf{i}_{J_{m-1}^c}} \sum_{i_m} h_{\mathbf{i}}\Big)^r\Big)^{p/r}\Big]. \tag{2.23}$$

Now, to estimate the last term, we note that

$$E_{J_{m-1}}\Big[E_m \max_{\mathbf{i}_{J_{m-1}}} E_{J_{m-1}^c}\Big(\sum_{\mathbf{i}_{J_{m-1}^c}} \sum_{i_m} h_{\mathbf{i}}\Big)^r\Big]^{p/r}$$

$$\leq E_{J_{m-1}}\Big[E_{J_{m-1}^c\cup\{m\}}\Big(\sum_{\mathbf{i}} h_{\mathbf{i}}\Big)^r\Big]^{p/r} \tag{2.24}$$

$$\le K_{p/r,1,|J_{m-1}|}$$

$$\times \sum_{J\subset J_{m-1}} E_J \max_{\mathbf{i}_J}\Big[E_{J_{m-1}\setminus J\cup J^c_{m-1}\cup\{m\}}\Big(\sum_{\mathbf{i}_{J_{m-1}\setminus J\cup J^c_{m-1}\cup\{m\}}} h_{\mathbf{i}}\Big)^r\Big]^{p/r},$$

which follows by the version of Corollary 2.2 for L^r ((2.8") for $p/r > 1$). Now (2.22), (2.23) and (2.24) complete the induction step. □

To deal with the term corresponding to $J = \emptyset$ in Proposition 2.6, we apply Paley-Zygmund as above, but now with $r < p$ replacing $1 < p$. The conclusion is:

Theorem 2.7. *There is a constant $K_{r,p,m}$ such that for $0 < r < p \le 1$, $m < \infty$, and $h_{\mathbf{i}} \ge 0$ with integrable p-th powers,*

$$\frac{1}{(2^{p+1}K_{r,p,m})^{1/(p-r)}}t_0^p \vee \sum_{\substack{J\subseteq\{1,\ldots,m\}\\ J\neq\emptyset}} \Big[E_J \max_{\mathbf{i}_J}\Big(E_{J^c}\Big(\sum_{\mathbf{i}_{J^c}} h_{\mathbf{i}}\Big)^r\Big)^{p/r}\Big]$$

$$\le E\Big(\sum_{\mathbf{i}} h_{\mathbf{i}}\Big)^p \tag{2.25}$$

$$\le 2K_{r,p,m}\Big\{2^{p/r}t_0^p + \sum_{\substack{J\subseteq\{1,\ldots,m\}\\ J\neq\emptyset}} \Big[E_J \max_{\mathbf{i}_J}\Big(E_{J^c}\Big(\sum_{\mathbf{i}_{J^c}} h_{\mathbf{i}}\Big)^r\Big)^{p/r}\Big]\Big\},$$

where

$$t_0 = \inf\Big[t : \Pr\Big\{\sum_{\mathbf{i}} h_{\mathbf{i}} > t\Big\} \le \frac{1}{2}(2^{p+1}K_{r,p,m})^{-1/(p-r)}\Big].$$

Hence, the p-th moment of a U-statistic of order m can be estimated by partial moments of maxima (or sums) of conditional moments of U-statistics of lower order plus the p-power of a quantile of the original U-statistic.

2.4. Canonical kernels and moments of order $1 \le p \le 2$, *or kernels h separately symmetric in each of the coordinates and* $0 < p < 1$. The canonical case reduces to the positive case by means of inequality (2.18), as before. The convexity part of inequality (2.18) fails for $p < 1$, but in this case, if h is symmetric separately in each of the coordinates, we can still randomize by products of independent Rademacher variables, and recursive application of Khinchin's inequality still reduces this case to nonnegative h. We leave the resulting statements to the reader in order to avoid repetition.

2.5. Regular (undecoupled) general U-statistics. If $h_{\mathbf{i}}(\mathbf{x}) = h_{\mathbf{i}\circ s}(\mathbf{x}\circ s)$ for any permutation s of $\{1,\ldots,m\}$ and $h_{\mathbf{i}} = 0$ if $\mathbf{i}$ has repeated indices, and if the sequences $\{X_i^{(j)} : i = 1, \ldots, n\}$ are independent copies of each other, then the decoupling inequalities of de la Peña and Montgomery-Smith (1995), together with the decoupling inequality for maxima in Hitczenko (1988), in combination with the previous inequalities give moment inequalities for the generalized U-statistics

$$\sum_{\mathbf{i}} h_{i_1,\ldots,i_m}(X_{i_1}, \ldots, X_{i_m})$$

where $\{X_i\}$ is a sequence of independent random variables, at the cost of vastly increasing the numerical constants (see e.g. Giné and Zinn (1992) for a similar application of the decoupling inequalities). We omit the resulting statements.

2.6. Comparison with previous results. We have already noted, below the statement of Theorem 2.3, that the inequalities there are better than the Hoffmann-Jørgensen type inequalities for U-statistics in Giné and Zinn (1992) in that they represent a decomposition into simpler quantities. Also, as mentioned in the Introduction, Ibragimov and Sharakhmetov (1998, 1999) obtained, except for constants, Proposition 2.1 and its analogue for canonical kernels for $m = 2$ and announced the result for general m; the final results in the present article for $p > 1$ in the nonnegative case (Theorem 2.3) and for $p > 2$ in the canonical case (Theorem 2.5), replacing some sums by maxima and lower moments by quantiles, seem to be more useful. As mentioned above, Corollary 2.2 restricted to $m = 2$ recovers inequalities (4.14) in Klass and Nowicki (1997). The inequalities in the last mentioned article for nonnegative kernels, $p < 1$ and $m = 2$ (the nonconvex case, inequalities (4.13) there) are different from our inequalities in Theorem 2.7 for $m = 2$, although they represent a similar level of decomposition of the p-th moment of the U-statistic. Basically, the difference is that they use inverses of truncated conditional moments whereas we use inverses of tail probabilities together with partial moments. This can be better seen by comparing Hoffmann-Jørgensen, which is Theorem 2.7 for $m = 1$, with their inequality for $m = 1$. The result of Klass and Nowicki (1997) can be described as the iteration of an inequality that follows from Hoffmann-Jørgensen, Paley-Zygmund ((2.13)) and (2.3), as follows. Given ξ_i, $i = 1, \ldots, n$, nonnegative, define v_0 as

$$v_0 = \sup\Big\{v \geq 0 : \sum E\Big(\frac{\xi_i}{v} \wedge 1\Big) \geq 1\Big\} \tag{2.26}$$

or, which is the same, v_0 is the largest number satisfying

$$v_0 = \sum E(\xi_i \wedge v_0). \tag{2.27}$$

Then, the inequality in question is:

Corollary 2.8. (Klass and Nowicki, 1997, Cor. 2.7) *Let* ξ_i, $1 = 1, \dots, n$, *be independent nonnegative random variables. Then, for all* $p > 0$,

$$E\Big(\sum \xi_i\Big)^p \simeq E \max \xi_i^p + v_0^p. \tag{2.28}$$

Proof. Since

$$\sum E(\xi_i \wedge \delta_0) = \sum E\xi_i I_{\xi_i < \delta_0} + \delta_0 \sum \Pr\{\xi_i \geq \delta_0\} \geq \delta_0,$$

it follows that $\delta_0 \leq v_0$. Therefore, if $p \leq 1$, inequality (2.3) and the definition of v_0 give

$$E\Big(\sum \xi_i\Big)^p \leq \Big(\sum E(\xi_i \wedge v_0)\Big)^p + \sum_j E\xi_i^p I_{\xi_i > v_0} \leq v_0^p + 2E \max_i \xi_i^p.$$

And if $p > 1$, Hoffmann-Jørgensen ((H)) and the previous inequality (with $p = 1$) give

$$E\Big(\sum \xi_i\Big)^p \lesssim \Big(E \sum \xi_i\Big)^p + E \max \xi_i^p \lesssim v_0^p + E \max \xi_i^p.$$

For the reverse inequality, if $p > 1$,

$$v_0^p = \Big(\sum E(\xi_i \wedge v_0)\Big)^p \leq E\Big(\sum \xi_i\Big)^p.$$

And if $p < 1$, following the proof of Lemma 2.2 in Klass and Nowicki (1997), we first observe that Paley-Zygmund and the first part of this proof give that for some universal constant C,

$$\begin{aligned} \Pr\Big\{\sum \xi_i \wedge v_0 > \frac{v_0}{2}\Big\} &\geq \frac{1}{4} \frac{\Big(E\sum(\xi_i \wedge v_0)\Big)^2}{E\Big(\sum(\xi_i \wedge v_0)\Big)^2} = \frac{1}{4} \frac{v_0^2}{E\Big(\sum(\xi_i \wedge v_0)\Big)^2} \\ &\geq \frac{C}{4} \frac{v_0^2}{E \max(\xi_i \wedge v_0)^2 + v_0^2} \geq \frac{C}{8}; \end{aligned}$$

therefore,

$$\begin{aligned} E\Big(\sum \xi_i\Big)^p &\geq E\Big(\sum(\xi_i \wedge v_0)\Big)^p \\ &\geq E\Big[\Big(\sum(\xi_i \wedge v_0)\Big)^p I_{\sum(\xi_i \wedge v_0) > v_0/2}\Big] \geq \frac{C}{8} \frac{v_0^p}{2^p}. \end{aligned}$$

□

In fact, if we bound t_0 by $t_0^p \leq 2E\Big(\sum(\xi_i \wedge t_0)\Big)^p$ and apply the above proof to the variables $\xi_i \wedge t_0$, Hoffmann-Jørgensen gives the following seemingly weaker inequality: letting $\tilde{v}_0$ be the parameter v_0 for the smaller variables $\xi_i \wedge t_0$ (note $\tilde{v}_0 \leq v_0$), then

$$E\Big(\sum \xi_i\Big)^p \simeq E \max \xi_i^p + \tilde{v}_0^p. \tag{2.22'}$$

3 Improved moment inequalities and exponential inequalities for $m = 2$

The right-hand side of inequality (2.19) for $m = 1$ is just

$$(3.1)\qquad E\Big|\sum \xi_i\Big|^p \le K^p \max\Big[p^p E\max \xi_i^p, p^{p/2}\Big(\sum E\xi_i^2\Big)^{p/2}\Big],\quad p \ge 2,$$

where ξ_i are independent mean–zero random variables. These inequalities were first obtained by Pinelis (1994). Part of their interest lies on the fact that they are basically equivalent to Bernstein's inequality up to constants. Here is how (3.1) (for all $p \ge 2$) implies Bernstein's inequality up to constants. Assume $\|\xi_i\|_\infty \le A < \infty$ for all i, and set $C^2 = \sum E\xi_i^2$. Then, (3.1) has the form

$$E\Big|\sum \xi_i\Big|^p \le K^p \max\big[p^p A^p,\ p^{p/2} C^p\big],\quad p \ge 2.$$

Let

$$p = \frac{x}{KeA} \wedge \Big(\frac{x}{KeC}\Big)^2$$

for any x for which $p \ge 2$. Then, by Markov's inequality, (3.1) gives, for these values of x,

$$\Pr\Big\{\Big|\sum \xi_i\Big| > x\Big\} \le \begin{cases} \frac{K^p p^p A^p}{x^p} \le e^{-p} & \text{if } p^p A^p \ge p^{p/2} C^p, \\ \frac{K^p p^{p/2} C^p}{x^p} \le e^{-p} & \text{otherwise.} \end{cases}$$

Hence,

$$(3.2)\qquad \Pr\Big\{\Big|\sum \xi_i\Big| > x\Big\} \le e^2 e^{-p} = e^2 \exp\Big\{-\frac{x}{KeA} \wedge \Big(\frac{x}{KeC}\Big)^2\Big\}$$

for all $x > 0$. Similarly, from the iteration (2.19) of the inequalities (3.1) we can obtain exponential inequalities for generalized decoupled U-statistics of any order. However, the inequalities we obtain, while better than the existing ones, are not of the best kind, as we will see below. We illustrate this comment by considering the case $m = 2$. In this case, inequality (2.19) is as follows:

$$\begin{aligned} E\Big|\sum_{i,j} h_{i,j}\Big|^p \le\ & K^p \max\Big[p^p \Big(\sum_{i,j} Eh_{i,j}^2\Big)^{p/2}, p^{3p/2} E_1 \max_i \Big(\sum_j E_2 h_{i,j}^2\Big)^{p/2}, \\ (3.3)\qquad & p^{3p/2} E_2 \max_j \Big(\sum_i E_1 h_{i,j}^2\Big)^{p/2}, p^{2p} E \max_{i,j} |h_{i,j}|^p\Big]. \end{aligned}$$

For bounded canonical kernels $h_{i,j}$ we define

$$(3.4)\qquad A = \max_{i,j} \|h_{i,j}\|_\infty,\quad C^2 = \sum_{i,j} Eh_{i,j}^2,$$

$$B^2 = \max_{i,j}\left[\Big\|\sum_i E_1 h_{i,j}^2(X_i^{(1)}, y)\Big\|_\infty, \Big\|\sum_j E_2 h_{i,j}^2(x, X_j^{(2)})\Big\|_\infty\right].$$

Then, we can proceed as in the deduction of (3.2) from (3.1), and easily obtain from (3.3) that there is a universal constant K such that

$$\Pr\left\{\Big|\sum_{i,j} h_{i,j}\Big| > x\right\} \le K\exp\left\{-\frac{1}{K}\min\left[\frac{x}{C}, \left(\frac{x}{B}\right)^{2/3}, \left(\frac{x}{A}\right)^{1/2}\right]\right\}. \tag{3.5}$$

This inequality also holds for regular canonical U-statistics by the decoupling inequalities of de la Peña and Montgomery-Smith (1995).

Inequality (3.5) is better than the Bernstein type inequality in Arcones and Giné (1993) as it is better for $x \le n^2 A$ and the probability is zero for $x \ge n^2 A$. Inequality (3.5) is suboptimal for small values of x, for which the exponent should be a constant times $-x^2$, just as in the Hanson and Wright (1971) exponential inequality for Gaussian chaos variables of order 2 (see also Ledoux and Talagrand (1991) and Latała (1999)). This suggests that inequality (2.9) is not of the best kind and can be improved.

Next we improve the Rosenthal type inequality (2.9) for $m = 2$ (that is, (3.3)) and deduce from it an exponential inequality for canonical U-statistics of order 2 which does detect the Gaussian portion of the tail probability.

First we show how Talagrand's (1996) extension of Prohorov's inequality to empirical processes, actually in Massart's (1999) version, produces an improved Rosenthal inequality for empirical processes. Then, we will use this inequality to estimate the terms resulting from conditionally applying inequality (3.1) to the U-statistic.

To describe Massart's version of Talagrand's inequality we must establish the setting and define some parameters. Let Z_i be independent random variables with values in some measurable space $(T, \mathcal{T})$, let $\mathcal{F}$ be a countable class of measurable real functions on T, and define

$$S := \sup_{f\in\mathcal{F}} \sum f(Z_i), \quad \sigma^2 = \sup_{f\in\mathcal{F}} \sum E(f(Z_i))^2, \quad a := \max_i \sup_{f\in\mathcal{F}} \big\|f(Z_i)\big\|_\infty.$$

Then,

$$\Pr\left\{|S| \ge 2E|S| + \sigma\sqrt{8x} + 34.5ax\right\} \le e^{-x} \tag{3.6}$$

for all $x > 0$. It follows easily from inequality (3.6) that

$$E|S|^p \le K^p\left[(E|S|)^p + p^{p/2}\sigma^p + p^p a^p\right] \tag{3.7}$$

for some universal constant $K < \infty$ and all $p \ge 1$, in fact, inequality (3.7) for all p large enough and inequality (3.6) for all $x > 0$ are equivalent up to constants. (We do not plan to keep track of constants in the derivation below and, therefore, we refrain from specifying a value for K in (3.7).)

Proposition 3.1. *Let $\{Z_i\}$ be as above, let $\mathcal{F}$ be a countable class of functions such that $Ef^2(Z_i) < \infty$ and $Ef(Z_i) = 0$ for all i. Then, in the notation from the previous paragraph,*

$$E|S|^p \le K^p\Big[(E|S|)^p + p^{p/2}\sigma^p + p^p E\max_i \sup_{f\in\mathcal{F}} \big|f(Z_i)\big|^p\Big] \tag{3.8}$$

for all $p \ge 1$, where K is a universal constant.

Proof. Set $F := \sup_{f\in\mathcal{F}} |f|$ and $M^p := 8\cdot 3^p E\max_i |F(Z_i)|^p$. Since the variables $f(Z_i)$ are centered, we can randomize by independent Rademacher variables ε_i, independent of the Z variables (at the price of increasing the value of the constant K). Set $\tilde{S} := \sup_f \big|\sum \varepsilon_i f(Z_i)\big|$. Then,

$$|\tilde{S}| \le \sup_f \Big|\sum \varepsilon_i f(Z_i) I_{F(Z_i)\le M}\Big| + \sup_f \Big|\sum \varepsilon_i f(Z_i) I_{F(Z_i)>M}\Big| := S_1 + S_2,$$

and notice that, since $ES_1^p \le 2^{p+1}E|S|^p$ (e.g., Lemmas 1.2.6 and 1.4.3 in de la Peña and Giné, 1999), inequality (3.7) gives

$$ES_1^p \le K^p\Big[(E|S|)^p + p^{p/2}\sigma^p + p^p M^p\Big].$$

To estimate ES_2^p we apply the original Hoffmann-Jørgensen inequality (from e.g., Ledoux and Talagrand (1991), (6.9) in page 156) to get

$$ES_2^p \le 2\cdot 3^p\big(t_0^p + E\max_i F(Z_i)^p\big),$$

where t_0 is any number such that $\Pr\{S_2 > t_0\} \le (8\cdot 3^p)^{-1}$. But the choice of M implies that we can take $t_0 = 0$ because

$$\Pr\big\{S_2 > 0\big\} = \Pr\big\{\max_i F(Z_i) > M\big\} \le \frac{1}{8\cdot 3^p},$$

proving the proposition. □

In what follows we will assume, just as above, that the kernels $h_{i,j}$, $i,j \le n$, are completely degenerate and define

$$D = \|(h_{i,j})\|_{L^2\to L^2} := \sup\Big\{E\sum_{i,j} h_{i,j}(X_i^{(1)}, X_j^{(2)}) f_i(X_i^{(1)}) g_j(X_j^{(2)}) \\ : E\sum_i f_i^2(X_i^{(1)}) \le 1, E\sum_j g_j^2(X_j^{(2)}) \le 1\Big\}. \tag{3.9}$$

Theorem 3.2. *There exists a universal constant $K<\infty$ such that, if $h_{i,j}$ are bounded canonical kernels of two variables for the independent random variables $X_i^{(1)}, X_j^{(2)}$, $i.j=1,\dots,n$, $n\in\mathbf{N}$, then*

$$E\Big|\sum_{1\le i,j\le n} h_{i,j}(X_i^{(1)},X_j^{(2)})\Big|^p \le K^p\Big[p^{p/2}\Big(\sum_{i,j}Eh_{i,j}^2\Big)^{p/2}+p^p\|(h_{i,j})\|_{L^2\to L^2}^p$$

$$+p^{3p/2}\Big[E_1\max_i\Big(\sum_j E_2h_{i,j}^2\Big)^{p/2}+E_2\max_j\Big(\sum_i E_1h_{i,j}^2\Big)^{p/2}\Big] \tag{3.10}$$

$$+p^{2p}E\max_{i,j}|h_{i,j}|^p\Big]$$

for all $p\ge 2$.

Inequality (3.10) is strictly better than the right-hand side inequality in (2.9) for $m=2$, that is, than (3.3).

Proof. Inequality (3.1) applied conditionally on the variables $X_i^{(1)}$ gives

$$E\Big|\sum_{i,j}h_{i,j}\Big|^p \le K^pE_1\Bigg(p^{p/2}\Big[\sum_j E_2\Big(\sum_i h_{i,j}\Big)^2\Big]^{p/2}+p^pE_2\sum_j\Big|\sum_i h_{i,j}\Big|^p\Bigg). \tag{3.11}$$

To bound the first summand at the right-hand side of (3.11) we first notice that

$$\Big[\sum_j E_2\Big(\sum_i h_{i,j}\Big)^2\Big]^{1/2}$$
$$=\sup\Big[\sum_i E_2\sum_j h_{i,j}(X_i^{(1)},X_j^{(2)})f_j(X_j^{(2)}) : E\sum_j f_j^2(X_j^{(2)})\le 1\Big],$$

where in fact, the sup is taken only over a countable subset of mean-zero vector functions $(f_1,\dots,f_n)$, dense in the unit ball of $L_2(\mathcal{L}(X_1^{(2)}))\times\cdots\times L_2(\mathcal{L}(X_n^{(2)}))$ for the seminorm $|(f_j)_{j\le n}|=\Big(\sum Ef_j^2(X_j^{(2)})\Big)^{1/2}$. [To see this, first apply duality in ℓ_2^n and then in $L_2(\mathcal{L}(X_j^{(2)}))$ for each j.] So we can apply (3.8) to $Z_i=(h_{i,j})_{j=1}^n$ with $f(Z_i)=E_2\sum_j h_{i,j}(X_i^{(1)},X_j^{(2)})f_j(X_j^{(2)})$. In this case, the right-hand side terms in (3.8) can be estimated as follows. The first term is

$$(E|S|)^2\le E|S|^2=E\Big[\sum_j E_2\Big(\sum_i h_{i,j}\Big)^2\Big]=E\sum_{i,j}h_{i,j}^2=C^2.$$

For the second we see that, since, by the previous duality argument,

$$\sum_i E_1\Big(E_2\sum_j h_{i,j}(X_i^{(1)},X_j^{(2)})f_j(X_j^{(2)})\Big)^2\le\|(h_{i,j})\|_{L^2\to L^2}^2=D^2,$$

it follows that $\sigma \le D$. The third term is

$$E \max_i \sup_f |f(Z_i)|^p = E_1 \max_i \sup_{E\sum f_j^2 \le 1} \Big[E_2 \sum_j h_{i,j}(X_i^{(1)}, X_j^{(2)}) f_j(X_j^{(2)})\Big]^p$$
$$\le E_1 \max_i \sup_{E\sum f_j^2 \le 1} \Big[\Big(E_2 \sum_j h_{i,j}^2\Big)^{1/2} \Big(E \sum_j f_j^2\Big)^{1/2}\Big]^p$$
$$= E_1 \max_i \Big(E_2 \sum_j h_{i,j}^2\Big)^{p/2}.$$

Thus, inequality (3.8) gives

$$(3.12) \quad p^{p/2} E\Big(\sum_j E_2\big(\sum_i h_{i,j}\big)^2\Big)^{p/2}$$
$$\le K^p \Big[p^{p/2} C^p + p^p D^p + p^{3p/2} E_1 \max_i \Big(E_2 \sum_j h_{i,j}^2\Big)^{p/2}\Big].$$

To estimate the second summand at the right-hand side of (3.11), we apply (3.1) once more and obtain

$$(3.13) \qquad p^p E_2 \sum_j E_1 \Big|\sum_i h_{i,j}\Big|^p$$
$$\le K^p \Big[p^{3p/2} E_2 \sum_j \Big(\sum_i E_1 h_{i,j}^2\Big)^{p/2} + p^{2p} E \sum_{i,j} |h_{i,j}|^p\Big].$$

Thus, to complete the proof of the theorem it suffices to replace the sum in j and the sum in i, j respectively by maxima in j and in i, j on the terms at the right-hand side of this inequality. But this is an easy application of inequality (2.6). For completeness sake, here it is. Applying (2.6) with $\alpha = 3$ and $p/2$ instead of p, the first term at the right of (3.13) bounds as:

$$p^{3p/2} E_2 \sum_j \Big(\sum_i E_1 h_{i,j}^2\Big)^{p/2}$$
$$\le 2^{1+3p/2}(1 + (p/2)^3) \Big[\Big(\frac{p}{2}\Big)^{3p/2} E_2 \max_j \Big(\sum_i E_1 h_{i,j}^2\Big)^{p/2} + C^p\Big],$$

which produces the conversion of the sum into a maximum without increasing the order of the multiplicative constant in front of C^p. The second term in (3.13) requires two steps. First, we apply (2.6) for $p/2$ and $\alpha = 4$, conditionally on $\{X_i^{(1)}\}$:

$$(3.14) \qquad p^{2p} E \sum_{i,j} |h_{i,j}|^p$$
$$\le 2^{2p+1}(1 + (p/2)^4) E_1 \sum_i \Big[\Big(\frac{p}{2}\Big)^{2p} E_2 \max_j |h_{i,j}|^p + \Big(\sum_j E_2 h_{i,j}^2\Big)^{p/2}\Big].$$

We apply (2.6) with respect to E_1, for $p/2$ and $\alpha = 0$, to the second term at the right-hand side of (3.14) and we obtain the bound

$$2^{2p+3}(1+(p/2)^4)\Big[E_1 \max_i \Big(\sum_j E_2 h_{i,j}^2\Big)^{p/2} + C^p\Big],$$

which is in terms of some of the quantities appearing at the right-hand side of (3.10) and with coefficients of lower order. As for the first term at the right of (3.14), we apply (2.6) with respect to E_1, again for $p/2$ and $\alpha = 4$, and get it bounded by

$$2^{4p+2}(1+(p/2)^4)^2\Big[\Big(\frac{p}{2}\Big)^{2p} E \max_{i,j} |h_{i,j}|^p + E_2\Big(\sum_i E_1 \max_j h_{i,j}^2\Big)^{p/2}\Big].$$

Here the first term coincides with the last one in (3.10), and the second term is dominated by

$$K^p E_2\Big[\sum_j\Big(\sum_i E_1 h_{i,j}^2\Big)\Big]^{p/2}.$$

Applying inequality (R_1) with respect to E_2, this is in turn dominated by

$$K^p\Big(\frac{p}{2}\Big)^{p/2} E_2 \sum_j\Big(\sum_i E_1 h_{i,j}^2\Big)^{p/2} + K^p C^p,$$

and the first summand has already been handled above (first term at the right of (3.13)). Collecting terms we obtain inequality (3.10). □

Theorem 3.2 implies the following moment inequality and exponential bound for bounded kernels.

Theorem 3.3. *There exist universal constants $K < \infty$ and $L < \infty$ such that, if $h_{i,j}$ are bounded canonical kernels of two variables for the independent random variables $X_i^{(1)}, X_j^{(2)}$, $i.j = 1, \ldots, n$, and if A, B, C, D are as defined in (3.4) and (3.9), then*

(3.15)
$$E\Big|\sum_{1\le i,j\le n} h_{i,j}(X_i^{(1)}, X_j^{(2)})\Big|^p \le K^p\Big[p^{p/2}C^p + p^p D^p + p^{3p/2} B^p + p^{2p} A^p\Big]$$

for all $p \ge 2$ and, equivalently,

$$\Pr\Big\{\Big|\sum_{i,j\le n} h_{i,j}(X_i^{(1)}, X_j^{(2)})\Big| \ge x\Big\}$$

$$\le L\exp\Big[-\frac{1}{L}\min\Big(\frac{x^2}{C^2}, \frac{x}{D}, \frac{x^{2/3}}{B^{2/3}}, \frac{x^{1/2}}{A^{1/2}}\Big)\Big] \tag{3.16}$$

for all $x > 0$.

The moment inequality is immediate from Theorem 3.2, and the equivalence with the exponential inequality follows just as (3.2) follows from (3.1) in one direction, and, in the other, by integration of tail probabilities.

Next we comment on the exponential inequality. For comparison purposes, let $h_{i,j}(X_i^{(1)}, X_j^{(2)}) = g_i g_j' x_{i,j}$ with g_i, g_j' independent standard normal. In this case,

$$C^2 = \sum_{i,j} x_{i,j}^2 \quad \text{and} \quad D = \sup\Big\{\sum_{i,j} u_i v_j x_{i,j} : \ \sum u_i^2 \le 1, \sum v_j^2 \le 1\Big\}$$

and the Gaussian chaos inequalities of Hanson and Wright (1971) and Latała (1999) yield the existence of universal constants $0 < k < K < \infty$ such that

$$\Pr\Big\{\Big|\sum_{i,j} h_{i,j}\Big| \ge K(Cx^{1/2} + Dx)\Big\} \le e^{-x}$$

and

$$\Pr\Big\{\Big|\sum_{i,j} h_{i,j}\Big| \ge k(Cx^{1/2} + Dx)\Big\} \ge k \wedge e^{-x}.$$

By the central limit theorem for canonical U-statistics, this implies that the coefficients of x^2 and x in (3.16) are correct (except for K). It is natural to have terms in smaller powers of x in (3.16) e.g., by comparison with Bernstein's inequality for sums of independent random variables. In fact, the term in $x^{1/2}$ cannot be avoided, at least up to logarithmic factors. To see this, consider the product V of two independent centered Poisson variables with parameter 1, which is the limit in law of $V_n = \sum_{i,j \le n} X_i^{(n)} Y_j^{(n)}$, where $X_i^{(n)}$ and $Y_j^{(n)}$ are centered Bernoulli random variables with parameter $p = 1/n$; then, for large x, the tail probabilities of V are of the order of $\exp(-x^{1/2} \log x)$, and therefore, so are those of V_n for large n. Also, note that the term in $x^{2/3}$ in the exponent corresponds, up to logarithmic factors, to the tail probabilities of the product of two independent random variables, one normal and the other centered Poisson.

If $X, Y, X_i^{(1)}, X_j^{(2)}$ are i.i.d., $h_{i,j} = h$ for all i, j, and h is completely degenerate, then the parameters defined by (3.4) and (3.8) become

$$A = \|h\|_\infty, \quad B^2 = n\big(\|E_Y h^2(x, Y)\|_\infty + \|E_X h^2(X, y)\|_\infty\big), \quad C^2 = n^2 E h^2$$

and

$$\begin{aligned} D &= n \sup\Big\{Eh(X,Y)f(X)g(Y) : \ Ef^2(X) \le 1, Eg^2(Y) \le 1\Big\} \\ &:= n\|h\|_{L_2 \mapsto L_2}, \end{aligned}$$

where $\|h\|_{L_2 \mapsto L_2}$ is the norm of the operator of $L_2(\mathcal{L}(X))$ with kernel h. Then, inequalities (3.15) and (3.16) become

Corollary 3.4. *Under the above assumptions, there exist universal constants $K < \infty$, $L < \infty$ such that, for all $n \in \mathbf{N}$ and $p \geq 2$,*

$$E\Big|\sum_{i,j\leq n} h(X_i^{(1)}, X_j^{(2)})\Big|^p \leq K^p\Big[p^{p/2}n^p(Eh^2)^{p/2} + p^p n^p \|h\|^p_{L_2\mapsto L_2} + p^{3p/2}n^{p/2}\big(\|E_Y h^2\|_\infty + \|E_X h^2\|_\infty\big)^{p/2} + p^{2p}\|h\|^p_\infty\Big] \tag{3.17}$$

and

$$\Pr\Big\{\Big|\sum_{i,j\leq n} h(X_i^{(1)}, X_j^{(2)})\Big| \geq x\Big\} \leq K\exp\Big[-\frac{1}{K}\min\Big(\frac{x^2}{n^2Eh^2}, \frac{x}{n\|h\|_{L_2\mapsto L_2}}, \frac{x^{2/3}}{\big[n(\|E_Y h^2\|_\infty + \|E_X h^2\|_\infty)\big]^{1/3}}, \frac{x^{1/2}}{\|h\|_\infty^{1/2}}\Big)\Big]. \tag{3.18}$$

Inequality (3.18) provides an analogue of Bernstein's inequality for degenerate U-statistics of order 2: note that inequalities (3.15), (3.16), (3.17) and (3.18) can all be 'undecoupled' using the result of de la Peña and Montgomery-Smith's (1995). It should also be noted that this exponential inequality for canonical U-statistics is strong enough to imply the sufficiency part of the law of the iterated logarithm for these objects: this can be seen by applying it to the kernels h_n in Steps 7 and 8 of the proof of Theorem 3.1 in Giné, Kwapień, Latała and Zinn (1999) (and using some of the computations there for the parameters C to D). Neither inequality (3.5) nor any of the previously published inequalities for U-statistics can do this.

Acknowledgement. We thank Stanisław Kwapień for several useful conversations.

References

Arcones, M. and Giné, E. (1993). Limit theorems for U-processes. *Ann. Probab.* **21**, 1494–1542.

de la Peña, V. and Giné, E. (1999). *Decoupling: From Dependence to Independence.* Springer-Verlag, New York.

de la Peña, V. and Montgomery–Smith, S. (1995). Decoupling inequalities for the tail probabilities of multivariate U-statistics. *Ann. Probab.* **23**, 806–816.

Figiel, T., Hitczenko, P., Johnson, W.B., Schechtman, G., and Zinn, J. (1997). Extremal properties of Rademacher functions with applications to the Khintchine and Rosenthal inequalities. *Trans. Amer. Math. Soc.* **349**, 997–1027.

GINÉ, E., KWAPIEŃ, S., LATAŁA, R., AND ZINN, J. (2000). The LIL for canonical U-statistics of order two. To appear.

GINÉ, E. AND ZHANG, C.-H. (1996). On integrability in the LIL for degenerate U-statistics. *J. Theoret. Probab* **9**, 385–412.

GINÉ, E. AND ZINN, J. (1983). Central limit theorems and weak laws of large numbers in certain Banach spaces. *Zeits. Wahrsch. v. Geb.* **62**, 323–354.

GINÉ, E. AND ZINN, J. (1992). On Hoffmann-Jørgensen's inequality for U-processes. *Probability in Banach Spaces 8* 80-91. Birkhäuser, Boston.

HANSON, D. L. AND WRIGHT, F. T. (1971). A bound on tail probabilities for quadratic forms in independent random variables. *Ann. Math. Statist.* **42**, 52–61.

HITCZENKO, P. (1988). Comparison of moments for tangent sequences of random variables. *Probab. Th. Rel. Fields* **78**, 223–230.

JOHNSON, W. B., SCHECHTMAN, G., AND ZINN, J. (1985). Best constants in moment inequalities for linear combinations of independent and exchangeable random variables. *Ann. Probab.* **13**, 234–253.

IBRAGIMOV, R. AND SHARAKHMETOV, SH. (1998). Exact bounds on the moments of symmetric statistics. In: Abstracts of the 7th Vilnius Conference on Probability Theory and Mathematical Statistics/ 22nd European Meeting of Statisticians, Vilnius, pp. 243–244.

IBRAGIMOV, R. AND SHARAKHMETOV, SH. (1999). Analogues of Khintchine, Marcinkiewicz-Zygmund and Rosenthal inequalities for symmetric statistics. *Scand. J. Statist.* **26**, 621–623.

KAHANE, J.-P. (1968). *Some Random Series of Functions.* Heath, Lexington, Massachusetts.

KLASS, M. AND NOWICKI, K. (1997). Order of magnitude bounds for expectations of Δ_2 functions of nonnegative random bilinear forms and generalized U-statistics. *Ann. Probab.* **25**, 1471–1501.

KWAPIEŃ, S. AND SZULGA, J. (1991). Hypercontraction methods in moment inequalities for series of independent random variables in normed spaces. *Ann. Probab.* **19**, 369–379.

KWAPIEŃ, S. AND WOYCZYŃSKI, W. (1992). *Random Series and Stochastic Integrals: Single and Multiple.* Birkhäuser, Boston.

LATAŁA, R. (1997). Estimation of moments of sums of independent random variables. *Ann. Probab.* **25**, 1502–1513.

LATAŁA, R. (1999). Tails and moment estimates for some type of chaos. *Studia Math.* **135**, 39-53.

LATAŁA, R. AND ZINN, J. (2000). Necessary and sufficient conditions for the strong law of large numbers for U-statistics. *Ann. Probab.*, to appear.

LEDOUX, M. AND TALAGRAND, M. (1991). *Probability in Banach Spaces: Isoperimetry and Processes.* Springer, New York.

MASSART, P. (2000). About the constants in Talagrand's concentration inequalities for empirical processes. *Ann. Probab.*, to appear.

PINELIS, I. (1994). Optimum bounds for the distributions of martingales in Banach spaces. *Ann. Probab.* **22**, 1679–1706.

TALAGRAND, M. (1996). New concentration inequalities in product spaces. *Invent. Math.* **126**, 505–563.

UTEV, S. A. (1985). Extremal problems in moment inequalities. In: *Limit Theorems in Probability Theory*, Trudy Inst. Math., Novosibirsk, 56–75 (in Russian).

Evarist Giné
Department of Mathematics
and Department of Statistics
University of Connecticut
Storrs, CT 06269
gine@uconnvm.uconn.edu

Rafał Latała
Institute of Mathematics
Warsaw University
Banacha 2
02-097 Warszawa, Poland
rlatala@mimuw.edu.pl

Joel Zinn
Department of Mathematics
Texas A&M University
College Station, TX 77843
jzinn@math.tamu.edu

A Multiplicative Inequality for Concentration Functions of n-Fold Convolutions

Friedrich Götze* and Andrei Yu. Zaitsev‡

ABSTRACT We estimate the concentration functions of n-fold convolutions of one-dimensional probability measures. The main result is a supplement to the results of Götze and Zaitsev (1998). We show that the estimation of concentration functions at arguments of bounded size can be reduced to the estimation of these functions at arguments of size $O(\sqrt{n})$ which is easier.

1 Introduction

Let us first introduce some notation. Let $\mathfrak{F}$ denote the set of probability distributions defined on the Borel σ-field of subsets of the real line $\mathbf{R}$, $\mathbf{L}(\xi) \in \mathfrak{F}$ the distribution of a random variable ξ, and $\mathbf{I}\{A\}$ the indicator function of an event A. For $F \in \mathfrak{F}$ the concentration function is defined by $Q(F, b) = \sup_x F\{[x, x+b]\}$, $b \geq 0$. For $F, H \in \mathfrak{F}$ we denote the corresponding distribution functions by $F(x)$, $H(x)$ and the characteristic functions by $\widehat{F}(t)$, $\widehat{H}(t)$. Let $E_a \in \mathfrak{F}$ be the distribution concentrated at a point $a \in \mathbf{R}$, $E = E_0$. Products and powers of measures will be understood in the convolution sense: $FH = F * H$, $H^m = H^{m*}$, $H^0 = E$. For $F = \mathbf{L}(\xi) \in \mathfrak{F}$ we shall use the notation $\overline{F} = \mathbf{L}(-\xi)$ and $\widetilde{F} = F\overline{F}$. The distribution $\widetilde{F}$ is called symmetrized. By $c(\cdot)$ we shall denote positive quantities depending on the arguments only. The symbol c will be used for absolute positive constants. Note that $c(\cdot)$, c can be different in different (or even in the same) formulas. The ends of proofs will be marked by $\square$.

The concentration functions were introduced and preliminarily estimated by Lévy (1937), see as well Doeblin (1939). We estimate the concentration functions of n-fold convolutions of one-dimensional probability measures. The bounds for these convolutions were obtained by Kolmogorov

*,‡Research supported by grant RFBR-DFG 99-01-04027 and by the SFB 343 in Bielefeld.

‡Research supported by the Russian Foundation of Basic Research (RFBR), grants 99-01-00112 and 96-15-96199.

(1956, 1958), Rogozin (1961a,b), Le Cam (1965), Esseen (1966, 1968), Kesten (1969, 1972), Morozova (1977), Postnikova and Yudin (1978), Arak (1981), Miroshnikov and Rogozin (1980, 1982), Hall (1983), Griffin, Jain and Pruitt (1984), Arak and Zaitsev (1988), Mukhin (1989), Nagaev and Khodzhabagyan (1996) and others.

The aim of this paper is to provide a supplement to a previous paper Götze and Zaitsev (1998) which is abbreviated in the following as GZ. We generalize Theorem 2 of that paper proving the following result.

Theorem 1. *Assume that a distribution* $F \in \mathfrak{F}$ *is represented in the form*

$$F = (1-p)\,U + p\,V, \qquad U, V \in \mathfrak{F}, \quad 0 < p < 1. \tag{1.1}$$

Let X *be a random variable with* $\mathbf{L}(X) = U$. *Suppose that*

$$0 < \sigma^2 = \mathbf{E}\,X^2 < \infty, \qquad \mathbf{E}\,X = 0, \tag{1.2}$$

and

$$b \geq \frac{\nu_n}{\sigma^2}, \qquad \text{where} \quad \nu_n = \mathbf{E}\,X^2 \min\{\,|X|, \sigma\sqrt{n}\,\}. \tag{1.3}$$

Let r, s *be integers,* $0 \leq r < s \leq n$. *Then, for any distribution* $H \in \mathfrak{F}$,

$$\begin{aligned} Q(H\,F^n, b) \;\leq\; & \mathbf{E}\,\frac{cb}{\sigma\sqrt{n-\mu}}\,Q(H\,V^\mu, \sigma\sqrt{n-\mu})\,\mathbf{I}\{\,r \leq \mu < s\,\} \\ & + \mathbf{E}\,Q(H\,U^{n-\mu}V^\mu, b)\,\mathbf{I}\{\,\mu \geq s\,\} \\ & + \min\Big\{\,1, \frac{cb}{\sigma\sqrt{n-r}}\,Q(H, \sigma\sqrt{n})\Big\}\,\mathbf{P}\,\{\,\mu < r\,\}, \end{aligned} \tag{1.4}$$

where μ *denotes a random variable having a binomial distribution with parameters* n *and* p. *Moreover,*

$$\begin{aligned} Q(H\,F^n, b) \;\leq\; & \frac{cb}{\sigma\sqrt{n\,(1-p)}}\,Q(H\,V^r, \sigma\sqrt{n}) \\ & + \min\Big\{\,1, \frac{cb}{\sigma\sqrt{n-r}}\,Q(H, \sigma\sqrt{n})\Big\}\,\mathbf{P}\,\{\,\mu < r\,\}. \end{aligned} \tag{1.5}$$

Theorem 2 of GZ (1998) is a particular case of our Theorem 1 appearing when $H = E$. The moment restrictions in Theorem 1 are imposed on the distribution U only. The distributions H and V are arbitrary and therefore the initial distribution F may have arbitrarily bad moment properties.

Taking in (1.5) $r = 0$, we obtain

Corollary 1. *Under the conditions of Theorem* 1

$$Q(H\,F^n, b) \leq \frac{cb}{\sigma\sqrt{n\,(1-p)}}\,Q(H, \sigma\sqrt{n}). \tag{1.6}$$

An important particular case of Theorem 1 and Corollary 1 appears when n is replaced by an integer $n\alpha$ and $H = F^{n(1-\alpha)}$ where $0 < \alpha < 1$. Then the bound (1.6) can be rewritten in the form

$$Q(F^n, b) \le \frac{cb}{\sigma\sqrt{n\alpha(1-p)}}\, Q\big(F^{n(1-\alpha)}, \sigma\sqrt{n\alpha}\big). \tag{1.7}$$

Theorem 1 and Corollary 1 provide in a sense "multiplicative inequalities" for concentration functions of convolutions. The inequality (1.7) reduces the problem of estimation of $Q(F^n, b)$ to that of $Q\big(F^{n(1-\alpha)}, \sigma\sqrt{n\alpha}\big)$. The trivial inequality

$$Q\big(F^{n(1-\alpha)}, \sigma\sqrt{n\alpha}\big) \le \Big(1 + \frac{\sigma\sqrt{n\alpha}}{b}\Big)\, Q(F^{n(1-\alpha)}, b) \tag{1.8}$$

(see (2.4)) together with (1.7) may return the problem to the initial stage (in particular, when $Q(F^n, b)$ decreases with some negative power of n). However, the inequality (1.8) may lead to a loss of precision. For instance, if F is concentrated on a one-dimensional lattice with step size $2\sigma\sqrt{n\alpha}$, we have

$$Q\big(F^{n(1-\alpha)}, \sigma\sqrt{n\alpha}\big) = Q(F^{n(1-\alpha)}, b), \qquad \text{for} \quad 0 \le b \le \sigma\sqrt{n\alpha}.$$

It is clear that it is much easier to estimate the concentration function for the large value $\sigma\sqrt{n\alpha}$ of the argument than for some fixed b.

Applying Theorem 1 and Corollary 1, we can use known classical bounds for concentration functions. We mention in this connection the papers by Rogozin (1961a), Esseen (1968), Kesten (1969), Arak (1981), Miroshnikov and Rogozin (1980), Hall (1983), Griffin, Jain and Pruitt (1984), and Petrov (1995). For example, the Esseen bound (1968) implies the following result (see the end of Section 2 for a proof).

Corollary 2. *Let the conditions of Theorem 1 be satisfied. Then, for any* $\delta > 0$,

$$Q(F^n, b) \le \frac{cb\,(\delta + \sigma)}{\delta\,\sigma\, n\sqrt{(1-p)\,D(\widetilde{F}, \delta\sqrt{n})}} \tag{1.9}$$

where

$$D(F, b) = \int_{-\infty}^{\infty} \min\{x^2 b^{-2}, 1\}\, F\{dx\}, \qquad F \in \mathfrak{F}, \quad b > 0. \tag{1.10}$$

Esseen (1968) (see Petrov (1976), inequality (2.7) of Chapter III) proved that

$$Q(F^n, b) \le \frac{c}{\sqrt{n D(\widetilde{F}, b)}}. \tag{1.11}$$

Corollary 2 implies the following result.

Corollary 3. *Let* $F \in \mathfrak{F}$, $b \geq 0$, $\delta > 0$. *Then*

$$Q(F^n, b) \leq \frac{c(F, b, \delta)}{n\sqrt{D(\widetilde{F}, \delta\sqrt{n})}}.$$

For the proof it suffices to note that after a shift every non-degenerate distribution can be represented in the form (1.1) with $p = 1/2$ and a non-degenerated U having bounded support. This yields the result for $b \geq \nu/\sigma^2$, where $\nu = \mathbf{E}\,|X|^3$. For $b < \nu/\sigma^2$ we can apply the result for $b = \nu/\sigma^2$, using the monotonicity of concentration functions. Corollary 3 is Theorem 1 from the paper of GZ (1998), which is a sharpening of a result of Esseen (1968) who showed that $Q(F^n, b) = o(n^{-1/2})$ as $n \to \infty$ iff the distribution F has an infinite second moment (see as well Morozova (1977)). However the proof of Corollary 3 in the present paper is somewhat easier and we obtain here a more explicit form of $c(F, b, \delta)$. For the connection of Corollary 3 with previous results about concentration functions, see GZ (1998).

Note that in the proof of Corollary 3 we use the inequality (1.11). Nevertheless, Corollary 3 can be considered as an improvement of the inequality (1.11). Indeed, the latter can be rewritten in the form

$$Q(F^n, b) \leq c\left(n \int \min\{\,x^2 b^{-2},\, 1\,\}\, \widetilde{F}\{dx\}\right)^{-1/2}. \tag{1.12}$$

Comparing (1.12) with the inequality

$$Q(F^n, b) \leq c(F, b)\left(n \int \min\{\,x^2 b^{-2},\, n\,\}\, \widetilde{F}\{dx\}\right)^{-1/2}, \tag{1.13}$$

which follows from Corollary 3 with $\delta = b$, we see that for any distributions with infinite variance the inequality (1.13) is sharper with respect to the order in n than (1.12). Note however that it is impossible to change $c(F, b)$ in (1.13) by some absolute constant c. The corresponding example is given by the distribution $F = F_n = \frac{1}{2}\, E_{-n} + \frac{1}{2}\, E_n$. It is well known that $Q(F_n^n, 1)$ behaves as $O(n^{-1/2})$ when $n \to \infty$. On the other hand,

$$\left(n \int \min\{\,x^2,\, n\,\}\, \widetilde{F}_n\{dx\}\right)^{-1/2} \leq \frac{c}{n}.$$

This implies that $c(F_n, 1) \geq c\sqrt{n}$ in (1.13).

Esseen (1968) proved that, for any $F \in \mathfrak{F}$ and $b > 0$,

$$Q(F, b) \leq cb \int_{|t| \leq b^{-1}} \bigl|\,\widehat{F}(t)\,\bigr|\, dt. \tag{1.14}$$

Applying (1.14) to the distribution F^n, we obtain

$$Q(F^n, b) \le cb \int_{|t| \le b^{-1}} \big|\widehat{F}(t)\big|^n \, dt. \tag{1.15}$$

On the other hand, using (1.7) and then (1.14), we see that

$$Q(F^n, b) \le \frac{cb}{\sqrt{1-p}} \int_{|t|\sigma\sqrt{n\alpha} \le 1} \big|\widehat{F}(t)\big|^{n(1-\alpha)} \, dt, \tag{1.16}$$

if $n\alpha$ is integer and $b \ge \mathbf{E}\,|X|^3/\sigma^2$. A comparison of (1.15) with (1.16) demonstrates an advantage of using the inequality (1.7).

Our proofs are based on non-uniform estimates in the Central Limit Theorem (CLT) and on elementary properties of concentration functions (see the proof of Lemma 1 in GZ (1998) and Zaitsev (1987, 1992)). In this respect our proofs differ from most of the previous papers, where Esseen's (1968) method of characteristic functions had been extensively used. One should note however that the CLT approach was applied in the seminal paper of Kolmogorov (1958). He used the uniform Berry–Esseen bound in the CLT since non-uniform ones were not known at that time.

2 Proofs

Lemma 1 (GZ (1998, Lemma 3)). *Let $\xi, \xi_1, \xi_2, \dots, \xi_n$ be i.i.d. random variables, $\mathbf{E}\,\xi = 0$, $W = \mathbf{L}(\xi_1 + \dots + \xi_n)$,*

$$B^2 = n\,\mathbf{E}\,\xi^2 > 0, \qquad \beta = n\,\mathbf{E}\,\xi^2 \min\{\,|\xi|,\, B\,\} < \infty, \tag{2.1}$$

and let $b \ge \frac{\beta}{B^2}$. Then, for any $G \in \mathfrak{F}$, we have

$$Q(WG, b) \le \frac{cb}{B}\, Q(G, B). \tag{2.2}$$

Note that the inequality (2.2) is a particular case of the inequality (1.6) appearing when $p = 0$.

We need the following well-known simple properties of concentration functions, which are valid for any $F, H \in \mathfrak{F}$ and $\gamma, \gamma_1, \gamma_2 > 0$:

$$Q(FH, \gamma) \le \min\{\, Q(F, \gamma),\, Q(H, \gamma)\,\}; \tag{2.3}$$

$$Q(F, \gamma_1) \le \big(1 + \lceil \gamma_1/\gamma_2 \rceil\big)\, Q(F, \gamma_2), \tag{2.4}$$

where $\lceil \cdot \rceil$ denotes the integer part of a number (see, e.g., Hengartner and Theodorescu (1973, Theorems 2.1.1 and 1.5.3)).

Proof of Theorem 1. It is known that HF^n can be written in the form

$$HF^n = \mathbf{L}(S_n), \qquad \text{where} \quad S_n = \zeta + \sum_{i=1}^{n} (1-\mu_i)\xi_i + \mu_i \eta_i,$$

and $\zeta,\ \xi_i,\ \eta_i,\ \mu_i$ are jointly independent random variables with

$$\mathbf{L}(\zeta) = H, \qquad \mathbf{L}(\xi_i) = U, \qquad \mathbf{L}(\eta_i) = V, \qquad \mathbf{L}(\mu_i) = (1-p)E + pE_1.$$

Define $\mu = \sum_{i=1}^{n} \mu_i$. Obviously, μ has binomial distribution with parameters n and p. Given a fixed value of μ, the random variable S_n has conditional distribution $HU^{n-\mu}V^{\mu}$. Hence, for any $x \in \mathbf{R}$ we have

$$\begin{aligned} \mathbf{P}\{S_n \in [x, x+b]\} &= \mathbf{E}\,\mathbf{I}\{S_n \in [x, x+b]\} \\ &= \mathbf{E}\,\mathbf{E}\{\mathbf{I}\{S_n \in [x, x+b]\} \,|\, \mu\} \\ &\le \mathbf{E}\,Q\big(HU^{n-\mu}V^{\mu}, b\big). \end{aligned}$$

Therefore,

$$Q(HF^n, b) \le \mathbf{E}\,Q\big(HU^{n-\mu}V^{\mu}, b\big). \tag{2.5}$$

In view of $\nu_{n-\mu} \le \nu_n$ and applying inequality (2.2) of Lemma 1 with $W = U^{n-\mu}$, $G = HV^{\mu}$, $B = \sigma\sqrt{n-\mu}$, we see that

$$\begin{aligned} \mathbf{E}\,Q\big(HU^{n-\mu}V^{\mu}, b\big) \le\ & \mathbf{E}\,\frac{cb}{\sigma\sqrt{n-\mu}}\, Q\big(HV^{\mu}, \sigma\sqrt{n-\mu}\big)\,\mathbf{I}\{r \le \mu < s\} \\ & + \mathbf{E}\,Q\big(HU^{n-\mu}V^{\mu}, b\big)\,\mathbf{I}\{\mu \ge s\} \\ & + \mathbf{E}\,Q\big(HU^{n-\mu}V^{\mu}, b\big)\,\mathbf{I}\{\mu < r\} \end{aligned} \tag{2.6}$$

and

$$\begin{aligned} &\mathbf{E}\,Q\big(HU^{n-\mu}V^{\mu}, b\big)\,\mathbf{I}\{\mu < r\} \\ &\qquad \le \min\Big\{1,\ \frac{cb}{\sigma\sqrt{n-r}}\, Q\big(H, \sigma\sqrt{n}\big)\Big\}\,\mathbf{P}\{\mu < r\}. \end{aligned} \tag{2.7}$$

The inequality (1.4) now follows from (2.5)–(2.7).

Using the relations $\mathbf{P}\{\mu = n\} = p^n$, (1.4) with $s = n$ and (2.3), we obtain

$$\begin{aligned} Q(HF^n, b) \le\ & \frac{cb}{\sigma}\, Q\big(HV^{r}, \sigma\sqrt{n}\big)\,\mathbf{E}\,\frac{1}{\sqrt{n-\mu}}\,\mathbf{I}\{r \le \mu < n\} \\ & + p^n\, Q\big(HV^{n}, b\big) \\ & + \min\Big\{1,\ \frac{cb}{\sigma\sqrt{n-r}}\, Q\big(H, \sigma\sqrt{n}\big)\Big\}\,\mathbf{P}\{\mu < r\}. \end{aligned} \tag{2.8}$$

Applying the Hölder inequality, we derive

$$\begin{aligned}\mathbf{E}\,\frac{1}{\sqrt{n-\mu}}\,\mathbf{I}\{r\le\mu<n\} &\le c\,\mathbf{E}\,\frac{1}{\sqrt{n-\mu+1}}\\ &\le c\Big(\mathbf{E}\,\frac{1}{n-\mu+1}\Big)^{1/2}\le\frac{c}{\sqrt{n\,(1-p)}}.\end{aligned}\tag{2.9}$$

Using (1.2) and (1.3), it is easy to see that

$$\begin{aligned}\nu_n &\ge \nu_1=\mathbf{E}\,|X|^3\,\mathbf{I}\{\,|X|\le\sigma\}+\sigma\,\mathbf{E}\,|X|^2\,\mathbf{I}\{\,|X|>\sigma\}\\ &\ge \big(\,\mathbf{E}\,|X|^2\,\mathbf{I}\{\,|X|\le\sigma\}\big)^{3/2}+\big(\,\mathbf{E}\,|X|^2\,\mathbf{I}\{\,|X|>\sigma\}\big)^{3/2}\\ &\ge c\sigma^3.\end{aligned}\tag{2.10}$$

The relations (1.3) and (2.10) together imply

$$b\ge\nu_n\,\sigma^{-2}\ge c\sigma.\tag{2.11}$$

According to (2.3), (2.4) and (2.11), we have

$$\begin{aligned}Q(HV^n,\,b)\le Q(HV^r,\,b) &\le \Big(\frac{b}{\sigma\sqrt{n}}+1\Big)\,Q(HV^r,\,\sigma\sqrt{n})\\ &\le \frac{cb}{\sigma}\,Q(HV^r,\,\sigma\sqrt{n}).\end{aligned}\tag{2.12}$$

Moreover,

$$p^n=\big(1-(1-p)\big)^n\le e^{-n(1-p)}\le\frac{c}{\sqrt{n\,(1-p)}}.\tag{2.13}$$

The inequality (1.5) follows from (2.8), (2.9), (2.12) and (2.13). □

Proof of Corollary 2. In view of (2.3), (2.11), we may assume without loss of generality that n is even and $n\ge 2$. Using (1.7) with $\alpha={}^1\!/_2$, (1.11), (2.3) and (2.4), we obtain

$$\begin{aligned}Q(F^n,\,b) &\le \frac{cb}{\sigma\sqrt{n(1-p)}}\;Q(F^{n/2},\,\sigma\sqrt{n})\\ &\le \frac{cb}{\sigma\sqrt{n(1-p)}}\;\big(1+\sigma\,\delta^{-1}\big)\,Q(F^{n/2},\,\delta\sqrt{n})\\ &\le \frac{cb}{\sigma\sqrt{n(1-p)}}\;\frac{1+\sigma\,\delta^{-1}}{\sqrt{n\,D(\widetilde{F},\,\delta\sqrt{n})}}\\ &= \frac{cb\,(\delta+\sigma)}{\delta\,\sigma\,n\sqrt{(1-p)\,D(\widetilde{F},\,\delta\sqrt{n})}}.\end{aligned}\tag{2.14}$$

□

References

ARAK, T. V. (1981). On the convergence rate in Kolmogorov's uniform limit theorem. I. *Theor. Probab. Appl.* **26**, 219–239.

ARAK, T. V., AND ZAITSEV, A. YU. (1988). Uniform limit theorems for sums of independent random variables. *Proc. Steklov Inst. Math.* **174**.

DOEBLIN, W. (1939). Sur les sommes d'un grand nombre des variables aléatoire indépendantes. *Bull Sci. Math.* **63**, 23–32; 35–64.

ESSEEN, C.-G. (1966). On the Kolmogorov–Rogozin inequality for the concentration function. *Z. Wahrscheinlichkeitstheor. verw. Geb.* **5**, 210–216.

ESSEEN, C.-G. (1968). On the concentration function of a sum of independent random variables. *Z. Wahrscheinlichkeitstheor. verw. Geb.* **9**, 290–308.

GÖTZE, F., AND ZAITSEV, A. YU. (1998). Estimates for the rapid decay of concentration functions of n-fold convolutions. *J. Theoret. Probab.* **11**, 715–731.

GRIFFIN, P. S., JAIN, N. C., AND PRUITT, W. E. (1984). Approximate local limit theorems outside domains of attraction. *Ann. Probab.* **12**, 45–63.

HALL, P. (1983). Order of magnitude of the concentration function. *Proc. Amer. Math. Soc.* **89**, 141–144.

HENGARTNER, W., AND THEODORESCU, R. (1973). *Concentration functions.* New York, Academic Press.

KESTEN, H. (1969). A sharper form of the Doeblin–Lévy–Kolmogorov–Rogozin inequality for concentration functions. *Math. Scand.* **25**, 133–144.

KESTEN, H. (1972). Sums of independent random variables—without moment conditions. *Ann. Math. Statist.* **24**, 701–732.

KOLMOGOROV, A. N. (1956). Two uniform limit theorems for sums of independent random variables. *Theor. Probab. Appl.* **1**, 384–394.

KOLMOGOROV, A. (1958). Sur les propriétés des fonctions de concentrations de M. P. Lévy. *Ann. Inst. H. Poincaré.* **16**, 27–34.

LE CAM, L. (1965). On the distribution of sums of independent random variables. *In: Bernoulli, Bayes, Laplace (anniversary volume)* Berlin, Springer, pp. 179–202.

LÉVY, P. (1937). *Théory de l'addition des variables aléatoires.* Paris, Gauthier–Villar.

MIROSHNIKOV, A. L., AND ROGOZIN, B. A. (1980). Inequalities for concentration functions. *Theor. Probab. Appl.* **25**, 176–180.

MIROSHNIKOV, A. L., AND ROGOZIN, B. A. (1982). Remarks on an inequality for the concentration function of sums of independent variables. *Theor. Probab. Appl.* **27**, 848–850.

MOROZOVA, L. N. (1977). Some bounds for the concentration functions of a sum of independent and identically distributed random variables. In: *Limit Theorems for Random Processes.* Fan, Tashkent pp. 85–91 (Russian).

MUKHIN, A. B. (1989). Local probabilities for sums of independent random variables. *Theor. Probab. Appl.* **34**, 616–624.

NAGAEV, S. V., AND KHODZHABAGYAN, S. S. (1996). On an estimate for the concentration function of sums of independent random variables. *Theor. Probab. Appl.* **41**, 560–569.

PETROV, V. V. (1976). *Limit theorems for sums of independent random variables.* Springer, Berlin.

PETROV, V. V. (1995). *Limit theorems of probability theory.* Oxford University Press, New York.

POSTNIKOVA, L. P., AND YUDIN, A. A. (1978). A sharper form of the inequality for the concentration function. *Theor. Probab. Appl.* **23**, 359–362.

ROGOZIN, B. A. (1961a). An estimate for concentration functions. *Theor. Probab. Appl.* **6**, 94–97.

ROGOZIN, B. A. (1961b). On the increase of dispersion of sums of independent random variables. *Theor. Probab. Appl.* **6**, 97–99.

ZAITSEV, A. YU. (1987). On the uniform approximation of distributions of sums of independent random variables. *Theor. Probab. Appl.* **32**, 40–47.

ZAITSEV, A. YU. (1992). Approximation of convolutions of probability distributions by infinitely divisible laws under weakened moment restrictions. *Zapiski Nauchnykh Seminarov LOMI* **194** 79–90 (Russian); English transl. in (1995). *J. Math. Sci.* **75**, 1922–1930.

Fakultät für Mathematik
Universität Bielefeld
Postfach 100131
33501 Bielefeld 1
Germany
goetze@mathematik.uni-bielefeld.de

St. Petersburg Branch
of Steklov Math. Inst.
Fontanka 27
St. Petersburg 191011
Russia
zaitsev@pdmi.ras.ru

On Exact Maximal Khinchine Inequalities

Iosif Pinelis

ABSTRACT Let $\varepsilon_1 \dots, \varepsilon_n$ be independent Rademacher random variables, and let $a_1, \dots, a_n$ be real numbers such that $a_1^2 + \dots + a_n^2 = 1$. Let (B_t) be a standard Brownian motion. One has the generalized moment identity

$$\mathsf{E}g\left(\max_{0\le t\le 1}|x+B_t|\right) = 2\sum_{n=1}^{\infty}(-1)^{n-1}\mathsf{E}g\left(x \vee \frac{|x+B_1|}{2n-1}\right),$$

where $x > 0$ and g is non-decreasing with $g(x) = 0$ and $\mathsf{E}g(|x+B_1|) < \infty$. This identity is used to show that the inequality

$$\mathsf{E}\max_{1\le k\le n}\left|\sum_{i=1}^{k} a_i\varepsilon_i\right|^p \le \mathsf{E}\max_{0\le t\le 1}|B_t|^p,$$

conjectured in [5], fails in a neighborhood of $p = 1$. However, this inequality is proved to hold for $p \ge 3$ in the case $a_1 = \dots = a_n = 1/\sqrt{n}$; the latter result is based on a "discrete" version of the above generalized moment identity.

1 Introduction

Let $\varepsilon_1, \varepsilon_2, \dots$ be independent Rademacher random variables (r.v.'s), so that

$$\mathsf{P}(\varepsilon_i = 1) = \mathsf{P}(\varepsilon_i = -1) = 1/2$$

for all i. Given any real numbers $a_1, a_2, \dots$, consider the weighted Rademacher sums and their maximum and absolute maximum:

$$S_0 := 0; \quad S_j := a_1\varepsilon_1 + \dots + a_j\varepsilon_j \text{ for } j \in \{1, 2, \dots\};$$
$$M_{\mathsf{Rad};\,n,+} := \max_{0\le j\le n} S_j;$$
$$M_{\mathsf{Rad};\,n} := \max_{0\le j\le n} |S_j|.$$

1991 Mathematics Subject Classification. Primary: 60E15, 60J65; Secondary: 60E05, 60G15, 60G50, 60J30

Keywords and phrases. Maximal Khinchine inequalities, exact comparison inequalities, Rademacher sums, Brownian motion, generalized moments, best constants, Bessel processes, multiple reflection, symmetrization, probabilistic identities.

In what follows (except for Section 4), we shall usually (unless specified otherwise) assume that n is fixed and

$$a_1^2 + \cdots + a_n^2 = 1. \tag{1.1}$$

In view of the Donsker invariance principle, it is natural to compare moments of the Rademacher sums with the corresponding moments of the Brownian motion. Let (B_t) be a standard Brownian motion. Let $T > 0$. Consider the maximum and the absolute maximum of the Brownian motion:

$$M_{T,+} := M_{\mathsf{Br};\,T,+} := \max_{0 \le t \le T} B_t;$$
$$M_T := M_{\mathsf{Br};\,T} := \max_{0 \le t \le T} |B_t|.$$

One has

$$\mathsf{E}|S_n|^p \le a(p) \tag{1.2}$$

for all $p > 0$, where $a(p) < \infty$ depends only on p and not on the a_i's – given (1.1). This inequality is usually referred to as the Khinchine inequality. It was proved by Khinchine [8] for the even natural numbers p. The best bound $a(p)$ in (1.2) is 1 if $0 < p \le 2$ and $\mathsf{E}|B_1|^p$ if $p > 2$, so that we have

$$\mathsf{E}|S_n|^p \le \mathsf{E}|B_1|^p \quad \forall p \ge 2; \tag{1.3}$$

see e.g. Haagerup [6]. The first case here, $0 < p \le 2$, is very easy; the second one, $p > 2$, is quite non-trivial, especially if $2 < p < 3$; the only proof in the latter case is due to Haagerup [6]. However, it is immediate from the Central Limit Theorem and the analogue of the Fatou Lemma for convergence in distribution (see, e.g., Theorem 5.3 in [1]) that (1.3) is *exact*, in the sense that $\sup \mathsf{E}|S_n|^p = \mathsf{E}|B_1|^p$, where the sup is taken over all n and all $a_1, \dots, a_n$ satisfying (1.1). Thus, (1.3) is an *exact comparison* inequality, comparing the power moments of the Rademacher sum S_n to the corresponding moments of the Brownian motion.

This paper is devoted mainly to "maximal" analogues of (1.3). The first such result is probably due to Paley and Zygmund [9]; see Lemma 6 there. They proved that $\mathsf{E}|S_\nu|^p \le a_{\max}(p)$, where ν is any random variable with values in the set $\{1, \dots, n\}$, $p > 0$, and $a_{\max}(p) < \infty$ depends only on p. Obviously, this is equivalent to

$$\mathsf{E}M_{\mathsf{Rad};\,n}^p \le a_{\max}(p). \tag{1.4}$$

In view of the Lévy inequality

$$\mathsf{P}(M_{\mathsf{Rad};\,n} > u) \le 2\mathsf{P}(|S_n| > u) \quad \forall u \ge 0, \tag{1.5}$$

we have

$$\mathsf{E}g(M_{\mathsf{Rad};\,n}) \le 2\mathsf{E}g(|S_n|) \tag{1.6}$$

for all nonnegative non-decreasing functions g. To deduce (1.6) from (1.5), first observe that (1.5) is a special case of (1.6), with the indicator $I_{(u,\infty)}$ in place of g; also, (1.6) holds with $I_{[u,\infty)}$ in place of g. Then represent any nonnegative non-decreasing function g as the pointwise limit of a non-decreasing sequence of linear combinations with nonnegative coefficients of such indicators $I_{(u,\infty)}$ and $I_{[u,\infty)}$.

In [10], a multidimensional version of the inequality

$$\mathsf{E}g(a_1\eta_1 + \cdots + a_n\eta_n) \le \mathsf{E}g(a_1\xi_1 + \cdots + a_n\xi_n) \tag{1.7}$$

was given, where $g\colon \mathbb{R} \to \mathbb{R}$ is any convex function with convex second derivative g''; $\eta_1, \ldots, \eta_n$ are independent r.v.'s with $\mathsf{E}\eta_i = 0$ and $\mathsf{P}(|\eta_i| \le 1) = 1$; and $\xi_1, \ldots, \xi_n$ are independent symmetric r.v.'s with $\mathsf{E}\xi_i^2 = 1$, for all i.[1]

In particular, letting $\eta_i = \varepsilon_i$ and $\xi_i \sim N(0,1)$, we have (cf. Eaton [3, 4])

$$\mathsf{E}g(S_n) \le \mathsf{E}g(B_1) \tag{1.8}$$

whenever g'' is convex. This may be considered as a generalized Khinchine inequality.

Combining (1.6) and (1.8), we obtain the following "maximal" analogue of the generalized Khinchine inequality (1.8):

$$\mathsf{E}g(M_{\mathsf{Rad};\,n}) \le 2\mathsf{E}g(|B_1|), \tag{1.9}$$

for all convex even functions $g\colon \mathbb{R} \to \mathbb{R}$ with convex second derivative g''. In particular, we have

$$\mathsf{E}M^p_{\mathsf{Rad};\,n} \le 2\mathsf{E}|B_1|^p \tag{1.10}$$

if $p = 2$ or $p \ge 3$; alternatively, combining (1.6) and (1.3), we have (1.10) for all $p \ge 2$.

Similarly, using the other Lévy inequality

$$\mathsf{P}(M_{\mathsf{Rad};\,n,+} > u) \le \mathsf{P}(|S_n| > u) \quad \forall u \ge 0, \tag{1.11}$$

we obtain

$$\mathsf{E}g(M_{\mathsf{Rad};\,n,+}) \le \mathsf{E}g(|B_1|), \tag{1.12}$$

[1] In [10], g was also supposed to be even. However, in view of (3.1) below, one may assume $\eta_i = \varepsilon_i$ $\forall i$, and then g may be symmetrized, as in the proof of (3.2).

for all convex even functions $g\colon \mathbb{R} \to \mathbb{R}$ with convex second derivative g'', and

$$\mathsf{E} M^p_{\mathsf{Rad};\, n,+} \le \mathsf{E}|B_1|^p \tag{1.13}$$

for all $p \ge 2$.

However, there is an essential difference between (1.9)–(1.10) and (1.12)–(1.13). Indeed, let $n \to \infty$ and $\max(|a_1|, \dots, |a_n|) \to 0$. Then, by the invariance principle, $M_{\mathsf{Rad};\, n,+}$ tends in distribution to $M_{\mathsf{Br};\, 1,+}$, which is well known to have the same distribution as $|B_1|$; thus, $M_{\mathsf{Rad};\, n,+}$ tends in distribution to $|B_1|$. Therefore, inequalities (1.12)–(1.13) are exact. On the other hand, inequalities (1.9)–(1.10) are hardly exact. Indeed, the limit version of inequality (1.5),

$$\mathsf{P}(M_{\mathsf{Br};\, 1} > u) < 2\mathsf{P}(|B_1| > u),$$

is strict for all real u (see (2.2) below), which implies

$$\mathsf{E} g(M_{\mathsf{Br};\, 1}) < 2\mathsf{E} g(|B_1|)$$

for all (say) strictly increasing g with $\mathsf{E} g(|B_1|) < \infty$.

It was conjectured in [5] that

$$\mathsf{E} M^p_{\mathsf{Rad};\, n} \le \mathsf{E} M^p_{\mathsf{Br};\, 1} \tag{1.14}$$

for all $p > 0$. Again by the invariance principle, this hypothetical inequality would be exact for all $p > 0$ for which it is true.

In this paper, we present the following points.

1. We show that conjecture (1.14) fails for $p = 1$ and some other p. The proof of this fact is based on Theorem 1 below, which provides an expression for generalized moments of the absolute maximum of the shifted Brownian motion $(x + B_t)_{0 \le t \le 1}$ with arbitrary moment functions. In turn, the proof of Theorem 1 is based on a multiple reflection argument. See Section 2.

2. We present inequalities that are less precise but more general than (1.14); these inequalities are based on a result of Hunt [7]. We also give a simpler proof of Hunt's inequality. See Section 3.

3. We prove (1.14) for $p \ge 3$ in the special case when $a_1 = \cdots = a_n = 1/\sqrt{n}$. This proof is based on Theorem 2, which provides an expression for generalized moments of the absolute maximum of the Rademacher sums, which is similar to that of Theorem 1, mentioned just above. See Section 4.

2 Generalized moments of the absolute maximum of the shifted Brownian motion and partial disproval of the conjecture

In this section, we shall show that conjecture (1.14) fails at least for $p = 1$ and also for p close to 2. Our main tool here is Theorem 1 below, which provides an expression for the expectation $\mathsf{E}g(M_{x,T})$ for arbitrary functions g, satisfying only some general conditions; here,

$$M_{x,T} := M_{\mathsf{Br};\,x,T} := \max_{0\le t\le T} |x + B_t|$$

is the maximum of the Bessel process $(|x + B_t|)_{0\le t\le T}$ of order 1; see [2] for a discussion of Bessel processes.

In turn, Theorem 1 is based on the following well-known fact about Brownian motion; see, e.g., formula (11.10) in [1].

Proposition 1. *Let I be an interval of the form (a_-, a_+), where $a_- < 0 < a_+$. Let I_0 be a Lebesgue-measurable subset of I. Then*

$$\mathsf{P}(B_t \in I\ \forall t \in [0,T]\ \&\ B_T \in I_0) = \sum_{k=-\infty}^{\infty} (-1)^k \mathsf{P}(B_T \in I_k), \tag{2.1}$$

where the I_k's are defined so that I_{k+1} is the reflection of I_k about $\alpha_k := a_+ + k(a_+ - a_-)$; that is, $I_{k+1} = 2\alpha_k - I_k$ for all $k \in \mathbb{Z}$.

Proof. The argument is based on multiple reflection of the Brownian motion. We present it here for the readers' convenience; we believe our presentation of the argument is simpler than in the sources known to us. Also, we present this proof because we shall use a similar argument in the "discrete" setting of Proposition 4 below. Introduce the events

$$\begin{aligned} A_0 &:= \{(B_t)_{0\le t\le T} \text{ hits neither } a_+ \text{ nor } a_- \ \&\ B_T \in I_0\}, \\ C_n^+ &:= \{(B_t)_{0\le t\le T} \text{ hits } a_+, \text{ then crosses } I \text{ at least } n \text{ times } \&\ B_T \in I_0\}, \end{aligned}$$

and the similarly defined C_n^-, for all $n \in \{0, 1, \dots\}$.

Observe that the left-hand side of (2.1) equals $\mathsf{P}(A_0)$. Hence, (2.1) follows by induction from the following identities:

$$\begin{aligned} \mathsf{P}(A_0) &= \mathsf{P}(B_T \in I_0) - \mathsf{P}(C_0^+ \cup C_0^-), \\ \mathsf{P}(C_n^+ \cup C_n^-) &= \mathsf{P}(C_n^+) + \mathsf{P}(C_n^-) - \mathsf{P}(C_n^+ \cap C_n^-), \\ C_n^+ \cap C_n^- &= C_{n+1}^+ \cup C_{n+1}^-, \\ \mathsf{P}(C_n^\pm) &= \mathsf{P}(B_T \in I_{\pm(n+1)}), \end{aligned}$$

for all $n \in \{0, 1, \dots\}$. The last of these identities follows from the reflection property of the Brownian motion. Indeed,

$$\mathsf{P}(C_n^+) = \mathsf{P}(B_T^{(n+1)} \in I_{n+1}) = \mathsf{P}(B_T \in I_{n+1}) \quad \forall n \in \{0, 1, \dots\},$$

where $(B_t^{(k)})$ is the standard Brownian motion obtained from $\left(B_t^{(0)}\right) := (B_t)$ by the k-fold reflection:

$$B_t^{(k+1)} := B_t^{(k)}\, I\{t \le \tau_k\} + (2\alpha_k - B_t^{(k)})\, I\{t > \tau_k\} \quad \forall t \ge 0,$$

where $\tau_k := \inf\{t > 0\colon B_t^{(k)} = \alpha_k\}$, $\forall k \in \{0, 1, \dots\}$. Similarly, $\mathsf{P}(C_n^-) = \mathsf{P}(B_T \in I_{-(n+1)})$. □

Corollary 1. *For all $x \in \mathbb{R}$ and all $b > |x|$,*

$$\mathsf{P}(M_{x,T} > b) = 2 \sum_{k=-\infty}^{\infty} \mathsf{P}(B_T \in I_{2k-1}) \tag{2.2}$$

$$= 2 \sum_{n=1}^{\infty} (-1)^{n-1} \mathsf{P}\left(\frac{|x + B_T|}{2n-1} > b\right), \tag{2.3}$$

where $I_m := 2bm + (-x - b, -x + b)$ for all $m \in \mathbb{Z}$.

Proof. Identity (2.2) is immediate from Proposition 1 with $I_0 := I := (-x - b, -x + b)$, because then (I_m) is a partition of $\mathbb{R}$ up to a countable set. Identity (2.3) follows because

$$\mathsf{P}\left(\frac{|x + B_T|}{2n-1} > b\right) - \mathsf{P}\left(\frac{|x + B_T|}{2n+1} > b\right) = \mathsf{P}(B_T \in I_n) + \mathsf{P}(B_T \in I_{-n})$$

for all natural n, which is used here only for $n = 1, 3, 5, \dots$. □

Theorem 1. *Let $g\colon [0, \infty) \to \mathbb{R}$ be non-decreasing and continuous at 0.*[2] *Then*

$$\mathsf{E}g(M_{x,T}) = g(|x|) + 2 \sum_{n=1}^{\infty} (-1)^{n-1} \left\{\mathsf{E}g\left(|x| \vee \frac{|x + B_T|}{2n-1}\right) - g(|x|)\right\} \tag{2.4}$$

for any real x such that $\mathsf{E}g(M_{x,T}) < \infty$.

Proof. Once this theorem is stated, its proof is an easy exercise, since the distribution of $M_{x,T}$ is described by Proposition 1.

W.l.o.g., $x \ge 0$ and $g(x) = 0$. Let $\mathcal{G}_x$ stand for the set of all non-decreasing and continuous at 0 functions $g\colon [0, \infty) \to \mathbb{R}$ with $g(x) = 0$.

It is immediate from Corollary 1 that formula (2.4) takes place for the indicators $I_{(b,\infty)}$ in place of g, for all $b > x$.

Hence, by the linearity in g, identity (2.4) holds for all linear combinations of such indicators $I_{(b,\infty)}$ in place of g. There is a non-decreasing sequence (g_k) in $\mathcal{G}_x$ of such linear combinations of indicators that converges to g almost everywhere on $[0, \infty)$.

Thus, it remains to perform a routine limit transition. It is based on the following simple fact:

Let us call an alternating series $\sum_{n=1}^{\infty} (-1)^{n-1} a_n$ *nice* if a_n converges monotonically to 0 as $n \to \infty$. Suppose that a sequence of nice alternating

[2] The necessity of the continuity of g at 0 in Theorem 1 was pointed out to me by G. Peskir.

series $\sum_{n=1}^{\infty}(-1)^{n-1}a_n^{(k)}$ converges term-wise as $k \to \infty$ to a nice alternating series $\sum_{n=1}^{\infty}(-1)^{n-1}a_n$; i.e., $a_n^{(k)} \to a_n$ as $k \to \infty$, for all n. Let $s^{(k)}$ and s denote the sums of these series, respectively. Then $s^{(k)} \to s$ as $k \to \infty$.

This fact follows because the sequence of series $\sum_{n=1}^{\infty}(-1)^{n-1}a_n^{(k)}$ converges uniformly in k; indeed, $\left|\sum_{n=N}^{\infty}(-1)^{n-1}a_n^{(k)}\right| \le \left|a_N^{(k)}\right| \to |a_N|$ as $k \to \infty$, for every N.

Since $|x+B_T| \le M_{x,T}$ and g is non-decreasing, $\mathsf{E}g(|x+B_T|) \le \mathsf{E}g(M_{x,T}) < \infty$. Hence, in view of the Lebesgue Theorem, the series $\sum_{n=1}^{\infty}(-1)^{n-1}\mathsf{E}g\left(x \vee \frac{|x+B_T|}{2n-1}\right)$ is nice (only here, and only for $x=0$, the continuity of g at 0 is used). The same is true with g_k in place of g. Moreover, again by the Lebesgue Theorem, $\mathsf{E}g_k(M_{x,T}) \to \mathsf{E}g(M_{x,T})$ and, for each n,

$$\mathsf{E}g_k\left(x \vee \frac{|x+B_T|}{2n-1}\right) \to \mathsf{E}g\left(x \vee \frac{|x+B_T|}{2n-1}\right)$$

as $k \to \infty$. The result now follows. □

Remark 1. It follows from the above proof that

$$\mathsf{E}g(M_{x,T}) < \infty \text{ iff } \mathsf{E}g(|x+B_T|) < \infty.$$

Remark 2. By the linearity, (2.4) remains true for functions g that can be represented as the difference of two non-decreasing functions. More exactly, let g be a real-valued function on $[0,\infty)$ that has locally bounded variation and is continuous at 0. Then (2.4) holds for all real x such that

$$\mathsf{E}g^*(|x+B_T|) < \infty,$$

where

$$g^*(u) := \underset{[0,u]}{\text{variation}}(g) := \sup\left\{\sum_{r=1}^{q}|g(u_r)-g(u_{r-1})| : q \in \{1,2,\dots\},\right.$$
$$\left. 0 = u_0 < u_1 < \cdots < u_q = u\right\}.$$

Corollary 2. *For any* $p > 0$,

$$\mathsf{E}M_T^p = c_p \mathsf{E}|B_T|^p, \tag{2.5}$$

where

$$c_p := 2\sum_{n=1}^{\infty}\frac{(-1)^{n-1}}{(2n-1)^p}. \tag{2.6}$$

This corollary is immediate from (2.4).

Note that for all $p > 0$,

$$\mathsf{E}|B_T|^p = \frac{2^{p/2}}{\sqrt{\pi}}\Gamma\left(\frac{p+1}{2}\right)T^{p/2}$$

and $c_1 = \frac{\pi}{2}$, so that we have the well-known result

$$\mathsf{E}M_T = \sqrt{\frac{\pi T}{2}}. \tag{2.7}$$

Similarly, in view of Remarks 1 and 2, we obtain from Theorem 1 the following identities:

$$\mathsf{E}\cosh(\lambda M_1) = 1 + 2\sum_{n=1}^{\infty}(-1)^{n-1}\left\{e^{\frac{\lambda^2}{2(2n-1)^2}} - 1\right\} \quad \forall \lambda \in \mathbb{C};$$

$$\mathsf{E}e^{\alpha M_1^2} = 1 + 2\sum_{n=1}^{\infty}(-1)^{n-1}\left\{\frac{2n-1}{\sqrt{(2n-1)^2 - 2\alpha}} - 1\right\} \quad \text{if } \operatorname{Re}\alpha < \frac{1}{2}.$$

Here is another corollary of the series formula of Proposition 1; it can be called a symmetrization identity for the Bessel process $(|x+B_t|)$, since the random variable $N_{x,T}$ below depends on (B_t) only via $(|B_t|)$.

Corollary 3. *$M_{x,T}$ and $N_{x,T} := \max(M_{0,T}, |x| + |B_T|)$ have the same distribution, for all real x.*

Proof. Once this corollary is stated, the proof is an easy exercise, since the series formula of Proposition 1 provides expressions for the distributions of both $M_{x,T}$ and $N_{x,T}$. In view of the symmetry of (B_t), we may and shall assume that $x > 0$. Obviously, $\mathsf{P}(M_{x,T} < y) = 0 = \mathsf{P}(N_{x,T} < y)$ if $y < x$. Let now $y > x$. Then, by (2.1),

$$\mathsf{P}(M_{x,T} < y) = \sum_{k\in\mathbb{Z}}(-1)^k\mu(I_k) \quad \text{and} \tag{2.8}$$

$$\mathsf{P}(N_{x,T} < y) = \sum_{k\in\mathbb{Z}}(-1)^k\mu(I_k^{\mathsf{symm}}), \tag{2.9}$$

where $\mu(A) := \mathsf{P}(B_T \in A)$ and

$$I_k := [-y-x, y-x] + 2ky \quad \text{and} \quad I_k^{\mathsf{symm}} := [x-y, y-x] + 2ky.$$

Hence, series (2.8) as an (infinite) linear combination of the values $F((2k-1)y - x)$, $k \in \mathbb{Z}$, where $F(u) := \mathsf{P}(B_T \le u)$. Similarly, series (2.9) as an (infinite) linear combination of the values $F((2k-1)y-x)$ *and* $F((2k-1)y+x)$, $k \in \mathbb{Z}$. Because $F((2k-1)y+x) = 1 - F((2l-1)y-x)$ for $l := 1-k$, one can represent series (2.9) as an (infinite) linear combination of the values $F((2k-1)y-x)$ *only*. It remains to perform the trivial verification that the coefficients of $F((2k-1)y-x)$ in the linear combination for $M_{x,T}$ are the same as those for $N_{x,T}$.

It may however be more interesting to know how the result of Corollary 3 may be obtained, rather than just verified. The crucial observation here is that

$$I_0 \setminus (-I_1) \overset{\text{a.s.}}{=} I_0^{\text{symm}}; \tag{2.10}$$

we write $A \overset{\text{a.s.}}{=} B$ if the Lebesgue measure of $(A \setminus B) \cup (B \setminus A)$ is 0. Using translation $u \mapsto u + 2y$ and reflection $u \mapsto (-u)$, we get

$$(-I_1) \setminus I_0 \overset{\text{a.s.}}{=} -I_1^{\text{symm}}. \tag{2.11}$$

Further apply the translations $u \mapsto u + 2ky$ to (2.10) and (2.11) to obtain

$$I_k \setminus (-I_{1-k}) \overset{\text{a.s.}}{=} I_k^{\text{symm}} \quad \text{and} \quad (-I_{1-k}) \setminus I_k \overset{\text{a.s.}}{=} -I_{1-k}^{\text{symm}} \quad \forall k \in \mathbb{Z}.$$

Taking into account now the symmetry of μ,

$$\mu(I_{1-k}) - \mu(I_k) = \mu(I_{1-k}^{\text{symm}}) - \mu(I_k^{\text{symm}}) \quad \forall k \in \mathbb{Z}.$$

It remains to rewrite the series in (2.8) and (2.9), respectively, as

$$\sum_{k=1}^{\infty} (-1)^{k-1}(\mu(I_{1-k}) - \mu(I_k)) \quad \text{and} \quad \sum_{k=1}^{\infty} (-1)^{k-1}(\mu(I_{1-k}^{\text{symm}}) - \mu(I_k^{\text{symm}})).$$

□

In conclusion of this section, we use Theorem 1 to show that conjecture (1.14) fails at least for some p.

Proposition 2. *Conjecture (1.14) fails at least for $p = 1$.*

Proof. Indeed, suppose (1.14) is true for $p = 1$. Take a positive real number x and a natural number $n \geq 2$, and let at this point

$$a_1 = x \text{ and } a_2 = \cdots = a_n = \frac{1}{\sqrt{n-1}},$$

instead of (1.1). Let now $n \to \infty$. Then (1.14), the invariance principle, and the symmetry and scaling properties of the Brownian motion will imply

$$\mathsf{E}M_{x,1}^p \leq \mathsf{E}M_{x^2+1}^p. \tag{2.12}$$

On the other hand, (2.4) yields

$$\mathsf{E}M_{x,1} = x + 2\sum_{n=1}^{\infty} (-1)^{n-1} \mathsf{E}\left(\frac{|x + B_1|}{2n-1} - x\right)_+,$$

where $u_+ := 0 \vee u$, as usual. Since we have a nice alternating series here (as in the proof of Theorem 1), its sum is no less than the sum of its first two terms. Hence, (2.12) with $p = 1$ will imply

$$x + 2\sum_{n=1}^{2}(-1)^{n-1}\mathsf{E}\left(\frac{|x+B_1|}{2n-1} - x\right)_+ \le \sqrt{\frac{\pi(x^2+1)}{2}}, \tag{2.13}$$

for all $x \geq 0$, in view of (2.7). However, direct computations show that (2.13) fails at $x = 1$. Thus, (1.14) fails for $p = 1$. □

Similarly, taking here $\sum\limits_{n=1}^{4}$ in place of $\sum\limits_{n=1}^{2}$, one can see that (2.12) fails for $p = 1.9985$ and $x = 0.45 + 0.8(2 - p)$.

Moreover, a Mathematica plot suggests that (2.12) fails for all $p \in (0, 1.9985)$ and $x = 0.45+0.8(2-p)$, while it seems to be true for $p = 2$ and all x. However, we will not formally prove here the two conjectures stated in this paragraph.

3 Hunt's inequalities and their applications

Let a function $F\colon \mathbb{R}^n \to \mathbb{R}$ be convex in each of its n arguments. Let $\eta_1, \dots, \eta_n$ be independent (but not necessarily identically distributed) r.v.'s with

$$\mathsf{E}\eta_1 = \cdots = \mathsf{E}\eta_n = 0 \quad \text{and} \quad \mathsf{P}(|\eta_1| \le 1) = \cdots = \mathsf{P}(|\eta_n| \le 1) = 1.$$

Let $\zeta_1, \dots, \zeta_n$ be independent symmetric (but not necessarily identically distributed) r.v.'s with

$$\mathsf{E}|\zeta_1| \ge 1, \dots, \mathsf{E}|\zeta_n| \ge 1.$$

The following result is due to Hunt [7].

Proposition 3.

$$\mathsf{E}F(\eta_1, \dots, \eta_n) \le \mathsf{E}F(\varepsilon_1, \dots, \varepsilon_n) \tag{3.1}$$

$$\le \mathsf{E}F(\zeta_1, \dots, \zeta_n). \tag{3.2}$$

Hunt's inequality (3.1) was used in [4] and then in [10]. G. Peskir has communicated to me the particular case of (3.2): $\mathsf{E}\max_{1\le k\le n} g\left(\sum_1^k a_i\varepsilon_i\right) \le \mathsf{E}\max_{1\le k\le n} g\left(\sum_1^k a_i\zeta_i\right)$, where g is convex. Later, he told me that inequality (3.2) also can be found in Hunt's paper [7].

Proof of Proposition 3. We present this proof, which is simpler than that in Hunt [7], for the readers' convenience. As Hunt notes, it is enough to consider the case $n = 1$; indeed, this is immediate by induction, using the Fubini Theorem. Let $\eta := \eta_1$, $\varepsilon := \varepsilon_1$, and $\zeta := \zeta_1$.

Then, by the convexity of F and the condition $\mathsf{P}(|\eta| \leq 1) = 1$,

$$F(\eta) \leq F(-1)\frac{1-\eta}{2} + F(1)\frac{1+\eta}{2};$$

applying now E and using $\mathsf{E}\eta = 0$, we get (3.1).

Next, in inequality $\mathsf{E}F(\varepsilon) \leq \mathsf{E}F(\zeta)$, we may assume that F is even (since ε and ζ are symmetric, we may replace here $F(u)$ by $F(u) + F(-u)$); then, since F is also convex, it is non-decreasing on $[0, \infty)$. Hence, using the Jensen inequality,

$$\mathsf{E}F(\varepsilon) = F(1) \leq F(\mathsf{E}|\zeta|) \leq \mathsf{E}F(|\zeta|) = \mathsf{E}F(\zeta).$$

□

Corollary 4. *Let, in addition, $F\colon \mathbb{R}^n \to \mathbb{R}$ be p-positive-homogeneous in the sense that*

$$F(tx_1, \ldots, tx_n) = |t|^p F(x_1, \ldots, x_n)$$

for some p and all real t and x_i's. Let $Z_1, \ldots, Z_n$ be iid standard normal. Then

$$\mathsf{E}F(\eta_1, \ldots, \eta_n) \leq \mathsf{E}F(\varepsilon_1, \ldots, \varepsilon_n) \leq \left(\frac{\pi}{2}\right)^{p/2} \mathsf{E}F(Z_1, \ldots, Z_n). \tag{3.3}$$

This follows because $\mathsf{E}|Z_i| = \sqrt{\frac{2}{\pi}}$.

Corollary 5. (Curvilinear maximal generalized moment comparison.) *Let $a(\cdot)$ and $b(\cdot)$ be any continuous real-valued functions defined on $[0,1]$. Let $g(\cdot)$ be any non-decreasing convex real-valued function defined on $\mathbb{R}$. Then*

$$\mathsf{E}g\left(\max_{0\leq k\leq n}\left|a\left(\frac{k}{n}\right) + b\left(\frac{k}{n}\right) S_k\right|\right) \leq \mathsf{E}g\left(\max_{0\leq t\leq 1}\left|a(t) + b(t)\sqrt{\frac{\pi}{2}}B_t\right|\right); \tag{3.4}$$

$$\mathsf{E}g\left(\max_{0\leq k\leq n}\left\{a\left(\frac{k}{n}\right) + b\left(\frac{k}{n}\right) S_k\right\}_+\right) \leq \mathsf{E}g\left(\max_{0\leq t\leq 1} \cdot \left\{a(t) + b(t)\sqrt{\frac{\pi}{2}}B_t\right\}_+\right). \tag{3.5}$$

In particular, for any $p \geq 1$,

$$\mathsf{E}\max_{0\leq k\leq n}\left|b\left(\frac{k}{n}\right) S_k\right|^p \leq \left(\frac{\pi}{2}\right)^{p/2} \mathsf{E}\max_{0\leq t\leq 1}|b(t)B_t|^p; \tag{3.6}$$

$$\mathsf{E}\max_{0\leq k\leq n}\left\{b\left(\frac{k}{n}\right) S_k\right\}_+^p \leq \left(\frac{\pi}{2}\right)^{p/2} \mathsf{E}\max_{0\leq t\leq 1}\{b(t)B_t\}_+^p. \tag{3.7}$$

An obvious advantage of inequalities (3.3) and (3.4)–(3.7) is their generality. The price that we have to pay is rather steep, though: having the factor $\sqrt{\frac{\pi}{2}}$ (or its power) on the right-hand sides of these inequalities. This loss of precision is due to the restriction $\mathsf{E}|\zeta_i| \ge 1$, to be compared with the condition $\mathsf{E}\xi_i^2 = 1$ for (1.7).

4 Identities and exact maximal inequalities for Rademacher sums with equal a_i's

In this section, assume that all the a_i's are the same. Also, it will be convenient here to remove the norm restriction (1.1). Instead, we shall assume that

$$a_1 = \cdots = a_n = 1.$$

Proposition 4. *Let $I := (a_-, a_+) \cap \mathbb{Z}$, where $a_- \in \mathbb{Z}$, $a_+ \in \mathbb{Z}$, and $a_- < 0 < a_+$. Let $I_0 \subseteq I$. Then*

$$\mathsf{P}(S_j \in I \ \forall j \in \{0, \dots, n\} \ \ \& \ \ S_n \in I_0) = \sum_{k=-\infty}^{\infty} (-1)^k \mathsf{P}(S_n \in I_k), \tag{4.1}$$

where the I_k's are defined so that I_{k+1} is the reflection of I_k about $\alpha_k := a_+ + k(a_+ - a_-)$; that is, $I_{k+1} = 2\alpha_k - I_k$ for all $k \in \mathbb{Z}$.

This proposition is well known and can be proved quite similarly to Proposition 1. For a proof based on different ideas, see e.g. [1], especially formula (11.8) there.

Corollary 6. *For all $y \in \{1, 2, \dots\}$,*

$$\mathsf{P}(M_{\mathsf{Rad};\,n} \ge y) = \sum_{m=1}^{\infty} (-1)^{m-1} \Big\{ \mathsf{P}\big(|S_n| \ge (2m-1)y\big) + \mathsf{P}\big(|S_n| - 1 \ge (2m-1)y\big) \Big\}. \tag{4.2}$$

Proof. It follows from Proposition 4 and the symmetry of the distribution of S_n that for all $y \in \{1, 2, \dots\}$,

$$\begin{aligned} &\mathsf{P}(M_{\mathsf{Rad};\,n} \ge y) - \mathsf{P}(|S_n| \ge y) \\ &\quad = -\mathsf{P}(M_{\mathsf{Rad};\,n} < y) + \mathsf{P}(|S_n| < y) \\ &\quad = \sum_{m=1}^{\infty} (-1)^{m-1} \mathsf{P}\big((2m-1)y + 1 \le |S_n| < (2m+1)y\big) \\ &\quad = \sum_{m=1}^{\infty} (-1)^{m-1} \big\{ \mathsf{P}\big(|S_n| - 1 \ge (2m-1)y\big) - \mathsf{P}\big(|S_n| \ge (2m+1)y\big) \big\}. \end{aligned}$$

Now (4.2) is immediate. □

Theorem 2. *For all functions* $g\colon \{0,1,\dots\} \to \mathbb{R}$ *with* $g(0) = 0$,

$$\mathsf{E}g(M_{\mathsf{Rad};\,n}) = \mathsf{E}H_g(|S_n|), \tag{4.3}$$

where

$$H_g(u) := \sum_{m=1}^{\infty} (-1)^{m-1} \left\{ g\left(\left\lfloor \frac{u}{2m-1} \right\rfloor\right) + g\left(\left\lfloor \frac{u-1}{2m-1} \right\rfloor_+\right) \right\},$$

$u \geq 0$, $\lfloor x \rfloor$ *stands for the integer part of* x.

Proof. This proof is quite similar to that of Theorem 1; it is even easier, because no limit transition is needed here. For $g = I_{[y,\infty)}$ and $y \in \{1,2,\dots\}$, identity (4.3) is equivalent to (4.2). The general case follows by the linearity of both sides of (4.3) in g. □

Proposition 5. *For* $g(u) \equiv u^p$ *and* $p \geq 3$,

$$H_g(u) \leq c_p g(u) \quad \forall u \in \{0,1,\dots\}, \tag{4.4}$$

where the constant c_p *is the same as the one given by (2.6).*

Proof. For $u \in \{1,2,\dots\}$ and $p \geq 1$,

$$H_g(u) - c_p g(u)$$

$$= \sum_{m=1}^{\infty} (-1)^{m-1} \left\{ \left\lfloor \frac{u}{2m-1} \right\rfloor^p + \left\lfloor \frac{u-1}{2m-1} \right\rfloor^p - 2\left(\frac{u}{2m-1}\right)^p \right\} \tag{4.5}$$

$$\leq -u^p + (u-1)^p + 2\sum_{k=1}^{\infty} \left\{ \left(\frac{u}{4k-1}\right)^p - \left\lfloor \frac{u-1}{4k-1} \right\rfloor^p \right\} \tag{4.6}$$

$$\leq -u^p + (u-1)u^{p-1} + 2p\sum_{k=1}^{\infty} \left(\frac{u}{4k-1}\right)^{p-1} \tag{4.7}$$

$$= -u^{p-1}b(p), \tag{4.8}$$

where

$$b(p) := 1 - 2p\sum_{k=1}^{\infty} \frac{1}{(4k-1)^{p-1}}.$$

Here, to obtain (4.6), note that only the terms of the series in (4.5) with even m's are nonnegative, so retain in (4.6) only these terms and the first one (corresponding to $m = 1$); moreover, use also an obvious upper bound for the terms with even m's. To obtain (4.7), use the elementary inequalities $b^p - a^p \leq pb^{p-1}(b-a)$ for $0 \leq a < b$ and $p \geq 1$, and

$$\frac{u}{m} - \left\lfloor \frac{u-1}{m} \right\rfloor = \frac{r+1}{m} \leq 1$$

for all natural numbers u and m, where $r \in \{0, 1, \dots, m-1\}$ is the residual of the division of $u-1$ by m.

Note that $b(p)$ increases in $p > 2$, since pa^{-p} decreases in $p \geq 1$ for $a \geq e$. Also,

$$b(3) = 1 + 3C_{\text{Catalan}} - \frac{3\,\pi^2}{8} > 0,$$

where

$$C_{\text{Catalan}} = \sum_{n=1}^{\infty} \frac{(-1)^{n-1}}{(2n-1)^2} = 0.9159\dots$$

is the Catalan constant. Hence, $b(p) > 0$ for all $p \geq 3$, and it remains to refer to (4.5)–(4.8). This completes the proof. □

Theorem 3. *For all* $p \geq 3$,

$$\mathsf{E}M^p_{\mathsf{Rad};\,n} \leq c_p \mathsf{E}|S_n|^p \tag{4.9}$$

and

$$\mathsf{E}M^p_{\mathsf{Rad};\,n} \leq \mathsf{E}M^p_{\mathsf{Br};\,n}. \tag{4.10}$$

Proof. Inequality (4.9) is immediate from Theorem 2 and Proposition 5. Inequality (4.10) follows from (4.9), (1.3), and (2.5). □

Remark. It follows from the proof of Proposition 5 that inequality (4.4) is strict for all $u \in \{1, 2, \dots\}$, and so, inequalities (4.9) and (4.10) are strict for all natural numbers n and all $p \geq 3$.

Inequality (4.10) (valid if $a_1 = \cdots = a_n = 1$) is equivalent to conjecture (1.14) in the special case when $a_1 = \cdots = a_n = 1/\sqrt{n}$.

On the other hand, it can be verified that (4.9) fails for $p = 0.9$ and $n = 2$. We conjecture that for some cutoff point $p_* \in (0.9, 1)$, we have (4.9) for all natural n and all $p > p_*$.

Acknowledgement. G. Peskir drew my attention to conjecture (1.14), suggested that the case of equal a_i's may be of interest, and provided a reference to [2].

References

[1] P. Billingsley, *Convergence of Probability Measures*, Wiley, New York, 1968.

[2] L. E. Dubins, L. A. Shepp, and A. N. Shiryaev, Optimal stopping rules and maximal inequalities for Bessel processes, *Theory Probab. Appl.* **38** (1993), 226–261.

[3] M. L. Eaton, A note on symmetric Bernoulli random variables, *Ann. Math. Statist.* **41** (1970), 1223–1226.

[4] M. L. Eaton, A probability inequality for linear combinations of bounded random variables, *Ann. Statist.* **2** (1974), 609–614.

[5] S. E. Graversen and G. Peskir, Extremal problems in the maximal inequalities of Khinchine, *Math. Proc. Cambridge Philos. Soc.* **123**(1) (1995), 169–177.

[6] U. Haagerup, The best constants in the Khinchine inequality, *Studia Math.* **70** (1982), 231–283.

[7] G. A. Hunt, An inequality in probability theory, *Proc. Amer. Math. Soc.* **6** (1955), 506–510.

[8] A. Khinchin, Über dyadische Brüche, *Math. Z.* **18** (1923), 109–116.

[9] R. E. A. C. Paley and A. Zygmund, On some series of functions, *Proc. Cambridge Philos. Soc.* **26**(1) (1930), 337–357.

[10] I. Pinelis, Extremal probabilistic problems and Hotelling's T^2 test under a symmetry condition, *Ann. Statist.* **22** (1994), 357–368.

Department of Mathematical Sciences
Michigan Technological University
Houghton, Michigan 49931
e-mail: `ipinelis@math.mtu.edu`

Strong Exponential Integrability of Martingales with Increments Bounded by a Sequence of Numbers

Jan Rosiński*

ABSTRACT A strong exponential bound for the probability tail of a martingale whose increments are uniformly bounded by a sequence of numbers is established. The exponent, which grows to infinity faster than a quadratic polynomial, is a Young function determined by that sequence. The related Orlicz norm of a martingale is estimated, some examples are discussed, and the optimality of the exponent is proven.

1 Introduction

Let $\{d_i\}_{i\geq 1}$ be a martingale difference sequence, that is, $\{d_i\}$ is a sequence of random variables adapted to some increasing filtration $\{\mathfrak{F}_i\}$ such that $E(d_i|\mathfrak{F}_{i-1}) = 0$, $i \geq 1$. Suppose that, for some $K \in (0, \infty)$,

(1.1) $$\|d_i\|_\infty \leq Ka_i, \qquad i \geq 1,$$

where $\{a_i\}_{i\geq 1}$ is a sequence of numbers such that

(1.2) $$a_1 \geq a_2 \geq \cdots > 0,$$

(1.3) $$\sum_{i=1}^{\infty} a_i^2 < \infty,$$

and

(1.4) $$\sum_{i=1}^{\infty} a_i = \infty.$$

*Work partially supported by an NSF grant.

Condition (1.3) guarantees that $f := \sum_{i=1}^{\infty} d_i$ exists a.s. If (1.4) does not hold, then $|f|$ is bounded and so its integrability is trivial. Our goal is to give a sharp bound on the tail probability $P(|f| > t)$ in the form

$$P(|f| > t) \leq C_1 \exp\big(- C_2 \psi(t/K)\big) \tag{1.5}$$

where $\psi = \psi_{\mathbf{a}}$ is an increasing convex function determined by the sequence $\mathbf{a} = \{a_i\}_{i \geq 1}$ ($C_1, C_2 > 0$ are constants). The bound (1.5) is well known for some special sequences $\{a_i\}_{i \geq 1}$. Namely, if $a_i = i^{-1}$, then

$$\psi(v) = \exp(v/4), \tag{1.6}$$

$C_1 = 16$, $C_2 = 1$ (Lemma 1.8 in Ledoux and Talagrand, (1991)). If $a_i = i^{-1/p}$, $1 < p < 2$, then

$$\psi(v) = t^q, \tag{1.7}$$

where $1/p + 1/q = 1$, $C_1 = 2$, $C_2 = C(q)$ (Lemma 1.7 in Ledoux and Talagrand, (1991)). Marcus (1987) considered the case of $a_i = 1/\delta(i)$ where δ is strictly increasing, regularly varying at infinity with index $\in (1/2, 1)$, and such that $q(i) := i/\delta(i)$ is strictly increasing and concave. He obtained bound (1.5) with

$$\psi(v) = q^{-1}(t/C) \tag{1.8}$$

for some constant C.

In this paper we will give a formula for $\psi = \psi_{\mathbf{a}}$ without any restriction on a_i's except of (1.2)–(1.4). We establish inequality (1.5) for such ψ and determine the integrability of f satisfying (1.1). The exact evaluation of $\psi_{\mathbf{a}}$ may be difficult or impossible in many cases but it may still be possible to find a simple lower bound for $\psi_{\mathbf{a}}$ that captures the rate of decay of the probability tail of $|f|$. We illustrate this idea by examples in Section 3.

2 The tail estimate and integrability

A convex even function $\phi : \mathbb{R} \to \mathbb{R}_+$ with $\phi(0) = 0$ is called a Young function. A Young function ϕ is said to be an N–function (*a nice Young function*) if $\lim_{u \to 0} \frac{\phi(u)}{u} = 0$ and $\lim_{u \to \infty} \frac{\phi(u)}{u} = \infty$. A function ψ is called a *complementary to* ϕ if

$$\psi(v) = \sup_{u \geq 0}(u|v| - \phi(u)), \quad v \geq 0. \tag{2.1}$$

If ϕ is an N-function then so is ψ, see Krasnoselsky and Rutitsky (1961).

Definition 2.1. $\psi = \psi_{\mathbf{a}}(v)$ *will denote the complementary function to*

$$\phi_{\mathbf{a}}(u) = \sum_{i \le N(u)} (2|u|a_i - 1) + \sum_{i > N(u)} u^2 a_i^2 \tag{2.2}$$

where $\mathbf{a} = \{a_i\}_{i \ge 1}$ *satisfies (1.2)–(1.4) and* $N(u) = \max\{i : |u|a_i \ge 1\}$ $(\max \emptyset = 0)$.

We first notice that $\phi_{\mathbf{a}}$ is an N–function. Indeed, $\phi_{\mathbf{a}}(u) = \int_0^{|u|} \phi'_{\mathbf{a}}(x)\,dx$, where

$$\phi'_{\mathbf{a}}(x) = 2\sum_{i=1}^{\infty} (a_i \wedge x a_i^2), \quad x \ge 0 \tag{2.3}$$

is a strictly increasing continuous function with $\phi'_{\mathbf{a}}(0) = 0$ and $\phi'_{\mathbf{a}}(\infty) = \infty$. We also note that

$$\phi'_{\mathbf{a}}(x) = 2\sum_{i \le N(x)} a_i + 2x \sum_{i > N(x)} a_i^2, \quad x \ge 0. \tag{2.4}$$

In the sequel the sequence $\mathbf{a} = \{a_i\}_{i \ge 1}$ will be fixed. Therefore, we will just *write* ψ *for* $\psi_{\mathbf{a}}$. The following lemma is useful for the evaluation of ψ.

Lemma 2.2. *(i)* $\psi(v) = \int_0^{|v|} \psi'(x)\,dx$, *where* ψ' *is the inverse function to* ϕ'. *Hence* ψ' *is an increasing continuous function with* $\psi'(0) = 0$ *and* $\psi'(\infty) = \infty$.
(ii) If $t = \phi'(s)$ *for some* $t, s \ge 0$, *then*

$$\psi(t) = st - \phi(s). \tag{2.5}$$

Proof. These are the standard facts that can be found in Chapter I, Sections 1, 2 in Krasnoselsky and Rutitsky (1961).

Theorem 2.3. *Assume (1.1)–(1.4). Then, for every* $t > 0$,

$$P(|f| \ge t) \le 2\exp\Big(-\frac{\psi(t/K)}{2}\Big) \tag{2.6}$$

where $\psi = \psi_{\mathbf{a}}$.

Proof. By homogeneity we may and do assume that $K = 1$. Fix $t > 0$ and choose $s > 0$ such that

$$\phi'(s) = t. \tag{2.7}$$

By (1.1) we have

$$|f| \le \sum_{i \le N(s)} a_i + \Big|\sum_{i > N(s)} d_i\Big|$$

($N(s)$ is given in Definition 2.1) and, by (2.4) and (2.7),

$$t = 2 \sum_{i \le N(s)} a_i + 2s \sum_{i > N(s)} a_i^2$$

Hence

$$\begin{aligned} P(|f| > t) &\le P(|\sum_{i>N(s)} d_i| > t/2) \\ &\le 2 \exp\Big(-\frac{t^2}{8q^2}\Big) \end{aligned}$$

from a generalized subgaussian inequality, where $q^2 = \sum_{i>N(s)} \|d_i\|_\infty^2$ (see Lemma 1.5 in Ledoux and Talagrand (1991)). Since

$$q^2 \le \sum_{i>N(s)} a_i^2 \le \frac{t}{2s},$$

we obtain

$$P(|f| > t) \le 2\exp(-st/4). \tag{2.8}$$

In view of (2.5), (2.2) and (2.7) we get

$$\begin{aligned} \psi(t) &= st - \phi(s) \\ &= st - 2s \sum_{i \le N(s)} a_i + N(s) - s^2 \sum_{i>N(s)} a_i^2 \\ &= s\Big(t - 2 \sum_{i \le N(s)} a_i - 2s \sum_{i>N(s)} a_i^2\Big) \\ &\quad + N(s) + s^2 \sum_{i>N(s)} a_i^2 \\ &= N(s) + st/2 - s \sum_{i \le N(s)} a_i \\ &\le st/2. \end{aligned}$$

Putting this bound in (2.8) completes the proof. Q.E.D.

Lemma 2.4.

$$\lim_{v \to \infty} \frac{\psi_{\mathbf{a}}(v)}{v^2} = \infty.$$

Proof. Fix $\epsilon \in (0, 1)$. Choose $n_0 \in \mathbb{N}$ such that $\sum_{i>n_0} a_i^2 < \epsilon$ and $i a_i^2 \le 1$ for $i > n_0$. Choose $x_0 > 0$ such that $N(x_0) > n_0$ and $a_1 n_0 x_0^{-1} < \epsilon$. Then for $x \geq x_0$ we have

$$\begin{aligned}\phi'(x) &\leq 2n_0 a_1 + 2\sum_{i=n_0+1}^{N(x)} a_i + 2x\epsilon \\ &\leq 4\epsilon x + 2\Big(\sum_{i=n_0+1}^{N(x)} a_i^2\Big)^{1/2}(N(x)-n_0)^{1/2} \\ &\leq 4\epsilon x + 2\epsilon^{1/2}(N(x))^{1/2}.\end{aligned}$$

Since $N(x) > n_0$, $x^{-1} \leq a_{N(x)} \leq (N(x))^{-1/2}$, which yields $N(x) \leq x^2$ for $x \geq x_0$. Consequently, $\phi'(x) \leq 6\epsilon^{1/2}x$ for $x \geq x_0$, implying

$$\psi'(y) \geq 6^{-1}\epsilon^{-1/2}y$$

for $y \geq y_0 = 6x_0$. The conclusion of this lemma immediately follows as $\epsilon^{-1/2}$ can be arbitrarily large. Q.E.D.

Theorem 2.5. *Assume (1.1)–(1.4). Then*

$$E\exp\{\psi(bf)\} \leq \frac{1+2b^2K^2}{1-2b^2K^2} \tag{2.9}$$

for every $0 < b < (\sqrt{2}K)^{-1}$, *where* $\psi = \psi_{\mathbf{a}}$.

Proof. First we will show that for every $c \geq 1$,

$$\psi(cv) \geq c^2\psi(v), \quad v \geq 0. \tag{2.10}$$

Indeed, it is easy to see that $\phi'(cu) \leq c\phi'(u)$ for every $u > 0$ and $c \geq 1$, which implies $\phi(cu) \leq c^2\phi(u)$. Then, by Young's inequality, we have

$$c^2uv \leq \phi(cu) + \psi(cv) \leq c^2\phi(u) + \psi(cv).$$

Hence

$$\psi(cv) \geq c^2(uv - \phi(u)).$$

Taking the supremum over $u > 0$ on the right hand side yields (2.10). Using (2.6) and (2.10) for $c = (bK)^{-1}$ we get

$$\begin{aligned}E\exp\{\psi(bf)\} &= 1 + \int_1^\infty P\big(|f| > b^{-1}\psi^{-1}(\log t)\big)\,dt \\ &\leq 1 + 2\int_1^\infty \exp(-2^{-1}\psi(K^{-1}b^{-1}\psi^{-1}(\log t)))\,dt \\ &\leq 1 + 2\int_1^\infty \exp\big(-2^{-1}(bK)^{-2}\log t\big)\,dt \\ &= \frac{1+2b^2K^2}{1-2b^2K^2}.\end{aligned}$$

Q.E.D.

Consider now the space $L^{\Psi}(dP)$ defined by the Young function $\Psi(u) = \Psi_{\mathbf{a}}(u) := \exp(\psi_{\mathbf{a}}(u)) - 1$.

$$L^{\Psi}(dP) = \{X : E[\Psi(c^{-1}X)] < \infty \text{ for some } c = c(X) > 0\}.$$

$L^{\Psi}(dP)$ is a Banach space (random variables equal a.s. are indistinguishable) with the norm

$$\|X\|_{\Psi} := \inf\{c > 0 : E[\Psi(c^{-1}X)] \leq 1\}. \tag{2.11}$$

Corollary 2.6. *Assume (1.1)–(1.4). Then* $f \in L^{\Psi}(dP)$ *and*

$$\|f\|_{\Psi} \leq \sqrt{6}K$$

where $\Psi = \Psi_{\mathbf{a}}$.

Proof. This follows immediately from Theorem 2.5 with $b = c^{-1}$. Q.E.D.

3 Examples

We will now give several examples showing how Theorem 2.3 covers previously studied cases as well as how it leads to new estimates.

Remark 3.1. The exact evaluation of $\psi = \psi_{\mathbf{a}}$ may be difficult or impossible in concrete cases. However, for many applications, it is enough to find a lower bound for ψ. Namely, if $\psi_1(t) \leq \psi(t)$ for $t \geq t_0$, then by Theorem 2.3 we have a bound

$$P(|f| \geq t) \leq 2\exp\Big(-\frac{\psi_1(t/K)}{2}\Big), \quad t \geq t_0 \tag{3.1}$$

which often may be sufficient.

To construct a function ψ_1 we may begin with some nonincreasing function $g : (0,\infty) \to (0,\infty)$ such that $a_i = g(i)$. Put

$$\frac{1}{2}\phi_1'(x) = g(1) \wedge xg(1)^2 + \int_1^{\infty} \big(g(t) \wedge xg(t)^2\big)\, dt. \tag{3.2}$$

Since $\phi_1' \geq \phi'$, $\phi_1 \geq \phi$, and from (2.1), $\psi_1(t) \leq \psi(t)$, $t \geq 0$. We may further simplify ψ_1 and use its lower bound in (3.1). This method will be applied in the following examples.

Example 3.2. *Let* $a_i = i^{-1}$, $i \geq 1$.

Taking $g(t) = t^{-1}$ in (3.2) gives

$$\phi_1'(x) = \begin{cases} 4x & 0 \le x < 1 \\ 4 + 2\log x & x \ge 1. \end{cases}$$

In view of Lemma 2.2 (i), $\psi_1' = (\phi_1')^{-1}$ which yields

$$\psi_1(v) = \begin{cases} v^2/8 & 0 \le v < 4 \\ 2e^{-2}\exp(v/2) & v \ge 4. \end{cases}$$

Consequently,

$$P(|f| \ge t) \le 2\exp\big\{ -e^{-2}\exp(\frac{t}{2K})\big\} \quad t \ge 4K, \tag{3.3}$$

provided $\|d_i\|_\infty \le Ki^{-1}$, $i \ge 1$. Notice that (3.3) is actually sharper than the inequality of Lemma 1.8 in Ledoux and Talagrand (1991) (see (1.6)–(1.5)). Comparing both parts of ψ_1, we get the following estimate valid for all $t \ge 0$

$$P(|f| \ge t) \le 3\exp\big\{ -e^{-2}\exp(\frac{t}{2K})\big\}. \tag{3.4}$$

Example 3.3. *Let $a_i = i^{-1/p}$, $i \ge 1$, $1 < p < 2$.*

Taking ϕ_1' from (3.2) with $g(t) = t^{-1/p}$ gives

$$\phi_1'(x) \le \phi_2'(x) := \begin{cases} \frac{4}{2-p}x & 0 \le x < 1 \\ \frac{2p}{(p-1)(2-p)}x^{p-1} & x \ge 1. \end{cases}$$

Using Lemma 2.2 (i) we obtain

$$\psi_2(v) = \begin{cases} \frac{(2-p)}{8}v^2 & 0 \le v < \frac{4}{2-p} \\ 2c_p v^q + 2/(p-1) & v \ge \frac{4}{2-p}, \end{cases}$$

where $1/p + 1/q = 1$, $c_p = (1-p/2)^{1/(p-1)}/(2q^q)$. By (3.1) we have

$$P(|f| \ge t) \le \exp\big(-c_p|t/K|^q\big) \tag{3.5}$$

for $t \ge \frac{4}{2-p}K$, provided $\|d_i\|_\infty \le Ki^{-1/p}$, $i \ge 1$.

Example 3.4. *Let $a_i = i^{-1}\log i$, $i \ge 3$, $a_1 = 1$, $a_2 = 1/2$.*

Take $g(t) = t^{-1}$ for $0 < t < e$ and $g(t) = t^{-1}\log t$ if $t \ge e$. Then

$$\frac{1}{2}\phi_1'(x) = 1 \wedge x + \int_1^e t^{-1} \wedge xt^{-2}\,dt + \int_e^\infty \{t^{-1}\log t\} \wedge \{xt^{-2}(\log t)^2\}\,dt.$$

Evaluating the integrals for $x = \frac{a}{\log a} \geq e$ we get

$$\phi_1'(\frac{a}{\log a}) = 3 + (\log a)^2 + \frac{2}{\log a}\{(\log a)^2 + 2\log a + 2\}$$

which yields

$$\phi_1'(\frac{a}{\log a}) \leq C(\log a)^2,$$

for $a \geq e$, where C is a numerical constant which may change from line to line in the following computations. Using Lemma 2.2 (i) we get

$$\psi_1'(v) \geq Cv^{-1/2}\exp(Cv^{1/2}), \quad v \geq v_0.$$

Hence $\psi_1(v) \geq 2\exp(Cv^{1/2}),\ v \geq v_0$. Consequently,

$$P(|f| \geq t) \leq 2\exp(-\exp[C(t/K)^{1/2}]) \tag{3.6}$$

for $t \geq v_0 K$, provided $\|d_i\|_\infty \leq K\frac{\log i}{i}$, $i \geq 3$.

Notice that (1.8) cannot be used in Example 3.4 because in this example $\{a_i\}$ is regularly varying with index 1 and (1.8) requires the index to be strictly less than 1. The fact that (1.8) does not extend to regularly varying sequences of index 1 is shown in the fact that ψ in (3.6) is of different order than q^{-1} (see also the conclusion of this section).

Example 3.5. *Let* $a_i = (i\log i)^{-1}$, $i \geq 3$, $a_1 = 1$, $a_2 = 1/2$.

Take $g(t) = t^{-1}$ for $0 < t < e$ and $g(t) = (t\log t)^{-1}$ if $t \geq e$. Then

$$\frac{1}{2}\phi_1'(x) = 1 \wedge x + \int_1^e t^{-1} \wedge xt^{-2}\,dt + \int_e^\infty \{(t\log t)^{-1}\} \wedge \{x(t\log t)^{-2}\}\,dt.$$

Evaluating the integrals for $x = a\log a \geq e$ we get

$$\phi_1'(a\log a) \leq 6 + 2\log(\log a)$$

which yields

$$\psi_1'(v) \geq e^{v/2-3}\exp(e^{v/2-3}), \qquad v \geq 6.$$

Since $\phi_1'(x) \leq 4x$ for $0 \leq x < 1$ and $\phi_1'(x) \leq 4 + 2\log x$ for $1 \leq x < e$, we obtain by Lemma 2.2(i) that $\psi_1'(v) \geq v/4$ for $0 \leq v < 4$ and $\psi_1'(v) \geq \exp(v/2 - 2)$ for $4 \leq v < 6$. This yields

$$\psi_1(v) \geq 2\exp(\exp(v/2 - 3)), \quad v \geq 6.$$

Consequently,

$$P(|f| \geq t) \leq 2\exp(-\exp[e^{-3}\exp(t/2K)]) \tag{3.7}$$

for $t \geq 6K$, provided $\|d_i\|_\infty \leq K\frac{1}{i\log i}$, $i \geq 3$.

Comparing (3.3), (3.6), and (3.7) we notice a huge difference in the rate of decay of the probability tail of $|f|$ being made by the logarithmic term in a_i. To verify whether we have obtained the correct rates of decay, consider $s_n = a_1 + \cdots + a_n$ and $f = \sum_{i=1}^{\infty} r_i a_i$, where $\{r_i\}$ is a Rademacher sequence. Notice that $P(|f| \geq s_n) \geq 2^{-n}$. We have $s_n \sim \log n$ in Example 3.2, $s_n \sim (\log n)^2/2$ in Example 3.4, and $s_n \sim \log(\log n)$ in Example 3.5. Applying (3.3) and (3.6), we see that in Examples 3.2 and 3.4

$$C_1 \log n \leq \log(-\log P\{|f| \geq s_n\}) \leq C_2 \log n$$

for large n and some positive constants C_1, C_2. However, in Example 3.5 we get

$$C_1 (\log n)^{1/2} \leq \log(-\log P\{|f| \geq s_n\}) \leq C_2 \log n$$

for large n and some positive constants C_1, C_2. Taking into account that the function $\exp(\exp(\exp x))$ increases very rapidly, (3.7) still seems to be the right bound.

Addendum

After this paper was completed and circulated, Stanisław Kwapień kindly informed us that a similar type exponential bound for martingales, as in Theorem 2.3, can be obtained by combining Theorem 5.1.2 in Kwapień and Woyczyński (1992) with a corollary in Gluskin and Kwapień (1995). Namely, under conditions of Theorem 2.3 one can get

$$P(|f| \geq t) \leq \exp(-F(ct, \mathbf{a})) \tag{3.8}$$

for some universal constant c and $t \geq K(1, \mathbf{a})$, where

$$K(s, \mathbf{a}) = \max\{\sum_{i \leq s} a_i, \ \sqrt{s}(\sum_{i > s} a_i^2)^{1/2}\}$$

and

$$F(t, \mathbf{a}) = \max\{s : \ K(s, \mathbf{a}) \leq t).$$

The basic difference between functions F and ψ is that different procedures are applied in their definitions, F is the inverse of K and ψ is Young complementary to ϕ. A direct relationship between these two functions is unclear at the first glance. However, function F is optimal in the sense that if $d_i = r_i a_i$, where $\{r_i\}$ is a Rademacher sequence, then the reverse inequality to (3.8) also holds (see Corollary in Gluskin and Kwapień (1995)). We will show that ψ is optimal as well. The optimality of F was proven using precise moment bounds for the Rademacher series. We will prove the optimality of ψ differently, using Kolmogorov's inequality (3.10).

Proposition 3.6. *Assume that* $\mathbf{a} = \{a_i\}$ *satisfies (1.2)–(1.4) and let* $d_i = r_i a_i$*, where* $\{r_i\}$ *is a Rademacher sequence. There exists a universal constant* C *such that*

$$P(|f| \geq t) \geq \exp\big(-\frac{\psi_{\mathbf{a}}(Ct)}{2}\big) \tag{3.9}$$

for all $t \geq t_0(\mathbf{a})$.

Proof. Kolmogorov's inequality states that there exists a universal constant $C_0 \geq 2$ such that for any square summable sequence $\{x_i\}$,

$$P(\sum_i r_i x_i \geq u) \geq \exp\big(-\frac{C_0 u^2}{\sum_i x_i^2}\big) \tag{3.10}$$

provided $u \geq C_0(\sum_i x_i^2)^{1/2}$ and $u \sup_i |x_i| \leq C_0^{-1} \sum_i x_i^2$ (see, e.g., page 90 in Ledoux and Talagrand (1991)).

Choose $s > 0$ such that

$$\phi'(s) = 3C_0 t. \tag{3.11}$$

Observe from the proof of Theorem 2.3 that

$$\psi(3C_0 t) = N(s) + s^2 \sum_{i > N(s)} a_i^2. \tag{3.12}$$

Consider the cases:
(A) $2\sum_{i \leq N(s)} a_i < s \sum_{i > N(s)} a_i^2$ and
(B) $2\sum_{i \leq N(s)} a_i \geq s \sum_{i > N(s)} a_i^2$.

Using (2.4) and (3.11) in case (A) we get

$$\begin{aligned} P(|f| \geq t) &= 2P(\sum_i r_i a_i \geq t) \geq P(\sum_{i > N(s)} r_i a_i \geq t) \\ &\geq P(\sum_{i > N(s)} r_i a_i \geq C_0^{-1} s \sum_{i > N(s)} a_i^2). \end{aligned} \tag{3.13}$$

We want to apply Kolmogorov's inequality (3.10) for $u = C_0^{-1} s \sum_{i > N(s)} a_i^2$ in (3.13). First we need to check that u satisfies the required conditions. Note that

$$u \geq C_0(\sum_{i > N(s)} a_i^2)^{1/2} \tag{3.14}$$

if and only if

$$s^2 \sum_{i > N(s)} a_i^2 \geq C_0^4.$$

Under (A) we get

$$\begin{aligned} s^2 \sum_{i>N(s)} a_i^2 &\geq s \sum_{i\leq N(s)} a_i + (s^2/2) \sum_{i>N(s)} a_i^2 \\ &\geq N(s) + (s^2/2) \sum_{i>N(s)} a_i^2 \\ &\geq \psi(3C_0 t)/2. \end{aligned}$$

Hence (3.14) holds when $t \geq (3C_0)^{-1}\psi^{-1}(2C_0^4) =: t_0(\mathbf{a})$. The other condition on u always holds,

$$\begin{aligned} u \sup_{i>N(s)} a_i &= C_0^{-1} s a_{N(s)+1} \sum_{i>N(s)} a_i^2 \\ &< C_0^{-1} \sum_{i>N(s)} a_i^2. \end{aligned}$$

Now Kolmogorov's inequality gives

$$P(|f| \geq t) \geq \exp(-C_0^{-1} s^2 \sum_{i>N(s)} a_i^2) \geq \exp(-2^{-1}\psi(3C_0 t)).$$

The case (B) is simple. We have

$$\begin{aligned} P(|f| \geq t) &= 2P(\sum_i r_i a_i \geq t) \geq P(\sum_{i\leq N(s)} r_i a_i \geq t) \\ &\geq P(\sum_{i\leq N(s)} r_i a_i \geq \sum_{i\leq N(s)} a_i) \\ &\geq \exp(-(\ln 2)N(s)) \geq \exp(-(\ln 2)\psi(3C_0 t)). \end{aligned}$$

By (2.10), $\psi(3C_0 t) \leq (3/4)^2 \psi(4C_0 t) \leq (2\ln 2)^{-1}\psi(4C_0 t)$. Thus (3.9) holds with $C = 4C_0$ for all $t \geq t_0(\mathbf{a})$. Q.E.D.

From the fact that both ψ and F are optimal we infer that these functions are equivalent in the following sense.

Corollary 3.7. *There exist universal constants c_1, c_2 such that*

$$\psi_{\mathbf{a}}(c_1 t) \leq F(t, \mathbf{a}) \leq \psi_{\mathbf{a}}(c_2 t)$$

for $t \geq t_1(\mathbf{a})$.

Remark 3.8. Let $\mathbf{a} = \{a_i\}$ be a sequence of positive numbers satisfying (1.3)–(1.4) but *do not assume monotonicity* (1.2). Define

$$\phi_{\mathbf{a}}(u) = \sum_{\{i:\ |u|a_i \geq 1\}} (2|u|a_i - 1) + \sum_{\{i:\ |u|a_i < 1\}} u^2 a_i^2.$$

Equivalently, define $\phi_{\mathbf{a}}(u) = \int_0^{|u|} \phi'_{\mathbf{a}}(x)\,dx$, where $\phi'_{\mathbf{a}}$ is given by (2.3). Since $\phi_{\mathbf{a}}$ is invariant under permutations of $\{a_i\}$ and f is invariant under permutations of $\{d_i\}$, we infer that the assumption (1.2) of monotonicity of $\{a_i\}$ can be completely removed from this paper. The advantage of working with $\psi_{\mathbf{a}}$ versus $F(\cdot, \mathbf{a})$ is that $\psi_{\mathbf{a}}$ is convex, hence it generates a homogeneous norm $\| \cdot \|_{\Psi_{\mathbf{a}}}$ (see (2.11), and it does not require the monotonicity of the sequence $\mathbf{a}$. The latter property is important when, for instance, the sequence $\mathbf{a}$ is chosen at random.

References

[1] Gluskin, E. D. and Kwapień, S. *Tail and moment estimates for sums of independent random variables with logarithmically concave tails.* Studia Math. **114** (1995), 303–309.

[2] Krasnoselsky M. A. and Rutitsky Y. B. *Convex functions and Orlicz spaces.* Noordhof, 1961.

[3] Kwapień, S. and Woyczyński, W. A. *Random series and stochastic integrals: single and multiple. Probability and its Applications.* Birkhäuser Boston, 1992.

[4] Ledoux, M. and Talagrand, M. *Probability in Banach spaces. Isoperimetry and processes.* Springer-Verlag, 1991.

[5] Marcus, M. B. *ξ-Radial Processes and Random Fourier Series.* Memoirs. Amer. Math. Soc. **368** 1987. Amer. Math. Soc., Providence, R.I.

Department of Mathematics
University of Tennessee
Knoxville, TN 37996
e-mail: `rosinski@math.utk.edu`

PART 2

General Empirical Process Theory

On Uniform Laws of Large Numbers for Smoothed Empirical Measures

Peter Gaenssler and Daniel Rost

ABSTRACT We consider function-indexed smoothed empirical measures on linear metric spaces and focus on uniform laws of large numbers (ULLN) comparable with Glivenko–Cantelli results in the non-smoothed case. Using the random measure process approach we are able to give a set of sufficient conditions for a ULLN which are different from the ones known in the literature and are more close to being necessary.

1 Introduction

Let X be a linear metric space endowed with its Borel σ-field $\mathcal{X}$ and let $\eta_j, j \in \mathbb{N}$, be independent identically distributed (i.i.d.) random elements (r.e.) in $(X, \mathcal{X})$ with law ν on $\mathcal{X}$, viewed as coordinate mappings on the product probability space $(\Omega, \mathcal{A}, \mathbb{P}) := (X^{\mathbb{N}}, \mathcal{X}^{\mathbb{N}}, \nu^{\mathbb{N}})$. Let ν_n be the empirical measure based on $\eta_1, ..., \eta_n$, i.e.

$$\nu_n := \frac{1}{n} \sum_{j \leq n} \delta_{\eta_j},$$

where δ_x denotes the Dirac measure in $x \in X$. ν_n is the nonparametric maximum likelihood estimator for ν.

If the underlying measure ν is "smooth" it is natural to use a "smoothed" version $\tilde{\nu}_n$ of ν_n as an estimator for ν, rather than the empirical measure itself. Following Yukich (1989) we consider "smoothing through convolution" as follows: Let $\mu_n, n \in \mathbb{N}$, be a sequence of probability measures on $\mathcal{X}$ and let

$$\tilde{\nu}_n := \nu_n \star \mu_n$$

be the so-called smoothed empirical measure based on $\eta_1, ..., \eta_n$, i.e.

$$\tilde{\nu}_n(B) := \int_X \int_X 1_B(x+y) \nu_n(dx) \mu_n(dy), \quad B \in \mathcal{X},$$

where 1_B denotes the indicator function of B.

Let $\mathcal{F}$ be a class of measurable functions $f : X \longrightarrow \mathbb{R}$ with measurable envelope $F : X \longrightarrow \mathbb{R}$ (i.e. $\sup_{f\in\mathcal{F}} |f(x)| \leq F(x)$ for all $x \in X$). For $f \in \mathcal{F}$ we put

$$\tilde{\nu}_n(f) := \int_X f \, d\tilde{\nu}_n,$$

assuming tacitly that integrals of functions $f \in \mathcal{F}$ exist. For $\tilde{\nu}_n(f)$ this is the case if $\int_X |f(x+y)|\mu_n(dy) < \infty$ for all $x \in X$. It is also tacitly assumed that suprema over $f \in \mathcal{F}$, like $\sup_{f\in\mathcal{F}} |\tilde{\nu}_n(f) - \nu(f)|$, are measurable.

The aim of this paper is to present uniform laws of large numbers (ULLN), i.e. sufficient conditions on $\mathcal{F}$ and the smoothing measures μ_n under which

$$\sup_{f\in\mathcal{F}} |\tilde{\nu}_n(f) - \nu(f)| \xrightarrow{\mathrm{L}_p} 0 \quad \text{as } n \longrightarrow \infty \tag{1.1}$$

for some $p \geq 1$, where $\xrightarrow{\mathrm{L}_p}$ denotes convergence with respect to the L_p - metric.

Concerning distributional results for soothed empirical processes, i.e. weak convergence of $\left(n^{1/2}(\tilde{\nu}_n(f) - \nu(f))\right)_{f\in\mathcal{F}}$ to a certain Gaussian process, we refer to the paper of the second author in the present volume.

Note that in the special case $X = \mathbb{R}^d, \mathcal{F} = \{1_{(-\infty,\underline{t}]} : \underline{t} \in \mathbb{R}^d\}$ and $\mu_n \equiv \delta_{\underline{0}}$ (the non-smoothed case), (1.1) implies the classical Glivenko-Cantelli theorem (cf. Gaenssler and Ziegler, 1994, Section 4). On the other hand, one should note the relationship between smoothed empirical measures and kernel density estimators.

Finally, we want to mention also a setting for studying estimators $\hat{F}_n$ (as already considered in Winter (1973) and Yamato (1973)) for a distribution function F (in the case $X = \mathbb{R}$) by defining

$$\hat{F}_n(t) := \frac{1}{n} \sum_{j\leq n} \mu_n(t - \eta_j), \ t \in \mathbb{R},$$

with a distribution function μ_n. $\hat{F}_n$ is then called a *smoothed* (or *perturbed*) *empirical distribution function*. Note that in general, i.e. for arbitrary linear metric spaces X, smoothing by convolution is its natural extension, since (replacing $(-\infty, t]$ by $B \in \mathcal{X}$)

$$\tilde{\nu}_n(B) = \frac{1}{n} \sum_{j\leq n} \int_X 1_B(\eta_j + y)\mu_n(dy) = \frac{1}{n} \sum_{j\leq n} \mu_n(B - \eta_j), \ B \in \mathcal{X}. \tag{1.2}$$

2 Main results

Taking up the general setting of the introduction, let $\tilde{\mathcal{F}}$ be the class of translates of all $f \in \mathcal{F}$, i.e.

$$\tilde{\mathcal{F}} := \{f_x : x \in X, f \in \mathcal{F}\},$$

where $f_x(y) := f(x+y), y \in X$. Consider the decomposition

$$\tilde{\nu}_n - \nu = \tilde{\nu}_n - \nu \star \mu_n + \nu \star \mu_n - \nu \tag{2.1}$$

of $\tilde{\nu}_n - \nu$ into the stochastic term $\tilde{\nu}_n - \nu \star \mu_n$ (noticing that $\mathbb{E}(\tilde{\nu}_n) = \nu \star \mu_n$) and the (non-stochastic) bias-term $\nu \star \mu_n - \nu$. (2.1) together with the assumption $\mathcal{F} = \tilde{\mathcal{F}}$ is essential for the following lemma.

Lemma 2.1. *Let X be a linear metric space and suppose that $\mathcal{F} = \tilde{\mathcal{F}}$ (i.e. $\mathcal{F}$ is closed under translation). Suppose further that*

$$\sup_{f\in\mathcal{F}} |\nu_n(f) - \nu(f)| \xrightarrow{L_p} 0 \quad \text{and} \quad \sup_{f\in\mathcal{F}} |\nu \star \mu_n(f) - \nu(f)| \longrightarrow 0.$$

Then

$$\sup_{f\in\mathcal{F}} |\tilde{\nu}_n(f) - \nu(f)| \xrightarrow{L_p} 0.$$

Proof. According to (2.1) it suffices to show that

$$\sup_{f\in\mathcal{F}} |\tilde{\nu}_n(f) - \nu \star \mu_n(f)| \leq \sup_{f\in\mathcal{F}} |\nu_n(f) - \nu(f)|.$$

For this, let $g \in \mathcal{F}$ be arbitrary; then

$$\begin{aligned}
|\tilde{\nu}_n(g) - \nu \star \mu_n(g)| &= |\nu_n \star \mu_n(g) - \nu \star \mu_n(g)| \\
&= \Big| \int_X \int_X g(x+y)\nu_n(dx)\mu_n(dy) - \int_X \int_X g(x+y)\nu(dx)\mu_n(dy) \Big| \\
&= \Big| \int_X \int_X g_y(x)\nu_n(dx)\mu_n(dy) - \int_X \int_X g_y(x)\nu(dx)\mu_n(dy) \Big| \\
&= \Big| \int_X (\nu_n(g_y) - \nu(g_y))\mu_n(dy) \Big| \leq \int_X \sup_{f\in\mathcal{F}} |\nu_n(f) - \nu(f)| \mu_n(dy) \\
&= \sup_{f\in\mathcal{F}} |\nu_n(f) - \nu(f)|.
\end{aligned}$$

□

Concerning the bias-term $\nu \star \mu_n - \nu = \nu \star \mu_n - \nu \star \delta_0$, one shows in the same way that

$$\sup_{f\in\mathcal{F}} |\nu \star \mu_n(f) - \nu \star \delta_0(f)| \leq \sup_{f\in\mathcal{F}} |\mu_n(f) - \delta_0(f)|,$$

where, e.g. for separable X and uniformly bounded equicontinuous classes, $\mathcal{F}$ $\sup_{f\in\mathcal{F}} |\mu_n(f) - \delta_0(f)| \longrightarrow 0$ if $\mu_n \longrightarrow \delta_0$ weakly (in the sense of weak convergence of probability measures in metric spaces); cf. Theorem 1.12.1 in van der Vaart and Wellner (1996).

The conditions of Lemma 2.1 are e.g. fulfilled if $X = \mathbb{R}, \mathcal{F} = \{1_{(-\infty,t]} : t \in \mathbb{R}\}$ and $\mu_n \longrightarrow \delta_0$ weakly and ν is a continuous probability measure on the Borel σ-field $\mathcal{B}$ in $\mathbb{R}$. The result in this special case goes back to Winter (1973) and Yamato (1973).
The disadvantage of Lemma 2.1 is that it holds only under the rather restrictive assumption $\mathcal{F} = \tilde{\mathcal{F}}$, a condition which cannot be dispensed with in general. It should be noted also that assuming the existence of a real-valued envelope F of $\mathcal{F}$, $\mathcal{F} = \tilde{\mathcal{F}}$ necessarily implies that $\mathcal{F}$ is uniformly bounded.

Now, not imposing the assumption $\mathcal{F} = \tilde{\mathcal{F}}$, there is a completely different way to obtain ULLN's for smoothed empirical measures via the *Random Measure Process approach* as presented in Gaenssler, Rost and Ziegler (1998): For this we note that $\tilde{\nu}_n(f)$ can be represented as

$$\tilde{\nu}_n(f) = \sum_{j\leq j(n)} w_{nj}(f) \cdot \xi_{nj}$$

with $j(n) := n, w_{nj}(f) := \int_X f(\eta_j + y)\, \mu_n(dy)$ and $\xi_{nj} := \frac{1}{n}$.
Processes $S_n = \big(S_n(f)\big)_{f\in\mathcal{F}}$ with

$$S_n(f) := \sum_{j\leq j(n)} w_{nj}(f) \cdot \xi_{nj},$$

where $(w_{nj})_{1\leq j\leq j(n), n\in\mathbb{N}}$ is a triangular array of random probability measures on a measurable space $(X, \mathcal{X})$ and $(\xi_{nj})_{1\leq j\leq j(n), n\in\mathbb{N}}$ is a triangular array of real-valued random variables such that

> for all $n \in \mathbb{N}$ the sequence of pairs
> $(w_{n1}, \xi_{n1}), ..., (w_{nj(n)}, \xi_{nj(n)})$ is independent.

are called *Random Measure Processes (RMP's)*. However, we do not assume that the pairs are identically distributed; also dependence within each pair is allowed.

In passing we remark that RMP's typically occur in the area of

- partial-sum processes, where e.g. $X = [0,1]^d, d \geq 1, w_{nj} := \delta_{\underline{j}/n}$ and $\xi_{nj} = \frac{1}{n^d} \cdot \xi_{\underline{j}}$ with $\underline{j}$ ranging over the d-dimensional lattice $\mathbb{N}^d$

or

- empirical processes, where $w_{nj} := \delta_{\eta_{nj}}$, η_{nj} being r.e. in an arbitrary sample space $(X, \mathcal{X})$ and $\xi_{nj} = \frac{1}{j(n)}$.

As shown in Gaenssler, Rost and Ziegler (1998), using again the decomposition (2.1) and a general ULLN for RMP's as presented there in Section 2, one gets (by a slight modification of the proof as presented there in Section 4) the following ULLN for smoothed empirical measures. Note that the pseudo-metric $\tilde{\nu}_n^{(1)}$ used in Theorem 2.2 is different from the one occurring in Gaenssler, Rost and Ziegler (1998) (cf. (3.9) there) which is essential for our purposes (cf. the proof of Theorem 2.6 below.

Theorem 2.2. *Let X be an arbitrary linear metric space and F be a measurable envelope of $\mathcal{F}$. Assume (with $w_{nj}(F) := \int\limits_X F(\eta_j + y)\,\mu_n(dy))$ that (2.2)–(2.5) hold, where (with $I(\,\cdot\,)$ standing for the indicator function $1_{\{\cdot\}}$)*

(2.2) $$\lim_{n\to\infty} \mathbb{E}\Big(w_{n1}(F)^p \cdot I(\frac{1}{n} w_{n1}(F) > \delta)\Big) = 0 \quad \textit{for all } \delta > 0\,,$$

(2.3) $$\sup_{n\in\mathbb{N}} \mathbb{E}\Big(w_{n1}(F) \cdot I(\frac{1}{n} w_{n1}(F) \le \delta_1)\Big) < \infty \quad \textit{for some } \delta_1 > 0\,,$$

(2.4) $$\big(N(\tau, \mathcal{F}, d_{\tilde{\nu}_n}^{(1)})\big)_{n\in\mathbb{N}} \textit{ is stochastically bounded for all } \tau > 0$$

where for $f, g \in \mathcal{F}$

$$d_{\tilde{\nu}_n}^{(1)}(f,g) := \int\limits_X \Big| \int\limits_X (f(x+y) - g(x+y))\mu_n(dy)\Big| \nu_n(dx)\,,$$

(2.5) $$\sup_{f\in\mathcal{F}} |\nu \star \mu_n(f) - \nu(f)| \longrightarrow 0\,.$$

Then

$$\sup_{f\in\mathcal{F}} |\tilde{\nu}_n(f) - \nu(f)| \xrightarrow{L_p} 0\,.$$

In (2.4) the so-called covering number $N(\,\cdot\,, \mathcal{F}, d_{\tilde{\nu}_n}^{(1)})$ of $\mathcal{F}$ with respect to the random pseudometric $d_{\tilde{\nu}_n}^{(1)}$ is defined by

$$N(\tau, \mathcal{F}, d_{\tilde{\nu}_n}^{(1)}) := \min\{m \in \mathbb{N} : \text{ there exist } f_1, \ldots, f_m \in \mathcal{F} \text{ such that for all } f \in \mathcal{F}\ d_{\tilde{\nu}_n}^{(1)}(f, f_i) \le \tau \text{ for some } f_i\}$$

and (2.4) means that for each $\rho > 0$ there exists an $M \equiv M(\tau, \rho) < \infty$ such that

$$\limsup_{n\to\infty} \mathbb{P}^*\big(N(\tau, \mathcal{F}, d_{\tilde{\nu}_n}^{(1)}) > M\big) < \rho$$

(where $\mathbb{P}^*$ denotes outer probability).

Looking at the case of *uniformly bounded* $\mathcal{F}$ (i.e. $\sup_{x\in X}\sup_{f\in\mathcal{F}} |f(x)| \le 1$) in the conditions of Theorem 2.2, we now then have (with envelope

$F \equiv 1$ of $\mathcal{F}$) $\{\frac{1}{n}w_{n1}(F) > \delta\} = \varnothing$ and $\{\frac{1}{n}w_{n1}(F) \leq \delta\} = \Omega$ for each $\delta > 0$ and large enough n; hence (2.2) and (2.3) are fulfilled. Furthermore it is known that (2.4) is fulfilled when $\mathcal{F}$ is a uniformly bounded *Vapnik-Chervonenkis subgraph class* (VCSG-class) (for the definition of VCSG-classes, see e.g. van der Vaart and Wellner (1996)). In the context of smoothed empirical measures one usually assumes $\mu_n \longrightarrow \delta_0$ weakly. Weak convergence of μ_n towards δ_0 does not follow from (2.5) nor does $\mu_n \longrightarrow \delta_0$ weakly imply (2.5) in general, as seen by the following example.

Example 2.3. *Take $X = \mathbb{R}, \mathcal{F} = \{1_{(-\infty,t]} : t \in \mathbb{R}\}, \nu = \delta_0$ and $\mu_n := \delta_{\frac{1}{n}}, n \in \mathbb{N}$. Here (2.2)–(2.4) hold, but (2.5) does not!*

However, if $\mathcal{F}$ is "smooth" one can deduce (2.5) from $\mu_n \longrightarrow \delta_0$ weakly. Assuming in addition that X is separable, we arrive at the following result.

Theorem 2.4. *Let X be a separable linear metric space and let $\mathcal{F}$ be a uniformly bounded equicontinuous VCSG-class. Suppose that $\mu_n \longrightarrow \delta_0$ weakly. Then*

$$\sup_{f\in\mathcal{F}} |\tilde{\nu}_n(f) - \nu(f)| \xrightarrow{L_p} 0 .$$

Proof. It suffices to verify (2.5): For each bounded and continuous $f : X \longrightarrow \mathbb{R}$ we have according to dominated convergence that

$$\begin{aligned} \nu \star \mu_n(f) &= \int_X \int_X f(x+y)\mu_n(dy)\nu(dx) \\ &= \int_X \int_X f_x(y)\mu_n(dy)\nu(dx) \longrightarrow \int_X f_x(0)\nu(dx), \end{aligned}$$

since $\mu_n \longrightarrow \delta_0$ weakly and $f_x : X \longrightarrow \mathbb{R}$ is also bounded and continuous for all $x \in X$, where

$$\int_X f_x(0)\nu(dx) = \int_X \int_X f(x+y)\delta_0(dy)\nu(dx) = \nu \star \delta_0(f) = \nu(f),$$

whence $\nu\star\mu_n \longrightarrow \nu$ weakly. Applying now Theorem 1.12.1 in van der Vaart and Wellner (1996) yields (2.5) □

In the non-smoothed case ($\mu_n \equiv \delta_0$) we have the following deep result (see Talagrand (1996), not requiring X to be a linear metric space.)

Theorem 2.5. *Let $(X, \mathcal{X}, \nu)$ be a complete probability space and $\mathcal{F}$ be a uniformly bounded class of measurable functions $f : X \longrightarrow \mathbb{R}$. Then the condition*

$$(2.6) \qquad \left(N(\tau, \mathcal{F}, d_{\nu_n}^{(1)})\right)_{n\in\mathbb{N}} \quad \textit{is stochastically bounded for all } \tau > 0$$

is necessary and sufficient for

$$\sup_{f\in\mathcal{F}} |\nu_n(f) - \nu(f)| \xrightarrow{L_p} 0.$$

Here $N(\,\cdot\,, \mathcal{F}, d_{\nu_n}^{(1)})$ is defined as in (2.4) replacing $d_{\tilde{\nu}_n}^{(1)}(f,g)$ by $d_{\nu_n}^{(1)}(f,g) := \nu_n(|f-g|)$ for $f, g \in \mathcal{F}$.

In view of this result it is tempting to see what can be obtained in the smoothed case. Up to now we were able to prove the following theorem.

Theorem 2.6. *Let X be a linear metric space endowed with its Borel σ-field $\mathcal{X}$ such that $(X, \mathcal{X}, \nu)$ is complete, and let $\mathcal{F}$ be a uniformly bounded class of measurable functions $f : X \longrightarrow \mathbb{R}$ which is closed under translations, i.e. $\tilde{\mathcal{F}} = \mathcal{F}$. Suppose that*

$$\sup_{f\in\mathcal{F}} |\mu_n(f) - f(0)| \longrightarrow 0.$$

Then the following statements are equivalent:

a) $\left(N(\tau, \mathcal{F}, d_{\nu_n}^{(1)})\right)_{n\in\mathbb{N}}$ *is stochastically bounded for all* $\tau > 0$.

b) $\left(N(\tau, \mathcal{F}, d_{\tilde{\nu}_n}^{(1)})\right)_{n\in\mathbb{N}}$ *is stochastically bounded for all* $\tau > 0$.

c) $\sup_{f\in\mathcal{F}} |\tilde{\nu}_n(f) - \nu(f)| \xrightarrow{L_1} 0$.

d) $\sup_{f\in\mathcal{F}} |\nu_n(f) - \nu(f)| \xrightarrow{L_1} 0$.

Proof. As remarked after the proof of Lemma 2.1, $\tilde{\mathcal{F}} = \mathcal{F}$ implies

$$(2.7) \qquad \sup_{f\in\mathcal{F}} |\nu \star \mu_n(f) - \nu(f)| \le \sup_{f\in\mathcal{F}} |\mu_n(f) - \delta_0(f)|,$$

and similarly

$$(2.8) \qquad \sup_{f\in\mathcal{F}} |\nu_n(f) - \nu_n \star \mu_n(f)| \le \sup_{f\in\mathcal{F}} |\mu_n(f) - \delta_0(f)|.$$

Now, for the proof of the theorem, note that

b) $\Longrightarrow$ *c)* follows from (2.7) and Theorem 2.2 (remember that (2.2) and (2.3) are satisfied since $\mathcal{F}$ is uniformly bounded).

c) $\Longrightarrow$ *d)* is immediate from (2.8) and the decomposition $\nu_n - \nu =$

$\nu_n - \nu_n \star \mu_n + \nu_n \star \mu_n - \nu$.
$d) \Longrightarrow a)$ is just Theorem 2.5.

It remains to prove
$a) \Longrightarrow b)$:
For this, let $\tau > 0$ and $g, h \in \mathcal{F}$ with $d_{\nu_n}^{(1)}(g,h) \leq \tau$. Then

$$\begin{aligned}
d_{\tilde{\nu}_n}^{(1)}(g,h) &= \int_X \Big| \int_X (g(x+y) - h(x+y))\mu_n(dy)\Big|\nu_n(dx) \\
&\leq \int_X \Big| \int_X (g(x+y) - g(x))\mu_n(dy)\Big|\nu_n(dx) + \int_X |g(x) - h(x)|\nu_n(dx) \\
&\quad + \int_X \Big| \int_X (h(x+y) - h(x))\mu_n(dy)\Big|\nu_n(dx) \\
&= \int_X \Big| \int_X (g_x(y) - g_x(0))\mu_n(dy)\Big|\nu_n(dx) + \int_X |g(x) - h(x)|\nu_n(dx) \\
&\quad + \int_X \Big| \int_X (h_x(y) - h_x(0))\mu_n(dy)\Big|\nu_n(dx) \\
&\leq 2 \cdot \int_X \sup_{f \in \mathcal{F}} \Big| \int_X (f(y) - f(0))\mu_n(dy)\Big|\nu_n(dx) + d_{\nu_n}^{(1)}(g,h) \\
&\leq 2 \cdot \sup_{f \in \mathcal{F}} |\mu_n(f) - f(0)| + \tau \\
&\leq 3\tau
\end{aligned}$$

for n large enough. This shows that for all $\tau > 0$ and large n (not depending on the sample!) we have

$$N(3\tau, \mathcal{F}, d_{\tilde{\nu}_n}^{(1)}) \leq N(\tau, \mathcal{F}, d_{\nu_n}^{(1)})$$

which completes the proof of $a) \Longrightarrow b)$ and the proof of the theorem. □

As already remarked earlier the condition $\sup_{f\in\mathcal{F}} |\mu_n(f) - f(0)| \longrightarrow 0$ is fulfilled for any sequence $(\mu_n)_{n\in\mathbb{N}}$ with $\mu_n \longrightarrow \delta_0$ weakly, if, in addition, X is separable and $\mathcal{F}$ an equicontinuous class of functions. But these assumptions seem to be rather severe. Indeed, when taking $\mu_n = \delta_0$ for all $n \in \mathbb{N}$, the statement of Theorem 2.6 reduces to the one of Theorem 2.5 which even holds without $\mathcal{F}$ being invariant under translation. It would be interesting to know whether one could delete $\mathcal{F} = \tilde{\mathcal{F}}$ in Theorem 2.6, or, keeping $\mathcal{F} = \tilde{\mathcal{F}}$, one could relax the condition $\sup_{f\in\mathcal{F}} |\mu_n(f) - f(0)| \longrightarrow 0$ to bias convergence. Most likely the answer to both questions is "no". We are still looking for some suitable counterexamples.

References

Gaenssler, P., Ziegler, K., 1994. "A uniform Law of Large Numbers for set-indexed Processes with applications to Empirical and Partial-Sum Processes." In: *Probability in Banach Spaces*, 9, J. Hoffmann-Jørgensen, J. Kuelbs, M.B. Marcus, (eds.), Birkhäuser, Boston, pp. 385-400.

Gaenssler, P., Rost, D., Ziegler, K., 1998. On Random Measure Processes with Application to Smoothed Empirical Processes. In: *High Dimensional Probability*, 1, Progress in Probability, Vol. 43, E. Eberlein, M. Hahn, M. Talagrand, (eds.), Birkhäuser Boston, Cambridge, MA, pp. 93–102.

Talagrand, M., 1996. The Glivenko-Cantelli Problem, Ten years later. *J. of Theoret. Probab.* **9**, 371–384.

van der Vaart, A.W., Wellner, J.A., 1996. *Weak convergence and empirical processes.* Springer Series in Statistics, Springer-Verlag, New York.

Winter, B.B., 1973. Strong uniform consistency of integrals of density estimators. *Canadian J. of Statist.* **1**, 247–253.

Yamato, H., 1973. Uniform convergence of an estimator of a distribution function. *Bull. Math. Statist.* **15**, 69–78.

Yukich, J.E., 1989. A note on limit theorems for perturbed empirical processes. *Stoch. Processes and Their Applications* **33**, 163–173.

Mathematical Institute
University of Munich
Theresienstrasse 39
D-80333 Munich, Germany
e-mail: `gaensler@rz.mathematik.uni-muenchen.de`
e-mail: `rost@rz.mathematik.uni-muenchen.de`

Weak Convergence of Smoothed Empirical Processes: Beyond Donsker Classes

Dragan Radulović and Marten Wegkamp

ABSTRACT Let $X_1, \dots, X_n$ be a sequence of independent random variables with common distribution P on the real line. We study smoothed empirical processes $\sqrt{n}(P_n * K_h - P)(f)_{f \in \mathcal{F}}$ based on a kernel estimate $P_n * K_h$, and indexed by a class of uniformly bounded functions $\mathcal{F}$. Assuming that P has a smooth density, we demonstrate that the smoothed empirical process converges weakly to a tight Gaussian limit for classes $\mathcal{F}$ that are not necessarily P-Donsker. In addition, we provide some theory on the bootstrap estimate of the distribution of the process. In particular, we show that the bias inherent to the kernel density estimation method can consistently be captured by the bootstrap.

1 Introduction

Let $X_1, \dots, X_n$ be a sequence of independent, identically distributed random variables with common probability measure P on $\mathbb{R}$. Let P_n be the empirical probability measure putting mass $1/n$ at each observation X_i. The theory of empirical processes studies the behaviour of the ordinary empirical process $\sqrt{n}(P_n - P)$, indexed by a family of sets, or more generally, a class of uniformly bounded functions $\mathcal{F}$. The boundedness condition could certainly be relaxed by an appropriate envelope assumption at cost of more technical proofs. We refrain from doing so merely for ease of exposition and in order to keep the proofs as simple as possible. Also, for this reason consider only the case of univariate variables. Generalization to the case where X_i takes values in arbitrary Euclidean spaces is fairly straightforward, see the remark following Theorem 1 below. We say that $\mathcal{F}$ is a P-Donsker class if the empirical process converges to a tight Gaussian limit. This problem is well understood, and necessary and sufficient conditions for this to happen can be found in Giné and Zinn (1984), Theorem 3.2, page 951. One condition is a refined stochastic equicontinuity condition of a symmetrized version of the empirical process, the other assumption states that $\mathcal{F}$ is P-pregaussian. Recall that a class $\mathcal{F}$ is called P-pregaussian if there exists a tight Gaussian process G_P in $l^\infty(\mathcal{F})$, which is indexed by $\mathcal{F}$. Tightness of the Gaussian process is not as innocent as it might seem at first glance. For example, it guarantees that G_P has uniformly continuous sample paths with respect to its intrinsic metric. We refer to Dudley (1999) and Van der Vaart and Wellner (1996) for more details about the theory of empirical processes.

If we have more information about the probability measure, we might want to replace the estimate P_n by a more appropriate one. What consequences would this have for the empirical process based on this new estimate?

Assuming that P has a smooth density, one could consider estimating P by the kernel estimator $P_{n,K}$, based on some kernel K, instead of the fully nonparametric estimator P_n. Yukich (1992), and Van der Vaart (1994) showed that the smoothed empirical process $\sqrt{n}(P_{n,K}-P)$ converges weakly to a tight Gaussian process as long as $\mathcal{F}$ is P-Donsker, and a few regularity conditions must hold. However, these authors failed to indicate the superiority of this new process over the ordinary empirical process. In fact, they showed that $P_{n,K}$ is closer to P_n than to P, as they prove that $\sqrt{n}|(P_{n,K} - P_n)(f)| \to 0$ in outer probability, uniformly in $f \in \mathcal{F}$. So why bother considering this smoothed empirical process? We will address this problem here.

This work is a continuation of the recent results obtained in Radulović and Wegkamp (1999). Assuming that P has a density $p(x)$, which satisfies a Lipschitz condition (and some boundedness assumptions), they construct a histogram based estimator $P_{n,H}$ and establish weak convergence of the empirical process $\sqrt{n}(P_{n,H} - P)(f)_{f\in\mathcal{F}}$ under sharp conditions. In fact, if $\mathcal{F}$ is a class of indicators of sets, then the condition imposed on $\mathcal{F}$, to wit, $\mathcal{F}$ be P-pregaussian, is necessary and sufficient.

In this paper, we revisit the setting as considered by Yukich (1992), and Van der Vaart (1994). That is, we assume that P has a smooth density (essentially two bounded derivatives), and show that the kernel based empirical process converges weakly to a tight Gaussian process, without requiring that the class $\mathcal{F}$ is P-Donsker. The regularity conditions are easily satisfied if P has a smooth density (essentially two bounded derivatives) and the class $\mathcal{F}$ is appropriately supported. In this case weak convergence is guaranteed if $\mathcal{F}$ is only P-pregaussian. This is the content of Theorem 2 below, generalizing the results obtained earlier in Radulović and Wegkamp (1999) for classes of sets to general classes of functions. This indicates that if indeed P is known to be smooth, one can take advantage of this fact. Of course a P-Donsker class is P-pregaussian, but the converse is not true. To our knowledge, Dudley and Strassen were the first to exhibit a counterexample (cf. Dudley and Strassen (1969), Section 3). For additional results, we refer to Araujo and Giné (1980), page 178, and Section 3.8, and Giné and Zinn (1984), Example 7.5. Another example can be found in Radulović and Wegkamp (1999), where $\mathcal{F}$ is a countable union of certain VC classes. More precisely, for each integer k, partition $[0, 1]$ into 2^k dyadic intervals, and let $\mathcal{A}_k$ be the collection of sets that can be written as a union of exactly k (out of 2^k) disjoint dyadic intervals. Define $\mathcal{A} = (\cup_{k=1}^{\infty}\mathcal{A}_k) \cup \{\emptyset\}$, and $\mathcal{F} = \{I_A : A \in \mathcal{A}\}$. It is not difficult to see that $\mathcal{F}$ is not even P-Glivenko-Cantelli, that is,

$$\sup_{f\in\mathcal{F}} |P_n f - Pf| \xrightarrow{\text{a.s.}} 0$$

does not hold, and consequently $\mathcal{F}$ fails to be P-Donsker. On the other hand, $\mathcal{F}$ is P-pregaussian for e.g. P =Unif[0, 1]. For these computations, we refer to Radulović and Wegkamp (1999).

In addition we prove the asymptotic consistency of the bootstrap estimator of the distribution of the smoothed empirical process. In particular, we show that one can consistently estimate the bias caused by the kernel estimate of P. This property has already been established for histogram based empirical processes, indexed by classes of sets by Radulović (1999). Again, we do not require that $\mathcal{F}$ be P-Donsker, contrasting the result obtained by Giné and Zinn (1990), which states that the bootstrapped ordinary empirical process works if and only if $\mathcal{F}$ is P-Donsker. We refer to Giné (1997) and Van der Vaart and Wellner (1996) for an overview of the bootstrap of (ordinary) empirical processes, and its statistical applications.

2 Main results

Let $X_1, X_2, \ldots$ be independent random variables with common probability measure P and density p on $\mathbb{R}$. Let $p_{n,h}$ be a kernel density estimate of p, that is,

$$p_{n,h}(x) = \frac{1}{nh}\sum_{i=1}^{n} K\left(\frac{x - X_i}{h}\right),$$

where the kernel K is a density with $\int xK(x)\,dx = 0$, $\int x^2K(x)\,dx < \infty$ and $\int K^2(x)dx < \infty$.

Define

$$P_{n,h}f = \int f(x)p_{n,h}(x)\,dx.$$

The bandwidth $h = h_n$ is taken such that $nh^2 \to \infty$ and $nh^4 \to 0$. Set $K_h(\cdot) = (1/h)K(\cdot/h)$, and observe that $P_{n,h}$ is a convolution measure, since we can write

$$P_{n,h}f = \int f(x)\int K_h(x-y)\,dP_n(y)\,dx = \int\int f(x+y)\,K_h(y)\,dy\,dP_n(x).$$

We have the following result:

Theorem 2.1. *Suppose that*

(A1) The mapping $y \mapsto Pf(X+y)$ *admits the expansion*

$$Pf(X+y) = Pf(X) + yC_1(f) + \frac{1}{2}y^2 C_2(y,f),$$

where $\sup_f |C_1(f)| < \infty$ *and* $\sup_f \sup_y |C_2(y,f)| < \infty$*;*

(A2) For each $f, g \in \mathcal{F}$, $\int P(f-g)^2(X+y)K_h(y)\,dy \leq CP(f-g)^2$ *for some* $C > 0$*;*

(A3) For each $f, g \in \mathcal{F}$, $\int (f-g)^2(x)\,dx \leq CP(f-g)^2$ *for some* $C > 0$*;*

(A4) $\mathcal{F}$ *is a* P*-pregaussian class of uniformly bounded functions. Then*

$$\sqrt{n}(P_{n,h} - P)(f)_{f\in\mathcal{F}} \xrightarrow{W} G_P(f)_{f\in\mathcal{F}},$$

where G_P *is a tight, centered Gaussian process with covariance function* $\rho_P(f,g) = Pfg - PfPg$.

The proof of the theorem can be found in Section 3. It is in spirit of Radulović and Wegkamp (1999), which is based in turn on the Giné and Zinn (1984) paper.

A few remarks are in order. First, the requirement that $\mathcal{F}$ be P-pregaussian is necessary for weak convergence. Yukich (1992) and Van der Vaart (1994) proved weak convergence of the smoothed process, provided $\mathcal{F}$ is P-Donsker with additional assumptions, basically A1. It should be noted that Van der Vaart (1994) assumes that $\mathcal{F}$ is translation invariant, which is a somewhat strong requirement. The above result nicely complements their results, as we show here that for classes $\mathcal{F}$ which are not P-Donsker, but satisfy the P-pregaussian condition, $P_n * K_h$ is closer to P, as it is to P_n. To see this, observe that under the conditions of Theorem 2.1,

$$\sup_{f\in\mathcal{F}} |\sqrt{n}(P_{n,h}f - Pf)|$$

remains stochastically bounded by the continuous mapping theorem. In contrast,

$$\sup_{f\in\mathcal{F}} |\sqrt{n}(P_n - P)(f)|$$

diverges for non P-Donsker classes $\mathcal{F}$.

Also, it is interesting to see that weak convergence of the histogram based process as considered in Radulović and Wegkamp (1999), requires no smoothness assumptions A1 and A2, except for the condition on the density of course. However, their main result that the histogram based empirical process $\sqrt{n}(P_{n,H} - P)$ converges weakly to a tight Gaussian process if and only if $\mathcal{F}$ is P-pregaussian, was proven only for classes of sets. Here we extend the theory to general uniformly bounded classes of functions.

For sake of simplicity we consider only probability measures on $\mathbb{R}$. For arbitrary Euclidean spaces, the same results hold true under more stringent conditions on h, v.i.z. $nh^d \to \infty$, and one needs to impose further restrictions on the kernel and bandwidth to ascertain that $\sqrt{n}\sup_f |\mathbb{E}P_{n,h}f - Pf| \to 0$.

Conditions A1 and A2 follow from either smoothness of the density p or functions f. Here, since imposing smoothness on f would force $\mathcal{F}$ to be Donsker, we take the first approach and demand p to be twice differentiable. There is a slight complication though. Condition A2 is in general implied if p stays away from 0 on its support, but this contradicts the needed smoothness in A1 of p. It is for this reason that we restrict the domain of the functions $f \in \mathcal{F}$ to a subset of $\mathbb{R}$ where $p(x) > \delta_0$ for some small $\delta_0 > 0$. This leads to the following result.

Theorem 2.2. *Suppose that p is twice differentiable. Let $\delta_0 > 0$ and $A \subset \mathbb{R}$ such that $p(x) > \delta_0$ for $x \in A$, $\sup_{x\in A} |p'(x)| < \infty$ and $sup_{x\in A}|p''(x)| < \infty$. Let $\mathcal{F}$ be a class of uniformly bounded functions which are zero outside the set A. Then a necessary and sufficient condition for weak convergence of $\sqrt{n}(P_{n,h} - P)(f)_{f\in\mathcal{F}}$ to $G_P(f)_{f\in\mathcal{F}}$ is that $\mathcal{F}$ is P-pregaussian.*

Proof. We prove the result by checking conditions A1, A2 and A3 of Theorem 2.1.

First, A3 is met since $p(x) > \delta_0$ for all $x \in A$ so that $\int f^2 \le \delta_0^{-1} Pf^2$.

Second,

$$Pf(X + y) = \int f(x)p(x - y)\, dy$$

$$= \int f(x)p(x)dx - y\int f(x)p'(x)dx + \frac{1}{2}y^2 \int f(x)p''(\xi)dx,$$

where ξ is between $x - y$ and x. Observe that

$$\int f(x)p'(x)dx \le \delta_0^{-1}\|f\|_\infty \sup_{x\in A} |p'(x)|$$

and

$$\int f(x)p''(\xi)dx \le \delta_0^{-1}\|f\|_\infty \sup_{x\in A} |p''(x)|.$$

This implies A1.

Finally, A2 is met since by similar arguments as in the proof of Theorem 1 below,

$$\int Pf^2(X+y)K_h(y)dy \leq \left\{1 + \frac{1}{2}\delta_0^{-1} \sup_{x\in A} |p''(x)| h^2 \int x^2 K(x)dx\right\} Pf^2$$

$$= (1 + o(1))\, Pf^2.$$

This proves the result. □

Next we discuss the role of the bandwidth h. It is known that for $nh \not\to \infty$, the consistency of the kernel density estimator p_n breaks down. However, $P_{n,h}f$ is still a consistent estimate of Pf. The results above use the fact that for smooth densities p the kernel estimator $P_{n,h}$ improves upon the naive estimator P_n. So one should not be surprised at all that for small bandwidths h, we are not able to establish weak convergence if the class $\mathcal{F}$ is not P-Donsker. For P-Donsker classes, we have the following theorem in terms of sufficient conditions on the bracketing numbers of $\mathcal{F}$. Recall that for any $\delta > 0$, a δ-bracket in $L_2(P)$ is a set of functions $[f_L, f_U]$ such that $f_L \leq f \leq f_U$ and $\int (f_U - f_L)^2 dP \leq \delta^2$. Then $B(\delta, L_2(P), \mathcal{F})$ is the minimal number of $\delta-$brackets needed to cover $\mathcal{F}$.

Theorem 2.3. *Suppose assumptions A1 and A2 hold true, and that the bandwidth satisfies $nh^4 \to 0$, and that $\mathcal{F}$ satisfies*

$$\int_0^1 \sqrt{\log B(x, L_2(P), \mathcal{F})}\, dx < \infty.$$

Then

$$\sqrt{n}(P_{n,h} - P)(f)_{f\in\mathcal{F}} \xrightarrow{W} G_P(f)_{f\in\mathcal{F}}.$$

The proof can be found at the end of Section 3.

This result shows, in particular, that the restrictive assumption A3 is not needed for establishing weak convergence of the smoothed empirical process, indexed by most P-Donsker classes $\mathcal{F}$. Assumption A3 is crucial, however, in our argument of the proof of Theorem 1. We want to stress that Theorem 1 shows that in certain settings, smoothing does something extra, and this is the main theme of this paper. It should be noted that the preceding theorem is similar in spirit to the results obtained by Yukich (1992) and Van der Vaart (1994). The proof is simpler, and we do not require that $\mathcal{F}$ be closed under translations. Also, the range of the bandwidths is less restrictive for this type of kernel in comparison with those treated in aforementioned papers.

We end this section by showing that the bootstrap can provide a consistent estimate of the distribution of the smoothed empirical process

$$\sqrt{n}(P_{n,h} - P).$$

Let $X_1^*, \dots, X_n^*$ be independent, identically distributed random variables with common probability P_n, given the original observations $X_1, \dots, X_n$. That is, we consider Efron's original bootstrap plan, and we do not resample from $P_{n,h}$ as might perhaps be expected.

Instead, let P_n^* be the empirical measure putting mass $1/n$ at each X_i^*, and construct the density $p_{n,h}^*$ by

$$p_{n,h}^*(x) = P_n^* K_h(x - \cdot) = \frac{1}{nh} \sum_{i=1}^{n} K\left(\frac{x - X_i^*}{h}\right).$$

The associated measure is denoted by $P_{n,h}^*$, i.e.,

$$P_{n,h}^* f = \int f(x) p_{n,h}^*(x)\, dx.$$

We will prove the following result: Provided the *centered* smoothed empirical process $\sqrt{n}(P_{n,\gamma} - \mathbb{E}P_{n,\gamma})(f)_{f\in\mathcal{F}}$ converges weakly to $G_P(f)_{f\in\mathcal{F}}$ for all $h' \le \gamma \le h$, the bootstrap empirical process $\sqrt{n}(P_{n,h}^* - P_{n,h'})(f)_{f\in\mathcal{F}}$ consistently estimates the distribution of the original smoothed empirical process $\sqrt{n}(P_{n,h} - P)(f)_{f\in\mathcal{F}}$. This means that we allow $\sup_{f\in\mathcal{F}} |\sqrt{n}(\mathbb{E}P_{n,h} - P)(f)| \to \infty$ to happen, since we can capture the bias term using the estimate $\sqrt{n}(P_{n,h} - P_{n,h'})$ by taking a sufficiently small bandwidth h'.

Before stating the results we introduce some more notation. Define for $h > 0$, and $f \in \mathcal{F}$ the bias

$$B(f,h) = |\mathbb{E}P_{n,h} f - Pf| = \left| Pf - \int Pf(X + y) K_h(y)\, dy \right|,$$

and let $B(h) = \sup_{f\in\mathcal{F}} B(f,h)$. We will formulate the result in terms of the bounded Lipschitz metric d_{BL}, which metrizes weak convergence of probability measures on $\mathbb{R}$. Recall that for probability measures P and Q on a separable Banach space,

$$d_{BL}(P,Q) = \sup\left\{ \left\| \int F d(P - Q) \right\| \;:\; \|F\|_\infty \le 1,\ \|F(x) - F(y)\| \le \|x - y\| \right\}.$$

Theorem 2.4. *Let $h \to 0$ and $h' \le h$. Suppose that the centered smoothed empirical process $\sqrt{n}(P_{n,\gamma} - \mathbb{E}P_{n,\gamma})(f)_{f\in\mathcal{F}}$ converges weakly to $G_P(f)_{f\in\mathcal{F}}$ for all $h' \le \gamma \le h$, and that $\sqrt{n}B(h') \to 0$. Then*

$$d_{BL}\left(\sqrt{n}(P_{n,h} - P), \sqrt{n}(P_{n,h}^* - P_{n,h'})\right) = o_P(1).$$

The proof of this result can be found in Section 4.

It is interesting to note that Theorem 2.4 holds under no additional conditions on P and/or $\mathcal{F}$. Of course, to be able to guarantee the required weak convergence of the centered empirical processes $\sqrt{n}(P_{n,\gamma} -$

$\mathbb{E}P_{n,\gamma})(f)_{f\in\mathcal{F}}$, one needs some smoothness on P. Also it should be noticed that $\sqrt{n}B(h') \to 0$ under assumption (A1) and $n(h')^4 \to 0$. However, (A1) might be relaxed, as we can take $h' << h$, and it is not required that $nh^4 \to 0$!

Supplementing the preceding consistency result, the accuracy of the bootstrap approximation of the distribution of $\sqrt{n}(P_{n,h} - P)(f)$ is of interest, especially in comparison with the Gaussian approximation in Theorem 2.1. We express this in terms of the d_3 distance between $\sqrt{n}(P_{n,h} - P)(f)$ and its bootstrap approximation $\sqrt{n}(P^*_{n,h} - P_{n,h'})(f)$ for fixed f. The d_3 distance is defined by

$$d_3(P,Q) = \sup\left\{\left|\int F d(P-Q)\right| \ : \ \|F^{(i)}\|_\infty \le 1, \ i = 0,1,2\right\},$$

for measures P and Q on $\mathbb{R}$, and like the bounded Lipschitz metric, it metrizes weak convergence.

Theorem 2.5. *Under the same conditions as in Theorem 2.4, we have*

$$\begin{aligned} &d_3\left(\sqrt{n}(P_{n,h} - P)(f), \sqrt{n}(P^*_{n,h} - P_{n,h'})(f)\right) \\ &= \mathcal{O}_P\left(\sqrt{n}B(f,h') + n^{-1/2} + \sqrt{L(f,h',h)}\right), \end{aligned}$$

where

$$L(f,h',h) = P\left\{\int f(X+y)K_{h'}(y)\,dy - \int f(X+y)K_h(y)\,dy\right\}^2.$$

The proof of this result can be found in Section 4 as well.

We have already seen that if p has two derivatives, the bias $B(f,h') = \mathcal{O}((h')^2)$. The term $L(f,h',h)$ tends to zero for any P-integrable function f. In fact, by taking $h = h'$, we simply have that $L \equiv 0$, and the speed of convergence is $\mathcal{O}_P(\sqrt{n}h^2 + n^{-1/2})$. This rate can be improved if we assume that f is Lipschitz, because one verifies easily that for $h' < h$

$$L(f,h',h) \le 4P\left(\int \{f(X+y) - f(X)\}K_h(y)\,dy\right)^2 = \mathcal{O}(h^2),$$

and $B(f,h') = \mathcal{O}(h')$ regardless of the smoothness of p, so that the rate of convergence in this case is $\mathcal{O}_P(\sqrt{n}h' + h + n^{-1/2})$, or even $\mathcal{O}_P(\sqrt{n}(h')^2 + h + n^{-1/2})$ if p is twice differentiable. For suitably chosen values of h' and h, this amounts to an improvement over the normal approximation of $\sqrt{n}(P_{n,h} - P)(f)$.

3 Proof of Theorems 2.1 and 2.3

Since $\mathcal{F}$ is uniformly bounded, we make the blanket assumption that $\|f\|_\infty \le 1$ for all $f \in \mathcal{F}$. We first prove Theorem 1 in several steps.

3.1 Bias

First, we deal with the bias term. Letting $g(f,y) = Pf(X+y)$, we see that condition A1 implies that there is a $C > 0$ independent of f and y such that

$$|g(f,y) - g(f,0) - yg'(f,0)| \leq \frac{1}{2}Cy^2.$$

Therefore, we have

$$\sup_{f\in\mathcal{F}} \sqrt{n}\,|\mathbb{E}P_{n,h}f - Pf|$$

$$= \sup_{f\in\mathcal{F}} \sqrt{n}\left|\int\int \{f(x+y) - f(x)\}\, K_h(y)\, dy\, dP(x)\right|$$

$$= \sup_{f\in\mathcal{F}} \sqrt{n}\left|\int \{Pf(X+y) - Pf(X)\}\, K_h(y)\, dy\right| \text{ by Fubini}$$

$$= \sup_{f\in\mathcal{F}} \sqrt{n}\left|\int yg'(f,0)K_h(y)\, dy + \frac{1}{2}\int y^2 g''(f,\xi)K_h(y)\, dy\right|.$$

Combining the first two displays, and the fact that $\int xK(x)\,dx = 0$ we conclude that for some $C > 0$,

$$\sup_{f\in\mathcal{F}} \sqrt{n}\,|\mathbb{E}P_{n,h}f - Pf| \leq \frac{1}{2}\sqrt{n}Ch^2 \int x^2K(x)\,dx = o(1), \tag{3.1}$$

since $nh^4 \to 0$.

This means that the bias is asymptotically negligible and we will consider in the sequel only the centered process $\sqrt{n}(P_{n,h} - \mathbb{E}P_{n,h})(f)_{f\in\mathcal{F}}$.

3.2 Finite dimensional convergence

Next, we show the finite dimensional convergence of

$$\sqrt{n}(P_{n,h} - \mathbb{E}P_{n,h})(f)_{f\in\mathcal{F}}.$$

We will show that for fixed $f \in \mathcal{F}$,

$$\int fd(P_{n,h} - \mathbb{E}P_{n,h}) \xrightarrow{W} \mathcal{N}\left(0, P(f-Pf)^2\right).$$

Define

$$\overline{f}_n(x) \equiv \overline{f}_n(f,x) = \int f(x+y)K_h(y)\,dy,$$

and observe that

$$\int f d(P_{n,h} - \mathbb{E}P_{n,h}) = \int \overline{f}_n d(P_n - P). \tag{3.2}$$

Since $\overline{f}_n(X_1), \ldots, \overline{f}_n(X_n)$ are iid. bounded random variables for fixed sample size n, we only need to verify that the variance of $\overline{f}_n(X_1)$ converges to $P(f - Pf)^2$. It is well known (cf. Devroye and Györfi (1985), page 7) that

$$\lim_{h \downarrow 0} P \left| \int g(x+y) K_h(y)\, dy - g(x) \right| = 0,$$

for all P integrable g and K. Applying this lemma for $g = f$ and $g = f^2$ yields that both $P\overline{f}_n \to Pf$ and $P\overline{f}_n^2 \to Pf^2$ as $n \to \infty$, and the fidi result is established.

3.3 Stochastic equicontinuity

It remains to establish stochastic equicontinuity of

$$\sqrt{n}(P_{n,h} - \mathbb{E}P_{n,h})(f)_{f \in \mathcal{F}},$$

i.e., for all $\tau > 0$

$$\lim_{\delta \downarrow 0} \limsup_{n \to \infty} \mathbb{P}^* \left\{ \sup_{P(f-g)^2 < \delta} \left| \sqrt{n} \int (f - g)\, d(P_{n,h} - \mathbb{E}P_{n,h}) \right| > \tau \right\} = 0,$$

where $\mathbb{P}^*$ stands for outer probability.

Since $\mathcal{F}$ is pregaussian and uniformly bounded, $(\mathcal{F}, L_2(P))$ is totally bounded by Sudakov's lower bound. (Refer to Theorem 2.16 in Giné and Zinn (1984).) Hence, given ε and n, there exist finitely many functions $g_1, \ldots, g_m \in \mathcal{F}$ such that $P(g_i - g_j)^2 > \varepsilon/(4\sqrt{n})$, so that

$$\sup_{f \in \mathcal{F}} \min_{1 \le i \le m} P(f - g_i)^2 \le \frac{\varepsilon}{\sqrt{n}}.$$

Consequently, for n large enough,

$$\mathbb{P}^* \left\{ \sup_{P(f-g)^2 \le \delta} \left| \sqrt{n} \int (f - g)\, d(P_{n,h} - \mathbb{E}P_{n,h}) \right| > 3\tau\varepsilon \right\}$$

$$\le 2\mathbb{P}^* \left\{ \sup_{P(f-g)^2 \le \varepsilon n^{-1/2}} \left| \sqrt{n} \int (f - g)\, d(P_{n,h} - \mathbb{E}P_{n,h}) \right| > \tau\varepsilon \right\}$$

$$+ \mathbb{P} \left\{ \max_{P(g_i - g_j)^2 \le 4\delta} \left| \sqrt{n} \int (g_i - g_j)\, d(P_{n,h} - \mathbb{E}P_{n,h}) \right| > \tau\varepsilon \right\} = I + II.$$

Observe that by the Cauchy-Schwarz inequality and condition A3 respectively

$$\left|\sqrt{n}\int (f-g)\,d(P_{n,h}-\mathbb{E}P_{n,h})\right|$$

$$\leq \sqrt{n}\|f-g\|_2\|p_{n,h}-\mathbb{E}p_{n,h}\|_2$$

$$\leq \sqrt{n}\left(CP(f-g)^2\right)^{1/2}\|p_{n,h}-\mathbb{E}p_{n,h}\|_2.$$

Moreover, since

$$\mathbb{E}\int |p_{n,h}-\mathbb{E}p_{n,h}|^2 = \int \mathrm{Var}(p_{n,h}(x))dx \leq \frac{1}{n}\int \mathbb{E}K_h^2(x-X_1)dx$$

$$= \frac{1}{nh}\int\int K^2(y)p(x-yh)dydx \leq \frac{1}{nh}\int K^2(x)dx,$$

we obtain

$$\sup_{P(f-g)^2\leq \varepsilon n^{-1/2}}\left|\sqrt{n}\int (f-g)\,d(P_{n,h}-\mathbb{E}P_{n,h})\right|$$

$$\leq \sqrt{n}\left(C\varepsilon n^{-1/2}\right)^{1/2}\|p_{n,h}-\mathbb{E}p_{n,h}\|_2 = \mathcal{O}_P\left(n^{-1/4}h^{-1/2}\right) = o_P(1),$$

as $nh^2\to\infty$. Hence $I\to 0$ as $n\to\infty$.

It remains to bound

$$II = \mathbb{P}\left\{\max_{P(g_i-g_j)^2\leq 4\delta}\left|\sqrt{n}\int (g_i-g_j)\,d(P_{n,h}-\mathbb{E}P_{n,h})\right| > \tau\varepsilon\right\}$$

$$= \mathbb{P}\left\{\max_{f\in\mathcal{F}(\delta)}\left|\sqrt{n}\int \overline{f}_n\,d(P_n-P)\right| > \tau\varepsilon\right\},$$

where

$$\mathcal{F}(\delta) = \left\{g_i-g_j : P(g_i-g_j)^2\leq 4\delta :\ i,j=1,\ldots,m\right\}.$$

Observe that

$$P\overline{f}_n^2 = \int\left(\int f(x+y)K_h(y)\,dy\right)^2 dP(x)$$

$$\leq \int\int f^2(x+y)K_h(y)\,dy\,dP(x) \qquad \text{by Cauchy-Schwarz}$$

$$= \int\int f^2(x+y)\,dP(x)K_h(y)dy \qquad \text{by Fubini}$$

$$\leq CPf^2 \qquad \text{by assumption A2.}$$

We will now argue as in Giné and Zinn (1984, pp. 952–953). Define the event A by

$$A = \left\{ \max_{f\in\mathcal{F}(\delta)} \frac{\sum_{i=1}^n \overline{f}_n^2(X_i)}{nPf^2} < 1 + C \right\}.$$

This event has probability tending to one by the following string of inequalities:

$$P(A^c) \leq m^2 \max_{f\in\mathcal{F}(\delta)} \mathbb{P}\left\{(P_n - P)\overline{f}_n^2 \geq (1+C)Pf^2 - P\overline{f}_n^2\right\}$$

$$\leq m^2 \max_{f\in\mathcal{F}(\delta)} \mathbb{P}\left\{(P_n - P)\overline{f}_n^2 \geq Pf^2\right\},$$

since $P\overline{f}_n^2 \leq CPf^2$, which by Bernstein's inequality and $\|f\|_\infty \leq 2$ is

$$\leq m^2 \max_{f\in\mathcal{F}(\delta)} \exp\left(-\frac{n}{2}\frac{(Pf^2)^2}{4P\overline{f}_n^2 + (4Pf^2)/3}\right)$$

$$\leq m^2 \max_{f\in\mathcal{F}(\delta)} \exp\left(-\frac{n}{8}\frac{Pf^2}{(C+1/3)}\right) \text{ since } P\overline{f}_n^2 \leq CPf^2$$

$$\leq m^2 \exp\left(-\frac{n\varepsilon}{32(C+1/3)\sqrt{n}}\right),$$

by construction of $f = g_i - g_j$.

Since $\mathcal{F}$ is pregaussian, there exists $c_n(\varepsilon) \to 0$ as $n \to \infty$ such that $\log(m) \leq c_n\sqrt{n}/\varepsilon$ as a result of Sudakov's lower bound. Hence

$$P(A) \geq 1 - \exp\left(\frac{2c_n\sqrt{n}}{\varepsilon} - \frac{\sqrt{n}\varepsilon}{32(C+1/3)}\right) \to 1, \text{ as } n \to \infty.$$

By a standard symmetrization argument (cf. Pollard (1984), p. 14) we obtain for n large enough,

$$II \leq 4\mathbb{P}\left\{ \max_{f\in\mathcal{F}(\delta)} |\sqrt{n}\int \overline{f}_n\,dP_n^0| \geq \frac{\tau\varepsilon}{4}\right\}$$

$$\leq 4\mathbb{P}\{A^c\} + \frac{16}{\tau\varepsilon}\mathbb{E}\mathbb{E}\left(\max_{f\in\mathcal{F}(\delta)} |\sqrt{n}\int \overline{f}_n\, dP_n^0|I_A|X_1,\ldots,X_n\right),$$

where $P_n^0 = \frac{1}{n}\sum_{i=1}^n \varepsilon_i\delta_{X_i}$ is the symmetrized measure of P_n, and $\varepsilon_1,\ldots,\varepsilon_n$ is a Rademacher sequence, independent of $X_1,\ldots,X_n$.

The conclusion that $II \to 0$ as $\delta \to 0$, $n \to \infty$ follows from the same arguments as in Giné and Zinn (1984, pp. 952–953), or Radulović and Wegkamp (1999). Since finite dimensional convergence and stochastic equicontinuity of a process imply weak convergence, the proof is complete. □

3.4 Proof of Theorem 2.3.

The restriction on the range of the bandwidth and Assumption A1 imply that the bias is asymptotically negligible. It suffices to show that

$$\sqrt{n}\int f\, d(P_{n,h} - \mathbb{E}P_{n,h}) = \sqrt{n}\int \overline{f}\, d(P_n - P)$$

is asymptotically equicontinuous.

Observe that we are simply dealing with a class of functions changing in the sample size n, and indexed by f, as

$$f \mapsto \overline{f} = \int f(x+y)K_h(y)dy.$$

Theorem 2.3 follows directly from Theorem 2.11.23 in Van der Vaart and Wellner (1996, p. 221) where we invoke Assumption A2, which takes care of their condition (2.11.21), and the entropy condition on $\mathcal{F}_n$. Notice that a δ^2-covering of brackets $(f_L^1, f_U^1),\ldots,(f_L^M, f_U^M)$ for $\mathcal{F}$ provides a $C\delta^2$-covering $(\overline{f}_L^1, \overline{f}_U^1),\ldots,(\overline{f}_L^M, \overline{f}_U^M)$ for $\mathcal{F}_n$ for every n, since pointwise

$$\overline{f}_L(x) - \overline{f}(x) = \int \{f_L(x+y) - f(x+y)\}K_h(y)\, dy \leq 0 \text{ if } f_L \leq f$$

and similarly,

$$\overline{f}_U(x) - \overline{f}(x) \geq 0 \text{ if } f_U(x) \geq f(x).$$

Assumption A2 relates the $L_2(P)$ bracketing numbers between $\mathcal{F}_n$ and $\mathcal{F}$. □

4 Proofs of Theorems 2.4 and 2.5

4.1 Proof of Theorem 2.4

We begin by introducing some notation. Let $Z_n(f) := \sqrt{n}(P_{n,h} - P)(f)$, $Z_n^*(f) := \sqrt{n}(P_{n,h}^* - P_{n,h'})(f)$. We also define the bias and centered parts of the process:

$$C_n(f) := \sqrt{n}(P_{n,h} - \mathbb{E}P_{n,h})(f), \; B_n(f) := \sqrt{n}(\mathbb{E}P_{n,h} - P)(f),$$

$$C_n^*(f) := \sqrt{n}(P_{n,h}^* - P_{n,h})(f)$$

and

$$B_n^*(f) := \sqrt{n}(P_{n,h} - P_{n,h'})(f).$$

Finally, let the set $\{g_1, ..., g_{K_\delta}\}$ be a $\delta-$cover for the class $\mathcal{F}$ under the $L_2(P)$ distance. Such a set exists since the assumptions of Theorem 4 imply that $\mathcal{F}$ is P-pregaussian. For given $f \in \mathcal{F}$, g_f will denote the closest g_k in the cover to f. For any bounded, Lipschitz function $H : l^\infty(\mathcal{F}) \to \mathbb{R}$, we obtain by adding and subtracting the appropriate terms

$$H(Z_n(f)) - H(Z_n^*(f)) = A_1 + A_2 + A_3 + A_4$$

where

$$A_1 = H(C_n(f) + B_n(f)) - H(C_n(g_f) + B_n(f))$$

$$A_2 = H(C_n(g_f) + B_n(f)) - H(C_n^*(g_f) + B_n(f))$$

$$A_3 = H(C_n^*(g_f) + B_n(f)) - H(C_n^*(g_f) + B_n^*(f)).$$

$$A_4 = H(C_n^*(g_f) + B_n^*(f)) - H(C_n^*(f) + B_n^*(f)).$$

By the definition of d_{BL} it is sufficient to show that $\mathbb{E}^*|A_i| = o_P(1)$ uniformly in H. Since H is Lipschitz, for A_3 we have

$$\mathbb{E}|A_3| \le \sup_{f\in\mathcal{F}} |B_n(f) - B_n^*(f)|$$

$$\le \sup_{f\in\mathcal{F}} |\sqrt{n}(P_n - P)(\overline{f}_h - \overline{f}_{h'})| + \sup_{f\in\mathcal{F}} |\sqrt{n}P(f - \overline{f}_{h'})|,$$

by adding and subtracting $\mathbb{E}B_n^*(f)$, and where

$$\overline{f}_h = \int f(x+y)K_h(y)dy$$

is as in Section 3. The second term on the right is equal to $\sqrt{n}B(f, h')$, and tends to zero by assumption. The first term tends to zero by the following argument. Invoke (3.2) to see that

$$\sqrt{n}(P_n - P)(\overline{f}_h - \overline{f}_{h'})$$

$$= \sqrt{n}(P_{n,h} - \mathbb{E}P_{n,h})f + \sqrt{n}(P_{n,h'} - \mathbb{E}P_{n,h'})f. \tag{4.1}$$

Consequently, the assumed weak convergence of

$$\sqrt{n}(P_{n,\gamma} - \mathbb{E}P_{n,\gamma})(f)_{f\in\mathcal{F}}$$

for $\gamma = h', h$ implies that

$$\sqrt{n}(P_n - P)(\overline{f}_h - \overline{f}_{h'})_{f\in\mathcal{F}}$$

is stochastically equicontinuous. Since for each n,

$$(\overline{f}_h - \overline{f}_{h'})(X_1), \dots , (\overline{f}_h - \overline{f}_{h'})(X_n)$$

are independent, identically distributed, bounded random variables, it is easily verified that $\sqrt{n}(P_n - P)(\overline{f}_h - \overline{f}_{h'})$ converges in probability to 0. This follows from the following calculation of the variance for $h' < h$

$$\operatorname{Var}(\overline{f}_h - \overline{f}_{h'})(X) \le P(\overline{f}_h - \overline{f}_{h'})^2 \tag{4.2}$$

$$\le 4P(\overline{f}_h - f)^2 \to 0, \text{ as } h \to 0.$$

These two facts – the established stochastic equicontinuity and finite dimensional convergence of $\sqrt{n}(P_n - P)(\overline{f}_h - \overline{f}_{h'})$ – entail that

$$\sup_{f\in\mathcal{F}} \left|\sqrt{n}(P_n - P)(\overline{f}_h - \overline{f}_{h'})\right| \to 0.$$

For A_2, we observe that the term B_n is not random so the estimate reduces to the finite dimensional convergence result of Theorem 5, whence

$$A_2 = \mathcal{O}_P(\sqrt{n}B(g_f, h') + n^{-1/2} + \sqrt{L(g_f, h', h)}) = o_P(1).$$

Next, $\mathbb{E}^*|A_4|$ is of order $o_P(1)$ provided that, in probability,

$$\lim_{\delta\to 0} \limsup_{n\to\infty} \mathbb{P}^* \left(\sup_{Pf^2\le\delta} \left| \frac{1}{\sqrt{n}} \sum_{j=1}^{n} (\overline{f}_h(X_j^*) - \mathbb{E}^*\overline{f}_h(X_1^*)) \right| > \varepsilon \right) = 0.$$

By the "in probability" in the above statement we mean an a.s. convergence along the subsequences. Because $\{X_i^*\}$ are resampled with replacement from $\{X_i\}$, we can invoke the Gine-Zinn result on bootstrapping empirical processes (cf. Giné and Zinn (1990)). In other words, the above is true provided that for all $\varepsilon > 0$

$$\lim_{\delta\to 0} \limsup_{n\to\infty} \mathbb{P}^* \left(\sup_{Pf^2\le\delta} \left| \frac{1}{\sqrt{n}} \sum_{j=1}^{n} (\overline{f}_h(X_j) - \mathbb{E}\overline{f}_h(X_1)) \right| > \varepsilon \right) = 0\,.$$

The above requirement is sufficient for $\mathbb{E}|A_1| = o(1)$ as well. Finally, the assumed weak convergence of the centered process C_n implies the stochastic equicontinuity requirement above, and the proof follows. It should be noticed that in the above argument the class $\mathcal{F}_n = \{\overline{f}_h : f \in \mathcal{F}\}$ changes with n, since h depends on n, in contrast with the Giné and Zinn (1990) set-up. However, a careful inspection of their proof reveals that the choice of the class $\mathcal{F}$ does not matter. Their argument relies on a chain of inequalities, each one holding with universal constants independent of n or class $\mathcal{F}$. □

4.2 Proof of Theorem 2.5

We use the same decomposition argument as in the preceding proof of Theorem 2.4. That is, we bound

$$d_3(\mathcal{L}(Z_n), \mathcal{L}(Z_n^*)) \leq d_3(\mathcal{L}(C_n), \mathcal{L}(C_n^*)) + d_3(\mathcal{L}(B_n), \mathcal{L}(B_n^*)). \tag{4.3}$$

It is well known that the first term is of the order $O_P(n^{-1/2})$ (see Radulović (1999)). We give here just a sketch of the proof. Let $Y_{i,n} := \overline{f}_h(X_i) - \mathbb{E}\overline{f}_h(X_j)$ and $Y_{i,n}^* := \overline{f}_h(X_i^*) - \mathbb{E}^*\overline{f}_h(X_i^*)$ and observe that $\|Y_{i,n}\| + \|Y_{i,n}^*\| \leq 4$, $\mathbb{E}Y_{i,n} = \mathbb{E}^*Y_{i,n}^* = 0$ and $|\mathbb{E}(Y_{i,n})^2 - \mathbb{E}^*(Y_{i,n}^*)^2| = O_P(n^{-1/2})$, together with a Lindeberg type of argument imply the above rate.

For the second term in (4.3) we employ the same calculation as in the proof of Theorem 4. Namely

$$d_3(L(B_n), L(B_n^*)) \leq |\sqrt{n}(P_n - P)(\overline{f}_h - \overline{f}_{h'})| + |\sqrt{n}P(f - \overline{f}_{h'})|\,.$$

Invoke the variance calculation (4.2) to conclude that the term $\sqrt{n}(P_n - P)(\overline{f}_{h'} - \overline{f}_h)$ is of order $\mathcal{O}\left(\sqrt{P(\overline{f}_h - \overline{f}_{h'})^2}\right)$. This completes the proof. □

Acknowledgment. We thank Andrew Barron and Philippe Barbe for stimulating discussions. We are also indebted to the referee for detailed comments and suggestions, which greatly improved our paper.

References

Araujo, A. and Giné, E. (1980). *The Central Limit Theorem for Real and Banach Valued Random Variables,* Wiley, New York.

Devroye, L. and Györfi, L. (1985). *Nonparametric Density Estimation: the L_1 View*, John Wiley & Sons.

Dudley, R. (1999). *Uniform Central Limit Theorems*, Cambridge.

Dudley, R. and Strassen, V. (1969). The Central Limit Theorem and epsilon-entropy, in: Lecture Notes in Mathematics, 87, pp. 224–231, Springer-Verlag.

Gaenssler, P., Rost, D. and Zieger, K. (1997). On random measure processes with application to smoothed empirical processes, preprint of the Mathematisches Institut, University of Munich, **60**.

Giné, E. (1997). Lectures on some aspects of the bootstrap. In *Lectures on Probability Theory and Statistics,* Ecole d'été de probabilités de Saint-Flour (1996), ed. P. Bernard. *Lecture Notes in Mathematics*, **1665**, 37–151, Springer.

Giné, E. & Zinn, J. (1984). Some limit theorems for empirical processes. *Annals of probability*, **12**, 929–998.

Giné, E. & Zinn, J. (1990). Bootstrapping general empirical measures. *Annals of probability*, **18**, 851–869.

Pollard, D. (1984). *Convergence of Stochastic Processes*, Springer-Verlag.

Radulović, D. (1998). Can we bootstrap even if CLT fails? *Journal of Theoretical Probability*, **17**, 813–830.

Radulović, D. (1999). A note on smoothed bootstrap, preprint.

Radulović, D. and Wegkamp, M. (1999). Necessary and Sufficient Conditions for Weak Convergence of Smoothed Empirical Processes, preprint Department of Statistics, Yale University.

van der Vaart, A. (1994). Weak convergence of smoothed empirical processes. *Scandinavian Journal of Statistics*, **21**, 501–504.

van der Vaart, A. & Wellner, J. (1996). *Weak Convergence and Empirical Processes*, Springer-Verlag.

Yukich, J.E. (1992). Weak convergence of smoothed empirical processes. *Scandinavian Journal of Statistics*, **19**, 271–279.

D. Radulović
Department of Mathematics
Princeton University
Princeton, NJ 08544
e-mail: `dradulov@princeton.edu`

M. Wegkamp
Department of Statistics
Yale University
New Haven, CT 06520
e-mail: `marten.wegkamp@yale.edu`

Limit Theorems for Smoothed Empirical Processes

Daniel Rost

ABSTRACT We present functional central limit theorems (FCLT) for smoothed empirical processes indexed by a class $\mathcal{F}$ of measurable functions defined on a linear metric space X via the random measure process (RMP) approach. We shall work under rather weak conditions and shall e.g. not need the rather restrictive assumption of $\mathcal{F}$ being invariant under translation. The results should be compared to those obtained by van der Vaart (1994) and Yukich (1992) who tackled the problem from a different angle.

1 Introduction

Taking up the setting and notation of the paper by Gaenssler and Rost (in this volume) the aim of this contribution is to present functional central limit theorems (FCLT) for the *smoothed empirical process* indexed by $\mathcal{F}$

$$\sqrt{n}(\tilde{\nu}_n - \nu) \equiv \left(\sqrt{n}(\tilde{\nu}_n(f) - \nu(f))\right)_{f \in \mathcal{F}},$$

i.e. sufficient conditions under which

$$(1.1) \qquad \sqrt{n}(\tilde{\nu}_n - \nu) \xrightarrow[sep]{\mathcal{L}} \mathbb{G} \qquad \textit{in } l^\infty(\mathcal{F}).$$

Here $\xrightarrow[sep]{\mathcal{L}}$ denotes weak convergence in the sense of Hoffmann-Jørgensen (see e.g. van der Vaart and Wellner (1996), section 1.3 for the definition) to a (Gaussian) limit process $\mathbb{G}$ whose paths are obtained in a separable subspace of $l^\infty(\mathcal{F})$, the space of all bounded functions on $\mathcal{F}$ endowed with the sup-norm. In all our theorems $U^b(\mathcal{F}, d_\nu^{(2)})$, the space of all bounded, uniformly $d_\nu^{(2)}$-continuous functions on $\mathcal{F}$ will serve as a separable subspace of $l^\infty(\mathcal{F})$; here

$$d_\nu^{(2)}(f,g) := \left(\nu(f-g)^2\right)^{1/2} \equiv \left(\int_X (f-g)^2 d\nu\right)^{1/2}, \quad f,g \in \mathcal{F}.$$

Note that $U^b(\mathcal{F}, d_\nu^{(2)})$ is separable if and only if $(\mathcal{F}, d_\nu^{(2)})$ is totally bounded. In this case a neccessary and sufficient condition for (1.1) to hold is convergence of the finite dimensional distributions (fidi convergence) together with the asymptotic equicontinuity condition (AEC) for $\sqrt{n}(\tilde{\nu}_n - \nu)$ with respect to $d_\nu^{(2)}$ (see e.g. Giné, (1996), chapter 2, for a reference).

Several FCLT for the smoothed empirical process have been established in the literature (cf. van der Vaart (1994) and Yukich (1992)). We will show how an FCLT with partly weaker conditions can be obtained by viewing as in Gaenssler and Rost (in the present volume) the smoothed empirical process as a special case of a *random measure process* (RMP).

It should be noted that many processes of statistical importance in fact turn out to be RMP. Examples are partial sum processes with fixed or random locations, sequential uniform empirical processes, empirical integrated regression functions, local empirical processes, just to mention a few (see also Gaenssler and Rost (1999) for a survey). Thus the RMP approach is a fairly general theory covering a lot of different examples and yet yielding surprisingly good results in the various cases. Dealing with integrals of functions $f \in \mathcal{F}$ we tacitly assume that they exist as real numbers, resp. random variables. Concerning the envelope F of $\mathcal{F}$, we also assume regularity conditions such as measurability and finiteness of $w_{nj}(F)$ and of $w_{nj}(F^2)$. Furthermore, (to avoid measurability considerations) we assume for simplicity that $\mathcal{F}$ is countable.

2 Main results

For the following results we assume that $\mathcal{F}$ has uniformly integrable L_2-entropy. This means that (cf. van der Vaart and Wellner (1996), condition (2.5.1), p. 127)

$$(2.1) \qquad \int_0^\infty \sup_Q \sqrt{\log N\big(\tau\,(Q(F^2))^{1/2}, \mathcal{F}, L_2(Q)\big)}\,d\tau < \infty$$

where the supremum is taken over all finite measures Q on $(X, \mathcal{X})$ with $0 < Q(F^2) < \infty$ and $L_2(Q)$ denotes the L_2-pseudometric with respect to Q, i.e. $L_2(Q)(f,g) := (Q(f-g)^2)^{1/2}$, $f, g \in \mathcal{F}$. For $\varepsilon > 0$ $N(\varepsilon, \mathcal{F}, L_2(Q))$ denotes the covering number, defined as the minimal number of $L_2(Q)$-balls of radius ε needed to cover $\mathcal{F}$. Based on an FCLT for RMP's (cf. Ziegler (1997), Theorem 6.1 together with Remark 6.2, or Gaenssler and Rost (1999)) and using the decomposition

$$(2.2) \qquad \sqrt{n}(\nu_n \star \mu_n - \nu) = \sqrt{n}(\nu_n \star \mu_n - \nu \star \mu_n) + \sqrt{n}(\nu \star \mu_n - \nu)$$

we obtain the following FCLT for smoothed empirical processes.

Theorem 2.1. *Let X be a linear metric space and let $\mathcal{F}$ have uniformly integrable L_2-entropy. Assume that the following conditions (2.3)–(2.5) are fulfilled:*

For each $\rho > 0$ there exists $\delta_n \equiv \delta_n(\rho), n \in \mathbb{N}$, $\delta_n \to 0$ such that

$$\limsup_{n\to\infty} \sqrt{n} \cdot \mathbb{E}\Big(w_{n1}(F)\, I(w_{n1}(F) > \delta_n\sqrt{n})\Big) \le \rho\,, \tag{2.3}$$

$$\limsup_{n\to\infty} \sup_{d_\nu^{(2)}(f,g)\le\alpha} \mathbb{E}\Big((w_{n1}(f) - w_{n1}(g))^2) \cdot I(w_{n1}(F) \le \delta_n\sqrt{n})\Big) \to 0 \text{ for } \alpha \to 0\,, \tag{2.4}$$

$$\sup_{n\in\mathbb{N}} \mathbb{E}\Big(w_{n1}(F^2) \cdot I(w_{nj}(F) \le \delta_n\sqrt{n})\Big) < \infty\,. \tag{2.5}$$

Suppose further that

$$\sup_{f\in\mathcal{F}} \sqrt{n}|\nu \star \mu_n(f) - \nu(f)| \longrightarrow 0\,. \tag{2.6}$$

Assume in addition that there exists a mean-zero Gaussian process $\bar{\mathbb{G}} = (\bar{G}(f))_{f\in\mathcal{F}}$ such that the finite dimensional distributions of $\sqrt{n}(\tilde{\nu}_n - \nu\star\mu_n)$ converge to those of $\bar{\mathbb{G}}$. Then there exists a mean-zero Gaussian process $\mathbb{G} = (G(f))_{f\in\mathcal{F}}$ with sample paths in $U^b(\mathcal{F}, d_\nu^{(2)})$ such that

$$\sqrt{n}(\nu_n \star \mu_n - \nu) \xrightarrow[sep]{\mathcal{L}} \mathbb{G} \qquad \textit{in } l^\infty(\mathcal{F}).$$

Note that compared with theorem 3.6 in Gaenssler, Rost and Ziegler (1998), the condition (3.11) there can be replaced with the present condition (2.4) which is essential in proving the following main Theorem 2.2.

Theorem 2.2 should be compared with the conditions used in van der Vaart (1994) and his overall assumption that $\mathcal{F}$ is invariant under translation, which means that $\mathcal{F} = \tilde{\mathcal{F}} := \{f_x : x \in X\}$ where $f_x(y) := f(x+y), y \in X$, an assumption which is also needed for Yukich's (1992) result. Their proofs are both based on showing stochastic equivalence of the empirical processes $\sqrt{n}(\nu_n - \nu)$ and the unbiased smoothed empirical processes $\sqrt{n}(\nu_n\star\mu_n - \nu\star\mu_n)$. Therefore the FCLT for smoothed empirical processes was linked to an FCLT for the empirical process with some conditions for proving stochastic equivalence and bias convergence. Note that $\mathcal{F} = \tilde{\mathcal{F}}$ automatically implies that $\mathcal{F}$ is uniformly bounded or else there would exist no real-valued envelope F. The condition $\mathcal{F} = \tilde{\mathcal{F}}$, however, does not come up when viewing smoothed empirical processes as special cases of RMP. In view of van der Vaart's (1994) result, note also that $\mathcal{F}$ being ν-Donsker does not always imply that $\tilde{\mathcal{F}}$ is ν-Donsker too, as can be seen by examples.

Theorem 2.2. *Let X be a linear metric space and $\mathcal{F}$ have uniformly integrable L_2-entropy. Assume that the following conditions (2.7) and (2.8)* *are fulfilled:*

$$\sup_{f\in\mathcal{F}} \int_X \Big[\int_X (f(x+y)-f(x))\mu_n(dy)\Big]^2 \nu(dx) \longrightarrow 0, \tag{2.7}$$

$$\sup_{f\in\mathcal{F}\cup\{F^3\}} \sqrt{n}\big|\nu\star\mu_n(f)-\nu(f)\big| \longrightarrow 0. \tag{2.8}$$

Then

$$\sqrt{n}(\tilde{\nu}_n-\nu) \xrightarrow[sep]{\mathcal{L}} \mathbb{G} \qquad \text{in } l^\infty(\mathcal{F}),$$

where $\mathbb{G}=(G(f))_{f\in\mathcal{F}}$ is a mean-zero Gaussian process with sample paths in $U^b(\mathcal{F},d_\nu^{(2)})$ and $cov(G(f),G(g))=\nu(f\cdot g)-\nu(f)\cdot\nu(g)$ for $f,g\in\mathcal{F}$.

In (2.8) it is, of course, tacitly assumed that $\nu\star\mu_n(F^3)$ and $\nu(F^3)$ exist.

Proof. We show (2.3) up to (2.5) , fidi convergence and then apply Theorem 2.1.

Let $\rho>0$. To find an appropriate sequence $(\delta_n)_{n\in\mathbb{N}}$ such that (2.3) – (2.5) hold, we consider the expression in (2.3) for fixed $\delta>0$. Then we have

$$\sqrt{n}\cdot\mathbb{E}\Big(w_{n1}(F)\,I(w_{n1}(F)>\delta\sqrt{n})\Big) \le \frac{1}{\delta}\,\mathbb{E}\Big(w_{n1}(F^2)\,I(w_{n1}(F)>\delta\sqrt{n})\Big) \longrightarrow 0 \quad \text{as } n\longrightarrow\infty$$

since the sequence of random variables $\big(w_{n1}(F^2)\big)_{n\in\mathbb{N}}$ is uniformly integrable (note that $\mathbb{E}(w_{n1}(F^2)^{3/2})\le\mathbb{E}(w_{n1}(F^3))=\nu\star\mu_n(F^3)\longrightarrow\nu(F^3)$ according to (2.8) , so $\sup_{n\in\mathbb{N}}\mathbb{E}(w_{n1}(F^2)^{3/2})<\infty$, hence $\big(w_{n1}(F^2)\big)_{n\in\mathbb{N}}$ is uniformly integrable). So we can find a sequence $(\delta_n)_{n\in\mathbb{N}}$ with $\delta_n>0$ and $\delta_n\to 0$ such that (2.3) holds.

As to (2.4): For $\alpha>0$ and $f,g\in\mathcal{F}$ such that $d_\nu^{(2)}(f,g)\le\alpha$,

$$\begin{aligned}
&\mathbb{E}\big((w_{n1}(f)-w_{n1}(g))^2\cdot I(w_{n1}(F)\le\delta_n\sqrt{n})\big)\\
&\le \int_X\Big[\int_X (f(x+y)-f(x)+f(x)-g(x)+g(x)-g(x+y))\mu_n(dy)\Big]^2\nu(dx)\\
&\le 9\cdot\int_X\Big[\int_X (f(x+y)-f(x))\mu_n(dy)\Big]^2\nu(dx)\\
&\quad+9\cdot\int_X\Big[\int_X (f(x)-g(x))\mu_n(dy)\Big]^2\nu(dx)\\
&\quad+9\cdot\int_X\Big[\int_X (g(x)-g(x+y))\mu_n(dy)\Big]^2\nu(dx)\\
&\le 18\sup_{h\in\mathcal{F}}\int_X\Big[\int_X (h(x+y)-h(x))\mu_n(dy)\Big]^2\nu(dx)+9\cdot d_\nu^{(2)}(f,g)^2.
\end{aligned}$$

So (2.7) gives (2.4).

As to (2.5): Here we have

$$\sup_{n\in\mathbb{N}} \mathbb{E}\Big(w_{n1}(F^2)\cdot I(w_{nj}(F)\le \delta_n\sqrt{n})\Big) \le \sup_{n\in\mathbb{N}} \mathbb{E}(w_{n1}(F^2))$$
$$\le \sup_{n\in\mathbb{N}} (\mathbb{E}(w_{n1}(F^3)))^{2/3} < \infty$$

because of (2.8).

Next we prove fidi convergence: Let $k \in \mathbb{N}, f_1, ..., f_k \in \mathcal{F}$ be fixed. We show that $\sum_{j=1}^n \zeta_{nj}/\sqrt{n}$ is asymptotically normal $\mathcal{N}_k(0,\Sigma_k)$ distributed where the ζ_{nj} are defined by

$$\zeta_{nj} = \begin{pmatrix} \int_X f_1(\eta_j+y)\mu_n(dy) - \nu\star\mu_n(f_1) \\ \vdots \\ \int_X f_k(\eta_j+y)\mu_n(dy) - \nu\star\mu_n(f_k) \end{pmatrix}, \; j=1,...,n,\; n\in\mathbb{N},$$

and the matrix Σ_k by

$$\Sigma_k = \big(\nu(f_i\cdot f_l) - \nu(f_i)\nu(f_l)\big)_{1\le i,l\le k}.$$

From the multi-dimensional central limit theorem for triangular arrays it suffices to show that

$$(+) \qquad \sum_{j=1}^n \mathbb{E}(||\frac{\zeta_{nj}}{\sqrt{n}}||^2 \cdot I(||\frac{\zeta_{nj}}{\sqrt{n}}|| > \varepsilon)) \longrightarrow 0 \quad \text{for all } \varepsilon > 0,$$

$$(++) \qquad \mathbb{V}_{n1} \longrightarrow \Sigma_k,$$

where $||\cdot||$ here denotes the Euclidian norm on $\mathbb{R}^k$ and $\mathbb{V}_{n1}$ is the covariance matrix of ζ_{n1}.

As to (+): We have

$$\sum_{j=1}^n \mathbb{E}\Big(||\frac{\zeta_{nj}}{\sqrt{n}}||^2 \cdot I(||\frac{\zeta_{nj}}{\sqrt{n}}|| > \varepsilon)\Big)$$
$$\le \sum_{i=1}^k \mathbb{E}\Big([w_{n1}(f_i) - \nu\star\mu_n(f_i)]^2 \cdot I((\sum_{i=1}^k |w_{n1}(f_i) - \nu\star\mu_n(f_i)|^2)^{\frac{1}{2}} > \varepsilon\sqrt{n})\Big)$$
$$\le k\,\mathbb{E}\big((w_{n1}(F) + \nu\star\mu_n(F))^2 \cdot I(w_{n1}(F) + \nu\star\mu_n(F) > \frac{\varepsilon}{\sqrt{k}}\sqrt{n})\big).$$

Now since

$$\big((w_{n1}(F) + \nu\star\mu_n(F))^2\big)_{n\in\mathbb{N}}$$

is uniformly integrable we have (+).

As to (++): For fixed $i, l \in \{1, ..., k\}$ the corresponding element of $\mathbb{V}_{n1}$ is

$$\int_X \Big[\int_X f_i(x+y)\mu_n(dy) - \nu \star \mu_n(f_i)\Big]\Big[\int_X f_l(x+y)\mu_n(dy) - \nu \star \mu_n(f_l)\Big]\nu(dx)$$

which is equal to (inserting $f_i(x)$ and $f_l(x)$):

$$\begin{aligned}
&\int_X \Big[\int_X f_i(x+y)\mu_n(dy) - f_i(x)\Big] \cdot \Big[\int_X f_l(x+y)\mu_n(dy) - f_l(x)\Big]\nu(dx) \\
&+ \int_X \big[f_i(x) - \nu \star \mu_n(f_i)\big] \cdot \Big[\int_X f_l(x+y)\mu_n(dy) - f_l(x)\Big]\nu(dx) \\
&+ \int_X \big[f_l(x) - \nu \star \mu_n(f_l)\big] \cdot \Big[\int_X f_i(x+y)\mu_n(dy) - f_i(x)\Big]\nu(dx) \\
&+ \int_X \big[f_l(x) - \nu \star \mu_n(f_l)\big] \cdot \big[f_i(x) - \nu \star \mu_n(f_i)\big]\nu(dx) \\
&= I_{n1} + I_{n2} + I_{n3} + I_{n4}.
\end{aligned}$$

The Cauchy-Schwarz inequality together with (2.7) yields $I_{n1} \longrightarrow 0$. Now

$$\begin{aligned}
|I_{n4} - (\nu(f_i f_l) - \nu(f_i)\nu(f_l))| &= |\nu \star \mu_n(f_i) - \nu(f_i)| \cdot |\nu \star \mu_n(f_l) - \nu(f_l)| \\
&\leq \sup_{f \in \mathcal{F}} \Big|\int_X \int_X (f(x+y) - f(x))\mu_n(dy)\nu(dx)\Big|^2 \\
&\leq \sup_{f \in \mathcal{F}} \int_X \Big[\int_X (f(x+y) - f(x))\mu_n(dy)\Big]^2 \nu(dx) \\
&\longrightarrow 0
\end{aligned}$$

according to (2.7). This gives $I_{n4} \longrightarrow \nu(f_i f_l) - \nu(f_i)\nu(f_l)$.

From this, the Cauchy-Schwarz inequality and (2.7) again, we also have $I_{n2} \longrightarrow 0$ and $I_{n3} \longrightarrow 0$. So we have proved that $\sum_{j=1}^{n} \zeta_{nj}/\sqrt{n}$ converges in distribution to $\mathcal{N}_k(0, \Sigma_k)$. Now we have verified all the conditions of Theorem 2.1 which now gives the result. □

As already mentioned $\mathcal{F}$ being invariant under translation implies that $\mathcal{F}$ is uniformly bounded. But for uniformly bounded $\mathcal{F}$ (2.8) reduces to (2.6), and so we get finally the following result (cf. Rost (1997), Theorem 3.2.4):

Theorem 2.3. *Let X be a linear metric space and let $\mathcal{F}$ be uniformly bounded having uniformly integrable L_2-entropy. Assume that the conditions (2.6) and (2.7) are satisfied. Then*

$$\sqrt{n}(\tilde{\nu}_n - \nu) \xrightarrow[sep]{\mathcal{L}} \mathbb{G} \quad in \quad l^\infty(\mathcal{F}),$$

where $\mathbb{G} = (G(f))_{f \in \mathcal{F}}$ *is a mean-zero Gaussian process with sample paths in* $U^b(\mathcal{F}, d_\nu^{(2)})$ *and* $cov(G(f), G(g)) = \nu(f \cdot g) - \nu(f) \cdot \nu(g)$ *for* $f, g \in \mathcal{F}$.

References

Gaenssler, P., Rost, D., 1999. *Empirical and partial-sum processes; revisited as random measure processes*, MaPhySto, Centre for Mathematical Physics and Stochastics, Lecture Notes no. 5, Aarhus.

Gaenssler, P., Rost, D., Ziegler, K., 1998. On Random Measure Processes with Application to Smoothed Empirical Processes. In: *High Dimensional Probability*, 1, Progress in Probability, Vol. 43, E. Eberlein, M. Hahn, M. Talagrand, (eds.), Birkhäuser Boston, Cambridge, MA, pp. 93–102,

Giné, E., 1996. *Lectures on some aspects of the bootstrap*, École d'Été de Calcul de Probabilités de Saint-Flour.

Rost, D., 1997. *Approximation results for smoothed empirical processes: A new approach*, Habilitationsschrift, Ludwig-Maximilians-Universität München.

van der Vaart, A.W., 1994. Weak convergence of smoothed empirical processes, *Scand. J. Statist.* **21**, 501–504.

van der Vaart, A.W., Wellner, J.A., 1996. *Weak convergence and empirical processes*, Springer Series in Statistics, Springer-Verlag New York.

Yukich, J.E., 1992. Weak convergence of smoothed empirical processes, *Scand. J. Statist.* **19**, 271–279.

Ziegler, K., 1997. Functional central limit theorems for triangular arrays of function-indexed processes under uniformly integrable entropy conditions, *J. of Multivariate Analysis* **62**, 233–272.

Mathematical Institute
University of Munich
Theresienstrasse 39
D-80333 Munich, Germany
e-mail: `rost@rz.mathematik.uni-muenchen.de`

Preservation Theorems for Glivenko-Cantelli and Uniform Glivenko-Cantelli Classes

Aad van der Vaart and Jon A. Wellner

ABSTRACT We show that the $P-$Glivenko property of classes of functions $\mathcal{F}_1, \dots, \mathcal{F}_k$ is preserved by a continuous function φ from R^k to R in the sense that the new class of functions

$$x \to \varphi(f_1(x), \dots, f_k(x)), \quad f_i \in \mathcal{F}_i, \quad i = 1, \dots, k$$

is again a Glivenko-Cantelli class of functions if it has an integrable envelope. We also prove an analogous result for preservation of the uniform Glivenko-Cantelli property. Corollaries of the main theorem include two preservation theorems of Dudley (1998a,b). We apply the main result to reprove a theorem of Schick and Yu (1999) concerning consistency of the NPMLE in a model for "mixed case" interval censoring. Finally a version of the consistency result of Schick and Yu (1999) is established for a general model for "mixed case interval censoring" in which a general sample space $\mathcal{Y}$ is partitioned into sets which are members of some VC-class $\mathcal{C}$ of subsets of $\mathcal{Y}$.

1 Glivenko-Cantelli theorems

Let $(\mathcal{X}, \mathcal{A}, P)$ be a probability space, and suppose that $\mathcal{F} \subset \mathcal{L}_1(P)$. For such a class of functions, let $\mathcal{F}_{0,P} \equiv \{f - Pf : f \in \mathcal{F}\}$. We also let $F_{\mathcal{F}}(x) \equiv \sup_{f \in \mathcal{F}} |f(x)|$, the *envelope function* of $\mathcal{F}$. If $|f| \le F$ for all $f \in \mathcal{F}$ with F measurable, then F is an envelope for $\mathcal{F}$.

Suppose that $X_1, X_2, \dots$ are i.i.d. P. The Glivenko-Cantelli theorems give conditions under which the empirical measure $\mathbb{P}_n$ converges uniformly to P over a class $\mathcal{F}$, either in probability (in which case we say that $\mathcal{F}$ is a *weak Glivenko-Cantelli class for P*) or almost surely:

$$\|\mathbb{P}_n - P\|_{\mathcal{F}} \equiv \sup_{f \in \mathcal{F}} |\mathbb{P}_n f - Pf| \to_{a.s.^*} 0\,;$$

in this case we say that $\mathcal{F}$ is a *strong Glivenko-Cantelli class for P*. Useful sufficient conditions for a class $\mathcal{F}$ to be a strong Glivenko-Cantelli class for

P are that it has an integrable envelope and either

$$\log N_{[]}(\epsilon, \mathcal{F}, L_1(P)) < \infty \qquad \text{for all } \epsilon > 0$$

or

$$\log N(\epsilon, \mathcal{F}_M, L_1(\mathbb{P}_n)) = o_P^*(n) \qquad \text{for every } \epsilon > 0 \text{ and } 0 < M < \infty .$$

Here $\mathcal{F}_M = \{f 1_{[F \le M]} : f \in \mathcal{F}\}$. In the second case some additional measurability conditions are necessary; see Van der Vaart and Wellner (1996), Chapter 2.4, pp. 122–126. In particular, the condition in Theorem 2.4.3, page 123, is shown by Giné and Zinn (1984) to be both necessary and sufficient, under measurability assumptions, for the class $\mathcal{F}$ to be a strong Glivenko-Cantelli class (see Talagrand (1996) for another proof). Talagrand (1987b) gives necessary and sufficient conditions for the Glivenko-Cantelli theorem without any measurability hypotheses.

Theorem 1. (Giné and Zinn, 1984). *Suppose that $\mathcal{F}$ is $L_1(P)$ bounded and nearly linearly supremum measurable for P; in particular this holds if $\mathcal{F}$ is image admissible Suslin. Then the following are equivalent:*
(a) $\mathcal{F}$ is a strong Glivenko-Cantelli class for P.
(b) $\mathcal{F}$ has an envelope function $F \in L_1(P)$ and the truncated classes $\mathcal{F}_M \equiv \{f 1\{F \le M\} : f \in \mathcal{F}\}$ satisfy

$$\frac{1}{n} E^* \log N(\epsilon, \mathcal{F}_M, L_r(\mathbb{P}_n)) \to 0 \quad \textit{for all } \epsilon > 0, \textit{ and for all } M \in (0, \infty)$$

for some (all) $r \in (0, \infty]$ where $\|f\|_{L_r(P)} \equiv \|f\|_{P,r} \equiv \{P(|f|^r)\}^{r^{-1} \wedge 1}$.

2 Preservation of the Glivenko–Cantelli property

Our goal in this section is to present several results concerning the stability of the Glivenko-Cantelli property of one or more classes of functions under composition with functions φ. A theorem which motivated our interest is the following result of Dudley (1998a).

Theorem 2. (Dudley, 1998a). *Suppose that $\mathcal{F}$ is a strong Glivenko-Cantelli class for P with $PF < \infty$, J is a possibly unbounded interval including the ranges of all $f \in \mathcal{F}$, φ is continuous and monotone on J, and for some finite constants c, d, $|\varphi(y)| \le c|y| + d$ for all $y \in J$. Then $\varphi(\mathcal{F})$ is also a strong Glivenko-Cantelli class for P.*

Dudley (1998a) proves this via the characterization of Glivenko-Cantelli classes due to Talagrand (1987b). Dudley (1998b) also uses Talagrand's characterization to prove the following interesting proposition.

Proposition 1. (Dudley, 1998b). *Suppose that $\mathcal{F}$ is a strong Glivenko-Cantelli class for P with $PF < \infty$, and g is a fixed bounded function*

($\|g\|_\infty < \infty$). Then the class of functions $g \cdot \mathcal{F} \equiv \{g \cdot f : f \in \mathcal{F}\}$ is a strong Glivenko-Cantelli class for P.

Yet another proposition in this same vein is:

Proposition 2. (Giné and Zinn, 1984). *Suppose that $\mathcal{F}$ is a uniformly bounded strong Glivenko-Cantelli class for P, and $g \in \mathcal{L}_1(P)$ is a fixed function. Then the class of functions $g \cdot \mathcal{F} \equiv \{g \cdot f : f \in \mathcal{F}\}$ is a strong Glivenko-Cantelli class for P.*

Given classes $\mathcal{F}_1, \ldots, \mathcal{F}_k$ of functions $f_i : \mathcal{X} \to R$ and a function $\varphi : R^k \to R$, let $\varphi(\mathcal{F}_1, \ldots, \mathcal{F}_k)$ be the class of functions

$$x \to \varphi(f_1(x), \ldots, f_k(x)),$$

where $f_i \in \mathcal{F}_i$, $i = 1, \ldots, k$. Theorem 2 and Propositions 1 and 2 are all corollaries of the following theorem.

Theorem 3. *Suppose that $\mathcal{F}_1, \ldots, \mathcal{F}_k$ are $P-$ Glivenko-Cantelli classes of functions, and that $\varphi : R^k \to R$ is continuous. Then $\mathcal{H} \equiv \varphi(\mathcal{F}_1, \ldots, \mathcal{F}_k)$ is $P-$ Glivenko-Cantelli provided that it has an integrable envelope function.*

Proof. We first assume that the classes of functions $\mathcal{F}_i$ are appropriately measurable. Let $F_1, \ldots, F_k$ and H be integrable envelopes for $\mathcal{F}_1, \ldots, \mathcal{F}_k$ and $\mathcal{H}$ respectively, and set $F = F_1 \vee \ldots \vee F_k$. For $M \in (0, \infty)$, define

$$\mathcal{H}_M \equiv \{\varphi(f)1_{[F \le M]} : f = (f_1, \ldots, f_k) \in \mathcal{F}_1 \times \mathcal{F}_2 \times \cdots \times \mathcal{F}_k \equiv \mathcal{F}\}.$$

Now

$$\|(\mathbb{P}_n - P)\varphi(f)\|_{\mathcal{F}} \le (\mathbb{P}_n + P)H1_{[F>M]} + \|(\mathbb{P}_n - P)h\|_{\mathcal{H}_M}.$$

The expectation of the first term on the right converges to 0 as $M \to \infty$. Hence it suffices to show that $\mathcal{H}_M$ is $P-$Glivenko-Cantelli for every fixed M. Let $\delta = \delta(\epsilon)$ be the δ of Lemma 2 below for $\varphi : [-M, M]^k \to R$, $\epsilon > 0$, and $\|\cdot\|$ the L_1-norm $\|\cdot\|_1$. Then for any $(f_j, g_j) \in \mathcal{F}_j$, $j = 1, \ldots, k$,

$$\mathbb{P}_n|f_j - g_j|1_{[F_j \le M]} \le \frac{\delta}{k}, \quad j = 1, \ldots, k$$

implies that

$$\mathbb{P}_n|\varphi(f_1, \ldots, f_k) - \varphi(g_1, \ldots, g_k)|1_{[F \le M]} \le \epsilon.$$

It follows that

$$N(\epsilon, \mathcal{H}_M, L_1(\mathbb{P}_n)) \le \prod_{j=1}^{k} N(\frac{\delta}{k}, \mathcal{F}_j 1_{[F_j \le M]}, L_1(\mathbb{P}_n)).$$

Thus $E^* \log N(\epsilon, \mathcal{H}_M, L_1(\mathbb{P}_n)) = o(n)$ for every $\epsilon > 0$, $M < \infty$. This implies that $E^* \log N(\epsilon, (\overline{\mathcal{H}}_M)_N, L_1(\mathbb{P}_n)) = o(n)$ for $(\overline{\mathcal{H}}_M)_N$ the functions $h1\{H \leq N\}$ for $h \in \overline{\mathcal{H}}_M$. Thus $\overline{\mathcal{H}}_M$ is strong Glivenko-Cantelli for P by Theorem 1. This concludes the proof that $\mathcal{H} = \varphi(\mathcal{F})$ is weak Glivenko-Cantelli. Because it has an integrable envelope, it is strong Glivenko-Cantelli by, e.g., Lemma 2.4.5 of Van der Vaart and Wellner (1996). This concludes the proof for appropriately measurable classes $\mathcal{F}_j$, $j = 1, \ldots, k$.

We extend the theorem to general Glivenko-Cantelli classes using separable versions as in Talagrand (1987a). (Also see van der Vaart and Wellner (1996), pp. 115–120 for a discussion.) As shown in the preceding argument, it is not a loss of generality to assume that the classes $\mathcal{F}_i$ are uniformly bounded. Furthermore, it suffices to show that $\varphi(\mathcal{F}_1, \ldots, \mathcal{F}_k)$ is weak Glivenko-Cantelli. We first need a lemma.

Lemma 1. *Any strong P-Glivenko Cantelli class $\mathcal{F}$ is totally bounded in $L_1(P)$ if and only if $\|P\|_{\mathcal{F}} < \infty$. Furthermore for any $r \in (1, \infty)$, if $\mathcal{F}$ has an envelope that is contained in $L_r(P)$, then $\mathcal{F}$ is also totally bounded in $L_r(P)$.*

Proof. A class that is totally bounded is also bounded. Thus for the first statement we only need to prove that a strong Glivenko-Cantelli class $\mathcal{F}$ with $\|P\|_{\mathcal{F}} < \infty$ is totally bounded in $L_1(P)$.

It is well-known that such a class has an integrable envelope. E.g. see Giné and Zinn (1984) or Problem 2.4.1 of van der Vaart and Wellner (1996) to conclude first that $P^* \|f - Pf\|_{\mathcal{F}} < \infty$. Next the claim follows from the triangle inequality $\|f\|_{\mathcal{F}} \leq \|f - Pf\|_{\mathcal{F}} + \|P\|_{\mathcal{F}}$. Thus it is no loss of generality to assume that the class $\mathcal{F}$ possesses an envelope that is finite everywhere.

If $\mathcal{F}$ is not totally bounded in $L_1(P)$, then there is $\varepsilon > 0$ and $\mathcal{G} \subset \mathcal{F}$ countable such that the packing number $D(\varepsilon, \mathcal{G}, L_1(P)) = \infty$. This is obvious, since if $\mathcal{F}$ is not totally bounded, then $D(\varepsilon, \mathcal{F}, L_1(P)) = \infty$ for some $\varepsilon > 0$; then take a countable set of $\varepsilon-$ separated points. So, now $\mathcal{G}$ has all the measurablity properties, is GC and $L_1(P)$-bounded, and moreover, $D(\varepsilon, \mathcal{G}, L_1) = \infty$. But this is a contradiction because, as shown in Giné and Zinn (1984), first half of page 984, $D(\varepsilon, \mathcal{G}, L_1)$ must be finite for all $\varepsilon > 0$. This completes the proof of the first part of the lemma.

If $\mathcal{F}$ has an envelope in $L_r(P)$, then $\mathcal{F}$ is totally bounded in $L_r(P)$ if the class $\mathcal{F}_M$ of functions $f1\{F \leq M\}$ is totally bounded in $L_r(P)$ for every fixed M. The class $\mathcal{F}_M$ is P-GC by Theorem 3 and hence this class is totally bounded in $L_1(P)$. But then it is also totally bounded in $L_r(P)$, because $P|f|^r \leq P|f|M^{r-1}$ for any f that is bounded by M and we can construct the ϵ-net over $\mathcal{F}_M$ in $L_1(P)$ to consist of functions that are bounded by M. □

Because a Glivenko-Cantelli class $\mathcal{F}$ with $\|P\|_{\mathcal{F}} < \infty$ is totally bounded in $L_1(P)$ by Lemma 1, it is separable as a subset of $L_1(P)$. A minor generalization of Theorem 2.3.17 in van der Vaart and Wellner (1996) shows that there exists a bijection $f \leftrightarrow \tilde{f}$ of $\mathcal{F}$ onto a class $\tilde{\mathcal{F}} \subset L_1(P)$ such that

- $f = \tilde{f}$ P–almost surely for every $f \in \mathcal{F}$.
- there exists a countable subset $\mathcal{G} \subset \tilde{\mathcal{F}}$ such that for every n there exists a measurable set $N_n \subset \mathcal{X}^n$ with $P^n(N_n) = 0$ such that for all $(x_1, \ldots, x_n) \notin N_n$ and $f \in \tilde{\mathcal{F}}$ there exists $\{g_m\} \subset \mathcal{G}$ such that $P|g_m - \tilde{f}| \to 0$ and $(g_m(x_1), \ldots, g_m(x_n)) \to (\tilde{f}(x_1), \ldots, \tilde{f}(x_n))$.

By an adaptation of a theorem due to Talagrand (1987a) (see Theorem 2.3.15 in van der Vaart and Wellner (1996)) a class $\mathcal{F}$ is weak Glivenko-Cantelli if and only if the class $\tilde{\mathcal{F}}$ is weak Glivenko-Cantelli and

$$\sup_{f \in \mathcal{F}} \mathbb{P}_n |f - \tilde{f}| \to 0$$

in outer probability. Construct a "pointwise separable version" $\tilde{\mathcal{F}}_i$ for each of the classes $\mathcal{F}_i$. The classes $\tilde{\mathcal{F}}_i$ possess enough measurability to make the preceding argument work; in particular "pointwise separable version" in the above sense is sufficient for the nearly linearly supremum measurable hypothesis of Giné and Zinn (1984) for both $\mathcal{F}_1, \ldots, \mathcal{F}_k$ and $\varphi(\mathcal{F}_1, \ldots, \mathcal{F}_k)$. Thus the class $\varphi(\tilde{\mathcal{F}}_1, \ldots, \tilde{\mathcal{F}}_k)$ is Glivenko-Cantelli for P.

Now by Lemma 2 there exists for every $\epsilon > 0$ a $\delta > 0$ such that

$$\mathbb{P}_n |f_j - \tilde{f}_j| < \frac{\delta}{k}, \qquad j = 1, \ldots, k,$$

implies

$$\mathbb{P}_n |\varphi(f_1, \ldots, f_k) - \varphi(\tilde{f}_1, \ldots, \tilde{f}_k)| < \epsilon\,.$$

The theorem follows. □

Lemma 2. *Suppose that $\varphi : K \to R$ is continuous and $K \subset R^k$ is compact. Then for every $\epsilon > 0$ there exists $\delta > 0$ such that for all n and for all $a_1, \ldots, a_n, b_1, \ldots, b_n \in K \subset R^k$*

$$\frac{1}{n} \sum_{i=1}^{n} \|a_i - b_i\| < \delta$$

implies

$$\frac{1}{n} \sum_{i=1}^{n} |\varphi(a_i) - \varphi(b_i)| < \epsilon\,.$$

Here $\|\cdot\|$ can be any norm on R^k; in particular it can be

$$\|x\|_r = \left(\sum_{i=1}^{k} |x_i|^r \right)^{1/r},$$

$r \in [1, \infty)$ or $\|x\|_\infty \equiv \max_{1 \le i \le k} |x_i|$ for $x = (x_1, \ldots, x_k) \in R^k$.

Proof. Let U_n be uniform on $\{1, \ldots, n\}$, and set $X_n = a_{U_n}$, $Y_n = b_{U_n}$. Then we can write

$$\frac{1}{n}\sum_{i=1}^n \|a_i - b_i\| = E\|X_n - Y_n\|$$

and

$$\frac{1}{n}\sum_{i=1}^n |\varphi(a_i) - \varphi(b_i)| = E|\varphi(X_n) - \varphi(Y_n)| .$$

Hence it suffices to show that for every $\epsilon > 0$ there exists $\delta > 0$ such that for all (X, Y) random vectors in $K \subset R^k$,

$$E\|X - Y\| < \delta \qquad \text{implies} \qquad E|\varphi(X) - \varphi(Y)| < \epsilon .$$

Suppose not. Then for some $\epsilon > 0$ and for all $m = 1, 2, \ldots$ there exists (X_m, Y_m) such that

$$E\|X_m - Y_m\| < \frac{1}{m}, \qquad E|\varphi(X_m) - \varphi(Y_m)| \ge \epsilon .$$

But since $\{(X_m, Y_m)\}$ is tight, there exists $(X_{m'}, Y_{m'}) \to_d (X, Y)$. Then it follows that

$$E\|X - Y\| = \lim_{m' \to \infty} E\|X_{m'} - Y_{m'}\| = 0$$

so that $X = Y$ a.s., while on the other hand

$$0 = E|\varphi(X) - \varphi(Y)| = \lim_{m' \to \infty} E|\varphi(X_{m'}) - \varphi(Y_{m'})| \ge \epsilon > 0 .$$

This contradiction means that the desired implication holds. □

Another potentially useful preservation theorem is one based on building up Glivenko-Cantelli classes from the restrictions of a class of functions to elements of a partition of the sample space. The following theorem is related to the results of Van der Vaart (1996) for Donsker classes.

Theorem 4. *Suppose that $\mathcal{F}$ is a class of functions on $(\mathcal{X}, \mathcal{A}, P)$, and $\{\mathcal{X}_i\}$ is a partition of $\mathcal{X}$: $\cup_{i=1}^\infty \mathcal{X}_i = \mathcal{X}$, $\mathcal{X}_i \cap \mathcal{X}_j = \emptyset$ for $i \ne j$. Suppose that $\mathcal{F}_j \equiv \{f 1_{\mathcal{X}_j} : f \in \mathcal{F}\}$ is $P-$Glivenko-Cantelli for each j, and $\mathcal{F}$ has an integrable envelope function F. Then $\mathcal{F}$ is itself $P-$Glivenko-Cantelli.*

Proof. Since

$$f = f\sum_{j=1}^\infty 1_{\mathcal{X}_j} = \sum_{j=1}^\infty f 1_{\mathcal{X}_j} ,$$

it follows that

$$E^*\|\mathbb{P}_n - P\|_{\mathcal{F}} \le \sum_{j=1}^\infty E^*\|\mathbb{P}_n - P\|_{\mathcal{F}_j} \to 0$$

by the dominated convergence theorem, since each term in the sum converges to zero by the hypothesis that each $\mathcal{F}_j$ is $P-$ Glivenko-Cantelli, and we have

$$E^*\|\mathbb{P}_n - P\|_{\mathcal{F}_j} \le E^*\mathbb{P}_n(F1_{\mathcal{X}_j}) + P(F1_{\mathcal{X}_j}) \le 2P(F1_{\mathcal{X}_j})$$

where $\sum_{j=1}^{\infty} P(F1_{\mathcal{X}_j}) = P(F) < \infty$. □

3 Preservation of the uniform Glivenko–Cantelli property

We say that $\mathcal{F}$ is a strong uniform Glivenko-Cantelli class if for all $\epsilon > 0$

$$\sup_{P \in \mathcal{P}(\mathcal{X},\mathcal{A})} Pr_P^* \left(\sup_{m \ge n} \|\mathbb{P}_m - P\|_{\mathcal{F}} > \epsilon \right) \to 0 \qquad \text{as } n \to \infty$$

where $\mathcal{P}(\mathcal{X}, \mathcal{A})$ is the set of all probability measures on $(\mathcal{X}, \mathcal{A})$. For $x = (x_1, \dots, x_n) \in \mathcal{X}^n$, $n = 1, 2, \dots$, and $r \in (0, \infty)$, we define on $\mathcal{F}$ the pseudo-distances

$$\begin{aligned} e_{x,r}(f,g) &= \left\{ n^{-1} \sum_{i=1}^{n} |f(x_i) - g(x_i)|^r \right\}^{r^{-1} \wedge 1}, \\ e_{x,\infty}(f,g) &= \max_{1 \le i \le n} |f(x_i) - g(x_i)|, \qquad f, g \in \mathcal{F}. \end{aligned}$$

Let $N(\epsilon, \mathcal{F}, e_{x,r})$ denote the $\epsilon-$covering number of $(\mathcal{F}, e_{x,r})$, $\epsilon > 0$. Then define, for $n = 1, 2, \dots$, $\epsilon > 0$, and $r \in (0, \infty]$, the quantities

$$N_{n,r}(\epsilon, \mathcal{F}) = \sup_{x \in \mathcal{X}^n} N(\epsilon, \mathcal{F}, e_{x,r}).$$

Theorem 5. (Dudley, Giné, and Zinn (1991)). *Suppose that $\mathcal{F}$ is a class of uniformly bounded functions such that $\mathcal{F}$ is image admissible Suslin. Then the following are equivalent:*
(a) $\mathcal{F}$ is a strong uniform Glivenko-Cantelli class.
(b)

$$\frac{\log N_{n,r}(\epsilon, \mathcal{F})}{n} \to 0 \qquad \textit{for all } \epsilon > 0$$

for some (all) $r \in (0, \infty]$.

For the definition of the image admissible Suslin property see Dudley (1984), sections 10.3 and 11.1. The following theorem gives natural sufficient conditions for preservation of the uniform Glivenko-Cantelli theorem.

Theorem 6. *Suppose that $\mathcal{F}_1, \ldots, \mathcal{F}_k$ are classes of uniformly bounded functions on $(\mathcal{X}, \mathcal{A})$ such that $\mathcal{F}_1, \ldots, \mathcal{F}_k$ are image-admissible Suslin and strong uniform Glivenko-Cantelli classes. Suppose that $\varphi : R^k \to R$ is continuous. Then $\mathcal{H} \equiv \varphi(\mathcal{F}_1, \ldots, \mathcal{F}_k)$ is a strong uniform Glivenko-Cantelli class.*

Proof. It follows from Lemma 2 that for any $\epsilon > 0$ there exists $\delta > 0$ such that for any $f_j, g_j \in \mathcal{F}_j$, $j = 1, \ldots, k$, $x \in \mathcal{X}^n$,

$$e_{x,1}(f_j, g_j) \le \frac{\delta}{k}, \quad j = 1, \ldots, k$$

implies that

$$e_{x,1}(\varphi(f_1, \ldots, f_k), \varphi(g_1, \ldots, g_k)) \le \epsilon\,.$$

It follows that

$$N_{n,1}(\epsilon, \mathcal{H}) \le \prod_{j=1}^{k} N_{n,1}(\frac{\delta}{k}, \mathcal{F}_j)\,,$$

and hence that

$$\frac{1}{n} \log N_{n,1}(\epsilon, \mathcal{H}) \le \sum_{j=1}^{k} \frac{1}{n} \log N_{n,1}(\frac{\delta}{k}, \mathcal{F}_j) \to 0$$

where the convergence follows from part (b) of Theorem 5 with $r = 1$. If we show that $\mathcal{H} = \varphi(\mathcal{F})$ is image admissible Suslin, then the conclusion follows from (b) implies (a) in Theorem 5. We give a proof of a slightly stronger statement in the following lemma. □

Lemma 3. *If $\mathcal{F}_1, \ldots, \mathcal{F}_k$ are image admissible Suslin via $(\mathcal{Y}_i, \mathcal{S}_i, T_i)$, $i = 1, \ldots, k$, and $\phi : R^k \to R$ is measurable, then $\varphi(\mathcal{F}_1, \ldots, \mathcal{F}_k)$ is image admissible Suslin via $(\prod_{i=1}^k \mathcal{Y}_i, \prod_{i=1}^k \mathcal{S}_i, T)$ for*

$$T(y_1, \ldots, y_k) = \varphi(T_1 y_1, \ldots, T_k y_k)\,.$$

Proof. There exist Polish spaces $\mathcal{Z}_i$ and measurable maps $g_i : \mathcal{Z}_i \to \mathcal{Y}_i$ which are onto. Then $g : \mathcal{Z}_1 \times \cdots \times \mathcal{Z}_k \to \mathcal{Y}_1 \times \cdots \times \mathcal{Y}_k$ defined by $g(z_1, \ldots, z_k) = (g_1(z_1), \ldots, g_k(z_k))$ is onto and measurable for $\mathcal{S}_1 \times \cdots \times \mathcal{S}_k$. So $(\prod_i \mathcal{Y}_i, \prod_i \mathcal{S}_i)$ is Suslin. It suffices to check that T is onto and ψ defined by the map

$$\psi(x, y_1, \ldots, y_k) = \varphi(T_1 y_1(x), \ldots, T_k y_k(x))$$

is measurable. Obviously T is onto, and ψ is measurable because each map $(x, y_i) \mapsto T_i y_i(x)$ is measurable on $\mathcal{X} \times \mathcal{Y}_i$ and hence on $\mathcal{X} \times \mathcal{Y}_1 \times \cdots \times \mathcal{Y}_k$, and hence

$$(x, y_1, \ldots, y_k) \mapsto (T_1 y_1(x), \ldots, T_k y_k(x))$$

is measurable from $\mathcal{X} \times \mathcal{Y}$ to R^k. □

4 Nonparametric maximum likelihood estimation: a general result

Now we prove a general result for nonparametric maximum likelihood estimation in a class of densities. The main proposition in this section is related to results of Pfanzagl (1988) and Van de Geer (1993), (1996). Suppose that $\mathcal{P}$ is a class of densities with respect to a fixed $\sigma-$finite measure μ on a measurable space $(\mathcal{X}, \mathcal{A})$. Suppose that $X_1, \ldots, X_n$ are i.i.d. P_0 with density $p_0 \in \mathcal{P}$. Let

$$\widehat{p}_n \equiv \operatorname{argmax} \mathbb{P}_n \log p\,.$$

For $0 < \alpha \leq 1$, let $\varphi_\alpha(t) = (t^\alpha - 1)/(t^\alpha + 1)$ for $t \geq 0$, $\varphi(t) = -1$ for $t < 0$. Then φ_α is bounded and continuous for each $\alpha \in (0, 1]$. For $0 < \beta < 1$ define

$$h_\beta^2(p, q) \equiv 1 - \int p^\beta q^{1-\beta} d\mu\,.$$

Note that $h_{1/2}(p, q) \equiv h(p, q)$ is the Hellinger distance between p and q, and by Hölder's inequality, $h_\beta(p, q) \geq 0$ with equality if and only if $p = q$ a.e. μ.

Proposition 3. *Suppose that $\mathcal{P}$ is convex. Then*

$$h_{1-\alpha/2}^2(\widehat{p}_n, p_0) \leq (\mathbb{P}_n - P_0)\left(\varphi_\alpha(\frac{\widehat{p}_n}{p_0})\right)\,.$$

In particular, when $\alpha = 1$ we have, with $\varphi \equiv \varphi_1$,

$$h^2(\widehat{p}_n, p_0) \equiv h_{1/2}^2(\widehat{p}_n, p_0) \leq (\mathbb{P}_n - P_0)\left(\varphi(\frac{\widehat{p}_n}{p_0})\right)\,.$$

Corollary 1. *Suppose that $\{\varphi(p/p_0) : p \in \mathcal{P}\}$ is a P_0 Glivenko-Cantelli class. Then for each $0 < \alpha \leq 1$, $h_{1-\alpha/2}(\widehat{p}_n, p_0) \rightarrow_{a.s.} 0$.*

Proof of Proposition 3. It follows from convexity of $\mathcal{P}$ and convexity of $t \rightarrow \varphi_\alpha(1/t)$ that

$$(1) \qquad 0 \leq \mathbb{P}_n \varphi_\alpha(\widehat{p}_n/p_0) = (\mathbb{P}_n - P_0)\varphi_\alpha(\widehat{p}_n/p_0) + P_0\varphi_\alpha(\widehat{p}_n/p_0)\,;$$

see van der Vaart and Wellner (1996) page 330, and Pfanzagl (1988), pp. 141–143. Now we show that

$$(2) \qquad P_0\varphi_\alpha(p/p_0) = \int \frac{p^\alpha - p_0^\alpha}{p^\alpha + p_0^\alpha} dP_0 \leq -\left(1 - \int p_0^\beta p^{1-\beta} d\mu\right)$$

for $\beta = 1 - \alpha/2$. Note that this holds if and only if

$$-1 + 2\int \frac{p^\alpha}{p_0^\alpha + p^\alpha} p_0 d\mu \leq -1 + \int p_0^\beta p^{1-\beta} d\mu\,,$$

or

$$\int p_0^\beta p^{1-\beta} d\mu \geq 2 \int \frac{p^\alpha}{p_0^\alpha + p^\alpha} p_0 d\mu .$$

But this holds if

$$p_0^\beta p^{1-\beta} \geq 2 \frac{p^\alpha p_0}{p_0^\alpha + p^\alpha} .$$

With $\beta = 1 - \alpha/2$, this becomes

$$\frac{1}{2}(p_0^\alpha + p^\alpha) \geq p_0^{\alpha/2} p^{\alpha/2} = \sqrt{p_0^\alpha p^\alpha} ,$$

and this holds by the arithmetic mean–geometric mean inequality. Thus (2) holds. Combining (2) with (1) yields the claim of the proposition.

□

5 Example: a result of Schick and Yu

Our goal in this section is to give another proof of the consistency result of Schick and Yu (1999) for the Non-Parametric Maximum Likelihood Estimator (NPMLE) $\widehat{F}_n$ for "mixed case" interval censored data. Our proof is based on the inequality of the preceding section, and is similar in spirit to results of Van de Geer (1993), (1996).

Suppose that Y is a random variable taking values in $R^+ = [0, \infty)$ with distribution function $F \in \mathcal{F} = \{\text{all df's } F \text{ on } R^+\}$. Unfortunately we are not able to observe Y itself. What we do observe is a vector of times $T_K = (T_{K,1}, \ldots, T_{K,K})$ where K, the number of times, is itself random, and the interval $(T_{K,j-1}, T_{K,j}]$ into which Y falls (with $T_{K,0} \equiv 0$, $T_{K,K+1} \equiv \infty$). More formally, we assume that K is an integer-valued random variable, and $\underline{T} = \{T_{k,j}, j = 1, \ldots, k, k = 1, 2, \ldots\}$, is a triangular array of "potential observation times", and that Y and $(K, \underline{T})$ are independent. Let $X = (\Delta_K, T_K, K)$, with a possible value $x = (\delta_k, t_k, k)$, where $\Delta_k = (\Delta_{k,1}, \ldots, \Delta_{k,k})$ with $\Delta_{k,j} = 1_{(T_{k,j-1}, T_{k,j}]}(Y)$, $j = 1, 2, \ldots, k+1$, and T_k is the kth row of the triangular array $\underline{T}$. Suppose we observe n i.i.d. copies of X; $X_1, X_2, \ldots, X_n$, where $X_i = (\Delta^{(i)}_{K^{(i)}}, T^{(i)}_{K^{(i)}}, K^{(i)})$, $i = 1, 2, \ldots, n$. Here $(Y^{(i)}, \underline{T}^{(i)}, K^{(i)})$, $i = 1, 2, \ldots$ are the underlying i.i.d. copies of $(Y, \underline{T}, K)$.

We first note that conditionally on K and T_K, the vector Δ_K has a multinomial distribution:

$$(\Delta_K | K, T_K) \sim \text{Multinomial}_{K+1}(1, \Delta F_K)$$

where

$$\Delta F_K \equiv (F(T_{K,1}), F(T_{K,2}) - F(T_{K,1}), \ldots, 1 - F(T_{K,K})) .$$

Suppose for the moment that the distribution G_k of $(T_K|K=k)$ has density g_k and $p_k \equiv P(K=k)$. Then a density of X is given by

$$(3) \qquad p_F(x) \equiv p_F(\delta_k, t_k, k) = \prod_{j=1}^{k+1} (F(t_{k,j}) - F(t_{k,j-1}))^{\delta_{k,j}} g_k(t_k) p_k$$

where $t_{k,0} \equiv 0$, $t_{k,k+1} \equiv \infty$. In general,

$$\begin{aligned} p_F(x) \equiv p_F(\delta_k, t_k, k) &= \prod_{j=1}^{k+1} (F(t_{k,j}) - F(t_{k,j-1}))^{\delta_{k,j}} \\ &= \sum_{j=1}^{k+1} \delta_{k,j} (F(t_{k,j}) - F(t_{k,j-1})) \qquad (4) \end{aligned}$$

is a density of X with respect to the dominating measure ν where ν is determined by the joint distribution of $(K, \underline{T})$, and it is this version of the density of X with which we will work throughout the rest of the paper. Thus the log-likelihood function for F of $X_1, \ldots, X_n$ is given by

$$\begin{aligned} \frac{1}{n} l_n(F|\underline{X}) &= \frac{1}{n} \sum_{i=1}^{n} \sum_{j=1}^{K^{(i)}+1} \Delta_{K,j}^{(i)} \log \left(F(T_{K^{(i)},j}^{(i)}) - F(T_{K^{(i)},j-1}^{(i)}) \right) \\ &= \mathbb{P}_n m_F \end{aligned}$$

where

$$\begin{aligned} m_F(X) &= \sum_{j=1}^{K+1} \Delta_{K,j} \log \left(F(T_{K,j}) - F(T_{K,j-1}) \right) \\ &\equiv \sum_{j=1}^{K+1} \Delta_{K,j} \log \left(\Delta F_{K,j} \right) \end{aligned}$$

and where we have ignored the terms not involving F. We also note that, with $P_0 \equiv P_{F_0}$,

$$P_0 m_F(X) = P_0 \left(\sum_{j=1}^{K+1} \Delta F_{0,K,j} \log \left(\Delta F_{K,j} \right) \right).$$

The Nonparametric Maximum Likelihood Estimator (NPMLE) $\widehat{F}_n$ is the distribution function $\widehat{F}_n(t)$ which puts all its mass at the observed time points and maximizes the log-likelihood $l_n(F|\underline{X})$. It can be calculated via the iterative convex minorant algorithm proposed in Groeneboom and Wellner (1992) for case 2 interval censored data.

By Proposition 3 with $\alpha = 1$ and $\varphi \equiv \varphi_1$ as before, it follows that

$$h^2(p_{\widehat{F}_n}, p_{F_0}) \le (\mathbb{P}_n - P_0)\left(\varphi(p_{\widehat{F}_n}/p_{F_0})\right)$$

where φ is bounded and continuous from R to R. Now the collection of functions

$$\mathcal{G} \equiv \{p_F : \; F \in \mathcal{F}\}$$

is easily seen to be a Glivenko-Cantelli class of functions: this can be seen by first applying Theorem 4 to the collections $\mathcal{G}_k$, $k = 1, 2, \ldots$ obtained from $\mathcal{G}$ by restricting to the sets $K = k$. Then for fixed k, the collections $\mathcal{G}_k = \{p_F(\delta, t_k, k) : \; F \in \mathcal{F}\}$ are P_0 Glivenko-Cantelli classes since $\mathcal{F}$ is a uniform Glivenko-Cantelli class, and since the functions p_F are continuous transformations of the classes of functions $x \to \delta_{k,j}$ and $x \to F(t_{k,j})$ for $j = 1, \ldots, k+1$, and hence $\mathcal{G}$ is $P-$Glivenko-Cantelli by Theorem 3. Note that the single function p_{F_0} is trivially P_0- Glivenko-Cantelli since it is uniformly bounded, and the single function $(1/p_{F_0})$ is also P_0- GC since $P_0(1/p_{F_0}) < \infty$. Thus by Proposition 2 with $g = (1/p_{F_0})$ and $\mathcal{F} = \mathcal{G} = \{p_F : \; F \in \mathcal{F}\}$, it follows that $\mathcal{G}' \equiv \{p_F/p_{F_0} : \; F \in \mathcal{F}\}$ is P_0-Glivenko-Cantelli. Finally another application of Theorem 3 shows that the collection

$$\mathcal{H} \equiv \{L_1(p_F/p_{F_0}) : \; F \in \mathcal{F}\} = \{\varphi(p_F/p_{F_0}) : \; F \in \mathcal{F}\}$$

is also P_0-Glivenko-Cantelli. When combined with Proposition 3, this yields the following theorem:

Theorem 7. *The NPMLE $\widehat{F}_n$ satisfies*

$$h(p_{\widehat{F}_n}, p_{F_0}) \to_{a.s.} 0\,.$$

To relate this result to a recent theorem of Schick and Yu (1999), it remains only to understand the relationship between their $L_1(\mu)$ and the Hellinger metric h between p_F and p_{F_0}.

Let $\mathcal{B}$ denote the collection of Borel sets in R. On $\mathcal{B}$ we define measures μ and $\widetilde{\mu}$ as follows: For $B \in \mathcal{B}$,

$$\mu(B) = \sum_{k=1}^{\infty} P(K = k) \sum_{j=1}^{k} P(T_{k,j} \in B | K = k)\,, \tag{5}$$

and

$$\widetilde{\mu}(B) = \sum_{k=1}^{\infty} P(K = k) \frac{1}{k} \sum_{j=1}^{k} P(T_{k,j} \in B | K = k)\,. \tag{6}$$

Let d be the $L_1(\mu)$ metric on the class $\mathcal{F}$; thus for $F_1, F_2 \in \mathcal{F}$,

$$d(F_1, F_2) = \int |F_1(t) - F_2(t)| d\mu(t)\,.$$

The measure μ was introduced by Schick and Yu (1999); note that μ is a finite measure if $E(K) < \infty$. Note that $d(F_1, F_2)$ can also be written in terms of an expectation as:

$$d(F_1, F_2) = E_{(K,\underline{T})}\left[\sum_{j=1}^{K} \left|(F_1(T_{K,j}) - F_2(T_{K,j})\right|\right]. \tag{7}$$

As Schick and Yu (1999) observed, consistency of the NPMLE $\widehat{F}_n$ in $L_1(\mu)$ holds under virtually no further hypotheses.

Theorem 8. (Schick and Yu). *Suppose that* $E(K) < \infty$. *Then* $d(\widehat{F}_n, F_0) \to_{a.s.} 0$.

Proof. We will show that Theorem 8 follows from Theorem 7 and the following lemma.

Lemma 4.

$$\frac{1}{2}\left\{\int |\widehat{F}_n - F_0| d\tilde{\mu}\right\}^2 \le h^2(p_{\widehat{F}_n}, p_{F_0}).$$

Proof. We know that

$$h^2(p_{\widehat{F}_n}, p_{F_0}) \le d_{TV}(p_{\widehat{F}_n}, p_{F_0}) \le \sqrt{2} h(p_{\widehat{F}_n}, p_{F_0})$$

where, with $y_{k,0} = -\infty$, $y_{k,k+1} = \infty$,

$$h^2(p_{\widehat{F}_n}, p_{F_0}) = \sum_{k=1}^{\infty} P(K=k) \sum_{j=1}^{k+1} \int \{[\widehat{F}_n(y_{k,j}) - \widehat{F}_n(y_{k,j-1})]^{1/2} \\ - [F_0(y_{k,j}) - F_0(y_{k,j-1})]^{1/2}\}^2 dG_k(y)$$

while

$$d_{TV}(p_{\widehat{F}_n}, p_{F_0}) = \sum_{k=1}^{\infty} P(K=k) \sum_{j=1}^{k+1} \int |[\widehat{F}_n(y_{k,j}) - \widehat{F}_n(y_{k,j-1})] \\ - [F_0(y_{k,j}) - F_0(y_{k,j-1})]| dG_k(y).$$

Note that

$$\sum_{j=1}^{k+1} \int |[\widehat{F}_n(y_{k,j}) - \widehat{F}_n(y_{k,j-1})] - [F_0(y_{k,j}) - F_0(y_{k,j-1})]| \\ = \sum_{j=1}^{k+1} |(\widehat{F}_n - F_0)(y_{k,j-1}, y_{k,j}]| \\ \ge \max_{1 \le j \le k+1} |\widehat{F}_n(y_{k,j}) - F_0(y_{k,j})|,$$

so integrating across this inequality with respect to $G_k(y)$ yields

$$\sum_{j=1}^{k+1} \int |[\widehat{F}_n(y_{k,j}) - \widehat{F}_n(y_{k,j-1})] - [F_0(y_{k,j}) - F_0(y_{k,j-1})]| \, dG_k(y)$$
$$\geq \max_{1\leq j\leq k} \int |\widehat{F}_n(y_{k,j}) - F_0(y_{k,j})| \, dG_{k,j}(y_{k,j})$$
$$\geq \frac{1}{k} \sum_{j=1}^{k} \int |\widehat{F}_n(y_{k,j}) - F_0(y_{k,j})| \, dG_{k,j}(y_{k,j}) .$$

By multiplying across by $P(K = k)$ and summing over k, this yields

$$d_{TV}(p_{\widehat{F}_n}, p_{F_0}) \geq \int |\widehat{F}_n - F_0| d\tilde{\mu} ,$$

and hence

$$h^2(p_{\widehat{F}_n}, p_{F_0}) \quad \geq \quad \frac{1}{2} \left\{ \int |\widehat{F}_n - F_0| d\tilde{\mu} \right\}^2 . \tag{8}$$

□

The measure $\tilde{\mu}$ figuring in Lemma 4 is not the same as the measure μ of Schick and Yu (1999) because of the factor $1/k$. Note that this factor means that the measure $\tilde{\mu}$ is always a finite measure, even if $E(K) = \infty$. It is clear that

$$\tilde{\mu}(B) \leq \mu(B)$$

for every Borel set B, and that $\mu \prec\prec \tilde{\mu}$. The following lemma (Lemma 2.2 of Schick and Yu (1999)) together with Lemma 4 shows that Theorem 7 implies the result of Schick and Yu once again:

Lemma 5. *Suppose that μ and $\tilde{\mu}$ are two finite measures, and that $g, g_1, g_2, \ldots$ are measurable functions with range in $[0, 1]$. Suppose that μ is absolutely continuous with respect to $\tilde{\mu}$. Then $\int |g_n - g| d\tilde{\mu} \to 0$ implies that $\int |g_n - g| d\mu \to 0$.*

Proof. Write

$$\int |g_n - g| d\mu = \int |g_n - g| \frac{d\mu}{d\tilde{\mu}} d\tilde{\mu}$$

and use the dominated convergence theorem applied to a.e. convergent subsequences. □

6 Example: generalizing the result of Schick and Yu

Our goal in this section is to give a generalization of the consistency result of Schick and Yu (1999).

Suppose that Y is a random variable taking values in $\mathcal{Y}$. Suppose that Y has distribution Q on the measurable space $(\mathcal{Y}, \mathcal{B})$. Unfortunately we are not able to observe Y itself. What we do observe is a vector of random sets $C_K = (C_{K,1}, \ldots, C_{K,K})$ where K, the number of sets is itself random, and the set $\{C_{K,j}\}_{j=1}^K$ form a partition of $\mathcal{Y}$: $C_{K,j} \cap C_{K,j'} = \emptyset$ if $j \neq j'$ and $\cup_{j=1}^K C_{K,j} = \mathcal{Y}$. More formally, we assume that K is an integer-valued random variable, and $\underline{C} = \{C_{k,j}, j = 1, \ldots, k, k = 1, 2, \ldots\}$, is a triangular array of "random sets", and that Y and $(K, \underline{C})$ are independent. Let $X = (\Delta_K, C_K, K)$, with a possible value $x = (\delta_k, c_k, k)$, where $\Delta_k = (\Delta_{k,1}, \ldots, \Delta_{k,k})$ with $\Delta_{k,j} = 1_{C_{k,j}}(Y)$, $j = 1, 2, \ldots, k$, and C_k is the k-th row of the triangular array $\underline{C}$. Suppose we observe n i.i.d. copies of X; $X_1, X_2, \ldots, X_n$, where $X_i = (\Delta_{K^{(i)}}^{(i)}, C_{K^{(i)}}^{(i)}, K^{(i)})$, $i = 1, 2, \ldots, n$. Here $(Y^{(i)}, \underline{C}^{(i)}, K^{(i)})$, $i = 1, 2, \ldots$ are the underlying i.i.d. copies of $(Y, \underline{C}, K)$.

We first note that conditionally on K and C_K, the vector Δ_K has a multinomial distribution:

$$(\Delta_K | K, C_K) \sim \text{Multinomial}_K(1, Q_K)$$

where

$$Q_K \equiv (Q(C_{K,1}), Q(C_{K,2}), \ldots, Q(C_{K,K})) .$$

Suppose for the moment that the distribution G_k of $(C_K | K = k)$ has density g_k and $p_k \equiv P(K = k)$. Then a density of X is given by

$$p_Q(x) \equiv p_Q(\delta_k, c_k, k) = \prod_{j=1}^{k} Q(c_{k,j})^{\delta_{k,j}} g_k(c_k) p_k . \tag{9}$$

In general,

$$p_Q(x) \equiv p_Q(\delta_k, c_k, k) = \prod_{j=1}^{k} Q(c_{k,j})^{\delta_{k,j}} = \sum_{j=1}^{k} \delta_{k,j} Q(c_{k,j}) \tag{10}$$

is a density of X with respect to the dominating measure ν where ν is determined by the joint distribution of $(K, \underline{C})$, and it is this version of the density of X with which we will work throughout the rest of the paper. Thus the log-likelihood function for Q of $X_1, \ldots, X_n$ is given by

$$\frac{1}{n} l_n(Q|\underline{X}) = \frac{1}{n} \sum_{i=1}^{n} \sum_{j=1}^{K^{(i)}} \Delta_{K,j}^{(i)} \log Q(C_{K^{(i)},j}^{(i)}) = \mathbb{P}_n m_Q$$

where

$$m_Q(X) = \sum_{j=1}^{K} \Delta_{K,j} \log Q(C_{K,j})$$

and where we have ignored the terms not involving Q. We also note that, with $P_0 \equiv P_{Q_0}$,

$$P_0 m_Q(X) = P_0 \left(\sum_{j=1}^{K} Q_0(C_{K,j}) \log Q(C_{K,j}) \right) .$$

The Nonparametric Maximum Likelihood Estimator (NPMLE) $\widehat{Q}_n$ is a probability measure $\widehat{Q}_n$ which maximizes the log-likelihood $l_n(Q|\underline{X})$. Here we will bypass the many interesting existence, characterization, and computational issues connected with the NPMLE $\widehat{Q}_n$, and focus instead on the issue of consistency once we have the NPMLE in hand.

By Proposition 3 with $\alpha = 1$ it follows that

$$h^2(p_{\widehat{Q}_n}, p_{Q_0}) \le (\mathbb{P}_n - P_0) \left(\varphi(p_{\widehat{Q}_n}/p_{Q_0}) \right) .$$

where φ is bounded and continuous. Now the collection of functions

$$\mathcal{G} \equiv \{p_Q : \ Q \in \mathcal{Q}\}$$

where $\mathcal{Q}$ is the collection of all probability distributions on $(\mathcal{Y}, \mathcal{B})$, is easily seen to be a Glivenko-Cantelli class of functions: this can be seen by first applying Theorem 4 to the collections $\mathcal{G}_k$, $k = 1, 2, \ldots$ obtained from $\mathcal{G}$ by restricting to the sets $K = k$. The next step is to show that the collections

$$\mathcal{G}_k \equiv \{p_Q(\cdot, \cdot, k) : \ Q \in \mathcal{Q}\}$$

are P_0-Glivenko-Cantelli for each k. To this end, suppose that $(\mathcal{Y}, \mathcal{B})$ is a measurable space and let $\mathcal{C}$ be a universal GC - class of sets in $\mathcal{Y}$ (i.e. $Q-$GC for every probability measure Q on $(\mathcal{Y}, \mathcal{B})$). Let $(T, \mathcal{T}, G)$ be a probability space, and let $t \mapsto C_t$ be a map with values in $\mathcal{C}$.

Lemma 6. *Suppose that the collection $\{\{t : y \in C_t\} : \ y \in \mathcal{Y}\}$ is $G-GC$. Then the collection of functions*

$$t \mapsto Q(C_t)$$

where Q ranges over all probability measures Q on $\mathcal{Y}$ is $G-GC$.

Remark. By Assouad (1983), Proposition 2.12, page 246, together with Corollary 1.10, page 241, it follows that $\{\{t \in T : y \in C_t\} : \ y \in \mathcal{Y}\}$ is a VC-class of subsets of T if and only if $\{C_t : t \in T\}$ is a VC-class of subsets of $\mathcal{Y}$. Thus if we assume that all the subsets $C = C_t$ arising in the partitions are elements of a VC-class $\mathcal{C}$, then, subject to being suitably measurable, the hypothesis of Lemma 6 is satisfied (and the class in question is even universal Glivenko-Cantelli).

Proof. For Q the Dirac measure at y (unit mass at y), the function $t \mapsto Q(C_t)$ becomes

$$t \mapsto 1_{C_t}(y)\,.$$

By assumption, the set of all such functions, with y ranging over $\mathcal{Y}$, is G–GC. Then so is the set of all functions

$$t \mapsto \frac{1}{m}\sum_{i=1}^{m} 1_{C_t}(y_i), \qquad y_1, \ldots, y_m \in \mathcal{Y}, m \in \mathbb{N}.$$

Let $Y_1, \ldots, Y_m$ be an i.i.d. sample from a given Q. Because $\mathcal{C}$ is Q–GC,

$$\sup_{C \in \mathcal{C}} \Big|\frac{1}{m}\sum_{i=1}^{m} 1\{Y_i \in C\} - P(C)\Big| \to 0 \qquad \text{a.s.}^*\,.$$

Then there exists a sequence $y_1, y_2, \ldots$ in $\mathcal{Y}$ such that

$$\sup_{C \in \mathcal{C}} \Big|\frac{1}{m}\sum_{i=1}^{m} 1\{y_i \in C\} - P(C)\Big| \to 0\,,$$

and consequently

$$\sup_{t \in T} \Big|\frac{1}{m}\sum_{i=1}^{m} 1\{y_i \in C_t\} - P(C_t)\Big| \to 0\,.$$

It follows that the maps $t \to Q(C_t)$ are the (uniform) limits of sequences of functions from a G–GC class. Hence they are G–GC. □

Note on measurability: It is assumed implicitly that the sets $\{t : y \in C_t\}$ are measurable in T for every $y \in \mathcal{Y}$. The proof shows that the maps $t \mapsto P(C_t)$ are then also measurable.

It follows that for fixed k, the collections $\mathcal{G}_k = \{p_Q(\delta, c_k, k) : Q \in \mathcal{Q}\}$ are P_0-Glivenko-Cantelli, and since the functions p_Q are continuous transformations of the classes of functions $x \to \delta_{k,j}$ and $x \to Q(c_{k,j})$ for $j = 1, \ldots, k$, and hence $\mathcal{G}$ is P–Glivenko-Cantelli by Theorem 3. Note that the single function p_{Q_0} is trivially P_0– Glivenko-Cantelli since it is uniformly bounded. Now another application of Theorem 3 shows that the collection

$$\mathcal{H} \equiv \{L_1(p_Q/p_{Q_0}) : \ Q \in \mathcal{Q}\} = \{\varphi(p_Q/p_{Q_0}) : \ Q \in \mathcal{Q}\}$$

is also P_0-Glivenko-Cantelli. When combined with Proposition 3, this yields the following theorem.

Theorem 9. *If all $C_{K,j} \in \mathcal{C}$, a VC collection of subsets of $\mathcal{X}$, then the NPMLE $\widehat{Q}_n$ satisfies*

$$h(p_{\widehat{Q}_n}, p_{Q_0}) \to_{a.s.} 0\,.$$

To obtain a statement analogous to the theorem of Schick and Yu (1999), let Σ denote a sigma algebra for the space $\mathcal{C}$ of subsets: we are assuming that $P(\cap_{j=1}^K [C_{K,j} \in \mathcal{C}]) = 1$. Thus $(\mathcal{C}, \Sigma)$ is a measurable space. On Σ we define the measure μ as follows:

$$\mu(B) = \sum_{k=1}^{\infty} P(K = k) \sum_{j=1}^{k} P(C_{k,j} \in B | K = k)\,, \qquad B \in \Sigma. \tag{11}$$

Note that

$$\begin{aligned} d_{TV}(p_{\widehat{Q}_n}, p_{Q_0}) &= \sum_{k=1}^{\infty} P(K = k) \sum_{j=1}^{k} \int |\widehat{Q}_n(c_{k,j}) - Q_0(c_{k,j})| dG_k(c) \\ &= \int |\widehat{Q}_n(c) - Q_0(c)|\, d\mu(c) \equiv d(\widehat{Q}_n, Q_0)\,. \end{aligned} \tag{12}$$

Here is a generalization of the theorem of Schick and Yu (1999).

Theorem 10. *The NPMLE $\widehat{Q}_n$ satisfies $d(\widehat{Q}_n, Q_0) \to_{a.s.} 0$.*

Proof. Theorem 10 follows from Theorem 9, (12) and the observation that

$$d(\widehat{Q}_n, Q_0) = d_{TV}(p_{\widehat{Q}_n}, p_{Q_0}) \le \sqrt{2} h(p_{\widehat{Q}_n}, p_{Q_0})\,.$$

□

Example 1. (Mixed case interval censoring in R^2). Suppose that $Y \sim Q_0$ takes values in R^{+2}, and the partitions $\{C_{K,j}\}_{j=1}^K$ are obtained as the natural rectangles formed from two increasing collections $0 \le S_{K_1,1} \le \dots \le S_{K_1,K}$ and $0 \le T_{K_2,1} \le \dots \le T_{K_2,K}$ on the two coordinate axes. Then $K = (K_1 + 1)(K_2 + 1)$, and all the elements of the partitions are elements of the VC-class of rectangles in R^2. It follows that the NPMLE $\widehat{Q}_n$ of Q_0 satisfies

$$\int_{s,t \in R^{+2}} |\widehat{Q}_n(s,t] - Q_0(s,t]| d\mu(s,t) \to 0 \qquad \text{a.s.}^*\,.$$

Acknowledgments. We owe thanks to Evarist Giné and Joel Zinn for suggesting the proof given for Lemma 1.

References

Assouad, P. (1983). Densité et dimension. *Annales de l'Institut Fourier* **33**, 233–282.

Dudley, R. M. (1984). *A Course on Empirical Processes. Ecole d'été de probabilités de St.-Flour, 1982.* Lecture Notes in Mathematics **1097**, Springer, Berlin, 1–142.

DUDLEY, R. M. (1998a). "Consistency of M-estimators and one-sided bracketing." In *Proceedings of the First International Conference on High-Dimensional Probability*. E. Eberlein, M. Hahn, and M. Talagrand, eds. Birkhäuser, Basel, pp. 33–58.

DUDLEY, R. M. (1998b). Personal communication.

DUDLEY, R. M., GINÉ, E., AND ZINN, J. (1991). Uniform and universal Glivenko-Cantelli classes. *J. Theoretical Probab.* **4**, 485–510.

GINÉ, E., AND ZINN, J. (1984). Some limit theorems for empirical processes. *Ann. Probability* **12**, 929–989.

GROENEBOOM, P. AND WELLNER, J. A. (1992). *Information Bounds and Nonparametric Maximum Likelihood Estimation.* DMV Seminar Band 19, Birkhäuser, Basel.

PFANZAGL, J. (1988). Consistency of maximum likelihood estimators for certain nonparametric families, in particular: mixtures. *J. Statist. Planning and Inference* **19**, 137–158.

SCHICK, A. AND YU, Q. (1999), Consistency of the GMLE with Mixed Case Interval-Censored Data, *Scand. J. Statist.***27**, 45–55.

TALAGRAND, M. (1987a). Measurability problems for empirical measures. *Ann. Probability* **15**, 204–212.

TALAGRAND, M. (1987b). The Glivenko-Cantelli problem. *Ann. Probability* **15**, 837–870.

TALAGRAND, M. (1996). The Glivenko-Cantelli problem, ten years later. *J. Theor. Prob.* **9**, 371–384.

VAN DE GEER, S. (1993). Hellinger consistency of certain nonparametric maximum likelihood estimators. *Ann. Statist.* **21**, 14–44.

VAN DE GEER, S. (1996). Rates of convergence for the maximum likelihood estimator in mixture models. *J. Nonparametric Statist.* **6**, 293–310.

VAN DER VAART, A. W. (1996). New Donsker classes. *Ann. Probability* **24**, 2128–2140.

VAN DER VAART, A. W., AND WELLNER, J. A. (1996). *Weak Convergence and Empirical Processes with Applications to Statistics,* Springer-Verlag, New York.

Aad van der Vaart
Department of Mathematics
Free University
De Boelelaan 1081a
1081 HV Amsterdam
The Netherlands
e-mail: aad@cs.vu.nl

Jon A. Wellner
Department of Statistics
University of Washington
P.O. Box 354322
Seattle, Washington 98195-4322
e-mail: jaw@stat.washington.edu

PART 3

Gaussian Processes

Continuité de certaines fonctions aléatoires gaussiennes à valeurs dans l_p, $1 \leq p < \infty$

X. Fernique

RÉSUMÉ : Soit $X = \{X_n(t), t \in T, n \in \mathbf{N}\}$ une fonction aléatoire gaussienne à valeurs dans $\mathbf{R}^{\mathbf{N}}$ et à composantes indépendantes ; on étudie l'existence d'une modification de X à valeurs dans $\ell_p, 1 \leq p < \infty$, et à trajectoires continues. On obtient en particulier une caractérisation pour $p = 2$ si $T = \mathbf{R}^m$ et si les composantes sont stationnaires ; cette caractérisation s'étend aussi à certains cas non stationnaires sur un espace métrique compact T arbitraire. Les techniques employées sont celles des mesures majorantes.

Introduction

Soit $X = \{X_n(t), t \in T, n \in \mathbf{N}\}$ une suite de fonctions aléatoires gaussiennes indépendantes à trajectoires continues sur un espace métrique T ; X définissant ainsi une fonction aléatoire gaussienne à valeurs dans $\mathbf{R}^{\mathbf{N}}$, on se propose de caractériser l'existence d'une modification de X à valeurs dans $l_p, 1 \leq p < \infty$ et à trajectoires continues ; le cas fondamental est celui où $p = 2$, les sommes du type $\sum |X_n(t)|^2$ interviennent en effet dans des calculs d'énergie ou de dispersion.

Dans le cas gaussien stationnaire sur $\mathbf{R}$ ou $\mathbf{R}^m$ et puisque de tels espaces $E = l_p$ ne contiennent pas c_0, on sait ([4], [5]) que X a une modification à trajectoires continues dans E si et seulement si $X(0)$ est p.s. à valeurs dans E et si de plus pour tout $y \in E'$, $\langle X, y\rangle$ a une modification à trajectoires localement bornées, c'est-à-dire si et seulement si

$$(0.1) \qquad \sum_n \mathbf{E}|X_n(0)|^p < \infty, \ \sup_{||y|| \leq 1} \mathbf{E} \sup_{0 \leq t \leq 1} \langle X(t), y\rangle < \infty.$$

Le problème est alors d'obtenir des conditions équivalentes à la condition (0.1) s'exprimant explicitement en fonctions simples des paramètres définissant la loi de X et donc des seules suites $\{a_n, d_n, n \in \mathbf{N}\}$ définies par

$$(0.2) \quad |a_n|^2 = \mathbf{E}|X_n(0)|^2, \quad |d_n(t)|^2 = \mathbf{E}|X_n(t) - X_n(0)|^2, n \in \mathbf{N}, \ t \in \mathbf{R}.$$

La situation dite d'Ornstein-Uhlenbeck au sens strict se présente quand les X_n ont tous même loi à des changements d'échelle près : $\{x_n, n \in \mathbf{N}\}$ est une suite i.i.d. de f.a. d'Ornstein-Uhlenbeck au sens usuel, $\{a_n, b_n, n \in \mathbf{N}\}$ est une suite de coefficients numériques et $X_n = \{a_n x_n(b_n t), t \in \mathbf{R}\}, n \in \mathbf{N}$. Ce cas particulier, important pour les applications, a été précédemment l'objet de nombreux travaux (cf. par exemple [7]). On a montré dans [3]) que la condition (0.1) est alors équivalente à

$$\sum_n |a_n|^p < \infty,$$

$$(0.3) \qquad \int_0^\infty t^{p-1} \left[\sup_{n \in \mathbf{N}} a_n \mathbf{I}_{\{b_n > \exp t^2\}}\right]^p dt < \infty \quad \text{si } p \geq 2,$$

$$\int_0^\infty t^{p-1} \left[\sum_n |a_n|^r \mathbf{I}_{\{b_n > \exp t^2\}}\right]^{1-p/2} dt < \infty, r = \frac{2p}{2-p} \quad \text{si } p < 2.$$

Cette équivalence n'est d'ailleurs pas liée à la loi particulière des x_n et elle subsiste sous la seule condition que cette loi commune soit associée à une distance d croissante *au sens large et non identiquement nulle* sur un intervalle $[0, \epsilon], \epsilon > 0$ (relire la preuve du théorème 0.5 de [2] pour corriger en ce sens son énoncé et donc celui du théorème 1.3 de [3]).

Dans la première partie de ce travail et pour étendre ces résultats, on supposera que $T = \mathbf{R}$ et que les différentes lois gaussiennes stationnaires des composantes indépendantes X_n n'ont pas de liens mutuels tout en restant associées à des distances d_n croissantes sur $[0, 1]$; dans ces conditions, en utilisant les techniques qui ont permis de conclure dans la situation d'Ornstein-Uhlenbeck au sens strict, on rappellera une condition suffisante et on présentera une condition nécessaire nouvelle (proposition 1) ; cette condition nécessaire est en fait, dans la situation d'Ornstein-Uhlenbeck au sens strict, identique aux conditions suffisantes (0.3) bien qu'en général, elle ne soit pas suffisante de sorte qu'une caractérisation générale relève nécessairement d'autres techniques.

Sous les mêmes hypothèses et à partir de techniques différentes plus adaptées, on présentera ensuite pour $p \leq 2$, une condition nécessaire et suffisante. On démontrera en particulier en 1.3.2 :

Théorème 1. *On note $X = \{X_n(t), t \in T, n \in \mathbf{N}\}$ une suite de fonctions aléatoires gaussiennes indépendantes sur $T = \mathbf{R}$; on suppose les composantes de X stationnaires et associées à des distances d_n croissantes au sens large sur $[0, 1]$.*
(a) Pour que X ait une modification à trajectoires continues dans l_2, il faut

et il suffit qu'il existe une fonction positive Δ sur $\mathbf{R}_+$ telle que

$$\sum_n |a_n|^2 < \infty, \int_0^\infty \Delta(t)dt < \infty, \tag{0.4}$$
$$\sup_{n\in\mathbf{N}} \int_0^\infty d_n^2[\exp(-t^2)]\frac{dt}{\Delta(t)} < \infty.$$

(b) Pour que X ait une modification à trajectoires continues dans $l_p, p < 2$, il faut et il suffit qu'il existe une fonction positive Δ sur $\mathbf{R}_+$ telle que

$$\sum_n |a_n|^p < \infty, \int_0^\infty \Delta(t)dt < \infty \tag{0.5}$$
$$\sum_n \left\{\int_0^\infty d_n^2[\exp(-t^2)]\frac{dt}{\Delta(t)}\right\}^{r/2} < \infty, \; r = \frac{2p}{2-p}.$$

Dans une seconde partie, on étendra l'étude en supprimant l'hypothèse de croissance des distances et même l'hypothèse de stationnarité ; on utilisera alors des mesures majorantes et leur adaptatation au problème étudié nous imposera d'affiner les techniques usuelles de leur maniement. Dans le cas stationnaire, on obtiendra en particulier :

Théorème 2. *On note $X = \{X_n(t), t \in T, n \in \mathbf{N}\}$ une suite de fonctions aléatoires gaussiennes indépendantes sur $T = \mathbf{R}^m$; on suppose les composantes de X stationnaires et la série $\sum_n |a_n|^2$ convergente. Dans ces conditions, les deux propriétés suivantes sont équivalentes :*
(a) X a une modification à trajectoires continues dans ℓ_2.

(b) Il existe sur $T' =]0,1]^m \times]0,1]$ muni de la mesure de Lebesgue $\lambda' = \lambda_m \times \lambda$ deux fonctions mesurables et positives Δ' et K' telles que
(b.1) $\int \Delta'(t,\tau)\, K'(t,\tau)\, d\lambda_m(t)\, d\lambda(\tau)$ *soit fini,*

(b.2) $\sup_{n\in\mathbf{N}} \int \frac{d_n^2(t)}{\Delta'(t,\tau)} K'(t,\tau) d\lambda_m(t)\, d\lambda(\tau)$ *soit fini,*

(b.3) la loi de K' sur (T', λ') soit égale à celle de $\frac{1}{2t\sqrt{\log h/t}}$ *sur $(]0,1], \lambda)$,*

$h = \exp(1/2)$.

La valeur de h dans la condition (b.3) est choisie de sorte que la fonction $2t\sqrt{\log h/t}, t \in]0,1]$ soit monotone.

1 Le cas stationnaire et des distances croissantes

On fixe $p \geq 1$, E est l'espace l_p, q est l'exposant conjugué de p et

$$B_q = \{y \in E' : \forall n \in \mathbf{N}, y_n \geq 0, \|y\|_q \leq 1\}$$

est la partie positive de la boule unité de E' ; on note

$$X = \{X_n(t), t \in \mathbf{R}, n \in \mathbf{N}\}$$

une suite de fonctions aléatoires gaussiennes stationnaires indépendantes sur $\mathbf{R}$. On suppose les d_n croissantes (au sens large) sur $[0,1]$, les X_n à trajectoires continues de sorte que les intégrales $\int_0^\infty d_n[\exp(-t^2)]\,dt$ sont convergentes et que les conditions (0.1) s'écrivent

(1.1)

$$\sum_n |a_n|^p < \infty, \quad M = \sup_{y \in B_q} \int_0^\infty \left\{ \sum_n |y_n|^2 d_n[\exp(-t^2)]^2 \right\}^{1/2} dt < \infty.$$

On se propose d'expliciter la seconde partie de cette condition.

1.1 Une condition nécessaire, une condition suffisante

On introduit la fonction d définie par

$$\forall t \geq 0, d(t) = \sup_{y \in B_q} \left[\sum_n |y_n|^2 d_n(t)^2 \right]^{1/2}, \tag{1.2}$$

on remarque qu'on a aussi

$$\forall t \geq 0, d(t) = \sup_{y \in B_2} \left[\sum_n |y_n|^p d_n(t)^p \right]^{1/p}; \tag{1.3}$$

la forme de d dépend d'ailleurs des valeurs de p ; on rappelle en effet que

$$\forall p \geq 2, \quad \sup_{y \in B_q} \left[\sum_n |y_n|^2 d_n^2 \right]^{1/2} = \sup_n d_n$$

$$\forall p \in [1, 2[, \quad \sup_{y \in B_q} \left[\sum_n |y_n|^2 d_n^2 \right]^{1/2} = \left[\sum_n |d_n|^r \right]^{1/r}, r = \frac{2p}{2-p}. \tag{1.4}$$

Proposition 1. *Pour que X ait une modification à trajectoires continues dans $E = l_p$,*
(a) il suffit que

$$\sum_n |a_n|^p < \infty, \quad \int_0^\infty d[\exp(-t^2)]dt < \infty; \tag{1.5}$$

(b) il faut que

$$\sum_n |a_n|^p < \infty, \quad \int_0^\infty d[\exp(-t^{2/p})]^p dt < \infty. \tag{1.6}$$

Démonstration : La preuve de (a) est immédiate, nous démontrons (b) par deux schémas différents ; le premier schéma ne conclut que si p est ≥ 2, mais il met en évidence une propriété un peu surprenante même dans le cas d'Ornstein-Uhlenbeck au sens strict. Pour simplifier, on posera :

$$H_n(t) = d_n[\exp(-t^2)], \quad H(t) = d[\exp(-t^2)], \quad n \in \mathbf{N},\, t \in \mathbf{R}.$$

Premier schéma de preuve : Supposons la condition (1.1) réalisée, alors en particulier,

$$M' = \sup_{y \in B_q} \int_0^\infty \left[\sup_{n \in \mathbf{N}} y_n H_n(t)\right] dt$$

est fini ; c'est avec la décroissance des H_n la seule hypothèse qu'on utilisera ici. Cette décroissance implique

$$\infty > M' \geq \sup_{y \in B_q} \sum_k 2^k \left[\sup_{n \in \mathbf{N}} y_n H_n(2^{k+1})\right].$$

Pour tout $k \in \mathbf{Z}$, on note $n(k)$ un entier tel que

$$2H_{n(k)}(2^{k+1}) \geq \sup_{n \in \mathbf{N}} H_n(2^{k+1})\,;$$

pour tout m, on pose $K_m = \{k : n(k) = m\}$ et on sépare $\mathbf{Z}$ suivant la partition des K_m :

$$M' \geq \sup_{y \in B_q} \sum_m \left[y_m \sum_{k \in \mathbf{K_m}} 2^k H_m(2^{k+1})\right] = \left\{\sum_m \left[\sum_{k \in \mathbf{K_m}} 2^k H_m(2^{k+1})\right]^p\right\}^{1/p};$$

il suffit maintenant d'utiliser la formule $(a+b)^p \geq a^p + b^p$ pour obtenir

$$M' \geq \left[\sum_m \sum_{k \in K_m} 2^{kp} H_m(2^{k+1})^p\right]^{1/p}$$

et en réordonnant cette somme où interviennent effectivement toutes les valeurs de l'indice k, en utilisant aussi la définition des K_m

$$2M' \geq \left[\sum_{k \in \mathbf{N}} 2^{kp} \sup_{n \in \mathbf{N}} H_n(2^{k+1})^p\right]^{1/p};$$

on réutilise maintenant la décroissance des H_n et pour tout $p \geq 1$, on obtient

$$[4M']^p \geq (1/p) \int_0^\infty t^{p-1} \left[\sup_{n \in \mathbf{N}} H_n(t)\right]^p dt.$$

C'est bien, mais seulement pour $p \geq 2$, la condition nécessaire annoncée.

Deuxième schéma de preuve : Supposons la condition (1.1) réalisée, alors en particulier la décroissance des H_n implique

$$\infty > M = \sup_{y\in B_q} \int_0^\infty \left[\sum_n |y_n|^2 H_n(t)^2\right]^{1/2} dt$$
$$\geq \sup_{y\in B_q} \sum_{k\in\mathbf{Z}} 2^k \left[\sum_n |y_n|^2 H_n(2^{k+1})^2\right]^{1/2}.$$

Pour linéariser ce dernier terme, nous introduisons l'ensemble

$$A = \{\alpha = (\alpha_n(k), k \in \mathbf{N}, n \in \mathbf{N}) : \forall k \in \mathbf{N}, \alpha(k) \in B_2\};$$

on obtient

$$M \geq \sup_{y\in B_q} \sup_{\alpha\in A} \sum_k 2^k \left[\sum_n y_n \alpha_n(k) H_n(2^{k+1})\right]$$

et on en déduit

$$M^p \geq \sup_{\alpha\in A} \sum_n \left[\sum_k 2^k \alpha_n(k) H_n(2^{k+1})\right]^p$$
$$\geq \sup_{\alpha\in A} \sum_n \left[\sum_k 2^{kp} |\alpha_n(k)|^p H_n(2^{k+1})^p\right]$$
$$\geq \sum_k 2^{kp} \sup_{\alpha\in B_2} \sum_n |\alpha_n|^p H_n(2^{k+1})^p.$$

La formule (1.3) fournit alors

$$M^p \geq \sum_k 2^{kp} H(2^{k+1})^p,$$

et finalement en utilisant la monotonie de H

$$\infty > M^p \geq 2^{(1-2p)} \int_0^\infty t^{p-1} H^p(t) dt,$$

la propriété (b) est ainsi établie pour toutes les valeurs de $p \geq 1$.

1.2 Remarques sur l'énoncé précédent

(a) Dans le cas d'Ornstein-Uhlenbeck au sens strict, on constate facilement que la condition nécessaire (1.6) de la proposition 1 est équivalente aux conditions (0.3) ; elle est alors nécessaire et suffisante ; ceci montre qu'au moins dans certains cas cette condition nécessaire qui est nouvelle est plus significative que la condition suffisante bien classique du même énoncé.

(b) Dans ce même cas et si de plus $p \geq 2$, le premier schéma de preuve ci-dessus montre que pour que X ait une modification à trajectoires continues, il (faut et il) suffit que

$$X(0) \in l_p \quad \text{p.s.}, \quad \sup_{y \in B_q} \int_0^\infty \left\{ \sup_{n \in \mathbf{N}} y_n d_n[\exp(-t^2)] \right\} dt < \infty. \tag{1.7}$$

Cette condition est apparemment nettement moins forte que les conditions (0.3) ; curieusement, elle leur est en fait équivalente.

(c) On peut comparer la condition suffisante (1.5) aux conditions du même type fournies par le théorème 11.21 de [8] et qui restent exploitées par la plupart des utilisateurs. L'utilisation de cet autre théorème nécessite trois hypothèses qui s'écriraient ici

$$\begin{gathered} \sum_n |a_n|^p < \infty, \quad \int_0^\infty d[\exp(-t^2)]dt < \infty, \\ \int_0^\infty \left\{ \sum_n d_n[\exp(-e^t)]^p \right\}^{1/p} dt < \infty. \end{gathered} \tag{1.8}$$

L'énoncé ci-dessus implique qu'en fait dans la situation particulière étudiée ici la troisième hypothèse est inutile. Toute technique de majoration, pour l'analyse infinitésimale des accroissements par exemple, qui introduit une telle intégrale est donc abusive : les évaluations associées ne peuvent être significatives que dans les cas simples où les termes correspondants sont en fait négligeables !

(d) Dans le cas simple (et qu'on peut étudier directement) où $p = 1$, les conditions (1.5) et (1.6) sont identiques, elles sont alors nécessaires et suffisantes.

Dans le cas d'Ornstein-Uhlenbeck au sens strict, on a déjà dit que la condition nécessaire (1.6) est aussi suffisante ; ce n'est pas la situation générale : supposer $X_n = 0$ si $n \geq 1$ et $d_0(t) = \{|\log t|^{1/2} \log\log 1/t\}^{-1} \wedge 1$; on a alors pour $p \geq 2$ et pour t assez grand, $t^{p-1}H^p(t) = t^{-1}[\log t]^{-p}$; la condition nécessaire (1.6) est donc vérifiée indépendamment de $p \geq 2$; par contre X n'a de modification à trajectoires continues dans aucun des l_p puisque $H(t) = d_0[\exp(-t^2)]$ est de l'ordre de $1/[t \log t]$ et n'est donc pas intégrable à l'infini.

Enfin même dans le cas d'Ornstein-Uhlenbeck au sens strict, la condition suffisante (1.5) n'est pas nécessaire : avec les notations de l'énoncé, prendre $a_n = 1/(n \log n), b_n = \exp n^2, n \geq 2$; alors pour $p \geq 2$, $H(t)$ est encore de l'ordre de $1/[t \log t]$ non intégrable à l'infini alors que X a une modification à trajectoires continues dans l_p.

1.3 Une condition nécessaire et suffisante

1.3.1 Un calcul de variations, $p > 1$

Dans ce calcul, on utilise les notations précédentes et on suppose qu'il existe un rang n_0 tel que $H_{n_0}(0)$ soit non nul. On fixe un entier $N \geq n_0$ et on pose

$$B_{N,q} = \left\{ y \in \mathbf{R}_+^N : \sum_{n=1}^{N} |y_n|^q \leq 1 \right\};$$

on définit les fonctions

$$K(y,t) = \left[\sum_{n=1}^{N} |y_n\, H_n|^2 \right]^{1/2} , y \in B_{N,q}, t \in \mathbf{R},$$

$$\Phi(y) = \int_0^\infty \left[\sum_{n=1}^{N} |y_n\, H_n|^2 \right]^{1/2} dt\, , y \in B_{N,q};$$

alors Φ atteint son maximum M non nul en au moins un point η du bord de $B_{N,q}$. On pose $K(t) = K(\eta, t)$, on a donc $M = \int_0^\infty K(t)dt$. On décompose $[1, N]$ suivant les valeurs des η_n :

$$N_0 = \{n : \eta_n = 0\}, \quad N_+ = \{n : \eta_n > 0\};$$

puisque M est non nul, N_+ n'est pas vide ; pour tout $j \in N_+$, on peut choisir comme variables principales sur le bord de $B_{N,q}$ les η_n, n différent de j, et Φ est alors continûment dérivable au voisinage de η relativement aux y_n, n différent de j et appartenant à N_+ ; pour ces valeurs de n, on obtient

$$|y_n|^{1-q} \frac{\partial \Phi}{\partial y_n}(y) = \int_0^\infty \left[|y_n|^{2-q} H_n(t)^2 - |y_j|^{2-q} H_j(t)^2 \right] \frac{dt}{K(y,t)}$$

$$\frac{\partial \Phi}{\partial y_n}(\eta) = 0\,;$$

cette formule indique alors que

$$\forall n \in N_+, \quad \int_0^\infty \left[|\eta_n|^{2-q} H_n(t)^2 \right] \frac{dt}{K(t)} = C, \text{ où } C \text{ est indépendant de } n\,;$$

en multipliant par $|\eta_n|^q$ et en sommant sur $n \in N_+$, on constate que $C = M$.

Le traitement des éléments de N_0 nécessite un traitement plus spécifique. Nous nous limiterons au cas $p \leq 2$. Fixons $n \in N_0$ et $j \in N_+$; pour tout $y \in B_2$, on aura

$$0 \leq \int_0^\infty \frac{K^2(\eta, t) - K^2(y,t)}{K(\eta, t) + K(y,t)}\, dt\,; \tag{1.9}$$

pour tout $u > 0$ assez petit, nous définissons $y = y(u)$ en posant

$$y_k = \eta_k \text{ si } k \notin \{n, j\}, \quad y_k = u, \quad y_j = \left[\eta_j^q - u^q\right]^{1/q} ;$$

la relation (1.9) fournit alors

$$\int_0^\infty \frac{H_n^2(t)\, dt}{K(\eta, t) + K(y, t)} \leq \left[\frac{\eta_j^2 - \left[\eta_j^q - u^q\right]^{2/q}}{u^2}\right] \int_0^\infty \frac{H_j^2(t)\, dt}{K(\eta, t) + K(y, t)} ;$$

on remarque alors que la fonction à intégrer au dernier terme est majorée indépendamment de y par $\dfrac{H_j^2}{K(\eta)}$ qui est intégrable et que la fonction à intégrer au premier membre tend vers $\dfrac{H_n^2}{2K(\eta)}$ quand u tend vers zéro. Le théorème de convergence dominée au second membre et le lemme de Fatou au premier fournissent donc en tenant compte des valeurs de p :

$$\text{si } p = 2, \quad \int_0^\infty \frac{H_n^2(t)\, dt}{K(\eta, t)} \leq \int_0^\infty \frac{H_j^2(t)\, dt}{K(\eta, t)} = M,$$

$$\text{si } p < 2 \text{ et donc } q > 2, \quad \int_0^\infty \frac{H_n^2(t)\, dt}{K(\eta, t)} = 0$$

dans ce deuxième cas, N_0 est donc inclus dans $\{n : H_n(0) = 0\}$. Pour résumer ce calcul :

(1.10) Si $p = 2$, alors pour tout $n \in [1, N]$, on a $\displaystyle\int_0^\infty H_n^2(t) \frac{dt}{K(t)} \leq M.$

Si $p < 2$, pour tout $n \in [1, N]$, on a :

$$0 \leq |\eta_n|^{q-2} = \left[\int_0^\infty K(t) dt\right]^{-1} \int_0^\infty H_n^2(t) \frac{dt}{K(t)}, \text{ et donc}$$

(1.11)

$$\left\{\sum_{n=1}^N \left[\int_0^\infty H_n^2(t) \frac{dt}{K(t)}\right]^{r/2}\right\}^{2/r} = \int_0^\infty K(t) dt = M < \infty, \; r = \frac{2p}{2-p}.$$

Si $p > 2$ et au vu du changement de signe des exposants, le calcul ne semble pas utilisable.

1.3.2 Une caractérisation, démonstration du théorème 1

Le calcul de variations ci-dessus permet de démontrer le théorème 1 énoncé dans l'introduction.

Démonstration : Nous démontrons d'abord la nécessité en supposant $p > 1$. Nous supposons la condition (0.1) réalisée et donc que la fonction

$$\Phi(y) = \int_0^\infty \left[\sum_n |y_n H_n|^2\right]^{1/2} dt, y \in B_q$$

est bornée par un nombre M ; pour tout entier N, nous notons Φ_N, M_N, K_N, les éléments introduits ci-dessus par le calcul de variations sur $B_{N,q}$; la suite $\{M_N, N \in \mathbf{N}\}$ est donc majorée par M et la suite $\{K_N, N \in \mathbf{N}\}$ est une suite de fonctions positives et décroissantes sur $\mathbf{R}_+$ telles que

$$\forall N \in \mathbf{N}, \quad K_N(0) \leq \sup_{y \in B_q} \sum_n y_n H_n(0) \leq 2[\sum_n |a_n|^p]^{1/p} ;$$

on peut donc en extraire une suite partielle convergeant vers une fonction Δ décroissante sauf peut-être sur un ensemble dénombrable. Le lemme de Fatou montre alors que Δ a les propriétés indiquées pour tout $p \in]1, 2]$. Si $p = 1$ et si la condition (0.1) est réalisée, la proposition 1 montre que la propriété (1.6) l'est aussi ; pour obtenir (0.5), il suffit alors de poser

$$\forall t \in \mathbf{R}_+, \; \Delta(t) = d[\exp(-t^2)] = \left[\sum_n |H_n(t)|^2\right]^{1/2} ;$$

la nécessité pour $p = 1$ en résulte.

Nous démontrons maintenant la suffisance, elle résulte des inégalités de Cauchy-Schwarz ; si les conditions du théorème sont réalisées, on a en effet pour tout $y \in B_q$,

$$\int_0^\infty \left[\sum_n |y_n H_n|^2\right]^{1/2} dt \leq \int_0^\infty \Delta(t)dt \cdot \sup_{y \in B_q} \sum_n |y_n|^2 \int_0^\infty H_n(t)^2 \frac{dt}{\Delta(t)} ;$$

le dernier terme s'évalue par (1.4) suivant que p est égal ou inférieur à 2 ; les hypothèses indiquent qu'il est fini, le théorème est démontré.

1.3.3 Remarque

La preuve de la suffisance donnée ci-dessus montre de manière étonnante que les majorations les plus simples peuvent quelquefois être remarquablement significatives.

2 Des situations plus générales

2.1 Les notations

On fixe $p \geq 1$; E, q et B_q ont les mêmes significations qu'au paragraphe précédent ; T est un espace métrique compact. On note $X = \{X_n(t), t \in$

$T, n \in \mathbf{N}\}$ une suite de fonctions aléatoires gaussiennes indépendantes sur T ; on fixe un élément t_0 de T et on détermine la loi de X à partir des paramètres $(a_n, d_n, n \in \mathbf{N})$ définis par

$$(2.1) \quad a_n^2 = \mathbf{E}|X_n(t_0)|^2, d_n^2(s,t) = \mathbf{E}|X_n(s) - X_n(t)|^2, n \in \mathbf{N}, (s,t) \subset T$$

Pour tout $y \in B_q$, on pose $d_y^2(s,t) = \sum_n y_n^2\, d_n^2(s,t)$.
On suppose la série $\sum_n |a_n|^p$ convergente de sorte que $X(t_0)$ a une modification à valeurs dans ℓ_p.

L'adaptation des techniques de la première partie de l'étude imposera l'hypothèse supplémentaire suivante :

Il existe un nombre $C < \infty$ tel que

$$(2.2) \qquad \forall n \in \mathbf{N}, \forall s \in T, \mathbf{E}|X_n(s)|^2 \le C^2 \mathbf{E}|X_n(t_0)|^2.$$

Une telle condition, vérifiée dans les cas stationnaires, permettra en effet d'analyser la situation en termes d'une seule mesure majorante sur T, même sans stationnarité.

2.2 Mesures majorantes, rappels et précisions

Les évaluations habituelles des fonctions aléatoires gaussiennes en termes de mesures majorantes privilégient la fonction $\sqrt{\log 1/x}, x \in]0,1]$. On préfère ici utiliser plutôt la fonction ψ_0 définie par

$$(2.3) \qquad \psi_0(x) = \sqrt{\log \frac{h}{x}} - \sqrt{\log h},\ x \in]0,1],\ h = \exp 1/2,$$

de sorte que ψ_0 soit positive, décroissante, convexe sur $]0,1]$, avec $\psi_0(1) = 0$. On utilisera aussi

$$(2.4) \qquad \begin{aligned} \Psi = &\ \{\psi = \theta \circ \psi_0 : \theta \text{ croissante et convexe (au sens large)}, \\ &\qquad\qquad\qquad \theta(0) = 0, \frac{\theta(x)}{x} \ge 1\} \\ \Psi' = &\ \{\psi = \theta \circ \psi_0 \in \Psi : \lim_{x \to \infty} \frac{\theta(x)}{x} = \infty\} \end{aligned}$$

On constate que pour tout $t \in]0,1]$,

$$\psi_0(t) \le \sqrt{\log 1/t} \le \psi_0(t) + 2^{-1/2}$$

et il en résulte que

$$(2.5) \qquad \forall (s,t) \subset]0,1], \quad \psi_0(s \times t) \le \psi_0(s) + \psi_0(t) + 2^{1/2}$$

Les énoncés usuels s'écrivent alors :

Rappel 1. *(a) Pour qu'une fonction aléatoire gaussienne F sur un espace métrique séparable S ait une modification bornée, il faut et il suffit qu'il existe une probabilité π sur S telle que*

$$\sup_{s\in S}\int \psi_0\circ\pi\{r\in S : d_F(r,s)\le u\}\, du<\infty.$$

(b) Pour que F ait une modification à trajectoires continues, il faut et il suffit
(b.1) qu'il existe une probabilité π sur S et un élément $\psi\in\Psi'$ tels que

$$\sup_{s\in S}\int \psi\circ\pi\{r\in S : d_F(r,s)\le u\}\, du<\infty.$$

Il faut et il suffit aussi
(b.2) qu'il existe une probabilité π sur S telle que les intégrales

$$\int \psi_0\circ\pi\{r\in S : d_F(r,s)\le u\}\, du, s\in S$$

soit majorées et uniformément convergentes (à l'origine).

Remarque. On sait qu'on peut écrire les conditions ci-dessus en termes des réarrangements croissants $\widetilde{d_F(.,s)}$ définis sur $([0,1],d\lambda)$ des différentes distances $d_F(.,s)$ définies sur (T,π) ; nous utiliserons aussi bien ces écritures.

C'est l'énoncé fondamental suivant qui permettra l'étude de situations non stationnaires :

Théorème 3. *On suppose que la série $\sum_n |a_n|^p$ est convergente et l'hypothèse (2.2) est vérifiée ; alors les deux propriétés suivantes sont équivalentes :*
(a) X a une modification à trajectoires continues dans ℓ_p.

(b) Il existe une probabilité π sur T telle que les intégrales

$$\int \psi_0\circ\pi\{s\in T : d_y(s,t)\le u\}du,\ y\in B_q, t\in T$$

soit majorées et uniformément convergentes .

Démonstration : (a) La preuve de la nécessité est basée sur le théorème 7.2.1 de [6] et l'introduction d'une fonction aléatoire gaussienne auxiliaire V définie sur T par

$$\forall n\in\mathbf{N}, V_n(t)=X_n(t)+\lambda_n\left[2C^2\mathbf{E}|X_n(t_1)|^2-\mathbf{E}|X_n(t)|^2\right]^{1/2},$$

où $\Lambda=(\lambda_n,n\in\mathbf{N})$ est une suite gaussienne normale indépendante de X et où le terme complémentaire est choisi pour que les $\mathbf{E}V_n(s)V_n(t)$ soient positifs ou nuls, que les $\mathbf{E}|V_n(t)|^2$ ne dépendent pas de t et que X ait une modification à trajectoires continues dans ℓ_p si et seulement si V a la même propriété ; remarquons que dans ces conditions, pour tout $(y, z) \subset B_q$, tout $(s, t) \subset T$ et tout $n \in \mathbf{N}$ on a :

$$\mathbf{E}\left[y_n V_n(t) - z_n V_n(s)\right]^2 \geq y_n^2 \left[\mathbf{E}|V_n(t)|^2 - \frac{(\mathbf{E}V_n(t)V_n(s))^2}{\mathbf{E}|V_n(s)|^2}\right]$$
$$\geq \frac{y_n^2}{2}\,\mathbf{E}[V_n(t) - V_n(s)]^2,$$

et par suite :

$$\mathbf{E}\left[\langle V(t), y\rangle - \langle V(s), z\rangle\right]^2 \geq d_y(s,t)^2/2. \tag{2.6}$$

Supposons alors que X ait une modification à trajectoires continues dans ℓ_p ; $V = \{\langle X(t), y\rangle, t \in T, y \in B_q\}$ a donc une modification continue ; nous notons π_0 la probabilité sur $T \times B_q$ associée à V au rappel 1(b.2) de sorte que les intégrales

$$\int \psi_0 \circ \pi_0\{\mathbf{E}|\langle V(t), y\rangle - \langle V(s), z\rangle|^2 \leq u^2\}\, du, t \in T, y \in B_q$$

soient majorées et uniformément convergentes. Soit π l'image de π_0 par la projection de $T \times B_q$ sur T ; pour tout $(t, y) \subset T \times B_q$ et tout $u \geq 0$, la propriété (2.6) implique

$$\pi_0\{\mathbf{E}|\langle V(t), y\rangle - \langle V(s), z\rangle|^2 \leq u^2\} \leq \pi\{s : d_y(s,t) \leq u\sqrt{2}\}$$

de sorte que la définition de π_0 implique que les intégrales

$$\int \psi_0 \circ \pi\{s : d_y(s,t) \leq u\}\, du, t \in T, y \in B_q$$

sont elles aussi majorées et uniformément convergentes : la propriété (b) est donc vérifiée.

(b) Supposons maintenant la propriété (b) vérifiée ; le rappel 1(b.2) associe aussi à la fonction aléatoire gaussienne $X(t_0)$ à trajectoires continues sur B_q une probabilité ρ sur B_q. Soit alors $\pi_0 = \pi \times \rho$, c'est une probabilité sur $T \times B_q$ et on vérifie facilement que c'est au sens du rappel 1(b.2) une mesure majorante pour $\{\langle X(t), y\rangle, t \in T, y \in B_q\}$ qui a donc une modification à trajectoires continues, c'est-à-dire que X a une modification à trajectoires continues dans ℓ_p, le théorème est établi.

2.3 Une condition nécessaire, une condition suffisante

On introduit la fonction d définie sur $T \times T$ par

$$d(s,t) = \sup_{y \in B_q} d_y(s,t). \tag{2.7}$$

On rappelle que d s'évalue par les formules (1.4).

En utilisant le théorème 3 et partiellement la preuve de la proposition 1, on démontre :

Proposition 2. *On suppose que la série $\sum_n |a_n|^p$ est convergente et que l'hypothèse (2.2) est vérifiée ; pour que X ait une modification à trajectoires continues dans $E = \ell_p$,*
(a) il suffit qu'il existe une probabilité π sur T telle que les intégrales

$$\int \psi_0 \circ \pi\{s; d(s,t) \leq u\} du, \; t \in T \tag{2.8}$$

soient majorées et uniformément convergentes ;

(b) il faut qu'il existe une probabilité π sur T telle que les intégrales

$$\int \sup_{n \in \mathbf{N}} [\psi_0 \circ \pi\{s : d_n(s,t)^p \leq u\}]^p \, du, t \in T, \tag{2.9}$$

soient bornées et uniformément convergentes.

Démonstration : L'affirmation (a) est immédiate (et bien connue dans le cas stationnaire). Pour prouver (b), on exprime les différentes intégrales intervenant dans les énoncés du théorème 3 et de la proposition pour t fixé dans T en termes des réarrangements croissants $\widetilde{d(t,.)}, \widetilde{d_n(t,.)}, \widetilde{d_y(t,.)}$ définis sur $(]0,1[, d\lambda)$ des différentes distances $d(t,.), d_n(t,.), d_y(t,.)$ définies sur (T, π). Pour tous $y \in B_q, n \in \mathbf{N}, t \in T$, le réarrangement de $d_y(t,.)$ est supérieur à celui de $y_n d_n(t,.)$; cet argument permet d'utiliser le premier schéma de preuve de la proposition 1 :

Pour simplifier, on posera :

$$H_n(t) = \widetilde{d_n}[\exp(-t^2)], \quad H(t) = \widetilde{d}[\exp(-t^2)], \quad n \in \mathbf{N}, \, t \in \mathbf{R}.$$

Supposons que X ait une modification à trajectoires continues dans $E = \ell_p$, alors le théorème 3 et l'argument ci-dessus montrent qu'en particulier,

$$M' = \sup_{y \in B_q} \int_0^\infty \left[\sup_{n \in \mathbf{N}} y_n H_n(t)\right] dt$$

est fini ; c'est avec la décroissance des H_n la seule hypothèse qu'on utilisera ici. Elle suffit comme dans la proposition 1 à montrer que

$$[4M']^p \geq (1/p) \int_0^\infty t^{p-1} \left[\sup_{n \in \mathbf{N}} H_n(t)\right]^p dt.$$

C'est bien la condition nécessaire annoncée.

Remarque sur l'énoncé précédent. Si les différents X_n sont stationnaires sur $\mathbf{R}$, on peut limiter l'étude à $T = [0,1]$ et choisir comme mesure majorante

π la mesure de Lebesgue λ ; avec ce choix, la condition nécessaire s'écrit plus explicitement :

Il faut que

$$\int \sup_{n \in \mathbf{N}} [\psi_0 \circ \lambda\{s : d_n(s)^p \leq u\}]^p \, du < \infty,$$

c'est-à-dire avec les notations usuelles,

$$\int_0^\infty \sup_{n \in \mathbf{N}} \left[\log \lambda^{-1}\{s : d_n(s) \leq u^{1/p}\}\right]^{p/2} du < \infty.$$

Si p est supérieur ou égal à 2, la proposition 2 étend alors exactement la proposition 1 en évitant l'hypothèse de croissance des distances.

2.4 Les caractérisations, $p = 2$

On se propose d'étendre à la situation plus générale la caractérisation établie au théorème 1 et en particulier de prouver le théorème 2 énoncé dans l'introduction.

Pour prouver le théorème 1, on utilisait essentiellement l'étude des variations d'intégrales du type $\int_0^\infty d(\exp[-x^2])\, dx$ en fonction de d. Dans la situation plus générale étudiée, les fonctionnelles dépendent des réarrangements et sont du type $\int_0^\infty \tilde{d}(\exp[-x^2])\, dx$; leurs variations en fonction de d sont apparemment moins régulières.

Il sera nécessaire pour analyser ces variations, de préciser les propriétés des mesures majorantes suivant des schémas esquissés précédemment dans [1].

2.4.1 D'autres propriétés des mesures majorantes

Comme précédemment, T est un espace métrique séparable et π est une probabilité sur T ; λ est la mesure de Lebesgue sur $]0,1]$.

La fonction ψ_0 et les ensembles Ψ et ψ' ont été définis en (2.3)-(2.4). On fixe un élément $\psi \in \Psi$; ce sera par exemple ψ_0. On note $\mu = \mu_\psi$ la mesure positive sur $]0,1]$ qui est la dérivée seconde de ψ de sorte que

$$\forall x \in]0,1],\ \psi(x) = \int (y-x)^+ \mu(dy); \tag{2.10}$$

on note $\phi = \phi_\mu$ la fonction définie par

$$\forall x \in]0,1], \phi(x) = \int_{[x,1]} \mu(dy) = \lim_{y \downarrow x} \left| \frac{d\psi}{dx}(y) \right| ; \tag{2.11}$$

on remarquera que la loi de ϕ sur $(]0,1], \lambda)$ est nécessairement diffuse.

Si $\psi = \psi_0$, on posera $\mu_{\psi_0} = \mu_0, \phi_{\mu_0} = \phi_0$; on aura donc sur $]0,1]$

$$(2.12) \qquad \mu_0(dx) = \frac{\log 1/x}{2x^2[\log h/x]^{3/2}}\, d\lambda(x), \qquad \phi_0(x) = \frac{1}{2x\sqrt{\log h/x}}\,.$$

On note $A = A(\pi, \mu)$ l'ensemble des fonctions K sur T qui sont de la forme

$$(2.13) \qquad K(t) = \int k(y,t)\mu(dy) \qquad \pi - p.s.$$

où $k = \{k(x,t), x \in]0,1], t \in T\}$ est mesurable sur $]0,1] \times T$ et telle que

$$(2.14) \qquad 0 \leq k \leq 1\,; \qquad \forall x \in]0,1], \qquad \int k(x,t)\,\pi(dt) = x.$$

On note enfin $B = B(\pi, \mu)$ l'ensemble des fonctions H sur (T, π) qui ont même loi que ϕ sur $(]0,1], \lambda)$.

Proposition 3. *Soient d une fonction mesurable positive sur (T,π) et $\tilde{d}$ son réarrangement croissant sur $(]0,1], \lambda)$; alors la fonction K_d définie sur T par*

$$(2.15) \quad K_d(t) = \mu\{y : \tilde{d}(y) > d(t)\} + \int_{\{\tilde{d}(y)=d(t)\}} \frac{y - \pi\{d < d(t)\}}{\pi\{d = d(t)\}}\ \mu(dy)$$

est un élément de A ; on a de plus :

$$(2.16) \quad \int \psi \circ \pi\{d \leq u\}\, du = \int d\, K_d\, d\pi = \inf\{\int d(t)K(t)\pi(dt), K \in A\}\,.$$

Démonstration : (a) On montre que K_d est un élément de A en constatant que la fonction k définie par

$$k(y,t) = \mathbf{I}_{\tilde{d}>d(t)} + \frac{y - \pi\{d < d(t)\}}{\pi\{d = d(t)\}}\mathbf{I}_{\tilde{d}=d(t)}$$

vérifie les propriétés (2.14) ; on remarquera pour cela que

$$\tilde{d}(y) = d(t) \quad \Longrightarrow \quad \pi\{d < d(t)\} \leq y \leq \pi\{d \leq d(t)\}.$$

(b) La propriété (2.16) résulte du lemme 1.4.1 de [1].

Proposition 4. *L'ensemble B est contenu dans A ; pour que la fonction K_d définie en (2.15) soit un élément de B, il faut et il suffit que la loi de d sur (T, π) soit diffuse.*

Démonstration : (a) On y utilisera la propriété suivante de la fonction ϕ sur $(]0,1], \lambda)$

$$(2.17) \qquad \lambda\{x : \phi(x) = \mu\{\phi \leq \phi(x)\} = 1$$

Soit H un élément de B ; la définition de B et l'équation (2.17) impliquent alors

$$(2.18) \qquad \pi\{t : H(t) = \mu\{\phi \leq H(t)\}\} = 1$$

on associe donc à H la fonction k définie sur $]0,1] \times T$ par

$$k(y,t) = \mathbf{I}_{\phi(y) \leq H(t)},$$

elle a les propriétés (2.14) de sorte que (2.18) montre que H appartient à A.

(b) Si la loi de d sur $(T, \pi\}$ n'est pas diffuse, celle de K_d n'est pas non plus diffuse et elle ne peut pas être égale à celle de ϕ ; les lois des éléments de Ψ sont en effet toutes diffuses.

(c) Si la loi de d est diffuse, le théorème de Fubini montre que

$$\lambda\{x : \mu\{\tilde{d} = \tilde{d}(x)\} > 0\} = \pi\{t : \mu\{\tilde{d} = d(t)\} > 0\} = 0$$

les inclusions

$$\{y : \tilde{d}(y) > \tilde{d}(x)\} \subset]x, 1] \subset [x, 1] \subset \{y : \tilde{d}(y) \geq \tilde{d}(x)\}$$

impliquent alors que

$$\lambda\left\{x : \mu\{\tilde{d} > \tilde{d}(x)\} = \mu\{\tilde{d} \geq \tilde{d}(x)\} = \phi(x)\right\} = 1$$

et ceci signifie exactement que ϕ est λ-p.s. le réarrangement décroisssant de K_d qui a donc même loi que ϕ.

Proposition 5. *(a) Soit D un ensemble de fonctions mesurables positives sur T ; on suppose que D est borné et que les intégrales $\int \psi \circ \pi\{d \leq u\} du, d \in D$ sont uniformément convergentes (à l'origine) ; dans ces conditions, l'ensemble des fonctions $d\, K_d, d \in D$ est équi-intégrable relativement à π.*

(b) Soient D un ensemble de fonctions mesurables positives sur (T, π) et $H_d, d \in D$ un sous-ensemble de B ; on suppose que l'ensemble des fonctions $d\, H_d, d \in D$ est équi-intégrable relativement à π ; dans ces conditions, les intégrales $\int \psi \circ \pi\{d \leq u\}\, du, d \in D$ sont majorées et uniformément convergentes (à l'origine).

(c) Soit $\{d_n, n \in \mathbf{N}\}$ une suite de fonctions mesurables positives sur T convergeant π-p.s. vers d ; on suppose que la loi de d est diffuse. Dans ces conditions, la suite $\{K_{d_n}, n \in \mathbf{N}\}$ des fonctions associées dans la proposition 2 converge π-p.s. vers K_d.

Démonstration : Démontrons (a) : On fixe $\epsilon > 0$ et on détermine η tel que

$$\forall d \in D, \quad \int_0^\eta \psi \circ \pi\{d \geq u\}\, du \leq \epsilon/2.$$

En particulier la décroissance de ϕ implique alors

$$\forall d \in D,\ \pi\{d \leq \eta\} = \lambda\{\tilde{d} \leq \eta\} \geq \overset{-1}{\psi}(\epsilon/2\eta),$$

c'est-à dire que

$$\forall d \in D, \quad \tilde{d}(t) > \eta \Longrightarrow t > \overset{-1}{\psi}(\epsilon/2\eta).$$

Dans ces conditions et en posant $M = \sup\{d(t), d \in D, t \in T\}$, on a pour toute partie mesurable $A \subset T$,

$$\begin{aligned}\forall d \in D, \int_A d(t)K_d(t)\pi(dt) &\leq \int_{d(t)\leq\eta} d(t)K_d(t)\pi(dt)\\ &\quad + \pi(A) \sup_{\tilde{d}(t)>\eta} \mathrm{ess}\{\tilde{d}(t)\mu\{\tilde{d} \geq \tilde{d}(t)\}\}\\ &\leq \epsilon/2 + \pi(A)\, M\, \phi \circ \overset{-1}{\psi}(\epsilon/2\eta).\end{aligned}$$

L'équi-intégrabilité s'ensuit.

Démontrons (b) : On fixe $\epsilon > 0$, on détermine successivement $\rho > 0$ et $M < \infty$ tels que

$$\begin{aligned}\pi(A) \leq \rho \quad &\Longrightarrow \quad \forall d \in D, \int_A d\, H_d\, d\pi \leq \epsilon/2,\\ \forall d \in D, \quad &\pi\{H_d > M\} = \lambda\{\phi > M\} = \rho\,;\end{aligned}$$

finalement on pose $\eta = \epsilon/[2M]$.

Dans ces conditions et en utilisant (2.16), on constate que pour tout $d \in D$,

$$\begin{aligned}\int_0^\eta \psi \circ \pi\{d \leq u\} du &\leq \int [d \wedge \eta]\, H_d\, d\pi\\ &\leq \int_{H_d>M} d\, H_d\, d\pi + \eta \int_{H_d \leq M} H_d\, d\pi \leq \epsilon/2 + \eta M \leq \epsilon\,;\end{aligned}$$

l'uniforme convergence des intégrales est donc établie.

La propriété (c) résulte immédiatement des propriétés de la convergence en loi.

2.4.2 Un calcul de variations, $p = 2$.

Les données du calcul

Les notations sont celles de 2.1 ; on note de plus $\{d_n, n \in \mathbf{N}\}$ une suite bornée convergeant uniformément vers 0 de fonctions continues positives sur T, non toutes nulles ; on note D l'ensemble de fonctions sur T défini par

$$D = \{d_y = \sqrt{\sum y_n^2\, d_n^2}\,; y = (y_n, n \in \mathbf{N}) \in B_2\}.$$

Pour tout $y \in B_2$, on pose

$$F(y) = \int \psi_0 \circ \pi\{d_y \leq u\}\, du$$

et on se propose d'analyser la situation où ces intégrales sont majorées et uniformément convergentes. On utilisera pour cela les résultats du paragraphe 2.4.1, ils ont montré en particulier que les variations de F étaient plus irrégulières si les lois des v.a. $d_y, y \in B_2$, n'étaient pas diffuses ; cette remarque justifie une régularisation préalable des données pour diffuser les lois.

La régularisation des données

On introduit l'espace produit $T' = T \times [0,1]$ qu'on munit de la mesure produit $\pi' = \pi \times \lambda$ et une suite strictement positive $(c_n, n \in \mathbf{N})$ tendant vers 0 ; pour tout $t' = (t, \tau) \in T'$ et tout $y \in B_2$, on pose

$$\forall n \in \mathbf{N}, d'_n(t') = \sqrt{d_n^2(t) + c_n^2 \tau^2},$$

$$c(y) = [\sum_n y_n^2 c_n^2]^{1/2}, \quad d'_y(t') = \sqrt{\sum_n y_n^2\, d'_n(t')^2}.$$

La suite $(d'_n, n \in \mathbf{N})$ est donc encore une suite bornée convergeant uniformément vers 0 de fonctions continues positives sur T'. Les inclusions

$$\{d_y \leq u/2,\ c(y)\tau \leq u/2\} \subset \{d'_y \leq u\} \subset \{d_y \leq u\}$$

et la propriété (2.5) montrent que

$$\begin{aligned} \psi_0 \circ \pi'\{d'_y \leq u\} &\leq \psi_0\{d_y \leq u/2,\ c(y)\tau \leq u/2\} \\ &\leq \psi_0\{d_y \leq u/2\} + \psi_0\{c(y)\tau \leq u/2\} + 2^{1/2}, \end{aligned}$$

il en résulte que les intégrales

$$F'(y) = \int \psi \circ \pi'\{d'_y \leq u\}\, du,\ y \in B_2,$$

resteront majorées et uniformément convergentes.

Enfin les lois des $d'_y, y \in B_2 \setminus \{0\}$, seront toutes diffuses ; en effet pour tout $y \in B_2 \setminus \{0\}$ et tout $m \in \mathbf{R}$,

$$\pi'\{d'_y = m\} = \int \pi\{d_y^2 = m^2 - c^2(y)\tau^2\} d\lambda(\tau)$$

et l'ensemble $\{\tau : \pi\{d_y^2 = m^2 - c^2(y)\tau^2\} > 0\}$ étant au plus dénombrable, sa mesure de Lebesgue est nulle.

Le calcul

Sur B_2 muni de la topologie faible, l'application : $y \longrightarrow d'_y$ à valeurs dans $\mathbb{C}[0,1]$ muni de la topologie uniforme est continue et l'ensemble des fonctions : $u \longrightarrow \psi_0 \circ \pi'\{d'_y \leq u\}, y \in B_2$, est équi-intégrable ; il en résulte que F' est continue sur le compact B_2 et atteint son maximum M (non nul) en au moins un point η. Pour tout $y \in B_2$, on note K_y l'élément de A associé à d'_y en (2.15) ; on a donc en particulier

$$M = \int d'_\eta(t')\, K_\eta(t')\, \pi'(dt') > 0.$$

On décompose $\mathbf{N}$ suivant les valeurs de η_n :

$$N_0 = \{n : \eta_n = 0\}, \qquad N_+ = \{n : \eta_n > 0\};$$

puisque M est non nul, N_+ n'est pas vide.

La fonction F' n'est sans doute pas dérivable au voisinage de η ; on a pourtant

$$F'(\eta) \leq \int d'_\eta(t')\, K_y(t')\, \pi'(dt')\,;$$

joint à la définition de η, ceci implique

$$0 \leq F'(\eta) - F'(y) \leq \int [d'_\eta(t') - d'_y('t)]\, K_y(t')\, \pi'(dt'). \tag{2.19}$$

Pour que ce calcul soit efficace, il faut que $\int d'_\eta(t')\, K_y(t')\pi'(dt')$ soit fini ; c'est en particulier le cas si deux au plus des composantes de y sont différentes de celles de η, ces composantes y_m et y_n différentes étant non nulles ; on a alors en effet

$$d'_\eta(t') \leq d'_y(t')[1 + \eta_n/y_n + \eta_m/y_m].$$

Supposons que N_+ ait au moins deux éléments différents m et n ; η_n est alors différent de 1 et η_m est non nul. Pour tout u positif assez petit, nous pouvons donc construire un élément $y(u)$ de B_2 en posant :

$$\forall k \notin \{n, m\}, y_k = \eta_k \;;\; y_n = \sqrt{\eta_n^2 + u^2}, y_m = \sqrt{\eta_m^2 - u^2} \quad ;$$

pour ce choix de y, la relation (2.19) fournit :

$$0 \leq \int \frac{d'_m(t')^2 - d'_n(t')^2}{d'_y(t') + d'_\eta(t')}\, K_y(t')\, \pi'(dt').$$

Quand u tend vers 0 et puisque la loi de d'_η est diffuse, d'_y et K_y convergent respectivement vers d'_η et K_η pour la convergence π'-p.s. ; par ailleurs,

$d'_m(t)^2$ est majoré par $[d'_y/y_m]^2$ et les $d'_y K_y$ sont équi-intégrables. Le passage à la limite fournit donc

$$0 \le \int \frac{d'_m(t')^2 - d'_n(t')^2}{d'_\eta(t')} K_\eta(t')\, \pi'(dt') \,;$$

en permutant m et n, on constate que l'intégrale ci-dessus est nulle et ceci signifie que $\int \frac{d'_m(t')^2}{d'_\eta(t')} K_\eta(t')\, \pi'(dt'), m \in N_+$ est constant et donc égal à M.

Fixons maintenant un élément m de N_+ et un élément n de N_0 ; pour tout u strictement positif assez petit, nous définissons un élément $y(u)$ de B_2 en posant :

$$\forall k \notin \{n, m\}, y_k = \eta_k \,;\, y_n = u, y_m = \sqrt{\eta_m^2 - u^2} \quad ;$$

pour ce choix de y, la même relation (2.19) fournit

$$\int \frac{d'_n(t')^2}{d'_y(t') + d'_\eta(t')} K_y(t')\, \pi'(dt') \le \int \frac{d_m(t)^2 + \tau^2}{d'_y(t') + d'_\eta(t')} K_y(t')\, \pi'(dt').$$

Quand u tend vers 0, le second membre se comporte bien puisque la fonction à intégrer est majorée par $d'_y(t')K_y(t')/(\eta_m y_m)$ équi-intégrable ; il tend donc vers $M/2$; au premier membre, les conditions d'équi-intégrabilité ne sont pas vérifiées, mais le lemme de Fatou s'applique et fournit :

$$\int \frac{d'_n(t')^2}{2d'_\eta(t')} K_\eta(t')\, \pi'(dt') \le M/2.$$

En regroupant, on constate donc que sous les hypothèses du calcul, on a

$$\text{(2.20)} \qquad \forall n \in \mathbf{N}, \qquad \int \frac{d_n^2(t)}{d'_\eta(t')} K_\eta(t')\, \pi'(dt') \le M.$$

Rappelons que la loi de d'_η étant diffuse, K_η est un élément de $B(\pi', \mu)$. Les conclusions de ce calcul nous fourniront dans les paragraphes suivants les caractérisations annoncées étendant les résultats du théorème 1.

2.4.3 La démonstration du théorème 2

On démontre en fait un énoncé un peu plus précis que le théorème 2 :

Théorème 4. *On note $X = \{X_n(t), t \in T, n \in \mathbf{N}\}$ une suite de fonctions aléatoires gaussiennes indépendantes sur $T = \mathbf{R}^m$; on suppose que les composantes de X sont stationnaires et que la série $\sum_n |a_n|^2$ est convergente. Dans ces conditions, les trois propriétés suivantes sont équivalentes :*
(a) X a une modification à trajectoires continues dans ℓ_2.

(b) Il existe sur $T' =]0,1]^m \times]0,1]$ *muni de la mesure de Lebesgue* $\lambda' = \lambda_m \times \lambda$ *deux fonctions mesurables et positives* Δ' *et* K' *telles que*

(b.1) $\sup_{n\in\mathbf{N}} \int \frac{d_n^2(t)}{\Delta'(t,\tau)} K'(t,\tau) d\lambda_m(t)\, d\lambda(\tau)$ *soit fini,*

(b.2) $\int \Delta'(t,\tau)\, K'(t,\tau)\, d\lambda_m(t)\, d\lambda(\tau)$ *soit fini,*

(b.3) la loi de K' *sur* (T', λ') *soit égale à celle de* $\frac{1}{2x\sqrt{\log h/x}}$ *sur* $(]0,1], \lambda)$.

(c) Il existe sur $T =]0,1]^m$ *muni de la mesure de Lebesgue* λ_m *deux fonctions mesurables et positives* Δ *et* K *telles que*

(c.1) $\sup_{n\in\mathbf{N}} \int \frac{d_n^2(t)}{\Delta(t)} K(t) d\lambda_m(t)$ *soit fini,*

(c.2) $\int \Delta(t)\, K(t)\, d\lambda_m(t)$ *soit fini,*

(c.3) la fonction K *sur* (T, λ_m) *appartienne à l'ensemble* $A(\lambda_m, \mu_0)$ *défini en (2.13)-(2.14) pour* $\psi = \psi_0$.

Démonstration : L'implication (a)$\Longrightarrow$(b) résulte du calcul précédent 2.4.2 et de sa conclusion (2.20) : sous l'hypothèse (a), poser $T =]0,1]^m, \pi = \lambda_m$, choisir $\Delta' = d'_\eta$, $K' = K_\eta$.

Pour démontrer (b)$\Longrightarrow$(c), on pose

$$K(t) = \int K'(t,\tau) d\lambda(\tau),\ \Delta(t) = \frac{1}{K(t)} \int \Delta'(t,\tau)\, K'(t,\tau)\, d\lambda(\tau);$$

on vérifie que K appartient à $A(\lambda, \mu_0)$: poser pour cela

$$k(y,t) = \int k'(y,t,\tau)\, d\lambda(\tau)$$

où les k' sont associés aux K' par (2.14) ; ceci établit (c.3). On vérifie aussi par l'inégalité de Cauchy-Schwarz que

$$\begin{aligned}\sup_n \int \frac{d_n^2(t)}{\Delta(t)} K(t) d\lambda_m(t) &\le \sup_n \int \frac{d_n^2(t)\, K^2(t)}{\int \Delta'(t,\tau)\, K'(t,\tau) d\lambda(\tau)} d\lambda_m(t)\\ &\le \sup_n \int d_n^2(t) \int \frac{K'(t,\tau)}{\Delta'(t,\tau)} d\lambda(\tau) d\lambda_m(t)\\ &\le \sup_n \int \frac{d_n^2(t)}{\Delta'(t,\tau)} K'(t,\tau)\, d\lambda_m(t) d\lambda(\tau) < \infty\,;\end{aligned}$$

(c.1) est donc vérifiée. On vérifie enfin que

$$\int \Delta(t)\, K(t)\, d\lambda_m(t) = \int \Delta'(t,\tau)\, K'(t,\tau)\, d\lambda_m(t) d\lambda(\tau) < \infty,$$

c'est (c.2) de sorte que (c) est vérifié. Il reste à prouver l'implication (c)$\Longrightarrow$(a) ; on utilise pour cela l'inégalité de Cauchy-Schwarz ; on a en effet pour tout $y \in B_2$,

$$\int \psi_0 \circ \lambda_m\{d_y \leq u\} du \leq \int d_y(t)\, K(t)\, d\lambda_m(t)$$

$$\leq \sqrt{\int \frac{\sum y_n^2\, d_n^2(t)}{\Delta(t)}\, K(t)\, d\lambda_m(t) \int \Delta(t)\, K(t)\, d\lambda_m(t)}$$

$$\leq \sqrt{\sup_n \int \frac{d_n^2(t)}{\Delta(t)}\, K(t)\, d\lambda_m(t) \int \Delta(t)\, K(t)\, d\lambda_m(t)}\,;$$

sous l'hypothèse (c), on a donc

$$\sup_{y \in B_2} \int \psi_0 \circ \lambda_m\{d_y \leq u\}\, du < \infty,$$

et ceci suffit à établir (a) ; le théorème 4 et donc le théorème 2 sont démontrés.

2.4.4 La caractérisation dans certains cas non stationnaires

Théorème 5. *On note $X = \{X_n(t), t \in T, n \in \mathbf{N}\}$ une suite de fonctions aléatoires gaussiennes indépendantes sur un espace métrique compact T ; on suppose que la série $\sum_n |a_n|^2$ est convergente et que la condition (2.2) est satisfaite. Dans ces conditions, les deux propriétés suivantes sont équivalentes :*

(a) X a une modification à trajectoires continues dans ℓ_2.

(b) Il existe une probabilité π sur T, il existe sur $T' = T\times]0,1]$ muni de la probabilité $\pi' = \pi \times \lambda$ deux familles $\Delta_t, K_t, t \in T$ de fonctions mesurables positives telles que :

(b.1) les intégrales

$$\int \frac{d_n^2(s,t)}{\Delta_t(s,\tau)}\, K_t(s,\tau)\, \pi'(ds, d\tau), t \in T, n \in \mathbf{N}$$

soient majorées,

(b.2) les fonctions

$$\Delta_t(s,\tau)\, K_t(s,\tau),\ t \in T,$$

soient équi-intégrables relativement à $\pi'(ds, d\tau)$,

(b.3) pour tout $t \in T$, la loi de K' sur (T', λ') soit égale à celle de $\dfrac{1}{2x\sqrt{\log h/x}}$ sur $(]0,1], \lambda)$.

Démonstration : Supposons la propriété (a) réalisée et notons π une probabilité sur T telle que (théorème 3) les intégrales

$$\int \psi_0 \circ \pi\{s \in T : d_y(s,t) \leq u\}\, du,\ y \in B_2, t \in T$$

soient majorées et uniformément convergentes. On utilise le calcul 2.4.2 pour les différentes valeurs de t. En appliquant les conclusions de 2.4.2, on constate que (b) est réalisée comme dans la preuve du théorème 2 ; la propriété (b.2) résultant plus particulièrement de la construction conjointe des Δ_t et des K_t et de la proposition 5(a).

Inversement supposons la propriété (b) réalisée, on montre alors, pour tout $t \in T$ et en utilisant l'inégalité de Cauchy-Schwarz comme dans la preuve du théorème 4, qu'il existe un nombre M indépendant de $t \in T$ tel que pour tout $t \in B_2$ et tout $S \subset T$,

$$\int_S d_y(s,t) K_t(s,\tau) \pi'(ds, d\tau) \leq M \int_S \Delta_t(s,\tau) K_t(s,\tau)\, \pi'(ds, d\tau)$$

de sorte que l'hypothèse (b.2) montre que les fonctions $d_y(s,t)K_t(s,\tau), t \in T, y \in B_2$, sont équi-intégrables relativement à $\pi'(ds, d\tau)$. La proposition 5(b) implique alors que les intégrales

$$\int \psi \circ \pi\{s \in T : d_y(s,t) \leq u\}\, du = \int \psi \circ \pi'\{(s,\tau) \in T' : d_y(s,t) \leq u\}\, du$$

sont majorées et uniformément convergentes ; le théorème 3 conclut alors à (a). Le théorème est démontré.

2.5 Remarques

(a) Les théorèmes 4 et 5 fournissent aussi des caractérisations pour l'existence de modification à trajectoires continues du chaos gaussien $\sum_n X_n^2$; on peut comparer ces caractérisations à la condition suffisante énoncée dans [8] au théorème 11.22 ; cette condition suffisante se compose de deux conditions intégrales portant respectivement sur deux distances δ_1, δ_2 qui dans notre cadre vérifient

$$\delta_1(s,t) = \Big\{\mathbf{E}\big[\sum_n X_n^2(s) - \sum_n X_n^2(t)\big]^2\Big\}^{1/2}$$
$$\leq \Big[\sum_n a_n^2\, d_n^2(s,t)\Big]^{1/2} + |F(s) - F(t)|,$$
$$c \sup_n \{d_n(s,t) \times [\mathbf{E}\,|X_n(s) + X_n(t)|^2]^{1/2}\} \leq \delta_2(s,t)$$
$$\leq C \sup_n [a_n \times d_n(s,t)],$$

où F est une fonction continue si le chaos a une modification continue. La condition énoncée dans [8] sur δ_1 est alors nécessaire dans les situations étudiées ici ; la condition énoncée sur δ_2 ne l'est pas.
(b) Les mêmes techniques de preuve étendraient les théorèmes 4 et 5 dans les espaces $\ell_p, 1 \leq p < 2$, avec une modification mineure due à la forme de (1.4) pour ces valeurs de p. Pour $p > 2$, les résultats restent insuffisants.

Références

[1] X. Fernique : Continuité et théorème central limite pour les transformées de Fourier des mesures aléatoires du second ordre, Z.Wahrscheinlichkeitstheorie verw. *Gebiete.* **42** (1978), 57–66.

[2] X. Fernique : Régularité de fonctions aléatoires gaussiennes stationnaires. *Probab. Th. Rel. Fields* **88** (1991), 521–536.

[3] X. Fernique : Sur la régularité des fonctions aléatoires d'Ornstein-Uhlenbeck à valeurs dans $l_p, p \in [1, \infty[$. *Ann. Prob.* **20** (1992), 1441–1449.

[4] X. Fernique : Analyse de fonctions aléatoires gaussiennes stationnaires à valeurs vectorielles. *C.R. Acad. Sci. Paris* **312** (1991), Série I, 927–929.

[5] X. Fernique : Sur les espaces de Fréchet ne contenant pas c_0. *Studia Math.* **101** :3 (1992), 299–309.

[6] X. Fernique : Fonctions aléatoires gaussiennes stationnaires, vecteurs aléatoires gaussiens. Publications C.R.M. Montréal, 1997.

[7] I. Iscoe, D. McDonald, M.B. Marcus, M. Talagrand et J. Zinn : On the continuity of ℓ_2-valued Ornstein-Uhlenbeck processes. *Annals of Probability* **18** (1990), 68–84.

[8] M. Ledoux, M. Talagrand : Probability in Banach spaces. Ergebnisse der Mathematik, série 3, vol. 23, Berlin-Heidelberg, New-York, Springer, 1991.

Institut de Recherche
Mathématique Avancée
Université Louis Pasteur et C.N.R.S.
7 rue René Descartes
67084 Strasbourg Cedex, France
e-mail : `fernique@math.u-strasbg.fr`

A Note on the Gaussian Correlation Conjecture

Wenbo V. Li* and Qi-Man Shao**

ABSTRACT We show that a special setting of the well-known Gaussian correlation conjecture, namely for sets of equal measure, can be useful in proving the existence of small ball constants. We hope this sheds light on the conjecture and points out new directions for useful partial results. A simple proof for another useful correlation inequality needed in our argument is also presented.

1 Introduction

Recently there has been a lot of interest and many studies of the so-called Gaussian correlation conjecture. See, for example, Hargé (1998), Hitczenko, Kwapien, Li, Schechtman, Schlumprecht and Zinn (1998), Schechtman, Schlumprecht and Zinn (1998), Lewis and Pritchard (1999), Li (1999), Shao (1999), Szarek and Werner (1999), and Vitale (1999). The conjecture states that for any two symmetric convex sets A and B in n-dimensional Euclidean space $\mathbb{R}^n$ and for any centered Gaussian measure μ on $\mathbb{R}^n$,

$$\mu(A \cap B) \geq \mu(A)\mu(B). \tag{1.1}$$

One of the equivalent formulations of this conjecture is that

$$\mathbb{P}\left(\max_{1\leq i\leq n} |\xi_i| \leq x\right) \geq \mathbb{P}\left(\max_{1\leq i\leq k} |\xi_i| \leq x\right) \mathbb{P}\left(\max_{k<i\leq n} |\xi_i| \leq x\right) \tag{1.2}$$

for any $x \geq 0$, $n \geq 2$, $1 \leq k \leq n$ and for any mean-zero jointly Gaussian random variables $\{\xi_i\}_{i=1}^n$. Other equivalent geometrical formulations and early history of the conjecture can be found in Schechtman, Schlumprecht and Zinn (1998). In particular, Pitt (1977) proved that (1.1) holds on $\mathbb{R}^2$. One of the most useful special cases is that (1.2) holds for $k = 1$, obtained independently by Khatri (1967) and Šidák (1967). The inequality in this case is called the Khatri-Šidák lemma, and it is a very useful tool in the study of the so-called small ball probabilities for Gaussian processes.

*Research partially supported by NSF Grant DMS-9972012.

**Research partially supported by NSF Grant DMS-9802451.

Let $\{X(t),\ 0 \le t \le 1\}$ be a continuous centered Gaussian process with $\sigma^2(t) = \mathbb{E}\, X^2(t)$ and $X(0) = 0$. The small ball problem under sup-norm for X is the determination of the rate

$$-\ln \mathbb{P}\Big(\sup_{0\le t\le 1} |X(t)| \le \sigma(x) \Big) \quad \text{as} \quad x \downarrow 0. \tag{1.3}$$

In the last few years there has been considerable progress on obtaining the small ball estimates for Gaussian processes. As established in Kuelbs and Li (1993)(see also Li and Linde (1999) for further developments), the behavior in (1.3) up to a constant is determined by the metric entropy of the unit ball of the reproducing kernel Hilbert space associated with $X(t)$, and vice versa. For other connections and applications of small ball probabilities, we refer readers to a recent survey paper by Li and Shao (2000).

When $\{X(t), 0 \le t \le 1\}$ is a fractional Brownian motion of order α $(0 < \alpha < 1)$, i.e., $\mathbb{E}\, X(t) = 0$, $\mathbb{E}\,(X(t+h) - X(t))^2 = h^{2\alpha}$, it was proved independently in Li and Linde (1998) and Shao (1999) that

$$\lim_{x\to 0} x \ln \mathbb{P}\left(\sup_{0\le t\le 1} |X(t)| \le x^\alpha \right) = -c_\alpha, \tag{1.4}$$

where $0 < c_\alpha < \infty$. The constants c_α play the role of the principle eigenvalues of certain operators in proper domains, since (1.4) can be rewritten as

$$\lim_{t\to\infty} t^{-1} \ln \mathbb{P}\,(\tau > t) = -c_\alpha, \tag{1.5}$$

where $\tau = \inf\{t : X(t) \notin [-1, 1]\}$ is the exit time. In the Brownian motion case, $\alpha = 1/2$, the constant in (1.5) is the principle eigenvalue of the Laplacian on the domain $[-1, 1]$ and is equal to $\pi^2/8$.

The proof given in Li and Linde (1998) is based on the following useful but special representation,

$$X(t) = \sqrt{a_\alpha}\,(W_\alpha(t) + Z_\alpha(t)), \quad 0 \le t \le 1,$$

where

$$a_\alpha = (2\alpha)^{-1} + \int_{-\infty}^{0} ((1-s)^{\alpha-1/2} - (-s)^{\alpha-1/2})^2 ds,$$

$$W_\alpha(t) = \int_0^t (t-s)^{\alpha-1/2} dW(s)$$

and

$$Z_\alpha(t) = \int_{-\infty}^{0} \{(t-s)^{\alpha-1/2} - (-s)^{\alpha-1/2}\} dW(s).$$

Furthermore, $W_\alpha(t)$ is independent of $Z_\alpha(t)$. The proof given in Shao (1999) is based on the following weaker correlation inequality:

$$\mathbb{P}\left(\max_{1\le i\le n} |\xi_i| \le x \right) \ge 2^{-k\wedge(n-k)} \mathbb{P}\left(\max_{1\le i\le k} |\xi_i| \le x \right) \mathbb{P}\left(\max_{k<i\le n} |\xi_i| \le x \right)$$

for any $1 \le k \le n$ and any mean-zero jointly Gaussian variables $\{\xi_i\}_{i=1}^n$.

In this paper, we show the usefulness of the following weaker correlation conjecture:

$$\mu(A \cap B) \ge \rho^2 \tag{1.6}$$

for any symmetric convex sets A and B such that $\mu(A) = \mu(B) = \rho$, which is the Gaussian correlation conjecture for sets of the same measure.

Theorem 1. *Let $\{X(t), 0 \le t \le 1\}$ be a fractional Brownian motion of order α $(0 < \alpha < 1)$. Then the weaker correlation conjecture (1.6) implies the existence of the limit in (1.4).*

We hope this note provides an interesting application and points out new directions for useful partial results. Furthermore, we present a very simple proof of another useful weaker correlation inequality given in Li (1999), see Lemma 2 in the next section. We include all the details so that readers with different backgrounds can go through the proofs easily.

2 Lemmas

The following lemma is standard in the small ball estimate and provides a nice example of the application of the Khatri-Šidák lemma and the chaining argument. The estimate is used to control the error term in the proof of the theorem.

Lemma 1. *Let $\{X(t), 0 \le t \le 1\}$ be a Gaussian process with $\mathbb{E}\, X(t) = 0$, $X(0) = 0$ and*

$$\mathbb{E}\,(X(t+h) - X(t))^2 \le h^{2\alpha}.$$

Then for any $0 < x < \Delta/4 \le 1/16$

$$\mathbb{P}\left(\max_{1 \le i \le 1/\Delta} |X(i\Delta)| \le x^\alpha\right) \ge \exp\left(-\frac{D_\alpha}{\Delta}\log(\frac{\Delta}{x})\right), \tag{2.1}$$

where $D_\alpha > 0$ is a constant depending only on α, $\log(x) = \ln(\max(e, x))$, while ln *is the natural logarithmic function.*

Proof. Let N be an integer such that

$$2^{N-1} < \Delta^{-1} \le 2^N. \tag{2.2}$$

To extend the process to $[1, 2]$, let $X(t) = X(1)$ for $1 \le t \le 2$. It is easy to see by a standard chaining argument that

$$\begin{aligned}\max_{1 \le i \le 1/\Delta} |X(i\Delta)| &\le \max_{1 \le i \le 2^N} |X(i\Delta)| \\ &\le \sum_{j=0}^{N} \max_{1 \le i \le 2^{N-j}} |X(i2^j\Delta) - X((i-1)2^j\Delta)|.\end{aligned}$$

Hence, using the Khatri-Šidák lemma in the last step,

$$\begin{aligned}
&\mathbb{P}(\max_{i\le 1/\Delta}|X(i\Delta)|\le x^\alpha)\\
\ge\ &\mathbb{P}\left(\sum_{j=0}^{N}\max_{1\le i\le 2^{N-j}}|X(i2^j\Delta)-X(i-1)2^j\Delta)|\le x^\alpha\right)\\
\ge\ &\mathbb{P}\Big(\max_{1\le i\le 2^{N-j}}|X(i2^j\Delta)-X((i-1)2^j\Delta)|\\
&\qquad\le(1-2^{-\alpha})2^{-\alpha j}x^\alpha),\ 0\le j\le N\Big)\\
\ge\ &\prod_{0\le j\le N}\ \prod_{1\le i\le 2^{N-j}}\mathbb{P}\left(|X(i2^j\Delta)-X(i-1)2^j\Delta)|\le(1-2^{-\alpha})2^{-\alpha j}x^\alpha\right).
\end{aligned}$$

Let Z be the standard normal random variable. Then

$$\begin{aligned}
&\mathbb{P}\left(|X(i2^j\Delta)-X(i-1)2^j\Delta)|\le(1-2^{-\alpha})2^{-\alpha j}x^\alpha\right)\\
=\ &\mathbb{P}\left(|Z|\le(1-2^{-\alpha})2^{-\alpha j}x^\alpha/(E(X(i2^j\Delta)-X((i-1)2^j\Delta))^2)^{\frac{1}{2}}\right)\\
\ge\ &\mathbb{P}\left(|Z|\le(1-2^{-\alpha})(x/\Delta)^\alpha 2^{-2\alpha j}\right)\\
\ge\ &(1-2^{-\alpha})(x/\Delta)^\alpha 2^{-2\alpha j-1}
\end{aligned}$$

where we used the fact that

$$\mathbb{P}(|Z|\le a)\ge a/2\ \text{ for }\ 0\le a\le 1/4.$$

Now putting all the estimates together yields

$$\begin{aligned}
&\mathbb{P}(\max_{i\le 1/\Delta}|X(i\Delta)|\le x^\alpha)\\
\ge\ &\exp\Big(\sum_{0\le j\le N}\ \sum_{1\le i\le 2^{N-j}}\{\ln(1-2^{-\alpha})+\alpha\ln(x/\Delta)-2\alpha j-1\}\Big)\\
\ge\ &\exp\Big(-2^{N+2}\{\log(2^\alpha/(2^\alpha-1))+\alpha\log(\Delta/x)-4\alpha-1\}\Big)\\
\ge\ &\exp\Big(-\frac{D_\alpha}{\Delta}\log(\frac{\Delta}{x})\Big),
\end{aligned}$$

as desired.

Remark. The use of a chaining argument is standard but crucial. Direct use of the Khatri-Šidák lemma in Lemma 1 without chaining gives only the estimate

$$\mathbb{P}(\max_{i\le 1/\Delta}|X(i\Delta)|\le x^\alpha)\ge\exp\Big(-\frac{C}{\Delta}\log(1/x)\Big),$$

for some constant $C>0$, which is not strong enough for what we need.

Our next lemma is a very useful weaker form of the Gaussian correlation conjecture. It is given in Li (1999) and is closely related to a result of

Schechtman, Schlumprecht and Zinn (1998). Here we present a simple proof based on Anderson's inequality and the idea of random shift.

Lemma 2. *Let μ be a centered Gaussian measure on a separable Banach space E. Then for any $0 < \lambda < 1$, any symmetric, convex sets A and B in E,*

$$\mu(A \cap B) \geq \mu(\lambda A)\mu((1-\lambda^2)^{1/2}B), \tag{2.3}$$

which is equivalent to

$$\mathbb{P}(X \in A, Y \in B) \geq \mathbb{P}\Big(X \in \lambda A\Big)\mathbb{P}\Big(Y \in (1-\lambda^2)^{1/2}B\Big) \tag{2.4}$$

for any centered joint Gaussian vectors X and Y in E.

Proof. Let $a = (1-\lambda^2)^{1/2}/\lambda$, and (X^*, Y^*) be an independent copy of (X, Y). It is easy to see that $X - aX^*$ and $Y + Y^*/a$ are independent. Thus, by Anderson's inequality

$$\begin{aligned} \mathbb{P}(X \in A, Y \in B) &\geq \mathbb{P}(X - aX^* \in A, Y + Y^*/a \in B) \\ &= \mathbb{P}(X - aX^* \in A)\mathbb{P}(Y + Y^*/a \in B) \\ &= \mathbb{P}\Big(X \in \lambda A\Big)\mathbb{P}\Big(Y \in (1-\lambda^2)^{1/2}B\Big), \end{aligned}$$

as desired.

Remark: A stronger result

$$\mu(A \cap B)\mu(\lambda^2 A + (1-\lambda^2)B) \geq \mu(\lambda A)\mu((1-\lambda^2)^{1/2}B)$$

was proved in Li (1999) following the arguments of Proposition 3 in Schechtman, Schlumprecht and Zinn (1998), where the case $\lambda = 1/\sqrt{2}$ was proved. The central and important fact used there is the rotational invariance of the Gaussian measure $\mu \times \mu$ for

$$(x, y) \mapsto (\lambda x + (1-\lambda^2)^{1/2}y, (1-\lambda^2)^{1/2}x - \lambda y).$$

It is of interest to compare these two related approaches.

The main difference between the Khatri-Šidák inequality and Lemma 2 in the applications to small ball probability is that the former provides only the rate (up to a constant) and the latter can preserve the rate together with the constant. The varying parameter λ plays a fundamental role in most of the applications we know so far, and the way we use Lemma 2 in the next section is typical. More applications of Lemma 2 are given in Li (1999).

3 Proof of theorem

From the works of Shao (1993) and Monrad and Rootzén (1995), there exist $0 < L \le l < \infty$ such that

$$\limsup_{x\to 0} x \ln \mathbb{P}\left(\sup_{0\le t\le 1} |X(t)| \le x^{\alpha}\right) = -L$$

and

$$\liminf_{x\to 0} x \ln \mathbb{P}\left(\sup_{0\le t\le 1} |X(t)| \le x^{\alpha}\right) = -l.$$

It suffices to show that $L \ge l$.

Let $\{x_k,\ k\ge 1\}$ and $\{y_k, k\ge 1\}$ be two sequences of positive numbers satisfying

$$x_k \downarrow 0,\ y_k \downarrow 0,\ y_k \ge k\, x_k, \tag{3.1}$$

$$\lim_{k\to\infty} y_k \ \ln\ \mathbb{P}\Big(\sup_{0\le t\le 1} |X(t)| \le y_k^{\alpha}\Big) = -L, \tag{3.2}$$

and

$$\lim_{k\to\infty} x_k \ \ln\ \mathbb{P}\Big(\sup_{0\le t\le 1} |X(t)| \le x_k^{\alpha}\Big) = -l. \tag{3.3}$$

For $0 < \epsilon < 1/2$, let $\Delta = \Delta_k = (1-\epsilon)^{1/\alpha} x_k/y_k$. We have for any $0 < \lambda < 1$

$$\begin{aligned}
&\mathbb{P}\Big(\sup_{0\le t\le 1} |X(t)| \le x_k^{\alpha}\Big)\\
\ge\ &\mathbb{P}\Big(\max_{0\le i\le 1/\Delta-1}\ \sup_{i\Delta\le t\le (i+1)\Delta} |X(t)| \le x_k^{\alpha},\ \max_{0\le i\le 1/\Delta} |X(i\Delta)| \le \epsilon x_k^{\alpha}\Big)\\
\ge\ &\mathbb{P}\Big(\max_{0\le i\le 1/\Delta-1}\ \sup_{i\Delta\le t\le (i+1)\Delta} |X(t)-X(i\Delta)| \le x_k^{\alpha}(1-\epsilon)\\
&\qquad \cdot \max_{0\le i\le 1/\Delta} |X(i\Delta)| \le \epsilon x_k^{\alpha}\Big)\\
\ge\ &\mathbb{P}\Big(\max_{0\le i\le 1/\Delta} |X(i\Delta)| \le (1-\lambda^2)^{1/2}\epsilon x_k^{\alpha}\Big)\\
&\qquad \cdot\mathbb{P}\Big(\max_{0\le i\le 1/\Delta-1}\ \sup_{i\Delta\le t\le (i+1)\Delta} |X(t)-X(i\Delta)| \le x_k^{\alpha}(1-\epsilon)\lambda\Big)
\end{aligned}$$

where we used Lemma 2 in the last step.

Let N be the integer defined in (2.2). Then

$$\begin{aligned}
&\mathbb{P}\Big(\max_{0\le i\le 1/\Delta-1}\sup_{i\Delta\le t\le(i+1)\Delta}|X(t)-X(i\Delta)|\le x_k^\alpha(1-\epsilon)\lambda\Big)\\
\ge\ &\mathbb{P}\Big(\max_{0\le i\le 2^N-1}\sup_{i\Delta\le t\le(i+1)\Delta}|X(t)-X(i\Delta)|\le x_k^\alpha(1-\epsilon)\lambda\Big)\\
\ge\ &\mathbb{P}\Big(\max_{0\le i\le 2^{N-1}-1}\sup_{i\Delta\le t\le(i+1)\Delta}|X(t)-X(i\Delta)|\le x_k^\alpha(1-\epsilon)\lambda\Big)\\
&\cdot\mathbb{P}\Big(\max_{2^{N-1}\le i\le 2^N-1}\sup_{i\Delta\le t\le(i+1)\Delta}|X(t)-X(i\Delta)|\le x_k^\alpha(1-\epsilon)\lambda\Big),
\end{aligned}$$

where we used the weaker correlation conjecture (1.6) in the last step. Now repeated application of the weaker correlation conjecture (1.6) together with the stationary increment and scaling properties implies

$$\begin{aligned}
&\mathbb{P}\Big(\max_{0\le i\le 1/\Delta-1}\sup_{i\Delta\le t\le(i+1)\Delta}|X(t)-X(i\Delta)|\le x_k^\alpha(1-\epsilon)\lambda\Big)\\
\ge\ &\left(\mathbb{P}\left(\sup_{0\le t\le\Delta}|X(t)|\le x_k^\alpha(1-\epsilon)\lambda\right)\right)^{2^N}\\
\ge\ &\left(\mathbb{P}\left(\sup_{0\le t\le 1}|X(t)|\le \lambda y_k^\alpha\right)\right)^{1+1/\Delta}.
\end{aligned}$$

Therefore by combining the above estimates and Lemma 1,

$$\begin{aligned}
\ln\mathbb{P}\Big(\sup_{0\le t\le 1}|X(t)|\le x_k^\alpha\Big)\ge&-\frac{D_\alpha}{\Delta}\ln\Big(\frac{\Delta}{x_k\epsilon^{1/\alpha}(1-\lambda^2)^{1/(2\alpha)}}\Big)\\
&+\Big(1+\frac{1}{\Delta}\Big)\ln\mathbb{P}\Big(\sup_{0\le t\le 1}|X(t)|\le\lambda y_k^\alpha\Big)
\end{aligned}$$

and thus by substituting in $\Delta=\Delta_k=(1-\epsilon)^{1/\alpha}x_k/y_k$,

$$\begin{aligned}
&x_k\ln\mathbb{P}\Big(\sup_{0\le t\le 1}|X(t)|\le x_k^\alpha\Big)\\
\ge\ &-\frac{D_\alpha\cdot y_k}{(1-\epsilon)^{1/\alpha}}\ \ln\Big(\frac{(1-\epsilon)^{1/\alpha}}{y_k\cdot\epsilon^{1/\alpha}(1-\lambda^2)^{1/(2\alpha)}}\Big)\\
&+\frac{x_k}{y_k}\Big(1+y_k(1-\epsilon)^{1/\alpha}x_k\Big)\cdot y_k\ln\mathbb{P}\Big(\sup_{0\le t\le 1}|X(t)|\le\lambda y_k^\alpha\Big).
\end{aligned}$$

Hence by taking $k\to\infty$, we obtain from (3.1), (3.2) and (3.3) that

$$-l\ge-\frac{L}{(1-\epsilon)^{1/\alpha}\lambda^{1/\alpha}},$$

which proves the theorem by letting $\epsilon\to 0$ and $\lambda\to 1$.

References

Hargé, G. (1998). Une inégalité de décorrélation pour la mesure gaussienne. *C. R. Acad. Sci. Paris Sér.* **326**, 1325–1328.

Hitczenko, P., Kwapien, S., Li, W.V., Schechtman, G., Schlumprecht, T. and Zinn, J. (1998). Hypercontractivity and comparison of moments of iterated maxima and minima of independent random variables. *Electronic Journal of Probability* Vol. **3**, Paper no. 2, 1–26.

Khatri, C. (1967). On certain inequalities for normal distributions and their applications to simultaneous confidence bounds. *Ann. Math. Stat.* **38**, 1853–1867.

Kuelbs, J. and Li, W.V. (1993). Metric entropy and the small ball problem for Gaussian measures. *J. Func. Anal.* **116**, 133–157.

Lewis, T. and Pritchard, G. (1999). Correlation measures. *Elect. Comm. in Probab.* **4**, 77–85.

Li, W.V. (1999). A Gaussian correlation inequality and its applications to small ball probabilities. *Elect. Comm. in Probab.* **4**, 111–118.

Li, W.V. and Linde, W. (1999). Approximation, metric entropy and small ball estimates for Gaussian measures. *Ann. Probab.* **27**, 1556–1578.

Li, W.V. and Linde, W. (1998). Existence of small ball constants for fractional Brownian motions. *C.R. Acad. Sci. Paris* **326**, 1329–1334.

Li, W.V. and Shao, Q.M. (2000). Gaussian processes: inequalities, small ball probabilities and applications. To appear in: *Stochastic Processes: Theory and Methods*, Handbook of Statistics, Vol. 19, C.R. Rao and D. Shanbhag, (eds.).

Monrad, D. and Rootzén, H. (1995). Small values of Gaussian processes and functional laws of the iterated logarithm. *Probab. Th. Rel. Fields* **101**, 173–192.

Pitt, L. (1977). A Gaussian correlation inequality for symmetric convex sets. *Ann. Probab.* **5**, 470–474.

Schechtman, G., Schlumprecht, T. and Zinn, J. (1998). On the Gaussian measure of the intersection. *Ann. Probab.* **26**, 346–357.

Šidák, Z. (1968). On multivariate normal probabilities of rectangles, their dependence on correlations. *Ann. Math. Statist.* **39**, 1425–1434.

Shao, Q.M. (1993). A note on small ball probability of Gaussian processes with stationary increments. *J. Theoret. Probab.* **6**, 595–602.

Shao, Q.M. (1999). A Gaussian correlation inequality and its applications to the existence of small ball constant, preprint.

Szarek, S. and Werner, E. (1999). A nonsymmetric correlation inequality for Gaussian measure. *J. Multi. Anal.* **68**, 193–211.

Vitale, R. (1999). Majorization and Gaussian Correlation. *Stat. Probab. Letters* **45**, 247–251.

Department of Mathematical Sciences
University of Delaware
Newark, DE 19716
E-mail: `wli@math.udel.edu`

Department of Mathematics
University of Oregon
Eugene, OR 97403
E-mail: `shao@math.uoregon.edu`

Probability Estimates for Lower Levels of Certain Gaussian Processes with Stationary Increments

Michael B. Marcus

ABSTRACT Let $\eta = \{\eta_x;\, x \in R\}$ be a mean–zero Gaussian process with covariance

$$E(\eta_x\, \eta_y) = \frac{1}{2}\left(\phi(x) + \phi(y) - \phi(x-y)\right).$$

Here

$$\phi(x) \stackrel{def}{=} \frac{1}{\pi}\int_{-\infty}^{\infty} \frac{1-\cos\lambda x}{\psi(\lambda)}\, d\lambda \qquad x \in R$$

where ψ is a non–negative even function on R. Note that ϕ is an even function and $\phi(0) = 0$. In this paper we assume that

1. $\phi(x+y) \le \phi(x) + \phi(y) \qquad$ for $x, y > 0$.
2. $\phi(x)$ is increasing on $[0,\infty)$.

We show that for all $\lambda \le 1/3$

$$P\left(\sup_{0\le x\le s} \eta(x) \le \lambda\right) \le \sqrt{\frac{2}{\pi}}\left(\frac{\lambda}{\phi^{1/2}(s)}\right)$$

and

$$P\left(\sup_{0\le |x|\le s} \eta(x) \le \lambda\right) \le C\left(\frac{\lambda}{\phi^{1/2}(s)}\right)^{3/2}$$

where C is an absolute constant.

1 Introduction

We obtain upper bounds for the probability that the supremum of a Gaussian process is below a small level. This is different from the small ball problem which considers the absolute value of the supremum of a Gaussian process. It seems that a lot less is known about the lower level problem than the small ball problem. Another innovation in these estimates is that the process is defined on the whole real line and we consider the supremum on a symmetric interval containing zero. This is not an artificial extension. It is necessary to make such an estimate in Bass, Eisenbaum and Shi (1999) and its generalization in Marcus (1999). The estimate involves a clever application of Slepian's lemma which reduces the problem to the consideration of the probability that planar Brownian motion spends a unit of time in a certain cone. This is done in Bass, Eisenbaum and Shi (1999) for fractional Brownian motion and generalized in Marcus (1999) to a larger class of Gaussian processes with stationary increments.

Let $\eta = \{\eta_x;\, x \in R\}$ be a mean–zero Gaussian process with covariance

$$E(\eta_x \eta_y) = \frac{1}{2}\left(\phi(x) + \phi(y) - \phi(x-y)\right), \tag{1.1}$$

so that, in particular

$$E(\eta_x - \eta_y)^2 = \phi(x-y). \tag{1.2}$$

Here

$$\phi(x) \stackrel{def}{=} \frac{1}{\pi}\int_{-\infty}^{\infty} \frac{1-\cos\lambda x}{\psi(\lambda)}\, d\lambda \qquad x \in R \tag{1.3}$$

where ψ is a non-negative even function on R satisfying

$$\int_0^{\infty} \frac{\lambda^2 \wedge 1}{\psi(\lambda)}\, d\lambda < \infty. \tag{1.4}$$

Note that when ψ varies regularly at zero with index $1 < \alpha \le 2$,

$$\phi(x) \sim c_\alpha \frac{1}{x\psi(1/x)} \qquad as \quad x \to \infty \tag{1.5}$$

where

$$c_\alpha = \frac{2}{\pi}\int_0^{\infty} \frac{1-\cos s}{s^\alpha}\, ds. \tag{1.6}$$

Furthermore, when $\psi(\lambda) = |\lambda|^\alpha$, $1 < \alpha \le 2$,

$$\phi(x) = c_\alpha |x|^{\alpha-1}. \tag{1.7}$$

That is, η is fractional Brownian motion with exponent $\alpha - 1$, including Brownian motion when $\alpha = 2$. Note that we define η on all of R. It is not restricted to R^+, as is usually the case. (We show in Lemma 1 that the right–hand side of (1.1) is positive definite).

Note that ϕ is an even function and $\phi(0) = 0$. In this paper we assume that

1. $\phi(x+y) \le \phi(x) + \phi(y)$ for $x, y > 0$.

2. $\phi(x)$ is increasing on $[0, \infty)$.

Theorem 1. *For all* $\lambda \le 1/3$

$$P\left(\sup_{0\le x\le s} \eta(x) \le \lambda\right) \le \sqrt{\frac{2}{\pi}}\left(\frac{\lambda}{\phi^{1/2}(s)}\right), \tag{1.8}$$

and

$$P\left(\sup_{0\le |x|\le s} \eta(x) \le \lambda\right) \le C\left(\frac{\lambda}{\phi^{1/2}(s)}\right)^{3/2} \tag{1.9}$$

where C is an absolute constant.

This result is proved in Section 2. Let us put it in perspective. The inequality in (1.8) is very easy to obtain. We will explain this in the proof of this theorem. Now consider the case when η is Brownian motion and denote it by B. One can see from (1.1) that $\{B(x), x \in [0,\infty)\}$ and $\{B(x), x \in (-\infty, 0]\}$ are independent. Therefore, since (1.8) holds for B, we get

$$P\left(\sup_{0\le |x|\le s} B(x) \le \lambda\right) \le \frac{2}{\pi}\left(\frac{\lambda}{\phi^{1/2}(s)}\right)^{2} \tag{1.10}$$

$$\le \frac{\lambda^2}{\pi s}, \tag{1.11}$$

since $\phi(s) = 2|s|$ in this case. This, of course, is a sharper result than (1.9). It is unlikely that the abrupt decrease to the three halves power from the square in (1.9), when we depart from Brownian motion, is necessary, but it is the best we can do. Another interesting problem is to find good lower bounds for the probabilities in (1.8) and (1.9). For Brownian motiion this is simple but, apparently, one can not even show that the lower bound is as big as some power of λ, for fractional Brownian motion.

In Section 2, Theorem 1 is proved. In Section 3 we show that there are many examples of Gaussian processes for which the hypotheses on ϕ are satisfied.

I would like to thank Wenbo Li for many interesting discussions on lower levels of Gaussian processes.

2 Proof of Theorem 1

We begin with several comments on the function ϕ.

Lemma 1. *$E(\eta_x \eta_y)$ is a covariance, i.e. it is positive definite.*

Proof. Note that

$$(2.1)\qquad \sum_{j,k=1}^{n} a_j a_k E(\eta_{x_j}\,\eta_{x_k})$$
$$= \frac{1}{\pi}\int_{-\infty}^{\infty} \frac{\left(\sum_{j=1}^{n} a_j((1-\cos\lambda x_j)+\sin\lambda x_j)\right)^2}{\psi(\lambda)}\,d\lambda$$
$$-\frac{2}{\pi}\sum_{j,k=1}^{n} a_j a_k \int_{-\infty}^{\infty} \frac{(1-\cos\lambda x_j)\sin\lambda x_k}{\psi(\lambda)}\,d\lambda.$$

The integrands in the last line of (2.1) are all odd. Therefore the last line of (2.1) is zero. This shows that the first line of (2.1) is greater than or equal to zero.

In the next lemma we record some elementary properties of the covariance of η that are implied by the hypotheses on ϕ. The proofs are elementary and are omitted.

Lemma 2. *When $xy > 0$, property 1. of ϕ is equivalent to*

$$(2.2)\qquad E(\eta_x\,\eta_y) \le \phi(x)\wedge\phi(y).$$

When $xy < 0$, property 1. of ϕ implies that

$$(2.3)\qquad E(\eta_x\,\eta_y) > 0$$

and property 2. of ϕ implies that

$$(2.4)\qquad E(\eta_x\,\eta_y) \le \frac{1}{2}(\phi(x)\wedge\phi(y)).$$

Proof of Theorem 1. It is more convenient to show that

$$(2.5)\qquad P\left(\eta(x)\le\sqrt{t}\lambda; 0\le|x|\le\phi^{-1}(t)\right)\le C\lambda^{3/2}$$

for an absolute constant C, from which (1.9) follows by letting $s=\phi^{-1}(t)$.

Let $\{W_1(x); x\ge 0\}$ and $\{W_2(x); x\ge 0\}$ be two independent real valued Brownian motions, with $W_1(0) = W_2(0) = 0$. Define the process $U = \{U(x); -\infty < x < \infty\}$ by

$$(2.6)\qquad U(x) \stackrel{def}{=} \begin{cases} W_1(\phi(x)) & x\ge 0\,,\\ \frac{1}{2}W_1(\phi(|x|))+\frac{\sqrt{3}}{2}W_2(\phi(|x|)) & x\le 0\,. \end{cases}$$

Consequently by (2.2) and (2.4), for all $x, y\in R$

$$\begin{aligned} EU^2(x) &= E\eta^2(x)\,,\\ EU(x)U(y) &\ge E\eta(x)\eta(y). \end{aligned}$$

Let $W_{1,2}(\cdot) \stackrel{def}{=} W_1(\cdot) + \sqrt{3}W_2(\cdot)$. By Slepian's Lemma

$$\begin{aligned}(2.7)\quad & P\left(\eta(x) \le \sqrt{t}\lambda; 0 \le |x| \le \phi^{-1}(t)\right) \\ & \le P\left(U(x) \le \sqrt{t}\lambda; 0 \le |x| \le \phi^{-1}(t)\right) \\ & \le P\left(W_1(\phi(x)) \le \sqrt{t}\lambda; W_{1,2}(\phi(x)) \le 2\sqrt{t}\lambda, 0 \le x \le \phi^{-1}(t)\right) \\ & \le P\left(W_1(y) \le \lambda, W_{1,2}(y) \le 2\lambda, 0 \le y \le 1\right) .\end{aligned}$$

Following Bass, Eisenbaum, and Shi (1999), let

$$\mathcal{S} \stackrel{def}{=} \{W_1(t) \le \lambda, W_{1,2}(t) \le 2\lambda, 0 \le t \le 1\}. \tag{2.8}$$

Note that

$$\begin{aligned}(2.9)\quad P(\mathcal{S}) &= P\left(W_1(t) < \lambda, W_1(t) + \sqrt{3}W_2(t) < 2\lambda, 0 \le t \le 1\right) \\ &= P((W_1(t), W_2(t)) \in \mathcal{C}, 0 \le t \le 1)\end{aligned}$$

where $\mathcal{C} \stackrel{def}{=} \{(x,y) : x < \lambda, x + \sqrt{3}y < 2\lambda\}$. I.e., starting from $(0,0)$, the planar Brownian motion (W_1, W_2) stays in $\mathcal{C}$ during $[0,1]$. Equivalently

$$\begin{aligned}P(\mathcal{S}) &= P(\text{ starting from } (\sqrt{4/3}\lambda, 0), \text{ the angular part of planar} \\ (2.10)\quad & \qquad \text{Brownian motion lies in} (-\pi/3, \pi/3) \text{ during } [0,1]\,) \\ &= P(\text{ starting from } (1,0), \text{ the angular part of planar Brownian} \\ & \qquad \text{motion lies in} (-\pi/3, \pi/3) \text{ during } [0, 3/(4\lambda^2)]\,).\end{aligned}$$

Following Bass, Eisenbaum and Shi (1999) again let $\{Z(t);\, t \ge 0\}$ be a planar Brownian motion, starting from $(1,0)$. It is known that with probability one, Z never hits the origin. Consequently there exists a continuous determination of $\theta(t)$, the total angle wound by Z around the origin up to time t (with, say, $\theta(0) = 0$). θ records the angle and keeps track of the number of times the Brownian path has wound around the origin, counting clockwise loops (-2π) and counterclockwise loops (2π). The following result is given in Bañuelos and Smits (1997): for any $a > 0$, there exists a constant $d = d(a) \in (0, \infty)$ such that for all $u > 0$,

$$P\left(\sup_{0 \le s \le u} |\theta(s)| < a\right) \le \frac{d}{u^{\pi/(4a)}}. \tag{2.11}$$

Using this in the last line of (2.10) we get

$$P(\mathcal{S}) \le d(\pi/3) \left(\frac{4}{3}\right)^{3/4} \lambda^{3/2} \tag{2.12}$$

which gives us (2.5) and hence (1.9).

The proof of (1.8) is simple. Let B denote standard Brownian motion. By (2.2)

$$
\begin{aligned}
(2.13)\qquad P\left(\sup_{0\le x\le s}\eta(x)\le\lambda\right) &\le P\left(\sup_{0\le x\le s}B(\phi(x))\le\lambda\right)\\
&\le P\left(\sup_{0\le y\le 1}B(y)\le\lambda/\phi^{1/2}(x)\right)\\
&\le \sqrt{\frac{2}{\pi}}\left(\frac{\lambda}{\phi^{1/2}(s)}\right).
\end{aligned}
$$

3 Examples

There are many examples of functions ψ for which ϕ satisfies properties 1. and 2. Let ξ be a symmetric infinitely divisible random variable with characteristic function

$$
(3.1)\qquad Ee^{i\lambda\xi}=e^{-\psi(|\lambda|)}.
$$

It follows from Markov process considerations that property 1. holds for this function ψ. (See Marcus (1999).) Also, it is easy to see that for any function ψ property 2. holds when $|\lambda|/\psi(|\lambda|)$ is decreasing as $|\lambda|$ increases.

In Marcus and Rosen (1993) we introduce functions ψ of the following form:

$$
(3.2)\qquad \psi(|\lambda|)\stackrel{def}{=}\int_{\alpha_1}^{\alpha_2}\lambda^u\,d\mu(u)
$$

$1\le\alpha_1<\alpha_2\le 2$, where μ is a finite measure on $[\alpha_1,\alpha_2]$. We show in Lemma 2, Marcus and Rosen (1993) that for ψ defined as in (3.2), the function ϕ defined in (1.3) is concave on $[0,\infty)$. Thus ϕ satisfies both properties 1. and 2. We show in Corollary 1, Marcus and Rosen (1993) and Corollary 3.1 Marcus (1999) that by different choices of the measure μ one can get a wide class of functions ϕ which vary regularly at zero and infinity.

References

R. Banuelos, and R.G. Smits (1997). Brownian motion in cones, *Prob. Theory Related Fields* **108**, 299–319.

R.F. Bass, N. Eisenbaum, and Z. Shi (1999). *The most visited sites of symmetric stable processes*, preprint.

M.B. Marcus (1999). *The most visited sites of certain Lévy processes*, preprint.

M.B. Marcus and J. Rosen (1993). ϕ-variation of the local times of symmetric Levy processes and stationary Gaussian processes In: *Seminar on Stochastic Processes*, Progress in Probability, vol. 33, Birkhäuser, Boston, 209–220, 1992.

Department of Mathematics
The City College of CUNY
New York, NY 10031
E-mail: `mbmarcus@earthlink.net`

PART 4

Strong Approximation and Embedding

Asymptotic Independence of the Local Empirical Process Indexed by Functions

Paul Deheuvels*, Uwe Einmahl and David M. Mason‡**

ABSTRACT We obtain independent joint strong approximations for multiple local empirical processes, which are independent of the approximation to the full empirical process. Applications include strong limit laws for iterated empirical processes.

1 Introduction

Einmahl and Mason (1997) introduced a notion of a local empirical process indexed by functions, generalizing an earlier version by Deheuvels and Mason (1994), defined for the uniform empirical process on the unit cube. Among other results, they obtained a strong approximation and a compact law of the iterated logarithm [LIL] for their general local empirical process. Our aim is to adapt their method of proof to derive joint strong approximations for local empirical processes at any number of distinct fixed locations. Moreover, we do this in a way so that the approximations are not only independent of each other, but also of a suitable strong approximation to the *full* empirical process. Establishing this last independence property is actually the most delicate part of the present construction.

Our study is motivated by previous work of Deheuvels and Mason (1994), who exploited an asymptotic independence property of the local uniform empirical process at different points in the unit interval to describe in their Theorem 3.3 the joint almost sure behavior of the Bahadur-Kiefer representation at these points. They accomplished this through strong limit laws for iterated empirical processes. We provide in Section 3 an example to show how such results are proved.

Before stating our results for multiple local empirical processes, we shall

*Research partially supported by a CNRS/NSF Grant.
**Research partially supported by an FWO Grant.
‡Research partially supported by an FWO Grant, a CNRS/NSF Grant and an NSF Grant.

first define what we mean by a local empirical process and then state the Einmahl and Mason (1997) strong Gaussian approximation to it when the index set is a class of bounded functions. Let us begin by recalling the basic notation and assumptions of Einmahl and Mason (1997).

Let $X_1, X_2, \ldots$, be a sequence of i.i.d. $R^d, d \geq 1$, valued random vectors with distribution $\mathbf{P}$ on the Borel sets $\mathcal{B}$ of R^d. Given any $x \in R^d$, we set for any invertible bimeasurable transformation $h : R^d \to R^d$ and any measurable subset $J \subseteq R^d$,

$$A(h) = x + hJ. \tag{1.1}$$

Let $(h_n)_{n\geq 1}$ be a sequence of invertible bimeasurable transformations from $R^d \to R^d$ and assume that with $A_n := A(h_n)$ and $a_n := \mathbf{P}(A_n), n \geq 1$,

$$a_n > 0, \text{ for } n \geq 1; \tag{A.i}$$

$$na_n \to \infty \text{ , as } n \to \infty; \tag{A.ii}$$

$$a_n \to 0, \text{ as } n \to \infty. \tag{A.iii}$$

Define the probability measure P_n on $(R^d, \mathcal{B})$,

$$P_n(B) = \mathbf{P}(x + h_n(J \cap B))/a_n, \;\; B \in \mathcal{B}. \tag{1.2}$$

Let $\mathcal{F}$ denote a class of square $\mathbf{P}$-integrable functions on R^d with supports contained in J. To avoid measurability problems we shall assume that there exists a countable subclass of $\mathcal{F}_c$ of $\mathcal{F}$ and a measurable set D with $P_n(D) = 0$ for all $n \geq 0$ such that for any $x_1, \ldots, x_m \in R^d - D$ and $f \in \mathcal{F}$ there exists a sequence $\{f_j\}_{j\geq 1} \subset \mathcal{F}_c$ satisfying

$$\lim_{j\to\infty} f_j(x_k) = f(x_k), \text{for } k = 1, \ldots, m; \tag{S.i}$$

$$\lim_{j\to\infty} P_n(f_j) = P_n(f), \text{ for each } n \geq 1; \tag{S.ii}$$

$$\lim_{j\to\infty} P_n(f_j^2) = P_n(f^2), \text{ for each } n \geq 1. \tag{S.iii}$$

Given any integer $n \geq 1$ and invertible bimeasurable transformation $h : R^d \to R^d$, we introduce the *local empirical process at x indexed by $\mathcal{F}$*

$$L_n(f, h) = \sum_{i=1}^{n} \frac{f(h^{-1}(X_i - x)) - Ef(h^{-1}(X_i - x))}{\sqrt{n\mathbf{P}(A(h))}}. \tag{1.3}$$

Notice when $h = I$, the identity mapping, $J = R^d$ and $x = 0$, $L_n(f,h)$ is the usual empirical process on R^d indexed by the class of functions $\mathcal{F}$, which we from now on write

$$\alpha_n(f) = \sum_{i=1}^{n} \frac{f(X_i) - Ef(X_i)}{\sqrt{n}}.$$

For integers $m \geq 1$ and $n \geq 1$ define the empirical process indexed by $\mathcal{F}$

$$\alpha_m^{(n)}(f) := \sum_{i=1}^{m} \frac{f(Y_i^{(n)}) - P_n(f)}{\sqrt{m}}, f \in \mathcal{F}, \tag{1.4}$$

where $Y_1^{(n)}, \ldots, Y_m^{(n)}$ are assumed to be i.i.d. P_n.

Let

$$\mathcal{F}' = \{f - g : f, g \in \mathcal{F}\},\ \mathcal{F}^2 = \{f^2 : f \in \mathcal{F}\}, \tag{1.5}$$

$$(\mathcal{F}')^2 = \{(f-g)^2 : f, g \in \mathcal{F}\} \text{ and } \mathcal{G} = \mathcal{F} \cup \mathcal{F}^2 \cup \mathcal{F}' \cup (\mathcal{F}')^2. \tag{1.6}$$

For any functional T defined on a subset $\mathcal{H}$ of the real valued functions on J we denote

$$\| T \|_{\mathcal{H}} := \sup\{| T(f) | : f \in \mathcal{H}\}. \tag{1.7}$$

We shall require the following additional assumption on the class of functions $\mathcal{F}$ and the sequence of probability measures $(P_n)_{n \geq 1}$. There exists a probability measure P_0 such that P_n converges weakly to P_0 and

$$\| P_n - P_0 \|_{\mathcal{G}} \to 0 \text{ as } n \to \infty. \tag{F.i}$$

We shall denote by

$$(B_0(f))_{f \in \mathcal{F}} \tag{1.8}$$

a P_0-Brownian bridge indexed by $\mathcal{F}$, that is, B_0 is a Gaussian process indexed by $\mathcal{F}$ with mean-zero and covariance function

$$\mathrm{Cov}(B_0(f), B_0(g)) = P_0(fg) - P_0(f)P_0(g), \quad f,\ g \in \mathcal{F}. \tag{1.9}$$

We shall assume that B_0 has uniformly ρ_0 continuous sample paths, where

$$\rho_0^2(f,g) = Var(B_0(f) - B_0(g)), \quad f,\ g \in \mathcal{F}. \tag{1.10}$$

Such a process B_0 exists whenever the class $\mathcal{F}$ is P_0-pregaussian. [For the definition of "P_0–pregaussian" refer to Ledoux and Talagrand (1991).] Let

Z be a standard normal random variable independent of B_0 and introduce the P_0-Brownian motion W_0 indexed by $\mathcal{F}$

$$W_0(f) := B_0(f) + ZP_0(f), \quad f \in \mathcal{F}. \tag{1.11}$$

Note that W_0 has covariance function $P_0(fg)$, $f, g \in \mathcal{F}$.

Einmahl and Mason (1997) established the following strong approximation and compact LIL for the local empirical process. We shall use the notation $Lx = \log(e \vee x)$ and $[x]$ = integer part of x.

Theorem 1. *Assume the measurability conditions (S.i), (S.ii) and (S.iii) along with (A.i), (A.ii) and (A.iii). Also assume that as $n \to \infty$*

$$a_n \sim d_n \text{ with } d_n \searrow 0, nd_n \nearrow \infty \text{ and } nd_n/LLn \to \infty. \tag{A.iv}$$

Further assume (F.i) and

$$| f | \leq K \text{ for all } f \in \mathcal{F} \text{ and some } K \geq 1. \tag{F.ii}$$

In addition, assume for each $n \geq 1$ and $m \geq n$, both

$$f \circ h_n^{-1} \text{ and } f \circ h_m^{-1} \circ h_n \in \mathcal{F}; \tag{F.iii}$$

there exists a sequence of positive constants $(b_n)_{n\geq 1}$ such that for all f, $g \in \mathcal{F}$ and $n \geq 1$

$$\int_{R^d} f(h_n^{-1}z)g(h_n^{-1}z)dP_0(z) = b_n \int_{R^d} f(z)g(z)dP_0(z), \tag{F.iv}$$

where $a_n/b_n \to 1$, as $n \to \infty$. Further assume that with $k(n) = [na_n]$

$$\| \alpha_{k(n)}^{(n)} \|_{\mathcal{F}} / \sqrt{LLn} \to_P 0, \text{ as } n \to \infty; \tag{F.v}$$

$$\mathcal{F} \text{ is } P_0\text{-pregaussian.} \tag{F.vi}$$

*Then one can construct $X_1, X_2, \ldots$, i.i.d. **P** and a sequence $W_1, W_2, \ldots$, of independent P_0 -Brownian motions indexed by $\mathcal{F}$ on the same probability space, such that with probability one, as $n \to \infty$,*

$$\sup_{f\in\mathcal{F}} |L_n(f, h_n) - \frac{1}{\sqrt{nb_n}} \sum_{i=1}^{n} W_i(f \circ h_n^{-1})|/\sqrt{LLn} \to 0. \tag{1.12}$$

(Note that our statement of condition (F.iv) above corrects a misprint in expressions (1.25) of Einmahl and Mason (1997) and (F.iv) of Einmahl and Mason (1998).)

Our aim here is to establish the following generalization of Theorem 1.

Theorem 2. *Assume that there exist $X_1, X_2, \dots$, i.i.d.* **P** *and independent* **P** *-Brownian bridges $B_1, B_2, \dots$, indexed by $\mathcal{F}$ on the same probability space, such that, with probability one,*

$$\sup_{f\in\mathcal{F}} |\alpha_n(f) - \frac{1}{\sqrt{n}}\sum_{i=1}^{n} B_i(f)|/\sqrt{LLn} \to 0, \quad as\ n\to\infty. \tag{1.13}$$

Further, let $x_i + h_{i,n}J_i$, $(h_{i,n})_{n\geq 1}$, and $\mathcal{F}_i$, $i=1,\dots,M$, satisfy for each $i=1,\dots,M$, the conditions of Theorem 1, where we assume that $x_i + h_{i,n}J_i$ for $i=1,\dots,M$ are pairwise disjoint. Then we can construct a probability space on which (1.13) holds and which includes M independent sequences $(W_{i,n})_{n\geq 1}$, $i=1,\dots,M$, of independent $P_{i,0}$ -Brownian motions indexed by $\mathcal{F}_i$, with $P_{i,0}$ being the measure in (F.i) that corresponds to i, jointly independent of $(B_n)_{n\geq 1}$, such that, with probability one, for each $i=1,\dots,M$, as $n\to\infty$,

$$\sup_{f\in\mathcal{F}_i} |L_{i,n}(f,h_{i,n}) - \frac{1}{\sqrt{nb_{i,n}}}\sum_{m=1}^{n} W_{i,m}(f\circ h_{i,n}^{-1})|/\sqrt{LLn} \to 0, \tag{1.14}$$

where $L_{i,n}(f,h_{i,n})$ is the local empirical process with accompanying constant $b_{i,n}$ corresponding to $x_i + h_{i,n}J_i$, $(h_{i,n})_{n\geq 1}$, and $\mathcal{F}_i$.

As in Einmahl and Mason (1998) it is possible to obtain a version of Theorem 2 for local empirical processes indexed by certain classes of unbounded functions.

We next state a compact LIL for the full empirical process and an arbitrary number M of local empirical processes. Let $\mathbb{B}$ denote the class of all bounded functionals on $\mathcal{F}\times\mathcal{F}_1\times\cdots\times\mathcal{F}_M$ equipped with the supremum norm

$$d(\Phi,\Psi) := \sup_{\overline{f}\in\mathcal{F}\times\mathcal{F}_1\times\cdots\times\mathcal{F}_M} |\Phi(\overline{f}) - \Psi(\overline{f})|, \quad \Phi,\Psi\in\mathbb{B},$$

where $\overline{f} = (f, f_1, \cdots, f_M)$. Further let $\mathcal{K}$ be the unit ball of the reproducing kernel Hilbert space pertaining to the R^{M+1} valued Gaussian process indexed by $\mathcal{F}\times\mathcal{F}_1\times\cdots\times\mathcal{F}_M$, defined for $\overline{f}\in\mathcal{F}\times\mathcal{F}_1\times\cdots\times\mathcal{F}_M$ to be

$$\langle B_1(f), W_{1,1}(f_1), \dots, W_{M,1}(f_M)\rangle .$$

(For the definition of the term "reproducing kernel Hilbert space" refer to Kuelbs (1976) or Ledoux and Talagrand (1991).)

Our next result gives as a corollary of Theorem 2 the compact law of the iterated logarithm for the local empirical process. Its proof is a consequence of the strong approximations (1.13) and (1.14) and the fact that the same result holds for the sequence of vector processes indexed by $\overline{f}\in\mathcal{F}\times\mathcal{F}_1\times\cdots\times\mathcal{F}_M$

$$\left\langle \frac{\sum_{i=1}^{n} B_i(f)}{\sqrt{2nLLn}}, \frac{\sum_{m=1}^{n} W_{1,m}(f_1\circ h_{1,n}^{-1})}{\sqrt{2nb_{1,n}LLn}}, \dots, \frac{\sum_{m=1}^{n} W_{M,m}(f_M\circ h_{M,n}^{-1})}{\sqrt{2nb_{M,n}LLn}} \right\rangle .$$

The proof, with obvious modifications, follows that of Corollary 1.1 of Einmahl and Mason (1997) and is based upon Theorem 4.1 of Carmona and Kôno (1976). Therefore we omit the details. A special case of this corollary is Theorem 1.2 of Deheuvels and Mason (1994).

Corollary 1. *In addition to the assumptions of Theorem 2, assume the following: there exists a* $C > 0$ *such that for each* $i = 1, \dots, M$, *we have for all* $f, g \in \mathcal{F}_i$ *such that* $P_{i,0}(f-g)^2 < \infty$, $n \geq 1$, *and* $2n \geq m \geq n$,

$$\int_{R^d} \left(f(h_{i,m}^{-1}(h_{i,n}(x)) - g(h_{i,m}^{-1}(h_{i,n}(x))\right)^2 dP_{i,0}(x) \leq CP_{i,0}(f-g)^2,$$

and for every compact set $A \subset R^d$ *and* $\delta > 0$ *there exists a* $q_0 > 1$ *such that for all* $1 < q < q_0$ *with* $n_k = [q^k]$, *uniformly in* $i = 1, \dots, M$,

$$\max_{n_k \leq m \leq n_{k+1}} \sup_{x \in A} |x - h_{i,m}^{-1}(h_{i,n_k}(x))| \leq \delta,$$

for all large enough k *depending on* A *and* $\varepsilon > 0$. *Then with probability one the sequence of vector processes indexed by* $\overline{f} \in \mathcal{F} \times \mathcal{F}_1 \times \dots \times \mathcal{F}_M$,

$$\left\langle \frac{\alpha_n(f)}{\sqrt{2LLn}}, \frac{L_{1,n}(f_1, h_{1,n})}{\sqrt{2LLn}}, \dots, \frac{L_{M,n}(f_M, h_{M,n})}{\sqrt{2LLn}} \right\rangle$$

is relatively compact in $\mathbb{B}$ *with a set of limit points equal to* $\mathcal{K}$.

Applications to Corollary 1 are provided in Section 3, which includes a preliminary study of a proposed estimator of the quantile density function.

2 Proof of Theorem 2

Our proof is divided into three steps. In the first step we show how one can construct for each given $n \geq 1$ two sequences of independent random vectors $X_1, \dots, X_n$ and $\widehat{X}_1, \dots, \widehat{X}_n$ which are "partially" independent, but where the corresponding empirical processes are still close to each other. In the second step we then use the first of these empirical processes (of size n) to construct the desired independent Wiener processes approximating the M local empirical processes. In the third step we use a blocking argument and the independence properties of the second empirical processes from Step 1 to establish Theorem 2.

For notational convenience from now on we shall assume without loss of generality that the class of functions $\mathcal{F}$ is uniformly bounded by 1.

We shall need the following Bernstein type inequality which follows from Doob's maximal inequality and the proof of Bernstein's inequality on page 14 of Dudley (1984).

Fact 1. *Let* $\xi_1, \dots, \xi_m$ *be independent mean zero random variables satisfying*

$$|\xi_i| \leq K, \ 1 \leq i \leq m.$$

Then for all $t \geq 0$

$$P\left(\max_{1\leq j\leq m} |\sum_{i=1}^{j} \xi_i| \geq t\right) \leq 2\exp\left(-t^2/\left(2B_m + t\frac{2K}{3}\right)\right),$$

where $B_m := \sum_{i=1}^{m} E\xi_i^2$.

Also we record an inequality for the empirical process which can be inferred from Lemmas 3.2.1 and 3.2.7 of Dudley (1984). For the definition of a *measurable cover function* used below, see section 3.1 of Dudley (1984).

Fact 2 *For any sequence* $(\alpha_k)_{k\geq 1}$ *of empirical processes indexed by a class of functions* $\mathcal{H}$ *bounded by a finite constant* $K > 0$ *and based upon a sequence of independent but not necessarily identically distributed random variables, the following inequality holds for all* $x \geq 0$ *and* $n \geq 1$:

$$P\left(\max_{1\leq m\leq n} ||\sqrt{m}\alpha_m(f)||^*_{\mathcal{H}} \geq x + 3E||\sqrt{n}\alpha_n(f)||^*_{\mathcal{H}}\right)$$

$$\leq 2\exp\left(-x^2/\left(12K^2n\right)\right) + 2\exp(-x/(4K)),$$

where $||\cdot||^*_{\mathcal{H}}$ *is a measurable cover function of* $||\cdot||_{\mathcal{H}}$.

Step 1. Given $n \geq 1$, consider the discrete random vector

$$\tau = (I_{A_0,}, \ldots, I_{A_M}),$$

where

$$A_i = x_i + h_{i,n}J_i \text{ for } i = 1, \ldots, M \text{ and } A_0 = R^d - A_1 - \ldots - A_M.$$

Assume that $x_i + h_{i,n}J_i$ for $i = 1, \ldots, M$ are pairwise disjoint. Let

$$\tau_j = (\tau_j^{(0)}, \ldots, \tau_j^{(M)}),\ 1 \leq j \leq n,$$

be independent copies of τ and set

$$\nu_i(m) = \sum_{j=1}^{m} \tau_j^{(i)},\ 1 \leq m \leq n,\ 0 \leq i \leq M.$$

Define for each $i = 1, \ldots, M$, the probability measure $P_{i,n}$ on $(R^d, \mathcal{B})$,

$$P_{i,n}(B) = \mathbf{P}(x_i + h_{i,n}(J_i \cap B))/a_{i,n}, B \in \mathcal{B},$$

where $a_{i,n} = P(A_i)$. Let

$$Y_j^{(i)},\ 1 \leq j \leq n,\ 1 \leq i \leq M,$$

be independent of the τ_j' s and assume that each $Y_j^{(i)}$ has the distribution $P_{i,n}$. Set

$$X_j^{(i)} = h_{i,n} Y_j^{(i)} + x_i,\ 1 \le j \le n,\ 1 \le i \le M.$$

Further define independent random vectors

$$X_j^{(0)},\ 1 \le j \le n,$$

independent of the τ_j' s, each having distribution equal to the conditional distribution of X given that $X \in A_0$. Then we obtain independent random vectors on R^d, each with the same distribution as X, by setting

$$X_j := \sum_{i=0}^{M} \tau_j^{(i)} X_{\nu_i(j)}^{(i)},\ 1 \le j \le n. \tag{2.1}$$

Let $(\alpha_m(f))_{f \in \mathcal{F}}$ be the empirical process sequence based on these random vectors, i.e.

$$\alpha_m(f) = \sum_{j=1}^{m} \{f(X_j) - \mu(f)\}/\sqrt{m},\ f \in \mathcal{F},\ 1 \le m \le n,$$

where $\mu(f) = Ef(X_1)$. It is easy to see that

$$\sqrt{m}\alpha_m(f) = \sum_{i=0}^{M} \sum_{j=1}^{\nu_i(m)} f(X_j^{(i)}) - m\mu(f). \tag{2.2}$$

Further note that

$$\mu(f) = \sum_{i=1}^{M} a_{i,n} \mu_i(f) + (1 - s_n)\mu_0(f), \tag{2.3}$$

where

$$s_n = \sum_{i=1}^{M} a_{i,n}$$

and

$$\mu_i(f) = Ef(X_1^{(i)}),\ 0 \le i \le M.$$

This enables us to rewrite (2.2) as follows: for $f \in \mathcal{F}$, $1 \le m \le n$,

(2.4)

$$\sqrt{m}\alpha_m(f) = \sum_{i=0}^{M} \sum_{j=1}^{\nu_i(m)} \{f(X_j^{(i)}) - \mu_i(f)\} + \sum_{i=0}^{M} \{\nu_i(m) - E\nu_i(m)\}\mu_i(f).$$

Let $\overline{X}_j^{(i)}$, $\overline{\tau}_j$, $1 \le j \le n$, $1 \le i \le M$, be independent random vectors, which are independent of the above sequences so that

$$\overline{X}_j^{(i)} =_d X_j^{(i)},\ \overline{\tau}_j =_d \tau_j,\ 1 \le j \le n,\ \ 1 \le i \le M.$$

Let $\widetilde{X}_j$ be the random vectors defined by (2.1) with $\overline{\tau}_j$ replacing τ_j, $1 \le j \le n$, and let $\widehat{X}_j$ be the random vectors defined in terms of $X_j^{(0)}, \overline{X}_j^{(1)}, \ldots, \overline{X}_j^{(M)}$, $\overline{\tau}_j$, $1 \le j \le n$, and denote the corresponding empirical processes by $(\widetilde{\alpha}_n(f))_{f\in\mathcal{F}}$ and $(\widehat{\alpha}_n(f))_{f\in\mathcal{F}}$, respectively.

Using (2.4) and the fact that $\sup_{f\in\mathcal{F}} |f(x)| \le 1$, we readily obtain,

$$\max_{1\le m\le n} ||\sqrt{m}\alpha_m(f) - \sqrt{m}\widetilde{\alpha}_m(f)||_{\mathcal{F}} \le 3\sum_{i=0}^{M} \max_{1\le m\le n} |\nu_i(m) - \overline{\nu}_i(m)| \tag{2.5}$$

and

$$\max_{1\le m\le n} ||\sqrt{m}\widetilde{\alpha}_m(f) - \sqrt{m}\widehat{\alpha}_m(f)||_{\mathcal{F}} \tag{2.6}$$

$$\le \max_{1\le m\le n} ||\sum_{i=1}^{M} \sum_{j=1}^{\overline{\nu}_i(m)} \{f(X_j^{(i)}) - f(\overline{X}_j^{(i)})\}||_{\mathcal{F}},$$

where the $\overline{\nu}_i(m)$ are sums of the first m of the $\overline{\tau}_j^{(i)}$. Set

$$\delta_n := 4E||\sum_{j=1}^{n}\{f(X_j) - \mu(f)\}||_{\mathcal{F}} + 4\sum_{i=1}^{M}\sqrt{na_{i,n}} + 4.$$

Using (2.5) and (2.6), we get

$$P\{\max_{1\le m\le n} ||\sqrt{m}\alpha_k(f) - \sqrt{m}\widehat{\alpha}_m(f)||_{\mathcal{F}} \ge x + 3\delta_n\}$$

$$\le P\{\sum_{i=0}^{M} \max_{1\le m\le n} |\nu_i(m) - \overline{\nu}_i(m)| \ge x/6\}$$

$$+P\{\max_{1\le m\le n} ||\sum_{i=1}^{M} \sum_{j=1}^{\overline{\nu}_i(m)} \{f(X_j^{(i)}) - f(\overline{X}_j^{(i)})\}||_{\mathcal{F}} \ge x/2 + 3\delta_n\} = p_n(1) + p_n(2).$$

Noticing that

$$\sum_{i=0}^{M} \{\nu_i(m) - \overline{\nu}_i(m)\} = 0,$$

we get

$$p_n(1) \le \sum_{i=1}^{M} P\{\max_{1\le m\le n} |\nu_i(m) - \overline{\nu}_i(m)| \ge \frac{x}{12M}\}$$

$$\leq 2\sum_{i=1}^{M} P\{|\nu_i(n) - \overline{\nu}_i(n)| \geq \frac{x}{12M}\},$$

which by Fact 1 is

$$\leq 4\sum_{i=1}^{M} \exp\left(-\frac{x^2}{576M^2 n a_{i,n}}\right) + 4M \exp\left(-\frac{x}{16M}\right)$$

$$\leq 4M \exp\left(-\frac{x^2}{576M^2 n s_n}\right) + 4M \exp\left(-\frac{x}{16M}\right).$$

Next observe that

$$p_n(2) \leq \sum_{i=1}^{M} P\{\nu_i(n) \geq 2na_{i,n}\}$$

$$+P\{\max_{1\leq k\leq M} \max_{1\leq m_i\leq 2na_{i,n}} ||\sum_{i=1}^{k}\sum_{j=1}^{m_i}\{f(X_j^{(i)}) - f(\overline{X}_j^{(i)})\}||_{\mathcal{F}} \geq x/2 + 3\delta_n\}$$

$$=: p_n(3) + p_n(4).$$

Using once more Fact 1, we conclude that

$$p_n(3) \leq 2\sum_{i=1}^{M} \exp(-\frac{2}{3}na_{i,n}) \leq 2M \exp(-\frac{2}{3}n\overline{a}_n),$$

where $\overline{a}_n := \min_{1\leq i\leq M} a_{i,n}$. To bound $p_n(4)$, we first use a version of the Lévy inequality for rectangular sums of B-valued random variables given in Lemma 1 of Li and Wu (1989) to get

$$p_n(4) \leq 4P\{||\sum_{i=1}^{M}\sum_{j=1}^{[2na_{i,n}]}\{f(X_j^{(i)}) - f(\overline{X}_j^{(i)})\}||_{\mathcal{F}} \geq x/2 + 3\delta_n\}.$$

(To formally apply their lemma we define for $1 \leq i \leq M$ and $1 \leq j \leq 2na_{i,n}$,

$$Z_{i,j}(f) = f(X_j^{(i)}) - f(\overline{X}_j^{(i)}),$$

and add a sufficient number of zero variables, if necessary, to obtain a rectangular sum.)

Next we claim

$$E||\sum_{i=1}^{M}\sum_{j=1}^{[2na_{i,n}]}\{f(X_j^{(i)}) - f(\overline{X}_j^{(i)})\}||_{\mathcal{F}} \leq \delta_n. \tag{2.7}$$

To verify (2.7), observe that

$$E\|\sum_{i=1}^{M}\sum_{j=1}^{[2na_{i,n}]}\{f(X_j^{(i)})-f(\overline{X}_j^{(i)})\}\|_{\mathcal{F}}$$

$$\le 2E\|\sum_{i=1}^{M}\sum_{j=1}^{\overline{\nu}_i(n)}\{f(X_j^{(i)})-f(\overline{X}_j^{(i)})\}\|_{\mathcal{F}}+4\sum_{i=1}^{M}E|\overline{\nu}_i(n)-na_{i,n}|+4$$

$$\le 4E\|\sum_{j=1}^{n}\{f(X_j)-\mu(f)\}\|_{\mathcal{F}}+4\sum_{i=1}^{M}\sqrt{na_{i,n}}+4.$$

Now we can apply Fact 2 to the empirical process based on the random vectors $(X_j^{(i)},\overline{X}_j^{(i)})$, $j=1,\dots,[2na_{i,n}]$, $i=1,\dots,M$, with $\mathcal{H}$ consisting of the functions $h(x,y)=f(x)-f(y)$, which are bounded by $K=2$, to obtain

$$p_n(4)\le 2\exp\left(-\frac{x^2}{384ns_n}\right)+2\exp\left(-\frac{x}{16}\right).$$

Putting everything together, we get

$$P\{\max_{1\le m\le n}\|\sqrt{m}\alpha_m(f)-\sqrt{m}\widehat{\alpha}_m(f)\|_{\mathcal{F}}\ge x+3\delta_n\}$$

(2.8)
$$\le 4(M+1)\left\{\exp\left(-\frac{x^2}{576M^2ns_n}\right)+\exp\left(-\frac{x}{16M}\right)\right\}+2M\exp(-\frac{2}{3}n\overline{a}_n).$$

We note for future reference that (1.13) in combination with $\mathcal{F}$ being clearly pre-gaussian, gives

$$\|\sum_{j=1}^{n}\{f(X_j)-\mu(f)\}\|_{\mathcal{F}}/\sqrt{nLLn}\to_P 0,\ \text{as } n\to\infty,$$

which by a standard argument based upon the Hoffmann-Jørgensen inequality implies that

$$E\|\sum_{j=1}^{n}\{f(X_j)-\mu(f)\}\|_{\mathcal{F}}/\sqrt{nLLn}\to 0,\ \text{as } n\to\infty.$$

This in conjunction with assumption (A.iv) yields

(2.9) $$\delta_n/\sqrt{nLLn}\to 0,\ \text{as } n\to\infty.$$

Step 2. Write for $i=1,\dots,M$, and $1\le m\le n$,

$$T_{i,m}^{(n)}=\sum_{j=1}^{m}\{f(Y_j^{(i)})-a_{i,n}P_{i,n}(f)\}.$$

As in Proposition 3.2 of Einmahl and Mason (1997) (see also the remark following relation (3.40) there) we can find for each $i = 1, \dots, M$, independent $P_{i,0}$-Brownian bridges $B_{i,j}, j \geq 1$ so that for any given $\delta > 0$ and for $n \geq m(\delta)$

$$P\left(\sqrt{na_{i,n}LLn}\max_{1\leq m\leq n}||T^{(n)}_{i,\nu_i(m)} - \sum_{j=1}^{m}\sqrt{a_{i,n}}B_{i,j}||_{\mathcal{F}_i} \geq A\delta\right)$$

$$\leq K_1\left\{na_{i,n}\exp(-K_2\sqrt{na_{i,n}LLn}) + (Ln)^{-1.5}\right\}$$

where K_1, K_2 are constants depending on δ only and A is an absolute constant. Due to the independence of the random vectors $Y_j^{(i)}, 1 \leq i \leq M, j \geq 1$, we can assume that all the Brownian bridges $B_{i,j}, 1 \leq i \leq M, j \geq 1$ are independent. Next an inspection of the proof of Proposition 2.1, Einmahl and Mason (1997) shows that one can find independent M-dimensional normal mean zero random vectors $\overline{Z}_j = (\overline{Z}_j^{(1)}, \dots, \overline{Z}_j^{(M)}), 1 \leq j \leq n$ with $\mathrm{cov}(\overline{Z}_j) = \Sigma_n - \tilde{\Sigma}_n$, where $\Sigma_n = \mathrm{diag}(a_{i,n})_{1\leq i\leq M}$ and $\tilde{\Sigma}_n = (a_{i,n}a_{j,n})_{1\leq i,j\leq M}$, so that for each $1 \leq i \leq M$,

$$P(\max_{1\leq k\leq n}|\nu_i(k) - ka_{i,n} - \sum_{j=1}^{k}\overline{Z}_j^{(i)}| \geq 10\delta\sqrt{na_{i,n}LLn})$$

$$\leq K_3\{\exp(-K_4\sqrt{n\overline{a}_nLLn}) + (Ln)^{-1.5}\},$$

where K_3, K_4 are positive constants and $\overline{a}_n = \min_{1\leq i\leq M} a_{i,n}$. Adding independent normal random vectors Δ_j with mean-zero and $\mathrm{cov}(\Delta_j) = \tilde{\Sigma}_n$ to the vectors $\overline{Z}_j$ we obtain independent mean-zero random vectors $Z_j = (Z_j^{(1)}, \dots, Z_j^{(M)})$ with $\mathrm{cov}(Z_j) = \Sigma_n$.

Employing a standard exponential inequality for normal random variables, we can conclude that for each $1 \leq i \leq M$,

$$P(\max_{1\leq k\leq n}|\sum_{j=1}^{k}\Delta_j^{(i)}| \geq \delta\sqrt{na_{i,n}LLn}) \leq 4(Ln)^{-\delta^2/2a_{i,n}}.$$

Given that Σ_n is a diagonal matrix it is evident that the variables $Z_j^{(i)}, 1 \leq i \leq M, 1 \leq j \leq n$ are independent. Thus defining

$$\overline{W}_{i,j}(f) := B_{i,j}(f) + Z_j^{(i)}P_{i,0}(f)/\sqrt{a_{i,n}}, f \in \mathcal{F}_i,$$

we obtain *independent* $P_{i,0}$- Brownian motions, $1 \leq i \leq M, 1 \leq j \leq n$.

Step 3. We are ready to construct our probability space. Define $n_k = 2^k$ for $k = 0, 1, 2, \dots$, and set $m_0 = 0$ and $m_k = 1 + \dots + 2^k$, for $k = 1, 2, \dots$ For $j = m_{k-1} + 1$ to $m_{k-1} + n_k = m_k$, define X_j, $\widehat{X}_j$, $Y_j^{(1)}, \dots, Y_j^{(M)}$, $j =$

$m_{k-1}+1,\dots,m_k$ as above with $n=n_k$. Notice that $\widehat{X}_j$, $Y_j^{(1)},\dots,Y_j^{(M)}$, $j=m_{k-1}+1,\dots,m_k$, are independent random variables. Set for $k=1,2,..$, and $m_{k-1}+1\le m\le m_k$,

$$\sqrt{m}\alpha_m^{(k)}(f) := \sum_{j=1+m_{k-1}}^{m} \{f(X_j)-\mu(f)\},$$

$$\sqrt{m}\widehat{\alpha}_m^{(k)}(f) := \sum_{j=1+m_{k-1}}^{m} \{f(\widehat{X}_j)-\mu(f)\}.$$

We get then from inequality (2.8) with $x=\varepsilon\sqrt{n_k LLn_k}$,

(2.10)

$$P\{\max_{m_{k-1}+1\le m\le m_k} ||\sqrt{m}\alpha_m^{(k)}(f)-\sqrt{m}\widehat{\alpha}_m^{(k)}(f)||_{\mathcal{F}} \ge \varepsilon\sqrt{n_k LLn_k}+3\delta_{n_k}\}$$

$$\le 4(M+1)\left\{\exp\left(-\frac{\varepsilon^2 LLn_k}{576M^2 s_{n_k}}\right)+\exp\left(-\frac{\varepsilon\sqrt{n_k LLn_k}}{16M}\right)\right\}$$

$$+2M\exp(-\frac{2}{3}n_k\overline{a}_{n_k}) =: p_k(\varepsilon).$$

We can readily verify that

$$\sum_{k=1}^{\infty} p_k(\varepsilon)<\infty.$$

Thus we get by the Borel-Cantelli lemma and (2.9) that as $k\to\infty$,

$$\max_{m_{k-1}+1\le m\le m_k} ||\sqrt{m}\alpha_m^{(k)}(f)-\sqrt{m}\widehat{\alpha}_m^{(k)}(f)||_{\mathcal{F}}/\sqrt{n_k LLn_k}\to 0. \tag{2.11}$$

Now proceed to use the same coupling argument as in Section 3 of Einmahl and Mason (1997), beginning with their relation (3.38), to define M independent sequences $(W_{i,n})_{n\ge1}$, $i=1,\dots,M$, of independent $P_{i,0}$-Brownian motions indexed by $\mathcal{F}_i$ such that for each $i=1,\dots,M$

$$\sup_{f\in\mathcal{F}_i} |L_{i,n}(f,h_{i,n})-\frac{1}{\sqrt{nb_{i,n}}}\sum_{m=1}^{n} W_{i,m}(f\circ h_{i,n}^{-1})|/\sqrt{LLn}\to 0,$$

where $L_{i,n}(f,h_{i,n})$ is the local empirical process with accompanying constant $b_{i,n}$ corresponding to $x_i+h_{i,n}J_i$, $(h_{i,n})_{n\ge1}$, and $\mathcal{F}_i$. The independence of $(W_{i,n})_{n\ge1}$, $i=1,\dots,M$, follows from Step 2.

Now enlarge the probability space if necessary to include independent $\mathbf{P}$-Brownian bridges $B_1,B_2,\dots$, indexed by $\mathcal{F}$ such that

$$\sup_{f\in\mathcal{F}} |\widehat{\alpha}_n(f)-\frac{1}{\sqrt{n}}\sum_{i=1}^{n} B_i(f)|/\sqrt{LLn}\to 0,\ \text{as } n\to\infty, \tag{2.12}$$

where

$$\widehat{\alpha}_n(f) = \frac{1}{\sqrt{n}} \sum_{j=1}^{n} \left\{ f(\widehat{X}_j) - \mu(f) \right\}.$$

This is possible by (1.13) in combination with a well-known coupling lemma (Lemma A.1) of Berkes and Philipp (1979). Next, using (2.11), it is evident that almost surely

$$||\widehat{\alpha}_n - \alpha_n||_{\mathcal{F}}/\sqrt{LLn} \to 0, \text{ as } n \to \infty,$$

thus giving (1.13). Moreover, since according to our construction, the random vectors $\widehat{X}_1$, $\widehat{X}_2, \ldots$, are independent of the $Y_j^{(1)}, \ldots, Y_j^{(M)}$, $j = m_{k-1}+1, \ldots, m_k$, $k \geq 1$, we can choose the sequence of Brownian bridges $B_1, B_2, \ldots$, to be independent of the Wiener processes $(W_{i,n})_{n\geq 1}$, $i = 1, \ldots, M$. □

3 Application to iterated empirical processes

Recently, there has been considerable interest in the study of iterated processes of the form $X_1(X_2(\cdot))$, where $X_1(\cdot)$ and $X_2(\cdot)$ denote random processes [see, e.g., Hu, Pierre Loti Viaud and Shi (1995) and the references therein]. In this section, we will show how to apply our results to the situation where $X_1(\cdot)$ and $X_2(\cdot)$ are functions of the uniform empirical and quantile processes. We begin by introducing some necessary notation and providing some historical remarks to motivate our results.

Let $U_1, U_2, \ldots$, be a sequence of i.i.d. uniform $(0,1)$ random variables. For any $n \geq 1$ denote by $\mathbf{U}_n(u) = n^{-1}\#\{U_i \leq u : 1 \leq i \leq n\}$, for $u \in R$, the uniform empirical distribution function [df] based upon $U_1, \ldots, U_n$ and by $\mathbf{V}_n(v) = \inf\{u \geq 0 : \mathbf{U}_n(u) \geq v\}$ for $v \in [0,1)$, with $\mathbf{V}_n(1) = 1$, the corresponding uniform quantile function [qf]. Let

$$\text{(3.1)} \qquad \alpha_n(u) = n^{1/2}(\mathbf{U}_n(u) - u) \quad \text{and} \quad \beta_n(v) = n^{1/2}(\mathbf{V}_n(v) - v),$$

for $u \in R$ and $v \in [0,1]$ denote, respectively, the uniform empirical and quantile processes. We recall in (3.2) below an equality [see, e.g., (1.6) in Shorack (1982)], showing that α_n and β_n (equivalently $\mathbf{U}_n$ and $\mathbf{V}_n$) are related through an iterated process structure of the type we have in mind. Let $I(u) = u$ denote the identity, and set $\|h\|_a^b = \sup_{a\leq u\leq b}|h(u)|$ for the sup-norm of a bounded function h on $[a,b]$. In this notation, we have

$$\text{(3.2)} \quad \left\|\beta_n + \alpha_n(I + n^{-1/2}\beta_n)\right\|_0^1 = n^{1/2}\left\|\mathbf{U}_n(\mathbf{V}_n) - I\right\|_0^1 = n^{-1/2} \quad \text{a.s.}$$

The fact that the right side of (3.2) tends to 0 is the starting point of the Bahadur-Kiefer [BK] theory [see Bahadur (1966), Kiefer (1967,1970), and, e.g., Deheuvels and Mason (1990)]. In view of (3.2), the fact that α_n is close (in a sense specified in (3.3)–(3.4) below) to $\alpha_n(I-n^{-1/2}\beta_n)$, leads to the basic BK representation, which consists in approximating β_n by $-\alpha_n$. Kiefer (1970) evaluated a rate for this approximation by showing that

$$(3.3) \qquad \limsup_{n\to\infty} n^{1/4}(Ln)^{-1/2}(LLn)^{-1/4}\|\beta_n+\alpha_n\|_0^1 = 2^{-1/4} \quad \text{a.s.}$$

Making use of a functional limit law [FLL] approach, Deheuvels and Mason (1992) gave a short proof of the Kiefer (1967) pointwise version of (3.3), that is, for each fixed $t_0 \in [0,1]$, almost surely,

$$(3.4) \quad \limsup_{n\to\infty} n^{1/4}(Ln)^{-3/4}|\beta_n(t_0)+\alpha_n(t_0)| = 2^{5/4}3^{-3/4}\big(t_0(1-t_0)\big)^{1/4}.$$

Later on Arcones and Mason (1997) generalized this approach to obtain BK representations for M-estimators.

Deheuvels (2000) introduced a family of multiply iterated processes to achieve a better uniform rate of approximation of β_n than that given by (3.3). This approach makes a recursive use of (3.2) and starts with an application of the BK representation to approximate β_n by $-\alpha_{n;1} := -\alpha_n$. In the next steps, for $k = 2, 3, \ldots$, given $\alpha_{n;k-1}$, the formal replacement of β_n by $-\alpha_{n;k-1}$ in $\alpha_n(I+n^{-1/2}\beta_n)$ leads to the further approximation of β_n by the iterated process $-\alpha_{n;k} := -\alpha_n(I-n^{-1/2}\alpha_{n;k-1}))$. Deheuvels (2000) used FLL based arguments to evaluate the rate of uniform approximation of β_n by $-\alpha_{n;k}$ for an arbitrary $k \geq 1$. For example, for $k = 2$ with $\alpha_{n;2} = \alpha_n(I-n^{-1/2}\alpha_n))$, he showed [see his Example 5.1] that, almost surely,

$$(3.5) \quad \limsup_{n\to\infty} n^{3/8}(Ln)^{-3/4}(LLn)^{-3/8}\|\beta_n+\alpha_n(I-n^{-1/2}\alpha_n)\|_0^1 = 2^{-3/8}.$$

Stute (1982) [see (3.3), p.100 in his paper] had anticipated in part this result by showing that the left side of (3.5) is almost surely a constant, but his methods were not adequate to derive the exact value of this quantity. We note also that Shorack (1982) introduced the use of FLL for proving (3.3).

Before we show by example how to apply our results to iterated processes, we shall establish two corollaries of our main results, which will provide the theoretical tools to handle such applications. The following notation will be needed.

Let $L \geq 1$ and $J \geq 1$ be fixed integers. Let $0 < t_1 \leq \ldots \leq t_J < 1$ be possibly equal points of $(0,1)$, and set, for convenience $t_0 = 0$ and $t_{J+1} = 1$. Consider L sequences $\{a_{n;\ell} : n \geq 1\}$, $\ell = 1, \ldots, L$ of positive constants fulfilling the following assumptions (H.1–2–3):

(H.1) for all $1 \leq \ell \leq L$: (i) $a_{n;\ell} \to 0$; (ii) $a_{n;\ell}\downarrow$; (iii) $na_{n;\ell}\uparrow$;
(H.2) for all $1 \leq \ell \leq L$: $na_{n;\ell}/LLn \to \infty$;
(H.3) for all $1 \leq \ell < L$: $a_{n;\ell+1}/a_{n,\ell} \to 0$.

Consider the increment functions of $s \in [-1,1]$, defined, for $1 \le j \le J$ and $1 \le \ell \le L$, by

$$(3.6)\quad g_{n;j,\ell}(t) = \big(2a_{n;\ell}LLn\big)^{-1/2}\big(\alpha_n(t_j + a_{n;\ell}s) - \alpha_n(t_j)\big), \quad s \in [-1,1].$$

Introduce the sequences

$$(3.7)\qquad g_{n;j,0} = \big(2LLn\big)^{-1/2}\alpha_n(t_j) \quad \text{for} \quad 0 \le j \le J+1,$$

$$(3.8)\qquad g_{n;0}(s) = (2LLn)^{-1/2}\alpha_n(s) \quad \text{for} \quad s \in [0,1].$$

Denote by $B[-1,1]$ the set of bounded functions on $[-1,1]$ endowed with the sup-norm $\|g\| = \|g\|_{-1}^{1} = \sup_{-1\le t\le 1}|g(t)|$. Let $AC_0[-1,1]$ stand for the subset of $B[-1,1]$ composed of absolutely continuous functions g such that $g(0)=0$. For each $g \in B[-1,1]$, set

$$(3.9)\qquad |g|_H = \Big\{\int_{-1}^{1} \dot g(s)^2 ds\Big\}^{1/2} \quad \text{if} \quad g \in AC_0[-1,1],$$
$$= \infty \qquad \text{otherwise,}$$

where for each $g \in AC_0[-1,1]$, the Lebesgue derivative of g is denoted by $\dot g = \frac{d}{ds}g$. Note for further use the inequality

$$(3.10)\qquad \|g\| \le |g|_H,$$

which readily follows from (3.9) and the Cauchy-Schwarz inequality.

The following two corollaries of Theorem 2 provide extensions of Theorem 1.2 and Lemma 3.4 in Deheuvels and Mason (1994), which are useful in the study of limit laws for iterated processes. We obtain the above-mentioned results of Deheuvels and Mason (1994) by setting $L=1$ in these corollaries.

Corollary 2. *Under the above assumptions, the sequence*

$$(3.11)\qquad G_n^* = \{g_{n;0}, g_{n;j,\ell} : 1 \le j \le J, 1 \le \ell \le L\}, \quad n \ge 1,$$

is almost surely relatively compact in $B[-1,1] \times B[-1,1]^{JL}$ endowed with the product topology. The corresponding limit set $\mathbb{L}^*$ consists of all

$$(3.12)\qquad G^* = \{g,\ g_{j,\ell} : 1 \le j \le J, 0 \le \ell \le L\},$$

with g and $g_{j,\ell} \in AC_0[-1,1]$ for $1 \le j \le J$, $1 \le \ell \le L$, and such that

$$(3.13)\qquad |g|_H^2 + \sum_{j=1}^{J}\sum_{\ell=1}^{L} |g_{j,\ell}|_H^2 \le 1 \quad \textit{with} \quad g(0) = g(1) = 1.$$

Corollary 3. *Under the above assumptions, the sequence*

$$G_n = \{g_{n;j,\ell} : 1 \le j \le J, 0 \le \ell \le L\}, \quad n \ge 1, \tag{3.14}$$

is almost surely relatively compact in $R^J \times B[-1,1]^{JL}$ endowed with the product topology. The corresponding limit set $\mathbb{L}$ consists of all

$$G = \{g_{j,\ell} : 1 \le j \le J, 0 \le \ell \le L\}, \tag{3.15}$$

with $g_{j,0} \in R$ and $g_{j,\ell} \in AC_0[-1,1]$ for $1 \le j \le J$, $1 \le \ell \le L$, and such that, with $g_{0,0} = g_{J+1,0} = 0$,

$$\sum_{j=1}^{J+1} \frac{\left(g_{j,0} - g_{j-1,0}\right)^2}{t_j - t_{j-1}} + \sum_{j=1}^{J} \sum_{\ell=1}^{L} |g_{j,\ell}|_H^2 \le 1, \tag{3.16}$$

with the conventions that, whenever $t_j - t_{j-1} = 0$ for some $2 \le j \le J$,

$$\begin{aligned} \frac{\left(g_{j,0} - g_{j-1,0}\right)^2}{t_j - t_{j-1}} &= 0 \quad \textit{whenever} \quad g_{j,0} - g_{j-1,0} = 0, \\ &= \infty \quad \textit{otherwise.} \end{aligned} \tag{3.17}$$

Proof. The proofs of Corollaries 2 and 3 being very similar, we shall give only details for Corollary 3. Choose any $0 < \delta < 1$. For each constant $g \in R$, set $g^{[\delta]} = g$, and for each function $g \in B[-1,1]$, set

$$\begin{aligned} g^{[\delta]}(s) &= g(s) - g(\delta) && \text{when} \quad \delta \le s \le 1, \\ &= 0 && \text{when} \quad -\delta < s < \delta, \\ &= g(s) - g(-\delta) && \text{when} \quad -1 \le s \le -\delta. \end{aligned}$$

In view of (3.14) and (3.15), set likewise

$$G_n^{[\delta]} = \{g_{n;j,\ell}^{[\delta]} : 1 \le j \le J, 0 \le \ell \le L\} \quad n \ge 1, \tag{3.18}$$

$$G^{[\delta]} = \{g_{j,\ell}^{[\delta]} : 1 \le j \le J, 0 \le \ell \le L\}. \tag{3.19}$$

Let $R^J \times B[-1,1]^{JL}$ be endowed with the topology generated by the norm

$$|G|_0 = \Big\{ \max_{1 \le j \le J} |g_{j,0}| \Big\} \vee \Big\{ \max_{1 \le j \le J} \max_{1 \le \ell \le L} \|g_{j,\ell}\| \Big\},$$

where $G = \{g_{j,\ell} : 1 \le j \le J, 0 \le \ell \le L\}$ is as in (3.15) with $g_{j,0} \in R$ and $g_{j,\ell} \in B[-1,1]$ for $1 \le j \le J$, $1 \le \ell \le L$. An application of the results of Kiefer (1972) (see, e.g., (1.4.1) p. 493 in Mason (1988), which also follows from Corollary 1 after some routine manipulations), shows via (3.6) that, for each $j = 1, \dots, J$ and $\ell = 1, \dots, L$, almost surely,

$$\begin{aligned} & \limsup_{n\to\infty} \|g_{j,\ell}^{[\delta]} - g_{j,\ell}\| \\ (3.20) \qquad = & \limsup_{n\to\infty} \left(2a_{n;\ell}LLn\right)^{-1/2} \sup_{|s|\le\delta a_{n;\ell}} |\alpha_n(t_j+s)-\alpha_n(t_j)| \\ = & \ \delta^{1/2}. \end{aligned}$$

Since $|g_{n;j,0}^{[\delta]} - g_{n;j,0}| = 0$ for $j = 1, \dots, J$, it follows from (3.18)–(3.20) that

$$(3.21) \qquad \limsup_{n\to\infty} |G_n^{[\delta]} - G_n|_0 = \delta^{1/2}, \quad \text{a.s.}$$

Recalling the definition (3.16) of $\mathbb{L}$, we infer readily from (3.10) that, likewise, for each $G \in \mathbb{L}$,

$$(3.22) \qquad |G^{[\delta]} - G|_0 \le \delta^{1/2}.$$

In view of (3.21)–(3.22) and the fact that we may choose $\delta \in (0,1)$ as small as desired, it is easy to see that the proof of the corollary boils down to showing that, for each $0 < \delta < 1$, the sequence $G_n^{[\delta]}$ is almost surely relatively compact with limit set equal to $\mathbb{L}^{[\delta]}$.

To complete the last part of our proof, we set

$$J(\delta) = [-1, -\delta] \cup [\delta, 1],$$

and for $1 \le j \le J$ and $1 \le \ell \le L$, and $n \ge 1$,

$$A_{j,\ell;n} = t_j + a_{n;\ell} J(\delta).$$

As follows from (H.1–3), the $\{A_{j,\ell;n} : 1 \le j \le J, 1 \le \ell \le L\}$ are eventually pairwise disjoint as $n \to \infty$. Therefore we may apply Corollary 1 to this case, from which we readily conclude that the sequence $G_n^{[\delta]}$ is almost surely relatively compact in $R^J \times B[-1,1]^{JL}$, with limit set equal to $\mathbb{L}^{[\delta]} = \{G^{[\delta]} : G \in \mathbb{L}\}$, as sought. □

We shall now apply our results to an instructive example. We shall describe the limiting behavior of the following statistic, defined for each $0 \le t_0, t_1 \le 1$, to be

$$(3.23) \qquad Q_n(t_0, t_1) = n^{1/2}\{\mathbf{U}_n(\mathbf{V}_n(t_0) + (t_1 - t_0)) - t_1\}.$$

To partially motivate the study of $Q_n(t_0, t_1)$, consider the version of this statistic defined in the general framework of an i.i.d. sequence $X_1, X_2, \dots$ of random variables with df $F(x) = P(X_1 \le x)$, for $x \in R$, and quantile function $Q(t) = \inf\{x : F(x) \ge t\}$, for $t \in (0,1)$. Denote by $F_n(x) = n^{-1}\#\{X_i \le x : 1 \le i \le n\}$, for $x \in R$, the empirical df based upon the first n observations, and by $Q_n(t) = \inf\{x : F_n(x) \geq t\}$, for $t \in (0,1)$, the empirical qf pertaining to F_n. Assuming that F has a continuous and positive density f, it is of interest to estimate the quantile density function $q(t) = Q'(t) = 1/f(Q(t))$ [see, e.g., Cheng (1995,1998) and the references therein]. A simple way to do this is first to consider a histogram-type estimator of $f(x)$, such as $f_n(x) = h^{-1}(F_n(x+h) - F_n(x))$ with a suitable bandwidth $h \neq 0$, then plugging into it an estimator of the quantile function, such as $Q_n(t)$, obtaining the natural estimator $f_n(Q_n(t))$ of $1/q(t)$. A systematic investigation of quantile density function estimators generated by such arguments is beyond the scope of this paper.

We shall limit ourselves to the observation that $\mathcal{Q}_n(t_0, t_1)$ is closely related to $f_n(Q_n(t_0))$ in the special case when F is the uniform df, with $F_n = \mathbf{U}_n$, $Q_n = \mathbf{V}_n$, $F = Q = I$ and $f(x) = 1$ for $x \in [0,1]$. By setting $t = t_0$, $h = t_1 - t_0$ and $f(Q(t_0)) = 1$, with $0 \leq t_0 < t_1 \leq 1$ in the previous definitions, we obtain readily from (3.2) and (3.23) that, in this case,

$$\begin{aligned} f_n(Q_n(t_0)) - f(Q(t_0)) &= (t_1 - t_0)^{-1}\Big(\big\{\mathbf{U}_n(\mathbf{V}_n(t_0) + (t_1 - t_0)) - t_1\big\} \\ &\qquad + \big\{t_0 - \mathbf{U}_n(\mathbf{V}_n(t_0))\big\}\Big) \\ &= (t_1 - t_0)^{-1}\Big(\mathcal{Q}_n(t_0, t_1) + O(1/n)\Big) \\ &= h^{-1}\Big(\mathcal{Q}_n(t_0, t_0 + h) + O(1/n)\Big) \quad \text{a.s.} \end{aligned}$$

Given this relation, it is obvious that a precise description of the limiting behavior of $\mathcal{Q}_n(t_0, t_0 + h)$, for various choices of the bandwith h, would be of interest within the general theory of density quantile estimation. This investigation will be carried out elsewhere. We shall give here a partial result along this line for the special case when t_0 and $t_1 = t_0 + h$ are fixed. In this case, we will show in the following corollary that $\mathcal{Q}_n(t_0, t_1)$ is closely approximated by $\alpha_n(t_1) - \alpha_n(t_0)$.

Corollary 4. *For any fixed* $0 \leq t_0, t_1 \leq 1$, *we have, almost surely,*

$$\begin{aligned} &\limsup_{n\to\infty} \pm n^{1/4}(LLn)^{-3/4}\Big\{\mathcal{Q}_n(t_0, t_1) - \big(\alpha_n(t_1) - \alpha_n(t_0)\big)\Big\} \\ &\qquad = 2\big(t_0(1-t_0)\big)^{1/4} 3^{-3/4} \quad \text{if} \quad t_0 \neq t_1, \\ &\qquad = 0 \qquad\qquad\qquad\qquad\quad \text{if} \quad t_0 = t_1. \end{aligned} \tag{3.24}$$

Remark. The approximation rate in (3.24) is unexpected in the sense that it is independent of the value of t_1 as long as the latter differs from that of t_0.

Proof. First, consider the case where $0 < t_1 < 1$. Since our definitions entail that $\mathbf{V}_n(0) = 0$, we see that, when $t_0 = 0$,

$$\mathcal{Q}_n(t_0, t_1) = \mathcal{Q}_n(0, t_1) - \big(\alpha_n(t_1) - \alpha_n(0)\big) = 0.$$

Likewise, since $\mathbf{V}_n(1) = 1$, we have, when $t_0 = 1$,

$$Q_n(t_0, t_1) = Q_n(1, t_1) - \big(\alpha_n(t_1) - \alpha_n(1)\big) = 0.$$

The above two equalities establish (3.24) for either $t_0 = 0 < t_1 < 1$ or $0 < t_1 < t_0 = 1$. We may therefore limit ourselves to the proof of (3.24) in the other cases by assuming from now on that

$$0 < t_0, t_1 < 1. \tag{3.25}$$

Towards this aim, we start by writing the following chain of equalities. Set

$$R_n(t_0, t_1) = Q_n(t_0, t_1) - \big(\alpha_n(t_1) - \alpha_n(t_0)\big).$$

Recalling the definition (3.23) of $Q_n(t_0, t_1)$, we see that

$$\begin{aligned} R_n(t_0, t_1) & \\ &= \alpha_n\big(\mathbf{V}_n(t_0) + (t_1 - t_0)\big) + \beta_n(t_0) - \big(\alpha_n(t_1) - \alpha_n(t_0)\big) \\ &= \alpha_n\big(t_1 + n^{-1/2}\beta_n(t_0)\big) - \alpha_n(t_1) + \big(\beta_n(t_0) + \alpha_n(t_0)\big). \end{aligned} \tag{3.26}$$

Also notice that from (3.2) and (3.5), we get

$$\begin{aligned} &\big|\alpha_n\big(t_1 + n^{-1/2}\beta_n(t_0)\big) - \alpha_n\big(t_1 - n^{-1/2}\alpha_n(t_0)\big)\big| \\ &= O\Big(n^{-3/8}(Ln)^{3/4}(LLn)^{1/8}\Big). \end{aligned} \tag{3.27}$$

We now replace, respectively, $\beta_n(t_0)$ and $\alpha_n\big(t_1 + n^{-1/2}\beta_n(t_0)\big)$ in (3.26) by $\alpha_n\big(t_0 - n^{-1/2}\alpha_n(t_0)\big)$ and $\alpha_n\big(t_1 - n^{-1/2}\alpha_n(t_0)\big)$, which in view of (3.5) and (3.27) yields

$$\begin{aligned} R_n(t_0, t_1) &= \Big(\alpha_n\big(t_1 - n^{-1/2}\alpha_n(t_0)\big) - \alpha_n\big(t_1\big)\Big) \\ &\quad - \Big(\alpha_n\big(t_0 - n^{-1/2}\alpha_n(t_0)\big) - \alpha_n\big(t_0\big)\Big) \\ &\quad + O\Big(n^{-3/8}(Ln)^{3/4}(LLn)^{1/8}\Big) \quad \text{a.s.} \end{aligned} \tag{3.28}$$

Notice that our arguments show that the "$O(\cdot)$" term in (3.28) holds uniformly over $0 \le t_0, t_1 \le 1$. By setting $t_0 = t_1$ in (3.28), we readily obtain the version of (3.24) corresponding to this special case. Thus, in view of (3.25), we may restrict ourselves to the case when

$$0 < t_0 \ne t_1 < 1. \tag{3.29}$$

To complete our proof, we use the following notation. Keeping in mind (3.29), set

$$x_n = \Big(2LLn\Big)^{-1/2} \alpha_n(t_0),$$

$$f_{n,0}(t) = \Big(2\ell_n LLn\Big)^{-1/2}\Big(\alpha_n(t_0+\ell_n t) - \alpha_n(t_0)\Big), \tag{3.30}$$

$$f_{n,1}(t) = \Big(2\ell_n LLn\Big)^{-1/2}\Big(\alpha_n(t_1+\ell_n t) - \alpha_n(t_1)\Big),$$

for $|t| \le 1$, where

$$\ell_n = n^{-1/2}(2LLn)^{1/2}.$$

We infer from Corollary 3 that the sequence $L_n := (x_n, f_{n,0}, f_{n,1})$, $n \ge 1$ is a.s. relatively compact in $R \times B[-1,1]^2$ with limit set $\mathbb{L}$ consisting of all $(x, f_0, f_1) \in R \times AC_0[-1,1]^2$ such that

$$\frac{x^2}{t_0(1-t_0)} + \int_{-1}^{1} \{\dot f_1(t)^2 + \dot f_2(t)^2\}dt \le 1, \quad f_1(0) = f_2(0) = 0. \tag{3.31}$$

Set $\psi_n = f_{n,0} - f_{n,1}$. Obviously, the sequence $K_n = (x_n, \psi_n)$ is a.s. relatively compact in $R \times B[-1,1]$ with limit set equal to $\mathbb{K}_1 = T(\mathbb{L})$, where T denotes the mapping defined by $T(x, f_0, f_1) = (x, \psi) := (x, f_0 - f_1)$. Thus it is straightforward to show that $\mathbb{K}_1$ consists of all $(x, \psi) \in R \times AC_0[-1,1]$ such that

$$\frac{x^2}{t_0(1-t_0)} + \frac{1}{2}\int_{-1}^{1} \dot\psi(t)^2 dt \le 1. \tag{3.32}$$

Now (3.28), (3.30) and (3.32), and an easy argument following the lines of (3.51)–(3.53) in Section 3.4 in Deheuvels and Mason (1994) lead to the conclusion that

$$\begin{aligned}
&\limsup_{n\to\infty} \pm n^{1/4}(2LLn)^{-3/4}\Big\{Q_n(t_0,t_1) - \big(\alpha_n(t_1) - \alpha_n(t_0)\big)\Big\} \\
= \;& \sup\Big\{ \pm (f_1(x) - f_2(x)) : (x, f_1, f_2) \in \mathbb{L}\Big\} \\
= \;& \sup\Big\{ \pm \psi(x) : (x, \psi) \in \mathbb{K}_1\Big\} \\
= \;& \sqrt{2}\big(t_0(1-t_0)\big)^{1/4} \sup_{0\le s\le 1} \sqrt{s(1-s^2)} = 2\big(t_0(1-t_0)\big)^{1/4} 3^{-3/4} \quad \text{a.s.,}
\end{aligned}$$

which completes the proof of (3.24) when $0 < t_1 < 1$. The proof of (3.24) when either $t_1 = 0$ or $t_1 = 1$ is achieved by similar arguments and will therefore be omitted. □

Acknowledgements. The second named author thanks the universities of Delaware and Paris and the third named author thanks the Free University Brussels and the University of Paris for their hospitality, while working on this paper.

References

Arcones, M. and Mason, D.M. (1997). A general approach to Bahadur-Kiefer expansions for M-estimators, *Math. Methods of Statistics*, **25**, 267–292.

Bahadur, R. (1966). A note on quantiles in large samples. *Ann. Math. Statist.* **37**, 577–580.

Berkes, I. and Philipp, W. (1979). Approximation theorems for independent and weakly dependent random vectors. *Ann. Probab.* **7**, 29–54.

Carmona, R. and Kôno, N. (1976). Convergence en loi et lois du logarithme itéré pour les vecteurs gaussiens. *Z. Wahrscheinlichkeitstheorie verw. Gebiete*, **36**, 241–267.

Cheng, C. (1998). A Berry-Essen theorem of quantile density estimation. *Statist. and Probab. Letters.* **39**, 255–262.

Cheng, C. (1995). Uniform consistency of generalized kernel estimators of quantile density. *Ann. Statist.* **23**, 2285–2291.

Deheuvels, P. (2000). Strong approximations of quantile processes by iterated Kiefer processes. *Ann. Probab.* (to appear).

Deheuvels, P. and Mason, D.M. (1990). Bahadur–Kiefer–processes. *Ann. Probability***18**, 669–697.

Deheuvels, P. and Mason, D.M. (1992). A functional LIL approach to pointwise Bahadur–Kiefer theorems. In: *Probability in Banach Spaces, 8: Proceedings of the 8th International Conference*, (R.M. Dudley, M.G. Hahn and J. Kuelbs, eds.) pp. 255–266, Birkhäuser, Boston, 1992.

Deheuvels, P. and Mason, D.M. (1994). Functional laws of the iterated logarithm for local empirical processes indexed by sets. *Ann. Probab.* **22**, 1619–1661.

Dudley, R.M. (1984). A course on empirical processes (École d'Été de Probabilités de Saint-Flour XII-1982) *Lecture Notes in Mathematics* **1097**, 2–141, (ed. P.L. Hennequin) Springer-Verlag, New York.

Einmahl, U. and Mason, D.M. (1997). Gaussian approximation of local empirical processes indexed by functions. *Probab. Th. Re. Fields* **107**, 283–301.

Einmahl, U. and Mason, D.M. (1998). Strong approximations to the local empirical processes. In: *Proceedings of the Oberwolfach Conference on High Dimensional Probability, Progress in Probability* **43**, (eds. E. Eberlein, M. Hahn and M. Talagrand) Birkhäuser, Boston, pp. 75–92.

Hu, Y., Pierre Loti Viaud, D. and Shi, Z. (1995). Laws of the iterated logarithm for iterated Wiener processes. *J. Theor. Probab.* **9**, 303–319.

Kiefer, J. (1967). On Bahadur's representation of sample quantiles. *Ann. Math. Statist.* **36**, 1323–1342.

Kiefer, J. (1970). Deviations between the sample quantile process and the sample d.f. In *Nonparametric Techniques in Statistical Inference* (M. Puri ed.). Cambridge Univ. Press, London, pp. 299–319.

Kiefer, J. (1972). Iterated logarithm analogues for sample quantiles when $p_n \downarrow 0$. *Proc. Sixth Berkeley Symp. Math. Statist. Probab.* **1**. Univ. of California Press, Berkeley, 227–244.

Kuelbs, J. (1976). The law of the iterated logarithm and related strong convergence theorems for Banach space valued random variables. (École d'Été de Probabilités de Saint-Flour V-1975) *Lecture Notes in Mathematics* **539**, (ed. P.L. Hennequin) Springer-Verlag, New York, 224–314.

Ledoux, M. and Talagrand, M. (1991). *Probability on Banach Spaces*, Springer-Verlag, Berlin.

Li, D. and Wu, Z. (1989). The law of the iterated logarithm for B-valued random variables with multidimensional indices. *Ann. Probab.* **17**, 760–774.

Mason, D. M. (1988). A strong invariance theorem for the tail empirical process. *Ann. Inst. Henri Poincaré* **24**, 491–506.

Shorack, G.R. (1982). Kiefer's theorem via the Hungarian construction. *Z. Wahrscheinlichskeit. verw. Gebiete.* **61**, 369–373.

Stute, W. (1982). The oscillation behaviour of empirical processes. *Ann. Probab.* **10**, 86–107.

Paul Deheuvels
L.S.T.A., Université Paris VI
7 avenue du Château
92340 Bourg-la-Reine
France
e-mail: `pd@ccr.jussieu.fr`

Uwe Einmahl
Departement Wiskunde
Vrije Universiteit Brussel
Pleinlaan 2
1050 Brussels, Belgium
e-mail: `ueinmahl@vub.ac.be`

Michael J. Klass
Food and Resource Economics
University of Delaware
206 Townsend Hall
Newark, Delaware 19717
e-mail: `davidm@udel.edu`

The Azéma-Yor Embedding in Brownian Motion with Drift

Goran Peskir*

ABSTRACT Let $B = (B_t)_{t\geq 0}$ be standard Brownian motion starting at zero, let $\mu > 0$ be given and fixed, and let ν be a probability measure on $\mathbf{R}$ having a strictly positive density F'. Then there exists a stopping time τ_* of B such that

$$(B_{\tau_*} + \mu\tau_*) \sim \nu$$

if and only if the following condition is satisfied:

$$D_\mu := \int_{\mathbf{R}} e^{-2\mu x}\nu(dx) \leq 1.$$

Setting in this case $C_\mu = -(2\mu)^{-1}\log(D_\mu)$, the following *explicit* formula is valid:

$$\tau_* = \inf\left\{ t > 0 \mid (B_t + \mu t) \leq h_\mu\Big(\max_{0\leq r\leq t}(B_r + \mu r)\Big)\right\}$$

where the map $s \mapsto h_\mu(s)$ for $s > C_\mu$ is expressed through its inverse by

$$h_\mu^{-1}(x) = -\frac{1}{2\mu}\log\left(\frac{1}{1-F(x)}\int_x^\infty e^{-2\mu t}dF(t)\right) \quad (x \in \mathbf{R})$$

and we set $h_\mu(s) = -\infty$ for $s \leq C_\mu$. This settles the question raised in [6]. In addition, it is proved that τ_* is pointwise the smallest possible stopping time satisfying $(B_{\tau_*} + \mu\tau_*) \sim \nu$ which generates stochastically the largest possible maximum of the process $(B_t + \mu t)_{t\geq 0}$ up to the time of stopping. This *minimax* property characterizes τ_* uniquely. The result recovers the Azéma-Yor solution of the Skorokhod embedding problem [1] by passing to the limit when $\mu \downarrow 0$. The condition on the existence of a strictly positive density is imposed for simplicity, and more general cases can be treated similarly. The line of arguments used in the proof can be extended to treat the case of more general *nonrecurrent* diffusions.

---

* Centre for Mathematical Physics and Stochastics, supported by the Danish National Research Foundation.

MR 1991 Mathematics Subject Classification: Primary 60G40, 60J65. Secondary 60J60, 60J25, 60G44.

Key words and phrases: The Skorokhod embedding problem, Brownian motion with drift, the Azéma-Yor embedding, stopping time, the barycentre function, the Hardy-Littlewood maximal function, non-recurrent diffusion, maximum process.

1 Introduction

This note is motivated by the following question raised in [6]: Given a Brownian motion with drift $X_t = B_t + \mu t$, and a probability measure ν on $\mathbf{R}$, find a stopping time τ_* of $(X_t)_{t\geq 0}$ satisfying $X_{\tau_*} \sim \nu$, and determine conditions on ν which make such a construction possible.

It is shown in [6] that under $\mu > 0$, the condition

$$\int_{\mathbf{R}} e^{-2\mu x} \nu(dx) = 1 \tag{1.1}$$

is sufficient to carry out an explicit construction of τ_*. Observe, however, that no $\nu \neq \delta_0$ with support in $\mathbf{R}_+$ satisfies (1.1).

The main purpose of this note is to show that the condition

$$\int_{\mathbf{R}} e^{-2\mu x} \nu(dx) \leq 1 \tag{1.2}$$

is necessary and sufficient to carry out an explicit construction of τ_*. In addition, some extremal properties of τ_* are revealed which makes it interesting for applications.

It may be noted that the simple convolution-type argument applied in the proof (following (3.26) below) can also be used to extend the construction given in [6] to the general case of (1.2). This procedure should also lead to the Azéma-Yor stopping time (3.3). The derivation presented below is entirely different and reveals in a clearer manner how (1.2) is actually found.

The problem of embedding the given law ν into a Markov process $X = (X_t)_{t\geq 0}$ has been a subject of many studies. This problem was initiated by Skorokhod [15] when X is Brownian motion, and it is usually referred to as the *Skorokhod embedding problem.* Rost [14] characterizes the existence of an embedding in a general Markov process. Falkner and Fitzsimmons [5] characterize the measures which can arise as stopped distributions of a transient right process for which semipolar sets are polar. These results, however, do not offer an explicit construction of the stopping time. Azéma and Yor [1] give an explicit construction of the stopping time when X is a recurrent diffusion. Bertoin and Le Jan [3] give yet another explicit construction of the stopping time when X is a Hunt process starting at a regular recurrent point. More general embedding problems for martingales are posed and solved by Rogers [13]. In view of these results, the most interesting feature in the problem above is the fact that the Brownian motion with drift *is not recurrent.*

The idea applied in the proof below is simple and well-known: The initial problem is first transformed into an analogous martingale problem by composing X with its scale function (it should be observed that the martingale problem has some new interesting features because the initial diffusion was not recurrent); we then proceed by deriving a differential equation for the optimal stopping boundary (this step is crucial in our approach); solving then the martingale problem, we also solve the initial problem. Note that, in essence, no new argument is used in the proof (if one compares it with [1] for instance) apart from the fact that we insist on deriving a differential equation for the optimal boundary; this line of arguments clarifies the proof and makes it also more tractable for a treatment of other *non-recurrent* diffusions. Actually, the main virtue of this note is to indicate that such extensions can be worked out along the same lines.

2 The Azéma-Yor embedding

Since our construction below extends the Azéma-Yor construction [1], we shall recall a few basic facts in this direction for comparison.

1. Let $B = (B_t)_{t \geq 0}$ be standard Brownian motion starting at zero, and let ν be a centered probability measure on $\mathbf{R}$. Assume for simplicity that ν has a strictly positive density F', and denote by $S_t = \max_{0 \leq r \leq t} B_r$ the maximum process associated with B. Then the Azéma-Yor solution [1] of the Skorokhod-embedding problem states that the stopping time

$$\tau_* = \inf \{ \, t > 0 \mid B_t \leq h_0(S_t) \, \} \tag{2.1}$$

satisfies $B_{\tau_*} \sim \nu$, if the map $s \mapsto h_0(s)$ is defined through its inverse by

$$h_0^{-1}(x) = \frac{1}{1 - F(x)} \int_x^\infty t \; dF(t) \tag{2.2}$$

for $x \in \mathbf{R}$. The map $x \mapsto h_0^{-1}(x)$ is the *barycentre function* of ν.

2. Using that $F(B_{\tau_*}) \sim U(0,1)$, and substituting $F(t) = v$ in (2.2), we see that $S_{\tau_*} = h_0^{-1}(B_{\tau_*})$ is equally distributed as the *Hardy-Littlewood maximal function* of ν :

$$H_0(u) = \frac{1}{1 - u} \int_u^1 F^{-1}(v) \; dv \tag{2.3}$$

being defined on the probability space $[0,1]$ with Lebesgue measure (see [7]). From this fact, and the well-known argument of Blackwell and Dubins [4], it follows that

$$P\big(S_\sigma \geq s\big) \leq P\big(S_{\tau_*} \geq s\big) \tag{2.4}$$

for all $s \geq 0$, whenever σ is a stopping time of B satisfying $B_\sigma \sim \nu$ and $E(S_\sigma) < \infty$. These facts were observed in [2], as well as that $E(S_{\tau_*}) < \infty$ if and only if $\int_0^\infty x \log x \; \nu(dx) < \infty$. Assuming then that this holds, it was shown in [11] (see also [16]) that if equality in (2.4) is attained for all $s \geq 0$, then $\tau_* \leq \sigma$ a.s. (Equality actually can be established here if one proceeds further by mimicking the proof of Proposition 3.3 below.) Observe that this result can also be restated as follows: If $\int_0^\infty x \log x \; \nu(dx) < \infty$ and σ

is any stopping time of B satisfying $B_\sigma \sim \nu$ and $S_\sigma \sim S_{\tau_*}$, then $\tau_* \le \sigma$ a.s. Thus, less formally, it says that τ_* is pointwise the smallest possible stopping time satisfying $B_{\tau_*} \sim \nu$ which generates stochastically the largest possible maximum of the process B up to the time of stopping. Another equivalent formulation (closer to terms of Hardy-Littlewood theory [7]) is that τ_* is pointwise the smallest possible stopping time satisfying $B_{\tau_*} \sim \nu$ and $S_{\tau_*} \sim H_0$, whenever $\nu \in L^+ \log L^+$ or $H_0 \in L^1$. The minimax property just described characterizes τ_* uniquely, and makes it also more interesting for applications, e.g. in option pricing theory (see [11]).

3. Yet another extremal property of τ_* follows from the argument of Monroe [9] upon the uniform integrability of $(B_{t\wedge\tau_*})_{t\ge 0}$ established by Azéma and Yor [2]: If $\sigma \le \tau_*$ is a stopping time of B satisfying $B_\sigma \sim \nu$, then $\sigma = \tau_*$ a.s.

3 The results and proof

The basic result of this note is contained in the following theorem. The explicit formulas (3.3)–(3.4) appearing below should be compared with their relatives (2.1)–(2.2) above. Observe that the latter corresponds to the limiting case $\mu = 0$ in the former. With a view of extending this result to more general non-recurrent diffusions, we point out the line of arguments used in the proof which makes the whole construction possible. Below in Proposition 3.2 and Proposition 3.3 we shall delve deeper into the structure of the stopping time (3.3) and reveal some of its extremal properties which make it interesting for applications.

Theorem 3.1. *Let $B = (B_t)_{t\ge 0}$ be standard Brownian motion starting at zero, let $\mu > 0$ be given and fixed, and let ν be a probability measure on $\mathbf{R}$ having a strictly positive density F'. Then there exists a stopping time τ_* of B such that*

$$(B_{\tau_*} + \mu\tau_*) \sim \nu \tag{3.1}$$

if and only if the following condition is satisfied:

$$D_\mu := \int_{\mathbf{R}} e^{-2\mu x}\, \nu(dx) \le 1. \tag{3.2}$$

Setting in this case $C_\mu = -(2\mu)^{-1}\log(D_\mu)$, the following explicit formula is valid:

$$\tau_* = \inf\left\{ t > 0 \;\middle|\; (B_t + \mu t) \le h_\mu\Big(\max_{0\le r\le t} (B_r + \mu r)\Big) \right\} \tag{3.3}$$

where the map $s \mapsto h_\mu(s)$ for $s > C_\mu$ is expressed through its inverse by

$$h_\mu^{-1}(x) = -\frac{1}{2\mu} \log \left(\frac{1}{1-F(x)} \int_x^\infty e^{-2\mu t} dF(t) \right) \quad (x \in \mathbf{R}) \tag{3.4}$$

and we set $h_\mu(s) = -\infty$ *for* $s \le C_\mu$.

Proof. 1. Set $X_t = B_t + \mu t$. Then $X = (X_t)_{t \ge 0}$ solves

$$dX_t = \mu\, dt + dB_t \tag{3.5}$$

with $X_0 = 0$. The scale function of X is given by

$$S(x) = \frac{1}{2\mu}\Big(1 - e^{-2\mu x}\Big) \tag{3.6}$$

for $x \in \mathbf{R}$. Thus the process

$$\begin{aligned} Z_t := S(X_t) &= \frac{1}{2\mu}\Big(1 - e^{-2\mu X_t}\Big) \\ &= \frac{1}{2\mu}\Big(1 - \exp\big(-2\mu B_t - 2\mu^2 t\big)\Big) \end{aligned} \tag{3.7}$$

is a continuous local martingale. It is easily verified that $Z = (Z_t)_{t \ge 0}$ solves

$$dZ_t = \big(1 - 2\mu Z_t\big)\, dB_t \tag{3.8}$$

with $Z_0 = 0$. Note that $Z_t < (2\mu)^{-1}$ with $Z_t \to (2\mu)^{-1}$ P-a.s. as $t \to \infty$. Observe that the diffusion coefficient in (3.8) takes value 0 at $(2\mu)^{-1}$; however, the solution Z started at zero never reaches $(2\mu)^{-1}$.

2. Let U be a random variable satisfying $U \sim F$, where F is the distribution function of ν. Then $V := S(U) \sim G$ with

$$G(x) = F\Big(S^{-1}(x)\Big) \tag{3.9}$$

for $x < (2\mu)^{-1}$. Suppose that τ_* is a stopping time of B satisfying $Z_{\tau_*} \sim V$. Then $X_{\tau_*} = S^{-1}(Z_{\tau_*}) \sim S^{-1}(V) = U \sim F$. This shows that the initial (diffusion) problem is reduced to the martingale problem of finding a stopping time τ_* of B satisfying

$$Z_{\tau_*} \sim V. \tag{3.10}$$

Note that $V < (2\mu)^{-1}$ so that $x \mapsto G(x)$ is strictly increasing and continuous on $]-\infty, (2\mu)^{-1}]$ with $G(-\infty) = 0$ and $G((2\mu)^{-1}) = 1$.

3. To solve the problem (3.10) we shall introduce the maximum process:

$$S_t = \max_{0 \le r \le t} Z_r \tag{3.11}$$

and consider the following stopping time:

$$\tau_g = \inf \{ t > 0 \mid Z_t \le g(S_t) \} \tag{3.12}$$

with a map $s \mapsto g(s)$ being defined on $[0, (2\mu)^{-1}]$. Motivated by the properties of G, it is natural to assume that $s \mapsto g(s)$ is an increasing C^1-function satisfying $g(0) = -\infty$ and $g\big((2\mu)^{-1}\big) = (2\mu)^{-1}$. Our main aim in the sequel is to show how to pick up a map $s \mapsto g_*(s)$ out of all admissible candidates just specified, so that (3.10) holds with $\tau_* = \tau_{g_*}$.

4. We show that $Z_{\tau_g} \sim V$ if and only if the inverse $x \mapsto g^{-1}(x)$ of the map $s \mapsto g(s)$ solves the following differential equation:

$$(g^{-1})'(x) - \frac{G'(x)}{1 - G(x)}\, g^{-1}(x) = -\frac{G'(x)}{1 - G(x)} x \tag{3.13}$$

for $x < (2\mu)^{-1}$. To prove this claim, we shall first verify that (cf. [8], [1])

$$F_g(s) := P\{S_{\tau_g} \le s\} = 1 - \exp\left(- \int_0^s \frac{dt}{t - g(t)} \right) \tag{3.14}$$

for all $0 \le s < (2\mu)^{-1}$. It is important to realize that this fact generally holds only for $s < (2\mu)^{-1}$, and depending on the magnitude of the left-hand derivative of $s \mapsto g(s)$ at $(2\mu)^{-1}$, the distribution function of S_{τ_g} may have a jump at $(2\mu)^{-1}$. Observe, however, that

$$\Big\{ S_{\tau_g} = (2\mu)^{-1} \Big\} = \Big\{ Z_{\tau_g} = (2\mu)^{-1} \Big\} = \big\{ \tau_g = \infty \big\}. \tag{3.15}$$

Hence the fact that $x \mapsto G(x)$ is continuous at $(2\mu)^{-1}$ will have as a consequence in our derivation below that the set in (3.14) is of P-measure zero. This is a quick way of establishing that $\tau_g < \infty$ P-a.s. upon a proper choice of the boundary $s \mapsto g(s)$ for $s < (2\mu)^{-1}$.

5. To derive (3.14), we may use the fact that

$$H(S_t) - (S_t - Z_t) H'(S_t) \tag{3.16}$$

is a continuous local martingale whenever $H \in C^2$, which is easily verified by Itô's formula. We shall apply this fact with $H_\varepsilon(s) = \int_0^s h_\varepsilon(r)\, dr$ where $h : [0, (2\mu)^{-1}[\; \to \mathbf{R}_+$ is any bounded C^1-function which is zero on $[0, \varepsilon]$. Then the process (3.16) is uniformly bounded and therefore it is a uniformly integrable martingale. Thus by the optional sampling theorem we get

$$E\Big(H_\varepsilon(S_{\tau_g}) \Big) = E\Big((S_{\tau_g} - Z_{\tau_g}) H'_\varepsilon(S_{\tau_g}) \Big). \tag{3.17}$$

Integrating by parts, the left-hand side in (3.17) becomes

$$
\begin{aligned}
(3.18) \qquad E\Big(H_\varepsilon(S_{\tau_g})\Big) &= \int_0^{1/2\mu} H_\varepsilon(s)\, dF_g(s) \\
&= H_\varepsilon(s)F_g(s)\Big|_0^{1/2\mu} - \int_0^{1/2\mu} F_g(s-)\, dH_\varepsilon(s) \\
&= H_\varepsilon(1/2\mu) - \int_0^{1/2\mu} F_g(s-)\, dH_\varepsilon(s) \\
&= \int_0^{1/2\mu} \Big(1 - F_g(s-)\Big) h_\varepsilon(s)\, ds.
\end{aligned}
$$

On the other hand, the right-hand side in (3.17) is equal to

$$
\begin{aligned}
(3.19) \qquad E\Big((S_{\tau_g} - Z_{\tau_g})H'_\varepsilon(S_{\tau_g})\Big) &= E\Big(\big(S_{\tau_g} - g(S_{\tau_g})\big)H'_\varepsilon(S_{\tau_g})\Big) \\
&= \int_0^{1/2\mu} \big(s - g(s)\big) h_\varepsilon(s)\, dF_g(s).
\end{aligned}
$$

Letting $\varepsilon \downarrow 0$ in (3.18) and (3.19), we find that

$$
(3.20) \qquad \int_0^{1/2\mu} \Big(1 - F_g(s-)\Big) h(s)\, ds = \int_0^{1/2\mu} \big(s - g(s)\big) h(s)\, dF_g(s)
$$

for all bounded C^1-functions $h : [0, (2\mu)^{-1}[\to \mathbf{R}_+$. This shows that

$$
(3.21) \qquad \frac{dF_g}{ds}(s) = \frac{1 - F_g(s)}{s - g(s)}
$$

for $0 < s < (2\mu)^{-1}$. The equation (3.21) is easily solved, and since clearly $F_g(0) = 0$, this leads to (3.14) above.

6. Suppose now that $Z_{\tau_g} \sim V$ for some g. Then by (3.15) we find that $P\{S_{\tau_g} = (2\mu)^{-1}\} = 0$, and therefore (3.14) holds for $s = (2\mu)^{-1}$ as well. This implies that

$$
\begin{aligned}
(3.22) \qquad 1 - G(x) &= P\big\{Z_{\tau_g} > x\big\} = P\big\{g(S_{\tau_g}) > x\big\} \\
&= P\big\{S_{\tau_g} > g^{-1}(x)\big\} = \exp\left(- \int_0^{g^{-1}(x)} \frac{dt}{t - g(t)} \right) \\
&= \exp\left(- \int_{-\infty}^{x} \frac{dr}{g'\big(g^{-1}(r)\big)\big(g^{-1}(r) - r\big)} \right)
\end{aligned}
$$

for all $x < (2\mu)^{-1}$ upon substituting $g(t) = r$. Differentiating over x in (3.22), we see that

$$
(3.23) \qquad \frac{G'(x)}{1 - G(x)} = \frac{1}{g'\big(g^{-1}(x)\big)\big(g^{-1}(x) - x\big)}
$$

for all $x < (2\mu)^{-1}$, and this equation is equivalent to (3.13). On the other hand, if g solves (3.13), or equivalently (3.23), then the final equality in (3.22) follows upon integrating in (3.23). This proves the claim above.

7. The general solution of (3.13) is given by

$$g^{-1}(x) = \frac{1}{1 - G(x)} \left(C - \int_{-\infty}^{x} t \, dG(t) \right) \tag{3.24}$$

where $C \in \mathbf{R}$ is a constant. Since $G((2\mu)^{-1}) = 1$, and we want $g^{-1}((2\mu)^{-1}) = (2\mu)^{-1}$, we see that the following identity must hold:

$$C = E(V) = \int_{-\infty}^{1/2\mu} t \, dG(t). \tag{3.25}$$

This gives the following explicit expression:

$$g_*^{-1}(x) = \frac{1}{1 - G(x)} \int_{x}^{1/2\mu} t \, dG(t) \tag{3.26}$$

for $x < (2\mu)^{-1}$. Hence we see that $g_*^{-1}(-\infty) = E(V)$, and thus $g_*\big(E(V)\big) = -\infty$, so everything agrees fine if $E(V) = 0$. Observe, however, that the whole preceding construction can be carried out in exactly the same way also for those functions g_* for which $0 \le E(V) < (2\mu)^{-1}$. The identity (3.14) holds then for all $s \in [E(V), (2\mu)^{-1}[$ with $F_g(E(V)) = 0$. The given τ_{g_*} is then described as follows: Let Z first hit $E(V)$, and then starting afresh, the time τ_{g_*} is obtained by adding that time onto the time needed for $Z_t \ge g_*(S_t)$ to happen. More formally, this can be described by $\tau_* = \tau_{s_*} + \tau_{g_*} \circ \theta_{\tau_{s_*}}$, where we set $s_* = E(V)$ and $\tau_{s_*} = \inf \{\, t > 0 \mid Z_t = s_* \,\}$. Observe that the more complicated form of this stopping time when $E(V) > 0$ has been formalised in terms of the initial process X by assigning the value $-\infty$ to h_μ in (3.3) above.

8. These considerations show that the condition

$$0 \le E(V) < (2\mu)^{-1} \tag{3.27}$$

is sufficient to solve the martingale problem (3.10). Note that the second inequality is trivial, and the first one is equivalently written as

$$E\Big(e^{-2\mu U}\Big) \le 1 \tag{3.28}$$

which is exactly the condition (3.2). However, this condition is necessary as well, as is easily seen from the fact that the process

$$\exp\Big(- 2\mu B_t - 2\mu^2 t \Big) \tag{3.29}$$

is a continuous local martingale (by the optional sampling theorem and Fatou's lemma). Thus the condition (3.2) is necessary and sufficient, and the martingale problem (3.10) is solved.

9. It remains to transfer this solution back to the initial problem. For this, note that from the definition of the stopping time τ_* and the process Z, we have

$$\begin{aligned}\tau_* &= \inf\{\, t > 0 \mid Z_t \le g_*(S_t)\,\} \\ &= \inf\{\, t > 0 \mid X_t \le (S^{-1} \circ g_* \circ S)(X_t^*)\,\}\end{aligned} \tag{3.30}$$

where we denote $X_t^* = \sup_{0 \le r \le t} X_r$ and define $g_*(s) = -\infty$ for $s \le E(V)$ if $E(V) > 0$. Setting $h_\mu = S^{-1} \circ g_* \circ S$, we see that

$$h_\mu^{-1}(x) = S^{-1}\Big(g_*^{-1}(S(x))\Big) = S^{-1}\Big(\frac{1}{1-F(x)}\int_x^\infty S(t)\; dF(t)\Big) \tag{3.31}$$

by means of (3.9). Now using that $S^{-1}(z) = -(2\mu)^{-1}\log(1-2\mu z)$, we end up with the formula (3.4). This can be then formally expressed in terms of the map $s \mapsto h_\mu(s)$ by assigning the value $-\infty$ as described above in the lines following (3.26). The proof is complete. □

Remarks:

1. It is easily verified by Jensen's inequality that under condition (3.2) any random variable $U \not\equiv 0$ satisfying $U \sim \nu$ must also satisfy $E(U) > 0$. This expectation may be $+\infty$. Moreover, it easily follows by the Burkholder-Davis-Gundy inequality and the optional sampling theorem (see e.g. [12]) that $E(U) < \infty$ if and only if $E(\tau_*) < \infty$; then $E(\tau_*) = \mu^{-1}E(U)$. In this case $E(\max_{0 \le t \le \tau_*} |X_t|) < \infty$ and the process $(X_{t\wedge\tau_*})_{t\ge 0}$ is uniformly integrable.

2. Observe that the function $x \mapsto h_\mu^{-1}(x)$ appearing in (3.4) may be viewed as an extension of the Hardy-Littlewood function (2.2) from the case of standard Brownian motion (when $\mu = 0$) to the case of Brownian motion with drift $\mu > 0$. It may be interesting to examine a role of this extension in the context of Hardy-Littlewood theory [7].

Our next aim is to explore the minimax property stated in Section 2 above in the case of Brownian motion with drift.

Proposition 3.2 (The minimax property). *Under the assumptions of Theorem 3.1, let σ be any stopping time of B satisfying*

$$(B_\sigma + \mu\sigma) \sim \nu. \tag{3.32}$$

Then the following inequality holds:

$$P\Big\{\max_{0\le t\le\sigma}(B_t + \mu t) \ge s\Big\} \le P\Big\{\max_{0\le t\le\tau_*}(B_t + \mu t) \ge s\Big\} \tag{3.33}$$

for all $s \ge 0$. Suppose, moreover, that ν satisfies

$$\int_0^\infty x\; \nu(dx) < \infty. \tag{3.34}$$

If σ is any stopping time of B satisfying (3.32) and

$$E\Big(\max_{0\le t\le\sigma}(B_t+\mu t)\Big)=E\Big(\max_{0\le t\le\tau_*}(B_t+\mu t)\Big)\,, \tag{3.35}$$

then the following inequality holds:

$$\tau_* \le \sigma \;\; P\text{-}a.s. \tag{3.36}$$

Remarks: 1. Observe from (3.33) and integration by parts that the left-hand expectation in (3.35) is always smaller than the right-hand one. Clearly, under (3.34) the condition (3.35) is equivalent to the fact that equality in (3.33) is attained for all $s \ge 0$, or in other words, that

$$\max_{0\le t\le\sigma}(B_t+\mu t)\sim\max_{0\le t\le\tau_*}(B_t+\mu t). \tag{3.37}$$

Thus, the result above shows that τ_* is *pointwise the smallest possible stopping time that generates stochastically the largest possible maximum* of the Brownian motion with drift up to the time of stopping. Observe that this minimax property characterizes τ_* uniquely, and that equality in (3.36) actually holds (see the proof of Proposition 3.3 below). Moreover, a slight modification of the argument below (if dealing with $(Z_t)_{t\ge0}$ instead of $(X_t)_{t\ge0}$ in part (II) of the proof) shows that (3.36) (with equality) is still valid under (3.37) even if (3.34) fails. The minimax property is of interest for applications whenever one wishes to stop the process at a prescribed law as soon as possible after the highest point feasible was reached (for more details see e.g. [11]).

2. Note that (3.33) holds without additional assumptions on the size of $\max_{0\le t\le\sigma}(B_t+\mu t)$ (compare it with Proposition 2.2 in [11]). This fact is intuitively clear as there is no recurrence.

Proof. (I): Set $X_t = B_t + \mu t$ and let $Z_t = S(X_t)$ where $x \mapsto S(x)$ is the scale function of X given by (3.6). Clearly, for (3.33) it is enough to prove that for all $0 < s < (2\mu)^{-1}$ we have

$$P\Big\{\max_{0\le t\le\sigma} Z_t\ge s\Big\}\le P\Big\{\max_{0\le t\le\tau_*} Z_t\ge s\Big\}. \tag{3.38}$$

Recall also from the proof above that $Z_\sigma \sim G$, and that

$$\tau_* = \inf\big\{\, t>0 \mid Z_t \le g_*(S_t)\,\big\} \tag{3.39}$$

where $s \mapsto g_*(s)$ for $s > E(V)$ is given through (3.26), and we set $g_*(s) = -\infty$ for $s \le E(V)$.

To prove (3.38) we will modify the well-known argument of Blackwell and Dubins [4]. Observe that our proof below in essence is the same as the proof of Proposition 2.2 in [11], and we include it here merely for completeness. In the sequel we use the notation (3.11).

By Doob's maximal inequality (see e.g. [12]), we have

$$sP\{S_{\sigma\wedge t} \geq s\} \leq \int_{\{S_{\sigma\wedge t}\geq s\}} Z_{\sigma\wedge t}\, dP \tag{3.40}$$

for all $t > 0$ with $s > 0$ given and fixed. Since $Z_t < (2\mu)^{-1}$ for all t, we may use Fatou's lemma, and by letting $t \to \infty$ in (3.40), we get

$$sP\{S_\sigma \geq s\} \leq \int_{\{S_\sigma\geq s\}} Z_\sigma\, dP. \tag{3.41}$$

Setting $A = \{S_\sigma \geq s\}$, we have

$$\begin{aligned}\int_A Z_\sigma\, dP &\leq \int_A (Z_\sigma \vee x)\, dP = \int_A \big(x + (Z_\sigma - x)^+\big)\, dP \\ &\leq xP(A) + \int (Z_\sigma - x)^+\, dP \\ &= xP(A) + \int_{\{Z_\sigma > x\}} Z_\sigma\, dP - xP\{Z_\sigma > x\} \\ &= x\Big(P(A) - P\{Z_\sigma > x\}\Big) + \int_{\{Z_\sigma > x\}} Z_\sigma\, dP\end{aligned} \tag{3.42}$$

for all $x \in \mathbf{R}$. Thus, if we choose x such that $P(A) - P\{Z_\sigma > x\} = 0$, or in other words, $G(x) = 1 - P(A)$, then by (3.41) and (3.42) we get

$$sP\{S_\sigma \geq s\} \leq \int_{\{Z_\sigma > x\}} Z_\sigma\, dP. \tag{3.43}$$

Since $G(Z_\sigma) \sim U(0,1)$ by our assumption (3.32), we have

$$\begin{aligned}\int_{\{Z_\sigma > x\}} Z_\sigma\, dP &= \int_{\{G(Z_\sigma)>G(x)\}} G^{-1}\big(G(Z_\sigma)\big)\, dP \\ &= \int_{G(x)}^1 G^{-1}(v)\, dv = \int_{1-P(A)}^1 G^{-1}(v)\, dv.\end{aligned} \tag{3.44}$$

On the other hand, since $G(Z_{\tau_*}) \sim U(0,1)$, we have

$$\begin{aligned}P\{S_{\tau_*} \geq s\} &= P\{g_*^{-1}(Z_{\tau_*}) \geq s\} \\ &= P\Big(\frac{1}{1-G(Z_{\tau_*})}\int_{G(Z_{\tau_*})}^1 G^{-1}(v)\, dv \geq s\Big) \\ &= \lambda\Big(u \in [0,1] \;\Big|\; \frac{1}{1-u}\int_u^1 G^{-1}(v)\, dv \geq s\Big) = 1 - u_*\end{aligned} \tag{3.45}$$

where u_* is the *smallest* u in $[0,1]$ for which

$$\frac{1}{1-u}\int_u^1 G^{-1}(v)\,dv \geq s \tag{3.46}$$

since the function on the left-hand side in (3.46) is increasing in u.

From (3.43) and (3.44) we now see that

$$sP(A) \leq \int_{1-P(A)}^1 G^{-1}(v)\,dv \tag{3.47}$$

and therefore $u_* \leq 1-P(A)$. This by (3.45) shows that

$$P\{S_\sigma \geq s\} = P(A) \leq 1-u_* = P\{S_{\tau_*} \geq s\} \tag{3.48}$$

and the proof of (3.38) is complete. Thus (3.33) is established.

(II): Due to its complexity and length, we shall only outline a proof of (3.36) which follows the line of arguments from [11]. Once the idea is properly understood, we believe that the text indicated below should be completed with no difficulties.

From Remark 1 following the proof of Theorem 3.1 above we know that under (3.34) the right-hand side expectation in (3.35) is finite. Motivated by facts from [11], consider the following optimal stopping problem:

$$\sup_\tau E\Big(\max_{0\leq t\leq \tau} X_t - \int_0^\tau c(X_t)\,dt\Big) \tag{3.49}$$

where the map $x \mapsto c(x)$ is defined by

$$c(x) = \frac{F'(x)S'(x)}{2\big(1-F(x)\big)S'\Big(S^{-1}\Big(\frac{1}{1-F(x)}\int_x^\infty S(t)\,dF(t)\Big)\Big)}. \tag{3.50}$$

Then it is possible to verify that the map $s \mapsto h_*(s)$ given through its inverse by the formula

$$h_*^{-1}(x) = S^{-1}\Big(\frac{1}{1-F(x)}\int_x^\infty S(t)\,dF(t)\Big) \quad (x \in \mathbf{R}) \tag{3.51}$$

is the maximal solution of the first-order nonlinear differential equation

$$g'(s) = \frac{S'\big(g(s)\big)}{2c\big(g(s)\big)\Big(S(s)-S\big(g(s)\big)\Big)}, \tag{3.52}$$

which stays strictly below the diagonal in $\mathbf{R}^2$, and thus by the maximality principle (see [10]) the stopping time τ_* is optimal for the problem (3.49).

(Observe that the maps in (3.51) and (3.4) coincide.) Then under the assumption (3.32) above, together with (3.35) being finite, one can extend the idea and argument used in Corollary 2.3 in [11] and prove first that

$$E\Big(\int_0^\sigma c(X_t)\,dt\Big) = E\Big(\int_0^{\tau_*} c(X_t)\,dt\Big) \tag{3.53}$$

as well that (3.36) then in essence follows from the optimality of τ_* within the problem (3.49). The key equation in this process is

$$\frac{1}{2}H''(x) + \mu H'(x) = c(x) \tag{3.54}$$

which is treated easily in this context. We omit all remaining details for simplicity. □

We shall conclude our considerations by examining yet another extremal property of τ_* which extends the argument of Monroe [9] to the Brownian motion with drift.

Proposition 3.3. *Under the assumptions of Theorem 3.1, let σ be any stopping time of B satisfying (3.32). If $\sigma \le \tau_*$, then $\sigma = \tau_*$ P-a.s.*

Proof. Set $X_t = B_t + \mu t$ and let $Z_t = S(X_t)$ where $x \mapsto S(x)$ is the scale function of X given by (3.6). Recall that $Z = (Z_t)_{t\ge 0}$ is a continuous local martingale satisfying $Z_t < (2\mu)^{-1}$ for all t. Choose a localization sequence of bounded stopping times $(\gamma_n)_{n\ge 1}$ for Z. Then $(Z_{t\wedge\gamma_n})_{t\ge 0}$ is a uniformly integrable martingale for all $n \ge 1$. Thus by the fact that $Z_{\tau_*} \sim Z_\sigma$ and the optional sampling theorem, we get

$$\begin{aligned} \int_{\{Z_{\tau_*}\ge z\}} Z_{\tau_*}\,dP &= \int_{\{Z_\sigma \ge z\}} Z_\sigma\,dP = \lim_{n\to\infty}\int_{\{Z_{\sigma\wedge\gamma_n}\ge z\}} Z_{\sigma\wedge\gamma_n}\,dP \\ &= \lim_{n\to\infty}\int_{\{Z_{\sigma\wedge\gamma_n}\ge z\}} Z_{\tau_*\wedge\gamma_n}\,dP = \int_{\{Z_\sigma\ge z\}} Z_{\tau_*}\,dP \end{aligned} \tag{3.55}$$

for all $z \in \mathbf{R}$. We now claim that this implies that $\{Z_{\tau_*} \ge z\} = \{Z_\sigma \ge z\}$ P-a.s. for all $z \in \mathbf{R}$, which in turn easily shows that $Z_{\tau_*} = Z_\sigma$ P-a.s.

For this, note by (3.32) and (3.55) that

$$\begin{aligned} &\int_{\{Z_{\tau_*}\ge z\}} (Z_{\tau_*}-z)\,dP = \int_{\{Z_\sigma\ge z\}} (Z_{\tau_*}-z)\,dP \\ &= \int_{\{Z_\sigma\ge z, Z_{\tau_*}\ge z\}} (Z_{\tau_*}-z)\,dP + \int_{\{Z_\sigma\ge z, Z_{\tau_*}< z\}} (Z_{\tau_*}-z)\,dP \\ &\le \int_{\{Z_{\tau_*}\ge z\}} (Z_{\tau_*}-z)\,dP \end{aligned} \tag{3.56}$$

with a strict inequality if $P\{Z_\sigma \ge z, Z_{\tau_*} < z\} > 0$. Thus $\{Z_\sigma \ge z\} \subseteq \{Z_{\tau_*} \ge z\}$ P-a.s., and since these two sets have equal P-probability, this proves the claim above.

Let now ρ be any stopping time of B satisfying $\sigma \le \rho \le \tau_*$. Then by the optional sampling theorem, and the fact that $Z_{\tau_*} = Z_\sigma$ P-a.s., we get

$$(3.57) \qquad (Z_\rho - n)^+ \le E\Big((Z_{\tau_*} - n)^+ \big| \mathcal{F}_\rho\Big) = E\Big((Z_\sigma - n)^+ \big| \mathcal{F}_\rho\Big) = (Z_\sigma - n)^+$$

for all $n \ge 1$. Letting $n \to \infty$ we see that $Z_\rho \le Z_\sigma$ P-a.s. Clearly, this is only possible if $\sigma = \tau_*$ P-a.s. The proof is complete. □

Remark: Observe that no uniform integrability condition is needed for this result (recall Remark 1 following Theorem 3.1). This is in contrast with the case when the drift is zero (recall Subsection 3 of Section 2) and can be intuitively explained by absence of recurrence (recall also Remark 2 stated after Proposition 3.2).

References

[1] AZÉMA, J. and YOR, M. (1979). Une solution simple au probleme de Skorokhod. *Sem. Probab.* XIII, *Lecture Notes in Math.* 721, Springer (90–115).

[2] AZEMA, J. and YOR, M. (1979). Le probleme de Skorokhod: Complements a l'expose precedent. *Sem. Probab.* XIII, *Lecture Notes in Math.* 721, Springer (625–633).

[3] BERTOIN, J. and LE JAN, Y. (1992). Representation of measures by balayage from a regular recurrent point. *Ann. Probab.* **20** (538–548).

[4] BLACKWELL, D. and DUBINS, L. (1963). A converse to the dominated convergence theorem. *Illinois J. Math.* **7** (508–514).

[5] FALKNER, N. and FITZSIMMONS, P.J. (1991). Stopping distributions for right processes. *Probab. Theory Related Fields* **89** (301–318).

[6] GRANDITS, P. (1998). Embedding in Brownian motion with drift. *Institute of Statistics, University of Vienna.* Preprint.

[7] HARDY, G. H. and LITTLEWOOD, J. E. (1930). A maximal theorem with function theoretic applications. *Acta Math.* **54** (81–116).

[8] LEHOCZKY, J. P. (1977). Formulas for stopped diffusion processes with stopping times based on the maximum. *Ann. Probab.* **5** (601–607).

[9] MONROE, I. (1972). On embedding right continuous martingales in Brownian motion. *Ann. Math. Statist.* **43** (1293–1311).

[10] PESKIR, G. (1997). Optimal stopping of the maximum process: The maximality principle. *Research Report* No. 377, *Dept. Theoret. Statist. Aarhus* (30 pp). *Ann. Probab.* **26**, 1998 (1614–1640).

[11] PESKIR, G. (1997). Designing options given the risk: The optimal Skorokhod embedding problem. *Research Report* No. 389, *Dept. Theoret. Statist. Aarhus* (18 pp). *Stochastic Process. Appl.* **81**, 1999 (25–38).

[12] REVUZ, D. and YOR, M. (1994). *Continuous Martingales and Brownian Motion.* (Second Edition) Springer-Verlag.

[13] ROGERS, L. C. G. (1993). The joint law of the maximum and terminal value of a martingale. *Probab. Theory Relat. Fields* **95** (451–466).

[14] ROST, H. (1971). The stopping distributions of a Markov process. *Invent. Math.* **14** (1–16).

[15] SKOROKHOD, A. (1965). *Studies in the theory of random processes.* Addison-Wesley, Reading.

[16] VAN DER VECHT, D. P. (1986). Ultimateness and the Azéma-Yor stopping time. *Sem. Probab.* XX, *Lecture Notes in Math.* 1204, Springer (375–378).

Institute of Mathematics
University of Aarhus
Ny Munkegade
8000 Aarhus, Denmark
e-mail: goran@imf.au.dk
http://www.imf.au.dk/~goran

(Department of Mathematics, University of Zagreb
Bijenička 30, 10000 Zagreb, Croatia)

A New Way to Obtain Estimates in the Invariance Principle

Alexander I. Sakhanenko*

ABSTRACT This paper presents a new simple method of obtaining estimates on rates of convergence in the invariance principle, which may be used in arbitrary separable linear spaces. This method is applied to one-dimensional and infinite dimensional martingales, among other examples.

1 Introduction

Let $Y_1, \ldots, Y_n$ be a sequence of random variables taking their values in a separable linear space $\mathcal{X}$ with the norm $\|\cdot\|$. Given the joint distribution of the random variables $X_1, \ldots, X_n$ with values in $\mathcal{X}$, suppose that this joint distribution is close, in some sense, to the joint distribution of the given random variables $Y_1, \ldots, Y_n$.

On the same probability space, where the random variables $Y_1, \ldots, Y_n$ are defined, we want to construct such random variables $X_1, \ldots, X_n$ that the distance

$$\Delta_n = \max_{1 \le k \le n} \left\| \sum_{m=1}^{k} X_m - \sum_{m=1}^{k} Y_m \right\| \tag{1.1}$$

would be small in some sense. The latter means that the value $\mathbf{P}(\Delta_n > x)$ is to be as small as possible for all $x > 0$ or for a specially chosen $x > 0$.

Solutions to this problem are usually the most difficult parts in deriving rates of convergence in the invariance principle. The main goal of this paper is to describe a new simple method to construct the random variables $X_1, \ldots, X_n$ and $Y_1, \ldots, Y_n$ as in (1.1) with a good rate of approximation.

This method is based upon the following simple idea. Suppose we have found smoothing random variables $V_0, V_1, \ldots, V_n$ that have specially chosen distributions and such that the differences

$$\delta_k = (X_k + V_k) - (Y_k + V_{k-1}), \quad k = 1, \ldots, n, \tag{1.2}$$

*This research is partially supported by the Grant of RFFI: 99-01-00561 and by the Grant of INTAS: 98-1625.

are small in some sense. In this case

$$(1.3) \qquad \forall k \quad \sum_{m=1}^{k} X_m - \sum_{m=1}^{k} Y_m = \sum_{m=1}^{k} \delta_m - V_k + V_0.$$

The representation (1.3) plays the key role in the method. For example, it follows from (1.3) that

$$(1.4) \qquad \Delta_n \le \|V_0\| + \max_{1\le k\le n} \|V_k\| + \max_{1\le k\le n} \|\delta_1 + \cdots + \delta_k\|.$$

In particular,

$$(1.5) \qquad \mathbf{P}(\Delta_n > x) \le \mathbf{P}(\max_{0\le k\le n} \|V_k\| > x/2) + \sum_{k=1}^{n} \mathbf{P}(\delta_k \ne 0).$$

We describe our construction in Section 2 and list the properties of the constructed random elements in Theorem 2.1. In Corollary 2.2, we estimate the probabilities $\mathbf{P}(\delta_m \ne 0)$ in (1.5) and, thus, find a very simple, but general, inequality for obtaining estimates in the invariance principles. Corollary 2.4 contains another inequality which may give better estimates but under additional assumptions.

In Section 3 we illustrate the possibilities of our method for one-dimensional and for infinite dimensional martingales. The remainder of the paper is devoted to proofs of the results stated in Sections 2 and 3.

Here are some conventions and notation. Later on we will assume that the linear space $\mathcal{X}$ is a complete separable metric space with respect to a specially chosen metric. We denote by $\mathcal{B}$ the corresponding σ-algebra of the Borel subsets of the space $\mathcal{X}$ and suppose that the function $(x, y) \mapsto \alpha x + \beta y$ is measurable with respect to $\mathcal{B} \times \mathcal{B}$ for all $\alpha, \beta \in (-\infty, \infty)$. The function $\|\cdot\|$ is treated as a real functional on $\mathcal{X}$ measurable with respect to $\mathcal{B}$ and such that

$$0 \le \|\alpha x + \beta y\| \le |\alpha|\,\|x\| + |\beta|\,\|y\| \qquad \forall x, y \in \mathcal{X}, \qquad \forall \alpha, \beta \in (-\infty, \infty).$$

We assume that all the distributions in this paper are probability measures defined on $(\mathcal{X}, \mathcal{B})$ or on a product of at most a countable number of $(\mathcal{X}, \mathcal{B})$ and intervals $[0,1]$ (see, for example, Loéve (1978), for details). The symbols $\mathcal{L}(Z)$, $\mathcal{L}(Z|Y)$ and $\mathcal{L}(Z,Y)$ correspond, respectively, to the (law of) distribution of the random element Z, the conditional (law of) distribution of Z with respect to the random element Y and the joint (law of) distribution of random elements Z and Y. In particular, we will write

$$\mathcal{L}(\widetilde{Z}) = \mathcal{L}(Z), \quad \mathcal{L}(\widetilde{Z}|Y) = \mathcal{L}(Z|Y) \quad \text{and} \quad \mathcal{L}(\widetilde{Z}, \widetilde{Y}) = \mathcal{L}(Z, Y)$$

whenever the corresponding distributions coincide.

Later on we denote by $\mathcal{L}(Z_1) \times \mathcal{L}(Z_2)$ the product of the distributions $\mathcal{L}(Z_1)$ and $\mathcal{L}(Z_2)$, where Z_1 and Z_2 may be random elements taking values in *arbitrary* spaces. Whereas, we denote by $\mathcal{L}(X) * \mathcal{L}(Y)$ the convolution of the distributions of the random variables X and Y taking their values in $\mathcal{X}$. Thus,

$$\mathcal{L}(Z_1,\, Z_2) = \mathcal{L}(Z_1) \times \mathcal{L}(Z_2) \quad \text{iff} \quad Z_1 \textit{ and } Z_2 \textit{ are independent}; \tag{1.6}$$

$$\mathcal{L}(X+Y) = \mathcal{L}(X) * \mathcal{L}(Y) \quad \text{if} \quad X \textit{ and } Y \textit{ are independent.} \tag{1.7}$$

Denote by $\mathbf{P}(\,\cdot\,|Y)$ and $\mathbf{E}(\,\cdot\,|Y)$ the conditional probability and the conditional expectation with respect to the random element Y. For brevity, sometimes we will write

$$\mathbf{P}_k(\,\cdot\,) = \mathbf{P}(\,\cdot\,|\vec{X}_k) \quad \text{and} \quad \mathbf{E}_k(\,\cdot\,) = \mathbf{E}(\,\cdot\,|\vec{X}_k) \quad \text{for} \quad \vec{X}_k = (X_1, \dots, X_k).$$

Let us point out that all the conditional distributions in the paper exist (see, for example, Loéve (1978) for proofs) because $\mathcal{X}^k$ are separable for all $k \geq 1$.

Finally, denote by $\rho(\cdot\,,\cdot)$ the distance in variation between distributions on $(\mathcal{X},\, \mathcal{B})$:

$$\rho\big(\mathcal{L}(X),\ \mathcal{L}(Y)\big) = \sup_{B \in \mathcal{B}} \big|\mathbf{P}(X \in B) - \mathbf{P}(Y \in B)\big|.$$

2 Description of the method

Given a sequence of mutually independent random variables $Y_1, Y_2, \dots$, there are many possibilities to construct random variables $X_1, X_2, \dots$ and $V_0, V_1, \dots$ which will satisfy the following three conditions:

i) *for all* $k \geq 1$ *the constructed random variables* $X_1, \dots, X_k$ *have the given distribution* $\mathcal{L}(X_1, \dots, X_k)$;

ii) *for all* $k \geq 0$ *the constructed random variable* V_k *has the given distribution* $\mathcal{L}(V_k)$;

iii) *the random variables* $V_0, Y_1, Y_2, \dots$ *are mutually independent and for all* $k \geq 1$ *the random elements* $\vec{X}_{k-1}$, V_k, $Y_{k+1}, Y_{k+2}, \dots$ *are also mutually independent.*

For example, we may take independent sequences $\{Y_1, Y_2, \dots\}$, $\{V_0, V_1, \dots\}$ and $\{X_1, X_2, \dots\}$. But such simply constructed $X_1, X_2, \dots$ and $V_0, V_1, \dots$ may not be appropriate for our purposes.

It is important to remark that conditions i), ii), iii) and definitions (1.6) and (1.7) allow us to define the distributions

$$\mathcal{L}(Y_k + V_{k-1}|\vec{X}_{k-1}) = \mathcal{L}(Y_k) * \mathcal{L}(V_{k-1}) \tag{2.1}$$

and

$$\mathcal{L}(X_k + V_k|\vec{X}_{k-1}) = \mathcal{L}(X_k|\vec{X}_{k-1}) * \mathcal{L}(V_k).$$

On the other hand, these conditions, in general, do not define the *joint* distribution $\mathcal{L}(Y_k + V_{k-1},\ X_k + V_k|\vec{X}_{k-1})$ uniquely. This fact makes us wish to construct the random variables $Y_k + V_{k-1}$ and $X_k + V_k$ with a specially chosen conditional joint distribution.

For each $k \geq 1$ put

$$\begin{aligned} \rho_k = \rho_k(\vec{X}_{k-1}) &\equiv \rho\big(\mathcal{L}(X_k|\vec{X}_{k-1}) * \mathcal{L}(V_k),\ \mathcal{L}(Y_k) * \mathcal{L}(V_{k-1})\big) \\ &= \sup_{B\in\mathcal{B}} \big|\, \mathbf{P}(X_k + V_k \in B|\vec{X}_{k-1}) - \mathbf{P}(Y_k + V_{k-1} \in B)\big|. \end{aligned} \tag{2.2}$$

Other useful representations for the values ρ_k may be found in Remark 2.6 at the end of the section.

Theorem 2.1. *Given a sequence of mutually independent random variables $Y_1, Y_2, \ldots$, it is possible to construct random variables $X_1, X_2, \ldots$, $V_0, V_1, \ldots$ which satisfy the conditions* i), ii), iii) *and such that the following assertion is fulfilled:*

iv) *for each* $k \geq 1$

$$\mathbf{P}(\delta_k \neq 0|\vec{X}_{k-1}) = \rho_k(\vec{X}_{k-1}) \quad a.s. \tag{2.3}$$

The next simple estimate for the invariance principle follows immediately from (1.5) and (2.3).

Corollary 2.2. *The random variables constructed in Theorem* 2.1 *satisfy the inequalities*

$$\mathbf{P}(\Delta_n > x) \leq \mathbf{P}\big(\max_{0\leq k\leq n} \|V_k\| > x/2\big) + \sum_{k=1}^{n} \mathbf{E}\,\rho_k \qquad \forall n \geq 1, \tag{2.4}$$

$$\mathbf{P}\big(\Delta_\infty > x\big) \leq \mathbf{P}\big(\sup_{k\geq 0} \|V_k\| > x/2\big) + \sum_{k\geq 1} \mathbf{E}\,\rho_k, \tag{2.5}$$

where

$$\Delta_\infty = \sup_{n\geq 1} \Delta_n = \sup_{k\geq 1} \Big\| \sum_{m=1}^{k} X_m - \sum_{m=1}^{k} Y_m \Big\|.$$

Remark 2.3. It is evident from (1.4), (2.4) and (2.5) that if

$$\forall k \geq 0 \qquad \mathbf{P}(\|V_k\| \leq x/2) = 1,$$

then

$$\forall n \leq \infty \qquad \mathbf{P}(\Delta_n > x) \leq \sum_{k=1}^{n} \mathbf{E}\,\rho_k.$$

This particular case of Corollary 2.2 was proved earlier in Sakhanenko (1987) by using a more complicated analytical method.

The next proposition may give better estimates than Corollary 2.2.

Corollary 2.4. *Suppose that for some non-random real numbers $\pi_1, \pi_2, \ldots$ the following condition is fulfilled;*

$$\forall k \geq 1 \qquad \rho_k \leq \pi_k \qquad a.s. \tag{2.6}$$

Then for all integers $m, n \geq 1$ and for all real $x, y, z > 0$ the random variables constructed in Theorem 2.1 satisfy the inequality

$$\begin{aligned} \mathbf{P}\big(\Delta_n > mx + (m-1)(y+z)\big) &\leq \Big(\sum_{k=1}^{n} \pi_k\Big)^m / m! \\ + \ \mathbf{P}(\max_{0 \leq k \leq n} \|V_k\| > x/2) &+ \mathbf{P}(\max_{1 \leq k \leq n} \|Y_k\| > y) + \mathbf{P}(\max_{1 \leq k \leq n} \|X_k\| > z). \end{aligned} \tag{2.7}$$

Moreover, inequality (2.7) *holds also for $n = \infty$ if we only replace in it all symbols $\max\limits_{l \leq k \leq n}$ by $\sup\limits_{k \geq l}$.*

Remark 2.5. If we want to apply inequalities, obtained in Corollaries 2.2 and 2.4, we need first to find appropriate distributions $\{\mathcal{L}(V_k)\}$ and estimate the values $\{\rho_k\}$ and $\{\mathbf{P}(\|V_k\| > x/2)\}$. In the one-dimensional case it is easy to do using the well-known idea of Lindeberg (see Example 1 in Section 3 and the proof of Lemma 3.1 in Section 6 for details). If the space $\mathcal{X}$ is infinite dimensional, then it is not easy to find appropriate distributions $\{\mathcal{L}(V_k)\}$ and estimate the $\{\rho_k\}$. (See, for example, Bentkus and Račkauskas (1984) or Paulauskas and Račkauskas (1989) where the analogous problems in CLT are investigated.) We consider below an infinite dimensional example (see Example 2 in Section 3) using an idea due to Bentkus (1985).

Remark 2.6. Assume now that the distributions $\mathcal{L}(X_k|\vec{X}_{k-1}) * \mathcal{L}(V_k)$ and $\mathcal{L}(Y_k) * \mathcal{L}(V_{k-1})$ have the densities $f_k(\,\cdot\,|\vec{X}_{k-1})$ and $f_k(\,\cdot\,)$ with respect to the same measure $\mu(\,\cdot\,)$ on $(\mathcal{X}, \mathcal{B})$. In this case (see, for example, Lindvall (1992))

$$\rho_k = \frac{1}{2} \int\limits_{\mathcal{X}} \big| f_k(x|\vec{X}_{k-1}) - f_k(x) \big| \, \mu(dx).$$

By the Radon–Nikodym Theorem such densities exist if we take the measure

$$\mu(B) = \mu_k(B|\vec{X}_{k-1}) = \mathbf{P}(X_k + V_k \in B|\vec{X}_{k-1}) + \mathbf{P}(Y_k + V_{k-1} \in B).$$

Remark 2.7. In this section we investigated infinite sequences $Y_1, Y_2, \ldots$ and $X_1, X_2, \ldots$. We may always assume that we are in this situation, since in the case of finite sequences $Y_1, \ldots, Y_N$ and $X_1, \ldots X_N$ we can take $X_k = Y_k = 0$ for all $k > N$.

3 Examples

Given $x > 0$, we will assume in this section that the distributions $\{\mathcal{L}(V_k)\}$ are such that

$$\forall k \geq 0 \qquad \mathcal{L}(V_k/x) = \mathcal{L}(V) \tag{3.1}$$

for a specially chosen distribution $\mathcal{L}(V)$.

Example 1. Let $\mathcal{X} = (-\infty, \infty)$ and assume we are given a sequence of mutually independent (and one-dimensional) random variables $Y_1, Y_2, \ldots$. Suppose that the given joint distribution $\mathcal{L}(X_1, X_2, \ldots)$ satisfies the following conditions with probability 1;

$$\mathbf{E}_{k-1} X_k = \mathbf{E}\, Y_k \quad \text{and} \quad \mathbf{E}_{k-1} X_k^2 = \mathbf{E}\, Y_k^2 < \infty \qquad \forall k \geq 1. \tag{3.2}$$

In Section 6 we prove

Lemma 3.1. *There exists a distribution $\mathcal{L}(V)$ such that $\mathbf{P}(|V| \leq 1/2) = 1$ and*

$$\rho_k \leq 16\, \mathbf{E}_{k-1} \min\left\{\frac{|X_k|^3}{x^3}, \frac{X_k^2}{x^2}\right\} + 16\, \mathbf{E} \min\left\{\frac{|Y_k|^3}{x^3}, \frac{Y_k^2}{x^2}\right\} \quad \forall k \geq 1$$

if conditions (3.1) *and* (3.2) *hold.*

Using Remark 2.3 and Lemma 3.1, we immediately obtain the following simple estimate in the invariance principle with an explicit bound for an absolute constant.

Corollary 3.1. *If conditions* (3.2) *hold, then for all $x > 0$ there exists a sequence of random variables $X_1, X_2, \ldots$ with the given joint distribution and such that for all $n \leq \infty$,*

$$\begin{aligned}\mathbf{P}(\Delta_n > x) &\leq 16 \sum_{k=1}^{n} \left(\mathbf{E} \min\left\{\frac{|X_k|^3}{x^3}, \frac{X_k^2}{x^2}\right\} + \mathbf{E} \min\left\{\frac{|Y_k|^3}{x^3}, \frac{Y_k^2}{x^2}\right\}\right) \\ &\leq 16\, x^{-r} \sum_{k=1}^{n} \left(\mathbf{E}\, |X_k|^r + \mathbf{E}\, |Y_k|^r\right) \quad \text{for all} \quad r \in [2, 3].\end{aligned}$$

Example 2. Now let $\mathcal{X}$ be the space of all infinite dimensional vectors $x = (x^{(1)}, x^{(2)}, \dots)$ with the norm

$$\|x\| = \sup_{i\geq 1} |x^{(i)}| < \infty.$$

As above, assume that we are given a sequence of mutually independent random variables $Y_1, Y_2, \dots$ defined on some probability space and taking their values in $\mathcal{X}$. On the same probability space, we want to construct random variables $X_1, X_2, \dots$ with the given joint distribution $\mathcal{L}(X_1, X_2, \dots)$ and such that the distance Δ_n, introduced in (1.1), will be small in some sense.

Suppose that the given joint distribution $\mathcal{L}(X_1, X_2, \dots)$ is such that the following conditions are fulfilled with probability 1;

$$\begin{gathered} \mathbf{E}_{k-1}X_k^{(i)} = \mathbf{E}\,Y_k^{(i)}, \qquad \mathbf{E}_{k-1}(X_k^{(i)})^2 = \mathbf{E}\,(Y_k^{(i)})^2 < \infty \\ \text{and} \qquad \mathbf{E}_{k-1}X_k^{(i)}X_k^{(j)} = \mathbf{E}\,Y_k^{(i)}Y_k^{(j)} \qquad \forall i, j, k \geq 1. \end{gathered} \tag{3.3}$$

If, in addition, $\mathbf{E}_{k-1}X_k^{(i)} = 0$ for all $i, k \geq 1$, then such a sequence $X_1, X_2, \dots$ may be naturally considered as an "infinite dimensional martingale".

For every infinite dimensional vector $x = (x^{(1)}, x^{(2)}, \dots)$ put

$$\|x\|_2 = \Big(\sum_{i\geq 1} (x^{(i)})^2\Big)^{1/2}.$$

The following statement is an infinite-dimensional analog of Lemma 3.1.

Lemma 3.2. *There exist a distribution $\mathcal{L}(V)$ and an absolute constant $C < \infty$ such that $\mathbf{P}(\|V\| \leq 1/2) = 1$ and*

$$\rho_k \leq C\,\mathbf{E}_{k-1} \min\left\{\frac{\|X_k\|_2^3}{x^3}, \frac{\|X_k\|_2^2}{x^2}\right\} + C\,\mathbf{E}\,\min\left\{\frac{\|Y_k\|_2^3}{x^3}, \frac{\|Y_k\|_2^2}{x^2}\right\} \tag{3.4}$$

for all $k \geq 1$, if conditions (3.1) *and* (3.3) *hold.*

As was shown in Sakhanenko (1987), Lemma 3.2 may be easily derived from a similar assertion (with $\|\cdot\|_2^3/x^3$ instead of $\min\{\|\cdot\|_2^3/x^3, \|\cdot\|_2^2/x^2\}$) proved in Bentkus (1985). But the latter work is difficult to access. For this reason and for the convenience of the readers in Section 7 we give a proof of Lemma 3.2.

If we substitute the estimate (3.4), obtained in Lemma 3.2, into Remark 2.3 and into Corollary 2.4, we immediately obtain the following two assertions.

Corollary 3.2. *If conditions* (3.3) *hold, then for all $x > 0$ there exists a sequence of random variables $X_1, X_2, \dots$ with the given joint distribution and such that for all $n \le \infty$,*

$$\mathbf{P}(\Delta_n > x) \le C \sum_{k=1}^{n} \Bigl(\mathbf{E} \min \Bigl\{ \frac{\|X_k\|_2^3}{x^3}, \frac{\|X_k\|_2^2}{x^2} \Bigr\} + \mathbf{E} \min \Bigl\{ \frac{\|Y_k\|_2^3}{x^3}, \frac{\|Y_k\|_2^2}{x^2} \Bigr\} \Bigr)$$

$$\le C x^{-r} \sum_{k=1}^{n} \left(\mathbf{E} \|X_k\|_2^r + \mathbf{E} \|Y_k\|_2^r \right) \quad \text{for all} \quad r \in [2,3], \tag{3.5}$$

where C is the same absolute constant as in Lemma 3.2.

Corollary 3.3. *Let conditions* (3.3) *be fulfilled and suppose that there exist non-random numbers $\{\pi_k(x)\}$ such that with probability* 1,

$$\mathbf{E}_{k-1} \min \Bigl\{ \frac{\|X_k\|_2^3}{x^3}, \frac{\|X_k\|_2^2}{x^2} \Bigr\} + \mathbf{E} \min \Bigl\{ \frac{\|Y_k\|_2^3}{x^3}, \frac{\|Y_k\|_2^2}{x^2} \Bigr\} \le \pi_k(x)$$

for each $k \ge 1$. Then for all integers $m, n \ge 1$ and for all real $y, z > 0$, the random variables $X_1, X_2, \dots$, constructed in Corollary 3.2, *satisfy the inequality*

$$\mathbf{P}\bigl(\Delta_n > mx + (m-1)(y+z)\bigr) \le \Bigl(C \sum_{k=1}^{n} \pi_k(x) \Bigr)^m / m!$$
$$+ \mathbf{P}\bigl(\max_{1 \le k \le n} \|Y_k\| > y\bigr) + \mathbf{P}\bigl(\max_{1 \le k \le n} \|X_k\| > z\bigr), \tag{3.6}$$

where C is the same absolute constant as in Lemma 3.2. *Moreover, inequality* (3.6) *holds also for $n = \infty$ if we replace in it all symbols $\max\limits_{1 \le k \le n}$ by $\sup\limits_{k \ge 1}$.*

Remark 3.4. We treat the results obtained in Corollaries 3.2 and 3.3 as estimates in the invariance principle for random variables taking values in the space $\mathcal{X}$ with norm $\|x\|$. But these estimates depend essentially on the values $\|X_k\|_2$ and $\|Y_k\|_2$. For this reason we may say that $\|x\|_2$ is "the second norm in the space $\mathcal{X}$". (See Bentkus and Račkauskas (1984) or Paulauskas and Račkauskas (1989) for other example of spaces in which it makes sense to introduce two different norms.)

Remark 3.5. Let us point out that the method, by which the random variables $X_1, X_2, \dots$ were constructed in Corollaries 3.1–3.3, depends very crucially on the chosen real number $x > 0$. It follows from the fact (see (3.1)) that the distributions of the smoothing random variables $\{V_k\}$ depend on x.

Remark 3.6. Earlier, in Sakhanenko (1987), Corollary 3.2 was proved by a more complicated analytical method, mentioned in Remark 2.3. In the particular case of mutually independent random elements $\{X_k\}$, a similar estimate was obtained independently by the same method, also in Bentkus and Liubinskas (1987). But Corollary 3.3 is impossible to derive by this analytical method.

Remark 3.7. If the random variables $\{X_k\}$ and $\{Y_k\}$ have moments only of order $r \le 3$ (see (3.5)), then the estimates, obtained in Corollaries 3.1–3.3, are unimprovable in some sense, because they are unimprovable in the one-dimensional case (see, for example, Sakhanenko (1974)). But if these random variables have moments of order $r > 3$, then the famous but complicated method of Komlós–Major–Tusnády and its more complicated generalizations (see Zaitsev (1998), for example) allow one to obtain incomparably better estimates for sequences of independent random variables. Moreover, the less complicated method of Einmahl (1987) also gives better estimates in case $r > 3$.

Remark 3.8. Due to the simplicity of our method, it works only under the additional assumptions (3.3). Whereas the more difficult method of Morrow and Philipp (1982) and Monrad and Philipp (1991) does not require such restricted assumptions.

4 Auxiliary lemmas

In this section we assume that the random variables V, X, Y, Z, α_0 and α are defined on the same probability space and that the variables V, X and Y take their values in the complete separable metric space $\mathcal{X}$. Whereas, α_0 and α are uniformly distributed on the interval $[0,1]$. The latter will be denoted, for brevity, as $\alpha \sim \mathcal{U}_{0,1}$.

Lemmas 4.1 and 4.3–4.5 are based on an idea of Skorohod (1976).

Lemma 4.1. *Let $\alpha_0 \sim \mathcal{U}_{0,1}$ be independent of the vector (X, V, Y). Then there exist functions $\psi_1(\cdot\,,\cdot)$, $\psi_2(\cdot\,,\cdot)$ and $\psi'(\cdot\,,\cdot\,,\cdot\,,\cdot)$ such that for the random variable $\alpha' = \psi'(X, V, Y, \alpha_0)$ the following statements hold:*

$$X = \psi_1(Y, \alpha'), \quad V = \psi_2(Y, \alpha'), \quad Y \text{ is independent of } \alpha' \sim \mathcal{U}_{0,1}.$$

Proof. We may treat the pairs (X, V) and (Y, Y) as random elements taking values in the complete separable metric space $\mathcal{X} \times \mathcal{X}$. Thus, Lemma 4.1 is a particular case of the main result from Skorohod (1976).

The following assertion is well known as Dobrushin's lemma (see, for example, Dobrushin (1970) or Lindvall (1992)).

Lemma 4.2. *For each pair of distributions $\mathcal{L}(X)$ and $\mathcal{L}(Y)$ there exist random variables $\widehat{X}$ and $\widehat{Y}$ such that*

$$\mathcal{L}(\widehat{X}) = \mathcal{L}(X), \quad \mathcal{L}(\widehat{Y}) = \mathcal{L}(Y) \quad \text{and} \quad \mathbf{P}(\widehat{X} \ne \widehat{Y}) = \rho\big(\mathcal{L}(X), \mathcal{L}(Y)\big).$$

Using Lemma 4.1, we may rewrite Lemma 4.2 in another known form.

Lemma 4.3. *Given a random variable X and a distribution $\mathcal{L}(Y)$, there exists a random variable $\widehat{Y}$ such that*

$$\mathcal{L}(\widehat{Y}) = \mathcal{L}(Y) \quad \text{and} \quad \mathbf{P}(X \neq \widehat{Y}) = \rho(\mathcal{L}(X), \mathcal{L}(Y)).$$

Now we prove the main assertion of the section.

Lemma 4.4. *There exist functions $\Psi_1(\cdot,\cdot,\cdot,\cdot,\cdot)$ and $\Psi_2(\cdot,\cdot,\cdot,\cdot,\cdot)$ such that the following statement is valid: if a random variable Y is independent of $\alpha \sim \mathcal{U}_{0,1}$, then*

$$\text{(4.1)} \quad \mathcal{L}(\widetilde{X}, \widetilde{V}) = \mathcal{L}(X) \times \mathcal{L}(V), \ \mathbf{P}(\widetilde{X} + \widetilde{V} \neq Y) = \rho(\mathcal{L}(X) * \mathcal{L}(V), \mathcal{L}(Y))$$

for

$$\text{(4.2)} \ \widetilde{X} = \Psi_1(Y, \alpha, \mathcal{L}(X), \mathcal{L}(V), \mathcal{L}(Y)), \quad \widetilde{V} = \Psi_2(Y, \alpha, \mathcal{L}(X), \mathcal{L}(V), \mathcal{L}(Y)).$$

Proof. Let random variables V and X be independent. Using Lemma 4.3, we may construct a random variable $\widehat{Y}$ such that

$$\text{(4.3)} \quad \mathcal{L}(\widehat{Y}) = \mathcal{L}(Y) \quad \text{and} \quad \mathbf{P}(X + V \neq \widehat{Y}) = \rho(\mathcal{L}(X) * \mathcal{L}(V), \mathcal{L}(Y)).$$

Now let $\alpha_0 \sim \mathcal{U}_{0,1}$ be independent of the random vector $(X, V, \widehat{Y})$. By Lemma 4.1, there exist functions $\psi_1(\cdot, \cdot)$, $\psi_2(\cdot, \cdot)$ and a random variable α' such that

$$\text{(4.4)} \quad X = \psi_1(\widehat{Y}, \alpha'), \quad V = \psi_2(\widehat{Y}, \alpha'), \quad \widehat{Y} \text{ is independent of } \alpha' \sim \mathcal{U}_{0,1}.$$

Put

$$\text{(4.5)} \qquad \widetilde{X} = \psi_1(Y, \alpha) \qquad \text{and} \qquad \widetilde{V} = \psi_2(Y, \alpha)).$$

Comparing (4.4) and (4.5), it is easy to see that the random vectors

$$(\widehat{Y}, \alpha', X, V) \qquad \text{and} \qquad (Y, \alpha, \widetilde{X}, \widetilde{V})$$

are identically distributed because the pairs $(\widehat{Y}, \alpha')$ and (Y, α) are identically distributed by assumption. This fact and (4.3) imply that all the statements in (4.1) are fulfilled for $\widetilde{X}$ and $\widetilde{V}$ defined by (4.5).

Note that we may construct such functions $\psi_1(\cdot, \cdot)$ and $\psi_2(\cdot, \cdot)$ for all distributions $\mathcal{L}(X)$, $\mathcal{L}(V)$ and $\mathcal{L}(Y)$. So, we may rewrite (4.5) in the form (4.2).

If we replace all distributions by conditional ones in Lemma 4.4, we obtain the following assertion.

Lemma 4.5. *Let a random element (Y, Z) be independent of $\alpha \sim \mathcal{U}_{0,1}$, where Z takes its values in $\mathcal{X}^k$ for some $k \geq 1$. Put*

$$\widetilde{X} = \Psi_1(Y, \alpha, \mathcal{L}(X|Z), \mathcal{L}(V), \mathcal{L}(Y|Z)),$$

$$\widetilde{V} = \Psi_2(Y, \alpha, \mathcal{L}(X|Z), \mathcal{L}(V), \mathcal{L}(Y|Z)).$$

Then

$$\mathcal{L}((\widetilde{X}, \widetilde{V})|Z) = \mathcal{L}(X|Z) \times \mathcal{L}(V),$$

$$\mathbf{P}(\widetilde{X} + \widetilde{V} \neq Y|Z) = \rho(\mathcal{L}(X|Z) * \mathcal{L}(V), \mathcal{L}(Y|Z)).$$

5 Proofs of results of Section 2

Suppose that we are given a set of mutually independent random variables

$$V_0,\ Y_1,\ Y_2,\ \ldots,\quad \alpha_1,\ \alpha_2,\ \ldots, \tag{5.1}$$

where V_0 has the desired distribution $\mathcal{L}(V_0)$ and $\alpha_k \sim \mathcal{U}_{0,1}$ for all $k \geq 1$. Put

$$X_1 = \Psi_1(Y_1 + V_0,\ \alpha_1,\ \mathcal{L}(X_1),\ \mathcal{L}(V_1),\ \mathcal{L}(Y_1) * \mathcal{L}(V_0)), \tag{5.2}$$

$$V_1 = \Psi_2(Y_1 + V_0,\ \alpha_1,\ \mathcal{L}(X_1),\ \mathcal{L}(V_1),\ \mathcal{L}(Y_1) * \mathcal{L}(V_0)). \tag{5.3}$$

For $k > 1$ we define the random variables $\{X_k,\ V_k\}$ recursively, by setting $k = 2,\ 3,\ \ldots$ in the following formulae:

$$X_k = \Psi_1(Y_k + V_{k-1},\ \alpha_k,\ \mathcal{L}(X_k|\vec{X}_{k-1}),\ \mathcal{L}(V_k),\ \mathcal{L}(Y_k) * \mathcal{L}(V_{k-1})), \tag{5.4}$$

$$V_k = \Psi_2(Y_k + V_{k-1},\ \alpha_k,\ \mathcal{L}(X_k|\vec{X}_{k-1}),\ \mathcal{L}(V_k),\ \mathcal{L}(Y_k) * \mathcal{L}(V_{k-1})). \tag{5.5}$$

Now we shall investigate the properties of the random variables constructed in (5.2)–(5.5).

Lemma 5.1. *For all $k > 1$ the sequence $X_1, \ldots, X_{k-1}, V_{k-1}$ is independent of the sequence $Y_k,\ Y_{k+1}, \ldots$ of mutually independent random variables.*

Proof. It follows from (5.2)–(5.5) that $X_1, \ldots, X_{k-1}, V_{k-1}$ are functions only of the variables $V_0, Y_1, \ldots, Y_{k-1}, \alpha_1, \ldots, \alpha_{k-1}$. This fact and the mutual independence of the random variables in (5.1) give us the assertion of Lemma 5.1.

Lemma 5.2. *If for some* $k > 1$

$$\mathcal{L}(X_1, \dots, X_{k-1}, V_{k-1}) = \mathcal{L}(X_1, \dots, X_{k-1}) \times \mathcal{L}(V_{k-1}), \tag{5.6}$$

then, for given k, *condition* (2.3) *holds and*

$$\mathcal{L}(X_1, \dots, X_k, V_k) = \mathcal{L}(X_1, \dots, X_k) \times \mathcal{L}(V_k). \tag{5.7}$$

Proof. Assumption (5.6) and Lemma 5.1 imply that condition (2.1) holds. Hence, we may rewrite (5.4) and (5.5) in the following way:

$$X_k = \Psi_1\big(Y_k + V_{k-1},\, \alpha_k,\, \mathcal{L}(X_k|\vec{X}_{k-1}),\, \mathcal{L}(V_k),\, \mathcal{L}(Y_k + V_{k-1}|\vec{X}_{k-1})\big), \tag{5.8}$$

$$V_k = \Psi_2\big(Y_k + V_{k-1},\, \alpha_k,\, \mathcal{L}(X_k|\vec{X}_{k-1}),\, \mathcal{L}(V_k),\, \mathcal{L}(Y_k + V_{k-1}|\vec{X}_{k-1})\big). \tag{5.9}$$

Lemma 4.5 and equalities (5.8)–(5.9) yield all the assertions of Lemma 5.2.

Proof of Theorem 2.1. Lemma 4.4 and definitions (5.2) and (5.3) imply that conditions (5.7) and (2.3) hold for $k = 1$. By induction on k, from Lemma 5.2 we have that properties (5.7) and (2.3) are satisfied for all $k \geq 1$. Hence, conditions i), ii) and iv) are fulfilled. After that, from (5.7) and Lemma 5.1 we obtain iii).

Remark 5.1. In the proof of Theorem 2.1 we introduced a set (5.1) of mutually independent random variables. It means that we have enlarged the initial probability space on which the given random variables Y_1, Y_2, $\dots$ were defined. It is well known (see, for example, Skorohod (1976)) that we do not need to enlarge the initial probability space in the case when it is "rich enough" in the following sense: on this probability space it is possible to define a random variable, uniformly distributed on the interval $[0, 1]$, which is independent of all given random variables.

Proof of Corollary 2.4. Denote by I_k the indicator of the event $\{\delta_k \neq 0\}$. It follows from (1.4) that

$$\Delta_n \leq \|V_0\| + \max_{1 \leq k \leq n} \|V_k\| + \max_{1 \leq k \leq n} \|\delta_k\| \cdot \sum_{k=1}^{n} I_k. \tag{5.10}$$

Thus, from (5.10) and (1.2) we obtain the inequality

$$\mathbf{P}\big(\Delta_n > x + (m-1)(x+y+z)\big) \leq \mathbf{P}\Big(\sum_{k=1}^{n} I_k > m-1\Big) \tag{5.11}$$

$$+\mathbf{P}\big(\max_{0 \leq k \leq n} \|V_k\| > x/2\big) + \mathbf{P}\big(\max_{1 \leq k \leq n} \|Y_k\| > y\big) + \mathbf{P}\big(\max_{1 \leq k \leq n} \|X_k\| > z\big).$$

Corollary 2.4 follows immediately from (5.11) and from the following auxiliary Lemma 5.3 for $l = 1$.

Lemma 5.3. *If conditions* (2.6) *are fulfilled, then for all integers* $m \geq 1$ *and for all* $1 \leq l \leq n$

$$\mathbf{P}_{l-1}\Big(\sum_{k=l}^{n} I_k > m-1\Big) \leq \Big(\sum_{k=l}^{n} \pi_k\Big)^m / m!. \tag{5.12}$$

Proof. The law of total probability gives

$$\mathbf{P}_{l-1}\Big(\sum_{k=l}^{n} I_k > m-1\Big) \leq \sum_{k=l}^{n} \mathbf{P}_{l-1}\Big(\delta_k \neq 0,\ \sum_{i=k+1}^{n} I_i > m-2\Big) \tag{5.13}$$

for all integers $m \geq 1$ and $1 \leq l \leq n$. On the other hand

$$\sum_{k=l}^{n} \pi_k \Big(\sum_{i=k+1}^{n} \pi_i\Big)^{m-1} / (m-1)! \leq \Big(\sum_{k=l}^{n} \pi_k\Big)^m / m!\,. \tag{5.14}$$

Inequalities (5.13) and (5.14) allow us to prove (5.12) using induction on m.

6 Proof of Lemma 3.1

Introduce the density

$$g(x) = 4\big((1-2|x|)^+\big)^2 - 2\big((1-4|x|)^+\big)^2. \tag{6.1}$$

We need the following auxiliary

Lemma 6.1. *Let random variables* X *and* Y *be such that*

$$\mathbf{E}\, X = \mathbf{E}\, Y \quad \text{and} \quad \mathbf{E}\, X^2 = \mathbf{E}\, Y^2 < \infty. \tag{6.2}$$

Suppose that a random variable V *has the density* $g(\cdot)$ *and is independent of the random variables* X *and* Y. *Then* $\mathbf{P}(|V| < 1/2) = 1$ *and*

$$\begin{aligned} \rho \ &\equiv\ \rho\big(\mathcal{L}(X) * \mathcal{L}(V),\, \mathcal{L}(Y) * \mathcal{L}(V)\big) \\ &\leq\ 16\,\mathbf{E}\, \min\big\{|X|^3,\ X^2\big\} + 16\,\mathbf{E}\, \min\big\{|Y|^3,\ Y^2\big\}\,. \end{aligned} \tag{6.3}$$

Proof. The assertion $\mathbf{P}(|V| < 1/2) = 1$ follows from the definition (6.1) of the density $g(\cdot)$ concentrated on the interval $(-1/2,\, 1/2)$. So we need only establish (6.3).

Remark 2.6 and the convolution formula yield

$$\rho = \frac{1}{2} \int_{-\infty}^{\infty} \big|\mathbf{E}\, g(x-X) - \mathbf{E}\, g(x-Y)\big|\, dx. \tag{6.4}$$

For

$$\varepsilon_0(x,y) = g(x-y) - g(x) + yg'(x) - y^2 g''(x)/2 \tag{6.5}$$

we immediately obtain

$$\mathbf{E}\, g(x-Y) = g(x) - g'(x)\,\mathbf{E}\, Y + g''(x)\,\mathbf{E}\, Y^2/2 + \mathbf{E}\,\varepsilon_0(x,Y), \tag{6.6}$$

$$\begin{aligned} \mathbf{E}\, g(x-X) &= g(x) - g'(x)\,\mathbf{E}\, X + g''(x)\,\mathbf{E}\, X^2/2 + \mathbf{E}\,\varepsilon_0(x,X) \\ &= g(x) - g'(x)\,\mathbf{E}\, Y + g''(x)\,\mathbf{E}\, Y^2/2 + \mathbf{E}\,\varepsilon_0(x,X). \end{aligned} \tag{6.7}$$

The last equality in (6.7) is a corollary of conditions (6.2). Using (6.4), (6.6) and (6.7), we have

$$\rho = \frac{1}{2}\int_{-\infty}^{\infty} \big|\mathbf{E}\,\varepsilon_0(x,X) - \mathbf{E}\,\varepsilon_0(x,Y)\big|\, dx \le \mathbf{E}\,\varepsilon_1(X) + \mathbf{E}\,\varepsilon_1(Y), \tag{6.8}$$

where

$$\varepsilon_1(y) = \frac{1}{2}\int_{-\infty}^{\infty} \big|\varepsilon_0(x,y)\big|\, dx. \tag{6.9}$$

Assertion (6.3) follows now from (6.8), and the inequality

$$\varepsilon_1(y) \le 16 \min\left\{|y|^3,\, y^2\right\}, \tag{6.10}$$

which will be proved below.

Lemma 6.2. *For the function* $g(\cdot)$ *defined in* (6.1), *inequality* (6.10) *holds.*

Proof. By definition (6.1), the function $g(x)$ has continuous derivative $g'(x)$ and its second derivative $g''(x)$ has the following simple form,

$$g''(x) = 32\,\mathbf{I}_{[-1/2,1/2]}(x) - 64\,\mathbf{I}_{[-1/4,1/4]}(x), \tag{6.11}$$

where $\mathbf{I}_A(\cdot)$ denotes the indicator of the set A. Put

$$\varepsilon_2(x,y) = g''(x-y) - g''(x), \quad \varepsilon_3(y) = \int_{-\infty}^{\infty} \big|\varepsilon_2(x,y)\big|\, dx. \tag{6.12}$$

Definitions (6.5), (6.12) and a Taylor expansion yield

$$\begin{aligned} \varepsilon_0(x,y) &= \int_0^1 (1-t)\, y^2 g''(x-ty)\, dt - y^2 g''(x)/2 \\ &= \int_0^1 (1-t)\, y^2 \varepsilon_2(x,ty)\, dt. \end{aligned} \tag{6.13}$$

Equalities (6.9), (6.12) and (6.13) imply the inequality

$$\varepsilon_1(y) \le \frac{1}{2} \int_0^1 (1-t)\, y^2 \varepsilon_3(ty)\, dt. \tag{6.14}$$

Using definitions (6.12) and (6.11), we obtain

$$\begin{aligned} \varepsilon_3(y) &\le \int_{-\infty}^{\infty} \left|g''(x-y)\right| dx + \int_{-\infty}^{\infty} \left|g''(x)\right| dx \\ &= 2 \int_{-\infty}^{\infty} \left|g''(x)\right| dx = 2 \int_{-1/2}^{1/2} 32\, dx = 64. \end{aligned}$$

So, by (6.14),

$$\varepsilon_1(y) \le \frac{1}{2} \int_0^1 (1-t) \cdot 64\, y^2\, dt = 16\, y^2. \tag{6.15}$$

On the other hand, the explicit formulae (6.11) and (6.12) give us the following representation,

$$\begin{aligned} \varepsilon_2(x,y) &= 32 \left(\mathbf{I}_{[-1/2+y,1/2+y]}(x) - \mathbf{I}_{[-1/2,1/2]}(x)\right) \\ &\quad - 64 \left(\mathbf{I}_{[-1/4+y,1/4+y]}(x) - \mathbf{I}_{[-1/4,1/4]}(x)\right). \end{aligned} \tag{6.16}$$

Using (6.16) and definition (6.12) it is not difficult to see that

$$\varepsilon_3(y) \le \int_{-1/2}^{-1/2+y} 32\, dx + \int_{1/2}^{1/2+y} 32\, dx + \int_{-1/4}^{-1/4+y} 64\, dx + \int_{1/4}^{1/4+y} 64\, dx = 192\, y$$

for any $y > 0$. Hence, by (6.14), we obtain that

$$\varepsilon_1(y) \le \frac{1}{2} \int_0^1 (1-t)\, t \cdot 192\, y^3\, dt = 16\, y^3$$

for $y > 0$. Similarly, $\varepsilon_1(y) \le 16\,|y|^3$ for $y < 0$ and, thus, for all y. The latter fact, together with (6.15), imply (6.10).

Proof of Lemma 3.1. It follows immediately from (2.2) and (3.1) that

$$\begin{aligned} \rho_k &= \rho\big(\mathcal{L}((X_k/x)|\vec{X}_{k-1}) * \mathcal{L}(V_k/x),\ \mathcal{L}(Y_k/x) * \mathcal{L}(V_{k-1}/x)\big) \\ &= \rho\big(\mathcal{L}((X_k/x)|\vec{X}_{k-1}) * \mathcal{L}(V),\ \mathcal{L}(Y_k/x) * \mathcal{L}(V)\big). \end{aligned} \tag{6.17}$$

Thus, to prove Lemma 3.1, we need only to put $\mathcal{L}(X) = \mathcal{L}((X_k/x)|\vec{X}_{k-1})$ and $\mathcal{L}(Y) = \mathcal{L}(Y_k/x)$ in Lemma 6.1 and to use conditions (3.2) instead of (6.2).

7 Proof of Lemma 3.2

Later on we suppose that $V^{(1)}, V^{(2)}, \ldots$ are independent and identically distributed random variables with the following density:

$$(7.1)\quad f(x) = C_0\big(((1/2-|x|)^+)^4 - 8((1/4-|x|)^+)^4\big), \qquad C_0 = 5\cdot 2^6/3.$$

In particular, $f(x) = 0$ for $|x| \geq 1/2$ and

$$(7.2)\quad f(x) = C_0\big(1/32 - 3x^2/2 + 6|x|^3 - 7x^4\big) \quad \text{for } |x| \leq 1/4.$$

It is easy to see that the derivatives $f'(x)$ and $f''(x)$ exist and are continuous for all $x \in (-\infty, +\infty)$, and that $f'''(x)$ exists and is continuous for all $x \neq 0$. Put

$$(7.3)\qquad \lambda_1(x) = \big(\ln f(x)\big)',\ \lambda_2(x) = \big(\ln f(x)\big)'',\ \lambda_3(x) = \big(\ln f(x)\big)'''$$

for $|x| < 1/2$ and let

$$(7.4)\qquad \lambda_1(x) = \lambda_2(x) = \lambda_3(x) = f(x) = 0 \quad \text{for } |x| \geq 1/2.$$

Using (7.1)–(7.4), it is not difficult to verify that the following expectations are finite:

$$(7.5)\quad C_1 = \mathbf{E}\,\lambda_1^2(V^{(1)}) = \int_{-1/4}^{1/4} \lambda_1^2(x)\, f(x)\, dx + 2\cdot 4^2 C_0 \int_{1/4}^{1/2} (1/2-x)^2\, dx,$$

$$(7.6)\quad C_2 = \mathbf{E}\,\lambda_2^2(V^{(1)}) = \int_{-1/4}^{1/4} \lambda_2^2(x)\, f(x)\, dx + 2\cdot 4^2 C_0 \int_{1/4}^{1/2} (1/2-x)^0\, dx,$$

$$(7.7)\quad C_3 = \mathbf{E}\,|\lambda_3(V^{(1)})| = \int_{-1/4}^{1/4} |\lambda_3(x)|\, f(x)\, dx + 2\cdot 8 C_0 \int_{1/4}^{1/2} (1/2-x)^1\, dx,$$

$$(7.8)\quad C_4 = \mathbf{E}\,\lambda_1^4(V^{(1)}) = \int_{-1/4}^{1/4} \lambda_1^4(x)\, f(x)\, dx + 2\cdot 4^4 C_0 \int_{1/4}^{1/2} (1/2-x)^0\, dx.$$

For n fixed later on, let R^n denote the n-dimensional Euclidean space, $\mu(\cdot)$ the n-dimensional Lebesgue measure and set

$$z = (z^{(1)}, \ldots, z^{(n)}),\quad |z|_n = \Big(\sum_{i=1}^n (z^{(i)})^2\Big)^{1/2}, \qquad \text{for } z \in R^n.$$

For $z, u \in R^n$ introduce the following notation:

(7.9) $$L_m(z,u) = \sum_{k=1}^{n} (-u^{(k)})^m \lambda_m(z^{(k)}), \quad m = 1,2,3,$$

(7.10) $$p(z) = \prod_{k=1}^{n} f(z^{(k)}), \quad p(t,z,u) = p(z - tu),$$

where $t \in [0,1]$. Let $p'(\cdot\,,\,\cdot\,,\cdot)$, $p''(\cdot\,,\,\cdot\,,\cdot)$ and $p'''(\cdot\,,\,\cdot\,,\cdot)$ be the derivatives of the function $p(\cdot\,,\,\cdot\,,\cdot)$ with respect to its first argument. It is easy to obtain that

(7.11) $$p'(t,z,u) = L_1(z - tu,\, u)\, p(z - tu),$$

(7.12) $$p''(t,z,u) = \big(L_2(z - tu,\, u) + L_1^2(z - tu,\, u)\big)\, p(z - tu),$$

(7.13) $$\begin{aligned} p'''(t,z,u) = \big(L_3(z - tu,\, u) &+ 3\,L_2(z - tu,\, u)\, L_1(z - tu,\, u) \\ &+ L_1^3(z - tu,\, u)\big)\, p(z - tu). \end{aligned}$$

Put

(7.14) $$\varepsilon_0(z,u) = p(z-u) - p(z) - p'(0,z,u) - p''(0,z,u)/2,$$

(7.15) $$\varepsilon_1(u) = \frac{1}{2} \int_{R^n} \big|\varepsilon_0(z,u)\big|\, \mu(dz).$$

Lemma 7.1. *There exists an absolute constant $C < \infty$ such that*

(7.16) $$\varepsilon_1(u) \le C \min\big\{|u|_n^3,\, |u|_n^2\big\} \qquad \text{for all} \qquad u \in R^n.$$

Proof. We will use the following two partial cases of the Taylor expansion formula:

$$\begin{aligned} &p(1) - p(0) - p'(0) - p''(0)/2 \\ &= \int_0^1 (1-t)\, p''(t)\, dt - p''(0)/2 = \int_0^1 (1-t)^2 p'''(t)\, dt/2. \end{aligned}$$

Hence, from definition (7.14), we have

(7.17) $$\varepsilon_0(z,u) = \int_0^1 (1-t)\, p''(t,z,u)\, dt - p''(0,z,u)/2$$

(7.18) $$= \int_0^1 (1-t)^2 p'''(t,z,u)\, dt/2.$$

Put

$$\beta(u) = \int\limits_{R^n} \left|p''(t,z,u)\right| \mu(dz), \quad \gamma(u) = \int\limits_{R^n} \left|p'''(t,z,u)\right| \mu(dz). \tag{7.19}$$

Equalities (7.12) and (7.13) allow us to rewrite (7.19) in the following way:

$$\beta(u) = \int\limits_{R^n} \left|p''(0, z-tu,\, u)\right| \mu(dz) = \int\limits_{R^n} \left|p''(0,z,u)\right| \mu(dz), \tag{7.20}$$

$$\gamma(u) = \int\limits_{R^n} \left|p'''(0, z-tu,\, u)\right| \mu(dz) = \int\limits_{R^n} \left|p'''(0,z,u)\right| \mu(dz). \tag{7.21}$$

Thus, from (7.15), (7.18) (7.19) and (7.21) we immediately obtain that

$$\varepsilon_1(u) \le \int\limits_0^1 (1-t)^2 \gamma(u)\, dt/4 = \gamma(u)/12, \tag{7.22}$$

and, in a similar way, (7.15), (7.17), (7.19) and (7.20) yield

$$\varepsilon_1(u) \le \int\limits_0^1 (1-t)\, \beta(u)\, dt/2 + \beta(u)/4 = \beta(u)/2. \tag{7.23}$$

Equalities (7.20) and (7.12) give us the formula

$$\beta(u) = \int\limits_{R^n} \left|L_2(z,u) + L_1^2(z,u)\right| p(z)\, \mu(dz). \tag{7.24}$$

Similarly, from (7.21) and (7.13) we have

$$\gamma(u) = \int\limits_{R^n} \left|L_3(z,u) + 3\, L_2(z,u)\, L_1(z,u) + L_1^3(z,u)\right| p(z)\, \mu(dz). \tag{7.25}$$

It is obvious that the function $p(\cdot)$, defined in (7.10), is the density of the random vector $V = (V^{(1)}, \ldots, V^{(n)})$. Hence, we may rewrite (7.24) and (7.25) in the following way:

$$\beta(u) = \mathbf{E}\left|L_2(V,u) + L_1^2(V,u)\right|, \tag{7.26}$$

$$\gamma(u) = \mathbf{E}\left|L_3(V,u) + 3\, L_2(V,u)\, L_1(V,u) + L_1^3(V,u)\right|. \tag{7.27}$$

The random variables $\lambda_1(V^{(1)}), \ldots, \lambda_1(V^{(n)})$ are independent and identically distributed with zero means. This fact, (7.5), (7.8) and (7.9) yield

$$\mathbf{E}\, L_1^2(V,u) = \sum_{k=1}^{n} (u^{(k)})^2 \mathbf{E}\, \lambda_1^2(V^{(k)}) = C_1 |u|_n^2, \tag{7.28}$$

$$\begin{aligned} \mathbf{E}\, L_1^4(V,u) &= 3\big(\mathbf{E}\, L_1^2(V,u)\big)^2 + \sum_{k=1}^{n} (u^{(k)})^4 \big(\mathbf{E}\, \lambda_1^4(V^{(k)}) - 3\left(\mathbf{E}\, \lambda_1^2(V^{(k)})\right)^2\big) \\ &\le \big(3C_1^2 + (C_4 - 3C_1^2)^+\big)|u|_n^4 \equiv C_5 |u|_n^4. \end{aligned} \tag{7.29}$$

The similar arguments, (7.6) and (7.9), give

$$\mathbf{E}\, L_2^2(V,u) \le C_2 |u|_n^4. \tag{7.30}$$

And it is evident from (7.7) and (7.9) that

$$\mathbf{E}\left|L_3(V,u)\right| \le \sum_{k=1}^{n} |u^{(k)}|^3 \mathbf{E}\left|\lambda_3(V^{(k)})\right| \le C_3 |u|_n^3. \tag{7.31}$$

Using (7.27) and (7.29)–(7.31), we obtain

$$\begin{aligned} \gamma(u) &\le \mathbf{E}\left|L_3(V,u)\right| + 3\big(\mathbf{E}\, L_2^2(V,u)\big)^{1/2}\big(\mathbf{E}\, L_1^2(V,u)\big)^{1/2} \\ &+ \big(\mathbf{E}\, L_1^4(V,u)\big)^{3/4} \le C_6 |u|_n^3, \quad C_6 = C_3 + 3C_2^{1/2}C_1^{1/2} + C_5^{3/4}. \end{aligned} \tag{7.32}$$

Similarly, by (7.26), (7.28) and (7.30) we have

$$\beta(u) \le \big(\mathbf{E}\, L_2^2(V,u)\big)^{1/2} + \mathbf{E}\, L_1^2(V,u) \le (C_1 + C_2^{1/2})|u|_n^2. \tag{7.33}$$

Finally, formulae (7.22), (7.23), (7.32) and (7.33) yield the desired inequality

$$\varepsilon_1(u) \le \min\big\{\beta(u)/2, \gamma(u)/12\big\} \le C \min\big\{|u|_n^3,\ |u|_n^2\big\}$$

with $C = \max\big\{(C_1 + C_2^{1/2})/2, C_6/12\big\}$.

Let $X = (X^{(1)}, \ldots X^{(n)})$ and $Y = (Y^{(1)}, \ldots, Y^{(n)})$ be n-dimensional random vectors, independent of the given vector $V = (V^{(1)}, \ldots, V^{(n)})$.

Lemma 7.2. *If*

$$\mathbf{E}\, X^{(i)} = \mathbf{E}\, Y^{(i)}, \qquad \mathbf{E}\,(X^{(i)})^2 = \mathbf{E}\,(Y^{(i)})^2 < \infty \quad \text{and} \quad \mathbf{E}\, X^{(i)}X^{(j)} = \mathbf{E}\, Y^{(i)}Y^{(j)} \tag{7.34}$$

for all $i, j = 1, \ldots, n$, then

$$\begin{aligned} \rho(n) &\equiv \rho\big(\mathcal{L}(X) * \mathcal{L}(V),\ \mathcal{L}(Y) * \mathcal{L}(V)\big) \\ &\le C\,\mathbf{E}\, \min\big\{|X|_n^3,\ |X|_n^2\big\} + C\,\mathbf{E}\, \min\big\{|Y|_n^3,\ |Y|_n^2\big\}, \end{aligned}$$

where C is the constant from Lemma 7.1.

Proof. Conditions (7.34) and definitions (7.9) imply the equalities

$$\mathbf{E}\,L_1(z,\,X) = \mathbf{E}\,L_1(z,\,Y) \quad \text{and} \quad \mathbf{E}\,L_2(z,\,X) = \mathbf{E}\,L_2(z,\,Y)$$

for all non-random z. Hence, using (7.11), (7.12) and (7.14) we obtain

$$\mathbf{E}\,p'(0,z,X) + \mathbf{E}\,p''(0,z,X)/2 = \mathbf{E}\,p'(0,z,Y) + \mathbf{E}\,p''(0,z,Y)/2,$$

$$\mathbf{E}\,\varepsilon_0(z,X) - \mathbf{E}\,\varepsilon_0(z,Y) = \mathbf{E}\,p(z-X) - \mathbf{E}\,p(z-Y). \tag{7.35}$$

By the convolution formula, the random vectors $X+V$ and $Y+V$ have the densities $\mathbf{E}\,p(z-X)$ and $\mathbf{E}\,p(z-Y)$, respectively. This fact and Remark 2.6 give us the equality

$$\rho(n) = \frac{1}{2}\int\limits_{R^n} \bigl|\mathbf{E}\,p(z-X) - \mathbf{E}\,p(z-Y)\bigr|\,\mu(dz). \tag{7.36}$$

Using (7.15), (7.35) and (7.36) we immediately obtain

$$\begin{aligned}
\rho(n) &\le \frac{1}{2}\int\limits_{R^n} \bigl|\mathbf{E}\,\varepsilon_0(z,X)\bigr|\,\mu(dz) + \frac{1}{2}\int\limits_{R^n} \bigl|\mathbf{E}\,\varepsilon_0(z,Y)\bigr|\,\mu(dz) \\
&\le \frac{1}{2}\mathbf{E}\int\limits_{R^n} \bigl|\varepsilon_0(z,X)\bigr|\,\mu(dz) + \frac{1}{2}\mathbf{E}\int\limits_{R^n} \bigl|\varepsilon_0(z,Y)\bigr|\,\mu(dz) \qquad (7.37)\\
&= \mathbf{E}\,\varepsilon_1(X) + \mathbf{E}\,\varepsilon_1(Y).
\end{aligned}$$

Lemma 7.2 follows from (7.16) and (7.37).

Denote by R^∞ the set of all infinite dimensional vectors $z = (z^{(1)},\, z^{(2)}\, \ldots)$. Let

$$z = (z^{(1)},\, z^{(2)},\, \ldots),\;\; z_{(n)} = (z^{(1)}, \ldots, z^{(n)}),\;\; |z|_n = |z_{(n)}|_n, \qquad \text{for } z \in R^\infty.$$

We will consider R^∞ as a metric space with the metric

$$d(z,u) = \sum_{k=1}^{\infty} 2^{-k} \frac{|z^{(k)} - u^{(k)}|}{1 + |z^{(k)} - u^{(k)}|}.$$

It is well known that in this case R^∞ is a complete separable metric space.

Let $X = (X^{(1)}, X^{(2)}, \ldots)$ and $Y = (Y^{(1)}, Y^{(2)}, \ldots)$ be infinite dimensional random vectors, independent of the given vector $V = (V^{(1)}, V^{(2)}, \ldots)$. Later on we will treat X, Y and V as random elements taking values in R^∞.

Lemma 7.3. *If conditions* (7.34) *hold for all* $i, j = 1, 2, \ldots,$ *then*

$$\begin{aligned}
\rho &\equiv \rho\bigl(\mathcal{L}(X) * \mathcal{L}(V),\, \mathcal{L}(Y) * \mathcal{L}(V)\bigr) \qquad (7.38)\\
&\le C\,\mathbf{E}\,\min\bigl\{\|X\|_2^3,\ \|X\|_2^2\bigr\} + C\,\mathbf{E}\,\min\bigl\{\|Y\|_2^3,\ \|Y\|_2^2\bigr\}.
\end{aligned}$$

Proof. Put

$$\xi = X + V, \quad \eta = Y + V, \quad \rho(n) = \rho\big(\mathcal{L}(\xi_{(n)}), \mathcal{L}(\eta_{(n)})\big). \tag{7.39}$$

It follows from Lemma 7.2 that

$$\begin{aligned} \rho(n) &\leq C\mathbf{E}\min\{|X|_n^3,\ |X|_n^2\} + C\mathbf{E}\min\{|Y|_n^3,\ |Y|_n^2\} \\ &\leq C\mathbf{E}\min\{\|X\|_2^3,\ \|X\|_2^2\} + C\mathbf{E}\min\{\|Y\|_2^3,\ \|Y\|_2^2\} \equiv \delta. \end{aligned} \tag{7.40}$$

By Dobrushin's Lemma 4.3, for each n there exists an n-dimensional random vector $\widetilde{\xi}_{(n)}$ such that

$$\mathcal{L}(\widetilde{\xi}_{(n)}) = \mathcal{L}(\xi_{(n)}), \quad \mathbf{P}(\widetilde{\xi}_{(n)} \neq \eta_{(n)}) = \rho(n). \tag{7.41}$$

Using Lemma A1 in Berkes and Philipp (1979), we derive from (7.39)–(7.41) that for all n there exist infinite dimensional random elements $\xi(n)$ and $\eta(n)$ such that

$$\mathcal{L}(\xi(n)) = \mathcal{L}(\xi), \quad \mathcal{L}(\eta(n)) = \mathcal{L}(\eta), \tag{7.42}$$

$$\mathbf{P}\big(\xi_{(n)}(n) \neq \eta_{(n)}(n)\big) = \rho(n) \leq \delta. \tag{7.43}$$

Note that $\mathcal{L}(\xi)$ and $\mathcal{L}(\eta)$ are two fixed distributions in the complete separable metric space R^∞. Hence, for every $\varepsilon > 0$ there exist compacts $K_1(\varepsilon)$ and $K_2(\varepsilon)$ such that

$$\mathbf{P}\big(\xi \notin K_1(\varepsilon)\big) < \varepsilon/2 \quad \text{and} \quad \mathbf{P}\big(\eta \notin K_2(\varepsilon)\big) < \varepsilon/2.$$

Thus,

$$\mathbf{P}\big((\xi(n), \eta(n)) \notin K_1(\varepsilon) \times K_2(\varepsilon)\big) \leq \mathbf{P}\big(\xi(n) \notin K_1(\varepsilon)\big) + \mathbf{P}\big(\eta(n) \notin K_2(\varepsilon)\big) =$$

$$= \mathbf{P}\big(\xi \notin K_1(\varepsilon)\big) + \mathbf{P}\big(\eta \notin K_2(\varepsilon)\big) < \varepsilon \qquad \forall n \quad \forall \varepsilon > 0.$$

So, the sequence of distributions $\mathcal{L}(\xi(n), \eta(n))$ is tight in the complete separable metric space $R^\infty \times R^\infty$. By Prokhorov's theorem, there exist a distribution $\mathcal{L}(\xi', \eta')$ and a subsequence $n' \to \infty$ such that

$$\mathcal{L}(\xi(n'), \eta(n')) \Longrightarrow \mathcal{L}(\xi', \eta') \quad \text{as} \quad n' \to \infty. \tag{7.44}$$

We shall now investigate the joint distribution of the resulting random variables ξ' and η'. First of all, note that

$$\forall n \quad \mathcal{L}(\xi) = \mathcal{L}(\xi(n)) = \mathcal{L}(\xi'), \quad \mathcal{L}(\eta) = \mathcal{L}(\eta(n)) = \mathcal{L}(\eta'),$$

by (7.42) and (7.44). Hence, from definitions (7.38) and (7.39) we have

$$\rho = \rho(\mathcal{L}(\xi), \mathcal{L}(\eta)) = \rho\big(\mathcal{L}(\xi'), \mathcal{L}(\eta')\big) \leq \mathbf{P}(\xi' \neq \eta'). \tag{7.45}$$

The last inequality is valid for all random elements (see, for example, Dobrushin (1970) or Lindvall (1992)).

Consider the sets

$$M(m) = \{x, y \in R^\infty : x_{(m)} \neq y_{(m)}\}. \tag{7.46}$$

It is obvious from (7.43) and (7.46) that

$$\forall n \geq m \ \ \mathbf{P}\big((\xi(n), \eta(n)) \in M(m)\big) = \mathbf{P}\big(\xi_{(m)}(n) \neq \eta_{(m)}(n)\big)$$

$$\leq \mathbf{P}\big(\xi_{(n)}(n) \neq \eta_{(n)}(n)\big) = \rho(n) \leq \delta. \tag{7.47}$$

But $M(m)$ is an open set in $R^\infty \times R^\infty$. This fact, (7.44) and (7.47) yield

$$\mathbf{P}\big((\xi', \eta') \in M(m)\big) \leq \limsup_{n' \to \infty} \mathbf{P}\big((\xi(n'), \eta(n')) \in M(m)\big) \leq \delta. \tag{7.48}$$

Introduce the set

$$M = \{x, y \in R^\infty : x \neq y\} = \bigcup_{m=1}^{\infty} M(m).$$

It is evident that

$$\mathbf{P}(\xi' \neq \eta') = \mathbf{P}\big((\xi', \eta') \in M\big) = \lim_{m \to \infty} \mathbf{P}\big((\xi', \eta') \in M(m)\big). \tag{7.49}$$

Thus, from (7.45), (7.49) and (7.48) we obtain the inequality $\rho \leq \delta$ which is equivalent to the assertion (7.38) of Lemma 7.3.

Proof of Lemma 3.2. It follows immediately from (2.2), (3.1) and (6.17) that

$$\rho_k \quad = \quad \rho\big(\mathcal{L}((X_k/x)|\vec{X}_{k-1}) * \mathcal{L}(V), \ \mathcal{L}(Y_k/x) * \mathcal{L}(V)\big).$$

Thus, to prove Lemma 3.1, it is sufficient to put $\mathcal{L}(X) = \mathcal{L}((X_k/x)|\vec{X}_{k-1})$ and $\mathcal{L}(Y) = \mathcal{L}(Y_k/x)$ in Lemma 7.3 and to use conditions (3.3) instead of (7.34).

Acknowledgement. The author would like to thank the two referees, David Mason and Lyudmila Sakhanenko for comments and suggestions that greatly improved the paper.

References

Bentkus V. (1985) Asymptotic analysis of sums of independent random elements in Banach spaces. *Doctor Dissertation*, Vilnius.

Bentkus V., Liubinskas K. (1987) Rates of convergence in the invariance principle in Banach spaces. *Lith. Math. J.*, **27**, 3, 205–213.

Bentkus V., Račkauskas A. (1984) Estimates of the distances between sums of independent random elements in Banach spaces. *Theory Probab. Appl.*, **29**, 1, 50–65.

Berkes I., Philipp W. (1979) Approximation theorems for independent and weakly dependent random vectors. *Ann. Probab.*, **7**, 1, 29–54.

Dobrushin R.L. (1970) Prescribing a system of random variables by conditional distributions. *Theory Probab. Appl.*, **15**, 3, 458–486.

Einmahl U. (1987) A useful estimate in the multidimensional invariance principle. *Probab. Th. Rel. Fields*, **76**, 81–101.

Lindvall T. (1992) *Lectures on the Coupling Method*, Wiley, New York.

Loéve M. (1978) *Probability theory*, Springer-Verlag, New York.

Morrow G., Philipp W. (1982) An almost sure invariance principle for Hilbert space valued martingales. *Trans. Amer. Math. Soc.*, **273**, 231–251.

Monrad D., Philipp W. (1991) Nearby variables with nearby conditional laws and a strong approximation theorem for Hilbert space valued martingales. *Probab. Th. Rel. Fields*, **88**, 381–404.

Paulauskas V. and Račkauskas A. (1989) *Approximation theory in the central limit theorem. Exact results in Banach spaces*, Kluwer Academic Publisher, Dordrecht.

Sakhanenko A. I. (1974) Estimates of the rate of convergence in the invariance principle. *Soviet Math. Dokl.*, **15**, 6, 1752–1755.

Sakhanenko A. I. (1987) Simple method of obtaining estimates in the invariance principle. *Lecture Notes in Math.*, Springer-Verlag, **1299**, 430–443.

Skorohod A.V. (1976) On a representation of random variables. *Theory Probab. Appl.*, **21**, 3, 628–632.

Zaitsev A.Yu. (1998) Multidimensional version of the results of Komlós, Major, Tusnády for vectors with finite exponential moments. *ESAIM: Probability and Statistics*, **2**, 41–108.

Departamento de Matemáticas
Universidad Autónoma Metropolitana – Iztapalapa
Av. Michoacán y la Purísima s/n
Col. Vicentina
09340 México, D.F., Mexico
e-mail: sakh@xanum.uam.mx; sakh@math.nsc.ru

PART 5

The Law of the Integrated Logarithm

On the Law of the Iterated Logarithm for Local Times of Recurrent Random Walks

Xia Chen

ABSTRACT We consider the law of the iterated logarithm (LIL) for the local time of one-dimensional recurrent random walks. First we show that the constants in the LIL for the local time and for its supremum (with respect to the space variable) are equal under a very general condition given in Jain and Pruitt (1984). Second we evaluate the common value of the constants, as the random walk is in the domain of attraction of a not necessarily symmetric stable law. The first problem relies on a special maximal inequality established in this paper and the second on the LIL for Markovian additive functionals given in the author's recent work.

1 Introduction and main results

Consider a random walk $\{S_n\}_{n\geq 0}$ on the lattice $\mathbf{Z}$

$$(1.1) \qquad S_0 = 0 \text{ and } S_n = \sum_{k=1}^{n} X_k \qquad n = 1, 2, \cdots$$

generated by an i.i.d. sequence $\{X, X_n\}_{n\geq 1}$ with integer values. We assume for convenience that the law of X is not supported on a proper subgroup of $\mathbf{Z}$. The local time $L_n(x)$ and its maxima L_n^* are defined as

$$(1.2) \quad L_n(x) = \sum_{k=0}^{n} I_{\{S_k=x\}} \text{ and } L_n^* = \sup_{y\in\mathbf{Z}} L_n(y) \qquad x \in \mathbf{Z},\ n = 1, 2, \cdots .$$

To make things interesting, we always consider the case when $\{S_n\}_{n\geq 0}$ is recurrent. This requires that $EX = 0$ whenever $E|X| < +\infty$. So the condition $EX = 0$ is assumed throughout without further mention.

Asymptotic properties for the local times have been extensively studied since the pioneering work by Chung and Hunt (1948), who established the first LIL for the local times of a symmetric simple random walk. Refer to Révész (1990) and the references therein for the historical development of this subject. Jain and Pruitt (1984) obtained some interesting LIL results for local times of the random walk under a very general condition, which we shall now describe.

Let $F(x)$ be the distribution function of X and write

$$G(x) = 1 - F(x) + F(-x) \quad \text{and} \quad K(x) = x^{-2} \int_{-x}^{x} y^2 dF(y)\,, \tag{1.3}$$

$$Q(x) = G(x) + K(x) = x^{-2} E\big(|X| \wedge x\big)^2 \tag{1.4}$$

for $x > 0$. The function Q is continuous and strictly decreasing for x large enough. Thus we can define the function $a(y)$ by

$$Q\big(a(y)\big) = \frac{1}{y} \tag{1.5}$$

for sufficiently large y, and $a(y) \uparrow \infty$. The basic assumption in Jain and Pruitt (1984) is

$$\limsup_{x \to \infty} \frac{G(x)}{K(x)} < 1 \tag{A}$$

which implies, in particular, that $E|X| < +\infty$. Under (A), the random walk $\{S_n\}_{n \geq 0}$ is recurrent, in which case Jain and Pruitt (1984) proved that there exist $0 < \theta_1, \theta_2 < \infty$ such that

$$\limsup_{n \to \infty} n^{-1} a\Big(\frac{n}{\log \log n}\Big) L_n(0) = \theta_1 \quad a.s.\ , \tag{1.6}$$

$$\limsup_{n \to \infty} n^{-1} a\Big(\frac{n}{\log \log n}\Big) L_n^* = \theta_2 \quad a.s. \tag{1.7}$$

The quantity on the left of (A) was introduced by Feller (1966) to describe the compactness and convergence of the normalized random walks. If X is in the domain of attraction of a stable law of index α, for example, then

$$\lim_{x \to \infty} \frac{G(x)}{K(x)} = \frac{2 - \alpha}{\alpha} \tag{1.8}$$

so that Jain and Pruitt's results include all cases when X is in the domain of attraction of a stable law of index $\alpha > 1$ and of zero mean. As pointed out by Jain and Pruitt, the class of the distributions described by (A) is much larger than this. Jain and Pruitt also pointed out that condition (A) excludes the case when the local time has a slowly varying increasing rate.

Jain and Pruitt (1984) then asked whether or not $\theta_1 = \theta_2$ under their condition. To this author's best knowledge, the first work related to this question is by Kesten (1965) who proved equality when the random walk is replaced by Brownian motion. By approximation, Kesten's result can be used to show that the equality holds for a symmetric simple random walk (Révész (1990)). As an application of large deviation theory, Donsker and Varadhan (1977) extended Kesten's observation to the case of symmetric stable processes with index $\alpha > 1$. In all mentioned cases, the common value of θ_1 and θ_2 is obtained.

Our first goal is to answer this question under condition (A). Indeed, we have

Theorem 1. *Under condition (A),*

$$\theta_1 = \theta_2 \tag{1.9}$$

where θ_1 and θ_2 are constants given by (1.6) and (1.7), respectively.

We are not able to determine the common value in (1.9) explicitly under condition (A). However, we have

Theorem 2. *Assume that*

$$\lim_{x\to\infty} \frac{G(x)}{K(x)} = \frac{2-\alpha}{\alpha} \quad \textit{and} \quad p \equiv \lim_{x\to\infty} \frac{1-F(x)}{G(x)} \tag{B}$$

exist for some $1 < \alpha < 2$. Then

(1.10)

$$\limsup_{n\to\infty} n^{-1} a\Big(\frac{n}{\log\log n}\Big) L_n(0) = \limsup_{n\to\infty} n^{-1} a\Big(\frac{n}{\log\log n}\Big) L_n^* = \Lambda(\alpha) \quad a.s.$$

where

(1.11)

$$\Lambda(\alpha) = \frac{\Gamma(1-1/\alpha)\Gamma(1/\alpha)}{\pi} \left(\frac{2\cos\big((\rho-1/2)\alpha\pi\big)}{\Gamma(3-\alpha)\sin\frac{(\alpha-1)\pi}{2}}\right)^{1/\alpha} (\alpha-1)^{\frac{2-\alpha}{\alpha}} \sin(\rho\pi),$$

$$\rho = \frac{1}{2} + \frac{1}{\alpha\pi} \arctan\Big((2p-1)\tan\big(\frac{\alpha\pi}{2}\big)\Big). \tag{1.12}$$

Theorem 3. *Assume that*

$$\lim_{x\to\infty} \frac{G(x)}{K(x)} = 0. \tag{C}$$

Then

(1.13)

$$\limsup_{n\to\infty} n^{-1} a\Big(\frac{n}{\log\log n}\Big) L_n(0) = \limsup_{n\to\infty} n^{-1} a\Big(\frac{n}{\log\log n}\Big) L_n^* = \sqrt{2} \quad a.s.$$

Remark. According to Chap. IX, Theorem 8.1, p.303 in Feller (1966), (B) and (C) are the necessary and sufficient conditions for F being in the domain of attraction of a stable law with $1 < \alpha < 2$ and $\alpha = 2$, respectively. Under (B) (or (C) with $\alpha = 2$),

$$G\big(a(n)\big) \sim \frac{2-\alpha}{2n} \quad \text{and} \quad K\big(a(n)\big) \sim \frac{\alpha}{2n}. \tag{1.14}$$

By Chap. XI, Theorem 8.2 and the comment on p. 305 in Feller (1966) there is a stable law $\mathfrak{L}$ with index α such that

$$S_n/a(n) \longrightarrow \mathfrak{L} \ \text{ in distribution.} \tag{1.15}$$

It turns out that when $1 < \alpha < 2$, the characteristic exponent of $\mathfrak{L}$ is of the form:

$$\begin{aligned}\Psi(\lambda) = {} & \frac{\Gamma(3-\alpha)}{2(\alpha-1)}\sin\frac{(\alpha-1)\pi}{2} \\ & \times|\lambda|^\alpha\Big(1 - i(2p-1)\mathrm{sgn}(\lambda)\tan\big(\tfrac{\alpha\pi}{2}\big)\Big)\end{aligned} \tag{1.16}$$

where $\lambda \in \mathbf{R}$. Hence (see, e.g., Section 2.6 in Zolotarev (1986))

$$\rho = 1 - \mathfrak{L}(0) = \lim_{n\to\infty} P\{S_n \geq 0\}. \tag{1.17}$$

Moreover, one can easily see that $1-1/\alpha \leq \rho \leq 1/\alpha$. The extreme points $\rho = 1/\alpha$ and $\rho = 1 - 1/\alpha$ correspond to the cases when $p = 0$ and $p = 1$, respectively. In particular $\cos\big((\rho-1/2)\alpha\pi\big) > 0$ and therefore the constant we obtain in Theorem 2 remains positive for all $0 \leq p \leq 1$. It is also interesting to see that the maximal limit value in Theorem 2 is achieved when $p = 1/2$, especially when X is symmetrically distributed — in which case Theorem 2 essentially belongs to (in the light of Theorem 1) Marcus and Rosen (1994). Bertoin (1995) obtains a law of the iterated logarithm for the local times of Lévy processes which are not necessarily symmetric. In Bertoin's result (Corollary 2), the constant of the LIL is given under a condition essentially the same as (B) (The author thanks the referee of this paper for pointing out Bertoin's contribution).

When X has a finite second moment σ^2, i.e.,

$$\sigma^2 \equiv E|X|^2 < +\infty, \tag{1.18}$$

one can easily see that $a(n) \sim \sqrt{n}\sigma$ as $n \to \infty$. Hence by Theorem 3 we have

$$\limsup_{n\to\infty} \frac{L_n(0)}{\sqrt{2n\log\log n}} = \limsup_{n\to\infty} \frac{L_n^*}{\sqrt{2n\log\log n}} = \frac{1}{\sigma} \qquad a.s. \tag{1.19}$$

2 Proof of Theorem 1

Clearly we have $\theta_1 \le \theta_2$. Hence we need to show

$$\theta_1 \ge \theta_2 \,. \tag{2.1}$$

Our approach is to compare the tail probabilities of $L_n(0)$ and L_n^* and we rely on the following inequalities.

Lemma 1. *For any integers $a, b > 0$ and $n \ge 1$,*

$$P\{L_n(x) \ge a\} \le P\{L_n(0) \ge a\} \qquad x \in \mathbf{Z}, \tag{2.2}$$

$$P\{L_n^* \ge a+b\} \le P\{L_n^* \ge a\}P\{L_n^* \ge b\}, \tag{2.3}$$

$$P\{L_n^* \ge a\} \le \Big(1 - P\{\max_{k\le n} |S_k| \ge b\}\Big)^{-1} P\{\sup_{|x|\le b} L_n(x) \ge a\}. \tag{2.4}$$

Proof. Define the stopping times as

$$\tau_1 = \inf\{k \ge 1; \quad S_k = x\}, \tag{2.5}$$

$$\tau_2 = \inf\{k \ge 1; \quad L_k^* \ge b\}, \tag{2.6}$$

$$\tau_3 = \inf\{k \ge 1; \quad |S_k| \ge b\}. \tag{2.7}$$

To prove (2.2), we may assume that $x \ne 0$. Hence

$$\begin{aligned} P\{L_n(x) \ge a\} &= P\Big\{\tau_1 \le n, \sum_{k=\tau_1}^{n} I_{\{S_k=x\}} \ge a\Big\} \\ &= \sum_{j=1}^{n} P\Big\{\tau_1 = j, \sum_{k=j}^{n} I_{\{S_k - S_j = 0\}} \ge a\Big\} \\ &= \sum_{j=1}^{n} P\{\tau_1 = j\}P\{L_{n-j}(0) \ge a\} \\ &\le P\{L_n(0) \ge a\}. \end{aligned}$$

The proof of (2.3) and (2.4) relies on the fact that for each $1 \le j \le n$, the random variable

$$\sup_{x\in\mathbf{Z}} \sum_{k=j}^{n} I_{\{S_k=x\}} \tag{2.8}$$

is independent of $\{X_1, \cdots, X_j\}$. Notice that $L^*_{\tau_2} = b$.

$$\begin{aligned} P\{L_n^* \geq a+b\} &= P\{\tau_2 \leq n,\ L_n^* - L^*_{\tau_2} \geq a\} \\ &\leq P\Big\{\tau_2 \leq n,\ \sup_{x\in\mathbf{Z}} \sum_{k=\tau_2}^{n} I_{\{S_k=x\}} \geq a\Big\} \\ &\leq P\{L_n^* \geq a\}P\{L_n^* \geq b\}. \end{aligned}$$

We shall now prove (2.4). Notice that

$$\begin{aligned} P\{L_n^* \geq a\} &\leq P\{\sup_{|x|\leq b} L_n(x) \geq a\} + P\{\sup_{|x|>b} L_n(x) \geq a\} \\ &= (I) + (II) \qquad \text{(say)} \end{aligned}$$

and that

(2.9)

$$(II) \leq P\Big\{\tau_3 \leq n,\ \sup_x \sum_{k=\tau_3}^{n} I_{\{S_k=x\}} \geq a\Big\} \leq P\{\max_{k\leq n}|S_k| \geq b\}P\{L_n^* \geq a\}.$$

Therefore

$$\Big(1 - P\{\max_{k\leq n}|S_k| \geq b\}\Big)P\{L_n^* \geq a\} \leq P\{\sup_{|x|\leq b} L_n(x) \geq a\} \tag{2.10}$$

which gives (2.4). Q.E.D.

To prove (2.1), we let

$$b_n = n\Big[a\Big(\frac{n}{\log\log n}\Big)\Big]^{-1} \quad \text{and} \quad c_n = a\Big(\frac{n}{\log\log n}\Big). \tag{2.11}$$

We let $\epsilon > 0$ be fixed but arbitrary. Applying (2.3) gives

$$P\{L_n^* \geq (\theta_1 + 4\epsilon)b_n\} \leq P\{L_n^* \geq 2\epsilon b_n\}P\{L_n^* \geq (\theta_1 + 2\epsilon)b_n\}. \tag{2.12}$$

Note (Lemma 3, Jain and Pruitt (1984)) that there is an $M > 0$ such that

$$P\{\max_{k\leq n}|S_k| \geq Ma(n)\} \leq \frac{1}{2}. \tag{2.13}$$

Thus by (2.4),

$$P\{L_n^* \geq 2\epsilon b_n\} \leq 2P\{\sup_{|x|\leq Ma(n)} L_n(x) \geq 2\epsilon b_n\}, \tag{2.14}$$

$$P\{L_n^* \geq (\theta_1 + 2\epsilon)b_n\} \leq 2P\{\sup_{|x|\leq Ma(n)} L_n(x) \geq (\theta_1 + 2\epsilon)b_n\}. \tag{2.15}$$

According to Lemma 11 in Jain and Pruitt (1984), there is a $\delta > 0$ and a $C > 0$ such that

$$P\{\sup_{|x-y|\le\delta c_n} |L_n(x) - L_n(y)| \ge \epsilon b_n\} \le \frac{C}{\log^2 n} \tag{2.16}$$

for all $n \ge 1$. On the other hand, notice that the interval $[-Ma(n), Ma(n)]$ can be covered by no more than $(K \log\log n)^\lambda$ intervals with diameters less than δc_n, where $\lambda > 0$ and $K > 0$ are constants. Hence

$$\begin{aligned}&P\{\sup_{|x|\le Ma(n)} L_n(x) \ge 2\epsilon b_n\}\\ &\le (K\log\log n)^\lambda \sup_{x\in\mathbf{Z}} P\{L_n(x) \ge \epsilon b_n\} + \frac{C}{\log^2 n},\end{aligned}$$

$$\begin{aligned}&P\{\sup_{|x|\le Ma(n)} L_n(x) \ge (\theta_1 + 2\epsilon) b_n\}\\ &\le (K\log\log n)^\lambda \sup_{x\in\mathbf{Z}} P\{L_n(x) \ge (\theta_1+\epsilon) b_n\} + \frac{C}{\log^2 n}.\end{aligned}$$

In view of (2.2) and (2.12),

$$\begin{aligned}&P\{L_n^* \ge (\theta_1 + 4\epsilon)b_n\}\\ &\le 4(K\log\log n)^{2\lambda} P\{L_n(0) \ge \epsilon b_n\} P\{L_n(0) \ge (\theta_1+\epsilon)b_n\}\\ &+C\frac{(\log\log n)^\lambda}{\log^2 n}\end{aligned}$$

where the constant $C > 0$ may differ from before. By examining the proof of Lemma 8 in Jain and Pruitt (1984) one can see that

$$P\{L_n(0) \ge \epsilon b_n\} \le \frac{1}{(\log n)^\delta} \tag{2.17}$$

eventually holds for some small constant $\delta > 0$. Hence,

$$\begin{aligned}&P\{L_n^* \ge (\theta_1 + 4\epsilon)b_n\}\\ &\le P\{L_n(0) \ge (\theta_1+\epsilon)b_n\} + C\frac{(\log\log n)^\lambda}{\log^2 n}\end{aligned} \tag{2.18}$$

for large n.

On the other hand, given a $\gamma > 1$ we take a subsequence $\{n_k\}$ such that

$$b_{n_k} \sim \gamma^k \qquad (k\to\infty). \tag{2.19}$$

By Lemma 3.3 in Chen (1999),

$$\sum_k P\{L_{n_k}(0) \ge (\theta_1+\epsilon)b_{n_k}\} < +\infty. \tag{2.20}$$

From (2.18),

$$\sum_k P\{L_{n_k}^* \ge (\theta_1 + 4\epsilon) b_{n_k}\} < +\infty. \tag{2.21}$$

Since $\gamma > 1$ and $\epsilon > 0$ are arbitrary, a standard argument via the Borel-Cantelli lemma gives

$$\limsup_{n\to\infty} L_n^*/b_n \le \theta_1 \quad a.s. \tag{2.22}$$

Hence, (2.1) is proved. Q.E.D.

3 Proof of Theorem 2 and Theorem 3

In view of Theorem 1, we only need to show

$$\limsup_{n\to\infty} n^{-1} a\Big(\frac{n}{\log\log n}\Big) L_n(0) = \Lambda(\alpha) \quad a.s. \tag{3.1}$$

for $1 < \alpha < 2$, and

$$\limsup_{n\to\infty} n^{-1} a\Big(\frac{n}{\log\log n}\Big) L_n(0) = \sqrt{2} \quad a.s. \tag{3.2}$$

for $\alpha = 2$.

The proof we present here appears as an application of the LIL (Chen (1999)) for additive functionals of Harris recurrent Markov chains. Let the stable law $\mathfrak{L}$ be given by (1.15). From (2.j) in Le Gall-Rosen (1991), as $n \to \infty$,

$$g(n) \equiv \sum_{k=1}^n P\{S_k = 0\} \sim p(0) \sum_{k=1}^n \frac{1}{a(k)}, \tag{3.3}$$

where $p(\cdot)$ is the density of $\mathfrak{L}$. Note that $a(y)$ is non-decreasing and varies regularly at infinity with index α^{-1}. As a routine exercise one can show that, as $n \to \infty$,

$$g(n) \sim p(0) \frac{n}{a(n)} \int_0^1 \frac{1}{x^{1/\alpha}} dx = p(0) \frac{\alpha}{\alpha - 1} \frac{n}{a(n)}. \tag{3.4}$$

In particular, $g(n)$ varies regularly at infinity with index $1 - 1/\alpha$. Viewing $\{S_n\}_{n\ge 0}$ as a recurrent Markov chain with the counting measure on $\mathbf{Z}$ as its invariant measure, we obtain from Theorem 2.4 in Chen (1999) that

(3.5)

$$\limsup_{n\to\infty} \Big(g\Big(\frac{n}{\log\log n}\Big) \log\log n\Big)^{-1} L_n(0) = (\alpha - 1)^{1/\alpha} \Gamma(1 - 1/\alpha) \quad a.s.$$

In view of (3.4),

$$\text{(3.6)}\quad \limsup_{n\to\infty} n^{-1}a\Big(\frac{n}{\log\log n}\Big)L_n(0)=\frac{\alpha\Gamma(1-1/\alpha)}{(\alpha-1)^{1-1/\alpha}}p(0)\quad a.s.$$

When $\alpha=2$, we have from (1.14) that

$$\text{(3.7)}\quad nP\{|X|>a(n)\}\longrightarrow 0 \quad\text{and}\quad E|X|^2I_{\{|X|\le a(n)\}}\sim\frac{a(n)^2}{n}\quad (n\to\infty).$$

By performing the truncation at the level $a(n)$ and by the classical method for the central limit theorem we can obtain that $\mathfrak{L}=N(0,1)$. In particular $p(0)=(2\pi)^{-1/2}$. Hence (3.2) follows from (3.5). We now assume $1<\alpha<2$.

To evaluate $p(0)$, we first verify that the characteristic exponent $\Psi(\lambda)$ of $\mathfrak{L}$ is given by (1.16) (This may be known, but we fail to find a complete statement of this result in the literature). By Theorem C.1 and (I.11) in Zolotarev (1986)

$$\text{(3.8)}\qquad P\{X>a_n\}\sim\frac{C_1}{n}\quad\text{and}\quad P\{X\le -a_n\}\sim\frac{C_2}{n}$$

where $C_1,C_2\ge 0$ are two parameters of the spectral function of $\mathfrak{L}$ given in that theorem. From the first part in (1.14) and the second part in (B) we have

$$\text{(3.9)}\qquad C_1+C_2=\frac{2-\alpha}{2}>0\quad\text{and}\quad p=\frac{2C_1}{2-\alpha}.$$

Hence (1.16) follows from the construction given in the proof of Theorem C.2 in Zolotarev (1986) and the fact that the law $\mathfrak{L}$ has zero mean.

We finally evaluate $p(0)$ in the case when $0<\alpha<2$. Write

$$\text{(3.10)}\qquad c=\frac{\Gamma(3-\alpha)}{2(\alpha-1)}\sin\frac{(\alpha-1)\pi}{2}.$$

Taking the inverse Fourier transformation gives

$$\begin{aligned}
p(0) &= \frac{1}{2\pi}\int_{-\infty}^{+\infty}\exp\Big\{-c|\lambda|^\alpha\Big(1-i(2p-1)\mathrm{sgn}(\lambda)\tan\big(\tfrac{\alpha\pi}{2}\big)\Big)\Big\}d\lambda\\
&= \frac{1}{\alpha\pi}\int_0^{+\infty}x^{1/\alpha-1}e^{-cx}\cos\Big[\big(c(2p-1)\tan\big(\tfrac{\alpha\pi}{2}\big)\big)x\Big]dx\\
\text{(3.11)}\quad &= \frac{\cos\Big(\alpha^{-1}\arctan\big((2p-1)\tan\big(\frac{\alpha\pi}{2}\big)\big)\Big)}{\alpha\pi c^{1/\alpha}\Big[1+(2p-1)^2\tan^2\big(\frac{\alpha\pi}{2}\big)\Big]^{1/2\alpha}}\Gamma(1/\alpha)\\
&= (\alpha\pi)^{-1}\Big[c^{-1}\cos\big((\rho-1/2)\alpha\pi\big)\Big]^{1/\alpha}\sin(\rho\pi)\Gamma(1/\alpha)\\
&= (\alpha\pi)^{-1}\Big[\frac{2(\alpha-1)\cos\big((\rho-1/2)\alpha\pi\big)}{\Gamma(3-\alpha)\sin\frac{(\alpha-1)\pi}{2}}\Big]^{1/\alpha}\sin(\rho\pi)\Gamma(1/\alpha)
\end{aligned}$$

where the third step follows from the fact that

$$(3.12) \qquad \int_0^{\infty} x^{t-1} e^{-ax} \cos(bx) dx = \frac{\sin\left(t \arctan\left(\frac{b}{a}\right)\right)}{(a^2+b^2)^{t/2}} \Gamma(t)$$

for all real numbers $t > -1$, $a > 0$ and b.

Bringing our computation of $p(0)$ back to (3.5) yields (3.1). Q.E.D.

4 Further remark

Since the LIL in a general Markovian context applied to Theorem 2 and Theorem 3 holds also in the case of non-discrete state space, we can achieve some results for non-lattice valued random walks, which take a form similar to (3.1) or (3.2). Here we consider a random walk $\{S_n\}_{n\geq 0}$ on $\mathbf{R}$. In the terminology of Revuz (1975), $\{S_n\}_{n\geq 0}$ is called *spread out*, if there is an integer $k \geq 1$ such that the kth convolution F^{*k} is not singular to the Lebesgue measure on $\mathbf{R}$, where F is the distribution of its i.i.d. increment. Revuz (1975) points out that this class is much larger than the class of absolutely continuous F. It is known (see, e.g., Sections 4 and 5, Chapter 3 in Revuz (1975)) that when viewed as a Markov chain with Lebesgue measure as its invariant measure, $\{S_n\}_{n\geq 0}$ is Harris recurrent if it is spread out.

We adopt all notations introduced in Section 1. By an argument almost identical to the one carried out in Section 3, we can prove

Theorem 4. *Assume that $\{S_n\}_{n\geq 0}$ is spread out and satisfies condition (B) for some $1 < \alpha < 2$. Then for any non-negative Lebesgue integrable function f on $\boldsymbol{R}$,*

$$(4.1) \qquad \limsup_{n\to\infty} n^{-1} a\Big(\frac{n}{\log\log n}\Big) \sum_{k=1}^{n} f(S_k) = \Lambda(\alpha) \int_{-\infty}^{\infty} f(x)dx \quad a.s.$$

where the constant $\Lambda(\alpha) > 0$ is given as in Theorem 2.

Theorem 5. *Assume that $\{S_n\}_{n\geq 0}$ is spread out and satisfies condition (C). Then for any non-negative Lebesgue integrable function f on $\boldsymbol{R}$,*

$$(4.2) \qquad \limsup_{n\to\infty} n^{-1} a\Big(\frac{n}{\log\log n}\Big) \sum_{k=1}^{n} f(S_k) = \sqrt{2} \int_{-\infty}^{\infty} f(x)dx \quad a.s.$$

References

Bertoin, J. (1995), Some applications of subordinators to local times of Markov processes, *Forum Math.* **7**, 629–644.

Chen, X. (1999), How often does a Harris recurrent Markov chain recur? *Ann. Probab.* **27**, 1324–1346.

Chung, K. L. and Hunt, C. A. (1949), On the zeros of $\sum_1^\infty \pm 1$, *Ann. Math.* **50**, 385–400.

Donsker, M. D. and Varadhan, S. R. S. (1977), On the law of the iterated logarithm for local times, *Comm. Pure. Appl. Math.* **XXX**, 707–753.

Feller, W. (1966), *An Introduction to Probability Theory and Its applications*, Vol. II, Wiley, New York.

Griffin, P. S., Jain, N. C. and Pruitt, W. E. (1984), Approximate local limit theorems for laws outside domains of attraction, *Ann Probab.* **12**, 45–63.

Jain, N. C. and Pruitt, W. E. (1984), Asymptotic behavior for the local time of a recurrent random walk, *Ann. Probab.* **12**, 64–85.

Kesten, H. (1965), An iterated logarithm law for local times, *Duke Math. J.* **32**, 447–456.

Le Gall, J.-F. and Rosen, J. (1991), The range of stable random walks, *Ann. Probab.* **19**, 650–705

Marcus, M. B. and Rosen, J. (1994), Law of the iterated logarithm for the local times of symmetric Lévy processes and recurrent random walks, *Ann. Probab.* **22**, 626–658.

Révész, P. (1990) *Random Walk in Random and Non-Random Environments*, World Scientific, London.

Revuz, D. (1975). *Markov Chains*, North-Holland, New York.

Zolotarev,V.M. (1986), *One-dimensional Stable Distributions*, Amer. Math. Soc., Providence, RI 1986.

Department of Mathematics
University of Tennessee
Knoxville, TN 37996
e-mail: xchen@math.utk.edu

A General Compact Law of the Iterated Logarithm in Banach Spaces

Uwe Einmahl* and James Kuelbs**

ABSTRACT We study a general compact law of the iterated logarithm, the related cluster set, and also the cluster sets determined by random series in Banach spaces.

1 Introduction

In the recent paper Einmahl-Kuelbs (1999), cluster sets for a generalized law of the iterated logarithm in Banach spaces were determined; here we examine the compact law of the iterated logarithm in this setting. We also study the related cluster sets for situations which generalize the independent coordinate results on $\mathbb{R}^d$ obtained in Einmahl (1995).

Let B denote a real separable Banach space with norm $\|\cdot\|$ and assume $X, X_1, X_2, \ldots$ are i.i.d. B-valued random vectors with $0 < E\|X\| < \infty$ and $E(X) = 0$. As usual, let $S_n = \sum_{j=1}^{n} X_j$ for $n \geq 1$, and write Lt to denote $\log(\max(t, e))$ for $t \geq 0$. The function $L(Lt)$ will be written as $L_2 t$, and B^* denotes the topological dual of B. For any sequence $\{x_n\} \subset B$, the set of its limit points is denoted by $C(\{x_n\})$ and is called the cluster set of $\{x_n\}$. If $A \subset B, x \in B$, the distance from x to A is given by $d(x, A) = \inf_{y \in A} \|x - y\|$.

The separability of B and the Hewitt-Savage zero-one law easily imply that for any sequence $\alpha_n \uparrow \infty$, with probability 1

$$C(\{S_n/\alpha_n\}) = A, \tag{1.1}$$

where A is non-random and depends only on $\{\alpha_n\}$ and the distribution of X (see Lemma 1 in Kuelbs (1981)). It is also immediate from the Hewitt-Savage zero-one law that if $A \subset B$ and $\alpha_n \uparrow \infty$, then the events determined by the statements

$$S_n/\alpha_n \quad \text{is relatively compact in} \quad B, \tag{1.2}$$

*Research partially supported by an FWO Grant.

**Research partially supported by NSF Grant 9703740.

and

$$\lim_n d(S_n/\alpha_n, A) = 0 \tag{1.3}$$

each have probability 0 or 1.

We say X satisfies the compact law of the iterated logarithm (CLIL) with respect to the sequence $\{\alpha_n\}$ if there is a compact subset A of B such that (1.1) and (1.3) both hold with probability 1. We combine (1.1) and (1.3) by writing $\{S_n/\alpha_n\} \twoheadrightarrow A$, and say $X \in$ CLIL with limit set A.

The classical choice for $\{\alpha_n\}$ is the sequence $\{(2nL_2n)^{1/2}\}$, and the CLIL is completely understood in this setting (see Ledoux-Talagrand (1991), pp. 213–216). In particular, the set A in (1.1) and (1.3) is then the unit ball of the reproducing kernel Hilbert space determined by the covariance structure of X.

If X fails to have a covariance function, there is still the possibility of an LIL result, and this is what we describe now.

Following Klass (1976) we associate with any real-valued random variable ξ satisfying $0 < E|\xi| < \infty$ a function $K(\cdot)$ which is defined as the inverse function of the strictly increasing function $G(\cdot)$ given for $y > 0$ by

$$G(y) = y^2 / \int_0^y E(|\xi| I(|\xi| > t))dt.$$

Since $E\|X\| < \infty$, we have $E|f(X)| < \infty$ for all $f \in B^*$, and for any $f \in B^*$ with $E|f(X)| > 0$ let K_f be the K-function corresponding to $f(X)$. For $y > 0$, define

$$\tilde{K}(y) = \sup\{K_f(y) : \|f\|_{B^*} \leq 1,\ E|f(X)| > 0\}, \tag{1.4}$$

and set for $n \geq 1$,

$$\gamma_n = \sqrt{2}\tilde{K}(n/L_2n)L_2n. \tag{1.5}$$

The generalized LIL for X with respect to $\{\gamma_n\}$ was studied in Einmahl (1993), and the related cluster set result of Einmahl-Kuelbs (1999) is the following theorem.

Theorem 1. *Suppose X is a mean-zero B-valued random vector such that $0 < E\|X\| < \infty$ and $\{\gamma_n\}$ is as in (1.5). Furthermore, assume*

$$S_n/\gamma_n \xrightarrow{\text{prob}} 0, \tag{1.6}$$

and

$$\sum_{n=1}^{\infty} P(\|X\| > \gamma_n) < \infty. \tag{1.7}$$

Then with probability 1,

$$\limsup_{n\to\infty} \|S_n/\gamma_n\| = 1, \tag{1.8}$$

and the cluster set $A = C(\{S_n/\gamma_n\})$ *is such that*

(1.9) A *is closed, non-empty, symmetric about 0, and*

(1.10) A *is star-like at 0.*

Remarks. (a) There is no loss of generality in assuming $E(X) = 0$ and $E\|X\| < \infty$ if we expect the CLIL (even the bounded LIL) to hold with respect to $\{\gamma_n\}$, since by (2.3) of Einmahl (1995) $\gamma_n/n \downarrow 0$, and hence the law of large numbers easily applies to yield $E(X) = 0$ and $E\|X\| < \infty$.

(b) The lim sup in (1.8) was obtained in Theorem 3 of Einmahl (1993) under (1.6) and (1.7). Theorem 2 there shows the lim sup in (1.8) is always greater than or equal to 1, and Theorem 1 of Einmahl (1993) shows that it is finite provided (1.7) holds and $\{S_n/\gamma_n\}$ is bounded in probability. The precise lim sup is unknown when $\{S_n/\gamma_n\}$ is only bounded in probability. This is even the case when there is a covariance structure available, although the results of Alexander (1989a-b) determine all possible cluster sets independent of $\{S_n/(2nL_2n)^{1/2}\}$ being stochastically bounded or not. Theorem 3 below shows that if X satisfies the CLIL, then (1.6) holds. It is also known from Lemma 6 in Einmahl (1993) that if X satisfies the bounded LIL and B is a Banach space of type 2, then (1.6) holds. Hence the distinction between $\{S_n/\gamma_n\}$ converging in probability to 0, or being bounded in probability, fails to be a problem in these two situations.

If B is an infinite dimensional Banach space and $A \subset \{x : \|x\| \leq 1\}$ satisfies (1.9) and (1.10), then Theorem 2 of Einmahl-Kuelbs (1999) shows there is a B-valued random vector X satisfying the conditions of Theorem 1 and such that $C(\{S_n/\gamma_n\}) = A$ with probability 1. Hence if A satisfies (1.9) and (1.10) and A is also compact with A a subset of the open unit ball of B, then (1.8) implies X does not satisfy the CLIL with respect to $\{\gamma_n\}$ even though $C(\{S_n/\gamma_n\}) = A$ is compact. This phenomenon suggests we try to understand the CLIL with respect to the $\{\gamma_n\}$. Our first result in this direction follows.

Theorem 2. *Let X be a mean-zero B-valued random vector with $0 < E\|X\| < \infty$, assume $\{\gamma_n\}$ is as in (1.5), and that (1.6) and (1.7) hold. Then the following are equivalent:*

(1.11)
- (*i*) X *satisfies the CLIL with respect to* $\{\gamma_n\}$.
- (*ii*) $\{S_n/\gamma_n\}$ *is relatively compact with probability 1.*
- (*iii*) *For each* $\epsilon > 0$ *there is a finite dimensional space* $F = F_\epsilon$ *such that with probability 1,* $\overline{\lim}_n d(S_n/\gamma_n, F) < \epsilon$.

Corollary 1. *Let X and $\{\gamma_n\}$ be as in Theorem 2 and assume for every $\epsilon > 0$ there exists $x_1, \ldots, x_N \in B$, $f_1, \ldots, f_N \in B^*$ such that with probability 1*

$$\overline{\lim}_n \|S_n/\gamma_n - \sum_{k=1}^{N} f_k(S_n/\gamma_n)x_k\| < \epsilon. \tag{1.12}$$

Then X satisfies the CLIL.

As a partial converse to the previous corollary we observe that the operator $T(x) = \sum_{k=1}^N f_k(x)x_k$, where the x_k's $\in B$ and f_k's are in B^*, is a finite rank operator from $B \to B$. If we assume that B has enough of these finite rank operators to approximate the identity operator on B, then we have a converse of Corollary 1.

Proposition 1. *Let X and $\{\gamma_n\}$ be as in Theorem 1 and assume X satisfies the CLIL with limit set A. If for each $\epsilon > 0$ there is a finite rank operator T on B, say $T(x) = \sum_{k=1}^N f_k(x)x_k$ as above, with $\sup_{x\in A} \|x - T(x)\| < \epsilon$, then (1.12) holds.*

Remark. A Banach space B is said to have the approximation property if for each compact set K and $\epsilon > 0$ there is a finite rank operator $T : B \to B$ such that

$$\sup_{x\in K} \|x - Tx\| < \epsilon. \tag{1.13}$$

Hence (1.12) is equivalent to the CLIL with respect to the $\{\gamma_n\}$ in all such spaces. Further references and information about the approximation property can be found in Day (1973) and Lindenstrauss-Tzafriri (1977). The main point for us is that many Banach spaces (but not all) have this property, which is more general than asking that the Banach space have a Schauder basis.

If we assume X has a covariance function, then $\gamma_n \sim (2\sigma^2 nL_2 n)^{1/2}$, where

$$\sigma^2 = \sup_{\|f\|_{B^*}\leq 1} E(f^2(X)) < \infty, \tag{1.14}$$

(see Einmahl (1993), p.2015). Furthermore, by Ledoux-Talagrand (1991), Theorem 8.6, the conditions (1.6) and (1.7) are then equivalent to X satisfying the CLIL, provided the canonical cluster set is compact. As we pointed out above, this equivalence does not hold for the general CLIL. However, the general CLIL does imply (1.6) and (1.7).

Theorem 3. *Let X be a mean zero random vector with $0 < E\|X\| < \infty$ and assume $\{\gamma_n\}$ is as in (1.5). Furthermore, assume X satisfies the CLIL* *with respect to $\{\gamma_n\}$, and with limit set A. Then (1.6) and (1.7) hold, the compact set A satisfies (1.9), (1.10), and* $\sup_{x\in A}\|x\| = 1$. *Moreover, each set of this type occurs as a cluster set for a suitable random vector $X : \Omega \to B$ satisfying the compact LIL with respect to the norming sequence $\{\gamma_n\}$.*

The proofs of Theorems 2 and 3, Corollary 1 and Proposition 1 will be given in the subsequent Section 2. Our results for random series are formulated and proven in Section 3.

2 Proofs

First we will prove Theorem 2, which is quite easy. From its proof it will be clear that the equivalence of (1.11–i) and (1.11–ii) requires only that the sequence $\{\gamma_n\} \uparrow \infty$. Furthermore, if we add the condition that X satisfies the bounded LIL with respect to $\{\gamma_n\}$ to (1.11–iii), then (1.11–i), (1.11–ii) and (1.11–iii) are still equivalent provided $\{\gamma_n\} \uparrow \infty$. We stated the theorem with $\{\gamma_n\}$, the constants of (1.5) satisfying (1.6) and (1.7), because this gives us the bounded LIL automatically. In addition, these $\{\gamma_n\}$ are natural generalizations of the classical constants when a covariance exists, and by Theorem 2 of Einmahl (1993) they provide the possibility of a non-degenerate cluster set, i.e. $\overline{\lim}_n\|S_n/\gamma_n\| \geq 1$ whenever X is centered and $0 < E\|X\| < \infty$. Of course, similar remarks also apply to Corollary 1, and Proposition 1.

Proof of Theorem 2. Now (1.11–i) implies (1.11–ii) by definition of the CLIL. Conversely, if (1.11–ii) holds then with probability 1 the cluster set

$$A = C(\{S_n/\gamma_n\})$$

is a non-empty, non-random compact subset of B (recall the earlier application of the Hewitt-Savage zero-one law). Thus X satisfies the CLIL with respect to $\{\gamma_n\}$ with limit set A. Moreover, if

$$\overline{\lim}_n d(S_n/\gamma_n, A) > 0 \tag{2.1}$$

with positive probability, then the probability is 1. Hence if this probability is not 0, then with probability 1, we have $\{S_n(\omega)/\gamma_n\}$ relatively compact with $C(\{S_n(\omega)/\gamma_n\}) = A$, yet (2.1) holds. This is a contradiction, and hence (2.1) holds with probability 0. Thus (1.11–ii) implies (1.11–i).

If (1.11–i) holds with compact limit set A, take $\epsilon > 0$ and choose $x_1, \dots, x_N \in A$ such that

$$A \subset \bigcup_{j=1}^{N} \{x : \|x - x_j\| < \epsilon\}. \tag{2.2}$$

Set $F = \text{span}\{x_1, \ldots, x_N\}$. Then $x_1, \ldots, x_N \in F \cap A$ and (2.2) implies

$$\overline{\lim}_n d\left(\frac{S_n}{\gamma_n}, \{x_1, \ldots, x_N\}\right) < \overline{\lim}_n d\left(\frac{S_n}{\gamma_n}, A\right) + \epsilon$$

with probability 1. Hence

$$\overline{\lim}_n d\left(\frac{S_n}{\gamma_n}, F\right) \le \overline{\lim}_n d\left(\frac{S_n}{\gamma_n}, \{x_1, \ldots, x_N\}\right) < \epsilon \tag{2.3}$$

since (1.11–i) implies $\overline{\lim}_n d(S_n/\gamma_n, A) = 0$. Thus (2.3) implies (1.11–iii).

Now assume (1.11–iii). Fix Ω_0, $P(\Omega_0) = 1$, where (1.11–iii) holds. Fix ϵ and take $y_n = y_n(\omega) \in F$ with $d\left(\frac{S_n(\omega)}{\gamma_n}, y_n(w)\right) < \epsilon/2$ for all sufficiently large $n \ge N(\omega)$ and $\omega \in \Omega_0$. Then $\|y_n\| < \|S_n/\gamma_n\| + \|y_n - \frac{S_n}{\gamma_n}\| < \|\frac{S_n}{\gamma_n}\| + \epsilon/2$, and since $\overline{\lim}_n \|S_n\|/\gamma_n = 1$ by (1.8), we have $\{y_n(\omega)\}$ relatively compact in F for $\omega \in \Omega_0$. Thus $\{S_n(\omega)/\gamma_n\}$ is totally bounded with probability 1 as ϵ can go to 0 through a countable set. Thus (1.11–ii) holds, and Theorem 2 is proven.

Proof of Corollary 1. Setting $F = \text{span}\{x_1, \ldots, x_N\}$, this is immediate from (1.12), in conjunction with (1.11–iii).

Proof of Proposition 1. Fix $\epsilon > 0$ and assume T is a finite rank operator $T(x) = \sum_{k=1}^N f_k(x)x_k$ such that $\sup_{x \in A} \|x - T(x)\| < \epsilon/2$, with A the limit set in the CLIL for X with respect to $\{\gamma_n\}$. Then with probability 1

$$\begin{aligned}\overline{\lim}_n \|S_n/\gamma_n \quad & - \textstyle\sum_{k=1}^N f_k(S_n/\gamma_n)x_k\| \\ &= \overline{\lim}_n \|(I - T)(S_n/\gamma_n)\| \\ &\le \overline{\lim}_n \{\|(I - T)(y_n)\| + \|(I - T)(y_n - S_n/\gamma_n)\|\},\end{aligned}$$

where $y_n = y_n(\omega) \in A$ and $\|S_n/\gamma_n - y_n\| = \delta_n \to 0$, since X satisfies the CLIL with respect to $\{\gamma_n\}$. Hence with probability 1,

$$\overline{\lim_n} \|(I - T)(S_n/\gamma_n)\| \le \epsilon/2 + \overline{\lim_n} \|I - T\| \cdot \delta_n < \epsilon,$$

and Proposition 1 is proven.

Proof of Theorem 3. If X satisfies the CLIL with respect to $\{\gamma_n\}$, where X is mean-zero, $0 < E\|X\| < \infty$, and $\{\gamma_n\}$ satisfies (1.5), then Theorem 1 of Einmahl (1993) implies (1.7).

Once we show (1.6) holds, the first part of Theorem 3 follows from Theorem 1 at the beginning of this paper. Hence it suffices to verify (1.6). For this, and also for the results on series in Section 3 below, it is useful to have the idea of a Schauder basis for a Banach space.

Let E denote an infinite dimensional real Banach space. The sequence $\{e_k\}$ in E is a Schauder basis for E if each $x \in E$ determines a unique sequence of real numbers $\{a_k\}$, depending on x, such that

$$\lim_N \|\sum_{k=1}^{N} a_k e_k - x\| = 0. \tag{2.4}$$

The coefficients $\{a_k\}$, being uniquely determined, are easily seen to determine linear coefficient functionals $f_k(x) = a_k$, $k \geq 1$, $x \in E$. Thus $f_k(x)$ is the unique coefficient of e_k in the series expansion of x, and the uniqueness immediately implies $f_k(e_j) = \delta_{j,k}$ for $k, j \geq 1$. Given a basis $\{e_k\}$, we define

$$\pi_N(x) = \sum_{k=1}^{N} f_k(x) e_k \quad \text{and} \quad Q_N(x) = x - \pi_N(x) \tag{2.5}$$

for $N \geq 1$, and point out that these operators are continuous on E. In fact, there exists a constant $C < \infty$ such that

$$\|\pi_N\| \leq C, \quad \|Q_N\| \leq C + 1 \tag{2.6}$$

for all $N \geq 1$, and since $\|f_n(x)e_k\| = \|\pi_{N+1}(x) - \pi_N(x)\| \leq 2C\|x\|$ for all $x \in E$, $N \geq 1$, the coefficient functionals $\{f_k\}$ are also continuous. In fact, replacing f_k by $f_k/\|f_k\|_{B^*}$ and e_k by $e_k\|f_k\|_{B^*}$ we may assume that coefficient functionals have norm 1. We do this in the remainder of the proof of Theorem 3. For further details we suggest the reader consult Day (1973), pp. 87–89.

To finish the proof of the first part of Theorem 3 we take $\Lambda : B \to C[0,1]$ to be a linear isometry, with the norm on $C[0,1]$ being the sup-norm $\|\cdot\|_\infty$. Λ exists because of the well-known theorem of Mazur. Let $Y = \Lambda(X), Y_j = \Lambda(X_j)$ for $j \geq 1$, and set $T_n = \sum_{j=1}^{n} Y_j = \Lambda(S_n)$. Then Λ a linear isometry implies the constants formed from Y in the way the γ_n in (1.5) are determined from X, yields exactly the same sequence $\{\gamma_n\}$. Hence it suffices to prove

$$T_n/\gamma_n \xrightarrow{\text{prob}} 0, \tag{2.7}$$

to obtain (1.6).

Now $C[0,1]$ with the sup-norm has a basis, which we denote by $\{u_k\}$. Hence consider π_N and Q_N as in (2.6). If X satisfies the CLIL in B with respect to $\{\gamma_n\}$, and with limit set A, then $\Lambda(A)$ is compact and for each $\epsilon > 0$, there exists an $N_0 = N_0(\epsilon)$ such that

$$\sup_{N \geq N_0} \sup_{x \in \Lambda(A)} \|Q_N(x)\|_\infty < \epsilon. \tag{2.8}$$

Hence with probability 1, $N \geq N_0$ implies

$$\overline{\lim}_n \|Q_N(T_n/\gamma_n)\|_\infty = \overline{\lim}_n \|Q_N(\Lambda(S_n/\gamma_n))\|_\infty < \epsilon, \tag{2.9}$$

where the inequality follows from (2.8) since $\{S_n/\gamma_n\} \twoheadrightarrow A$. Now

$$\pi_N(T_n/\gamma_n) = \sum_{j=1}^{N} (f_j \circ \Lambda)\left(\frac{S_n}{\gamma_n}\right) u_j$$

and since $f_j \circ \Lambda \in B^*$ has norm less than or equal to 1, $|(f_j \circ \Lambda)(X)| \leq \|X\|$ and (1.7) for $\|X\|$ implies

$$\sum_{n=1}^{\infty} P(|(f_j \circ \Lambda)(X)| > \gamma_n) < \infty \tag{2.10}$$

for $j = 1, \dots, N$. Since the real line is type 2, Lemma 6 of Einmahl (1993) with $c_n = \gamma_n$ implies $E|(f_j \circ \Lambda)\left(\frac{S_n}{\gamma_n}\right)| \to 0$ as $n \to \infty$. Here we use the properties (2.3) and (2.4) of Einmahl (1993) to be certain the sequence $c_n = \gamma_n$ satisfies the conditions of Lemma 6 there. Since this argument holds for $j = 1, \dots, N$, we thus have $\pi_N(T_n/\gamma_n) \overset{\text{prob}}{\longrightarrow} 0$ as $n \to \infty$. Combining this with (2.9) and recalling that $\epsilon > 0$ was arbitrary, we have $T_n/\gamma_n \overset{\text{prob}}{\longrightarrow} 0$. Thus Λ a linear isometry with $S_n/\gamma_n = \Lambda^{-1}(T_n/\gamma_n)$ implies (1.6) as claimed. Hence the first part of Theorem 3 is proven.

We now turn to the "moreover" part that for each given compact set $\tilde{A}$ as in the statement of Theorem 3, one can construct a suitable random vector $X : \Omega \to B$ which has a cluster set equal to $\tilde{A}$ and which satisfies the **compact** LIL with respect to the norming sequence $\{\gamma_n\}$. This will be accomplished by specifying a variable which leads to the given cluster set and which, at the same time, is such that the compact LIL holds true. We do not go through the details of the proof that the cluster set is $\tilde{A}$, as this can be done by using the proof in Einmahl and Kuelbs (1999). Actually the case $\tilde{A}$ compact is a bit easier than when the cluster set is merely closed. Hence what remains is to show the variable constructed actually satisfies the compact LIL. This we do by an application of Theorem 2. We assume again that the Banach space B has a Schauder basis. (This additional assumption can be removed by the same argument as above.)

Let us now recall the basic definitions. Our starting point was the observation that $\tilde{A}$ can be written as a closure of (at most) countably many open line segments, that is

$$\tilde{A} = \mathrm{cl}\left(\bigcup_{\mathrm{j}=1}^{\infty} \mathcal{L}_{\mathrm{j}}\right)$$

where

$$\mathcal{L}_j = \{tz_j : |t| < \sigma_j\}$$

for a suitable unit vector $z_j \in B$ and $0 \leq \sigma_j \leq \sigma = 1$.

We assume that $\sigma_1 = \sigma = 1$, and also $\sigma_j > 0, j \geq 2$. (This can always be accomplished by replacing all "trivial" line segments $\mathcal{L}_j = \emptyset$ by $\mathcal{L}_1$ if necessary.)

We set for $k \geq 1$, $m_k = 4^{k^4}, m_{k,0} = m_k$, and $m_{k,\ell+1} = m_{k,\ell} + k^3 + 4^{\sigma_{\ell+1}^2 k^4}$ for $0 \leq \ell \leq k-1$. Furthermore, we define $m_{k,k+1} = m_{k+1}$ and $n_{k,\ell} = m_{k,\ell} + k^3, 0 \leq \ell \leq k$.

We choose a symmetric and discrete random variable $Z : \Omega \to I\!R$, the truncated second moment function $H(t) = E(Z^2 I\{|Z| \leq t\}), t \geq 0$ of which satisfies,

$$H(t) = d_n, \ \exp(n) \leq t < \exp(n+1), n \geq 1,$$

where

$$\begin{aligned}
d_n &= 0, 0 \leq n < m_4 \quad \text{and for} \quad k \geq 4, \\
d_{m_k} &= \exp(k^3), \\
d_{m_{k,\ell}+j} &= \exp(k^3 + \ell k + j/k^2), 0 \leq j \leq k^3, \\
d_m &= \exp(k^3 + (\ell+1)k), n_{k,\ell} \leq m \leq m_{k,\ell+1}, 0 \leq \ell \leq k-1, \\
d_{m_{k,k}+j} &= \exp((2k^2 + 3k + 1)jk^{-3} + k^2 + k^3), 0 \leq j \leq k^3, \\
d_m &= \exp((k+1)^3), n_{k,k} \leq m \leq m_{k+1}.
\end{aligned}$$

Since $H(t)$ varies slowly at infinity, it follows that Z is in the domain of attraction of the normal distribution. Moreover we have shown that Z satisfies the conditions of the generalized LIL, i.e. if $Z_1, Z_2, Z_3, \ldots$ are independent copies of Z with probability 1,

$$\limsup_{n \to \infty} |\sum_{j=1}^{n} Z_j| / \overline{\gamma}_n = 1, \tag{2.11}$$

where $\overline{\gamma}_n = \sqrt{2}\ \overline{K}(n/L_2 n) L_2 n, n \geq 1$, and $\overline{K}$ is the K-function corresponding to Z. That is, $\overline{K}$ is the inverse function of $\overline{G}$ which is defined by

$$\overline{G}(t) = t^2 / (H(t) + tM(t))$$

for $t > 0$ and where $M(t) = E(|Z| I\{|Z| > t\})$. Next, recall that Z being in the domain of attraction of the normal distribution also implies

$$\lim_{t \to \infty} tM(t)/H(t) = 0. \tag{2.12}$$

Given Z, we define X by

$$X = \sum_{k=1}^{\infty} \sum_{\ell=1}^{k+1} z_{k,\ell} Z I\{\exp(m_{k,\ell-1}) < |Z| \leq \exp(m_{k,\ell})\}, \tag{2.13}$$

where $z_{k,\ell} = z_k, 1 \leq \ell \leq k$ and $z_{k,k+1} = z_1, k \geq 1$.

Further, recall that

$$\gamma_n \leq \overline{\gamma}_n, n \geq 1 \tag{2.14}$$

and also

$$\limsup_{n\to\infty} \overline{\gamma}_n/\gamma_n \leq 2. \tag{2.15}$$

We now show that given $\epsilon > 0$ we can construct a sequence of i.i.d. F−valued random vectors $\{X_i^{(\epsilon)} : i \geq 1\}$ so that with probability 1,

$$\limsup_{n\to\infty} \| \sum_{i=1}^{n} (X_i - X_i^{(\epsilon)})\|/\gamma_n \leq \epsilon, \tag{2.16}$$

where F is a finite-dimensional subspace of B. This clearly implies that with probability 1,

$$\limsup_{n\to\infty} d(S_n/\gamma_n, F) \leq \epsilon, \tag{2.17}$$

and Theorem 2 above implies that our random vector X satisfies a compact LIL.

To prove (2.16), choose a finite set of vectors $\{w_1, \dots, w_m\} \subset \tilde{A}$ so that

$$\sup_{x\in\tilde{A}} \min_{1\leq i\leq m} \|x - w_i\| \leq \epsilon^2/3600. \tag{2.18}$$

This is of course possible since we are assuming that the set $\tilde{A}$ is compact. Let F be the linear space spanned by $\{w_1, \dots, w_m\}$. Define a mapping $\pi_\epsilon : \{z_j : j \geq 1\} \to F$ by setting for any $j \geq 1$ such that $\sigma_j \geq \epsilon/60$, $\pi_\epsilon(z_j) = \sigma_j{}^{-1} w_i$, where w_i is the vector with the smallest index i so that $\|w_i - \sigma_j z_j\| \leq \epsilon^2/3600$ and $\pi_\epsilon(z_j) = 0$ otherwise. Let

(2.19)

$$X_i^{(\epsilon)} = \sum_{k=1}^{\infty} \sum_{\ell=1}^{k+1} \pi_\epsilon(z_{k,\ell}) Z_i I\{\exp(m_{k,\ell-1}) < |Z_i| \leq \exp(m_{k,\ell})\}, i \geq 1.$$

These are clearly finite-dimensional symmetric random vectors (with values in the linear space F) and it thus remains to prove (2.16).

Setting $\Delta_i := X_i - X_i^{(\epsilon)}, i \geq 1$, we obviously have,

$$\Delta_i = \sum_{k=1}^{\infty} \sum_{\ell=1}^{k+1} a_{k,\ell} Z_i I\{\exp(m_{k,\ell-1}) < |Z_i| \leq \exp(m_{k,\ell})\}, i \geq 1, \tag{2.20}$$

where $\|a_{k,\ell}\| \leq \epsilon/60$ if $\ell = k+1$ or $\sigma_\ell \geq \epsilon/60, 1 \leq \ell \leq k$. Otherwise, we have $\|a_{k,\ell}\| = 1$. To establish (2.16) it is now sufficient to prove that

$$\sum_{s=1}^{\infty} P\{\max_{1\leq n\leq n_s} \| \sum_{i=1}^{n} \Delta_i\| \geq \epsilon\gamma_{n_{s-1}}\} < \infty, \tag{2.21}$$

where $n_s := 2^s, s \geq 1$.

This is of course by the Lévy inequality equivalent to

$$\sum_{s=1}^{\infty} P\{\|\sum_{i=1}^{n_s} \Delta_i\| \geq \epsilon\gamma_{n_{s-1}}\} < \infty. \tag{2.22}$$

Let $\Delta_{s,i} := \Delta_i I\{|Z_i| \leq \overline{K}(n_s/L_2n_s)\}, 1 \leq i \leq n_s, s \geq 1$. Using the trivial inequality

$$\|\sum_{i=1}^{n_s}(\Delta_{s,i} - \Delta_i)\| \leq \sum_{i=1}^{n_s} |Z_i| I\{|Z_i| \geq \overline{K}(n_s/L_2n_s)\},$$

along with the fact that $n_s E[|Z|I\{|Z| \geq \overline{K}(n_s/L_2n_s)\}]/\gamma_{n_s} \to 0$ as $n \to \infty$, which follows from the definition of the $\overline{K}$-function in combination with relations (2.12) and (2.15), we can show by an adaptation of the double truncation argument used by Einmahl(1993) (pp. 2035–2038) that (2.22) is reducible to proving

$$\sum_{s=1}^{\infty} P\{\|\sum_{i=1}^{n_s} \Delta_{s,i}\| \geq (3\epsilon/4)\gamma_{n_{s-1}}\} < \infty. \tag{2.23}$$

Noticing that $E[\|\sum_{i=1}^{n_s} \Delta_{s,i}\|] = o(\gamma_{n_{s-1}})$ as $s \to \infty$, which follows by the same argument as used in Lemma 8, Einmahl and Kuelbs(1999), we can use inequality (3.1) of Einmahl(1993) to conclude that for large enough s,

$$\begin{aligned} &P\{\|\textstyle\sum_{i=1}^{n_s} \Delta_{s,i}\| \geq (3\epsilon/4)\gamma_{n_{s-1}}\} \\ &\leq 16\exp(-\epsilon^2\gamma_{n_{s-1}}^2/324\Lambda_s) + Cn_sE[|Z|^3 I\{|Z| \leq \overline{K}(n_s/L_2n_s)\}]/\gamma_{n_s}^3, \end{aligned}$$

where C is a positive constant and $\Lambda_s := n_sE[\|\Delta_{s,1}\|^2]$. Since the random variable Z satisfies the LIL with respect to the norming sequence $\{\overline{\gamma}_n\}$, we have

$$\sum_{n=1}^{\infty} P\{|Z| \geq \overline{\gamma}_n\} < \infty,$$

which in combination with (2.15) and Lemma 5 of Einmahl(1993) implies

$$\sum_{s=1}^{\infty} n_sE[|Z|^3 I\{|Z| \leq \overline{\gamma}_{n_s}\}]/\gamma_{n_s}^3 < \infty.$$

It thus remains to prove that

$$\sum_{s=1}^{\infty} \exp(-\epsilon^2\gamma_{n_{s-1}}^2/324\Lambda_s) < \infty. \tag{2.24}$$

Since $\|\Delta_i\| \leq |Z_i|, i \geq 1$ it easily follows that

$$\Lambda_s \leq n_sH(\overline{K}(n_s/L_2n_s)) \leq \overline{\gamma}_{n_s}^2/2L_2n_s.$$

Recalling (2.15) and the fact that γ_n/n is non-increasing we see that for large s,

$$\Lambda_s \leq 9\gamma^2_{n_{s-1}}/L_2 n_s. \tag{2.25}$$

Using the same notation as in Einmahl and Kuelbs(1999), let for $1 \leq \ell \leq k+1$,

$$J'_{k,\ell} := \{n : \exp(m_{k,\ell-1}) < \overline{K}(n/L_2 n) \leq \exp(n_{k,\ell-1}\},$$
$$J_{k,\ell} := \{n : \exp(n_{k,\ell-1}) < \overline{K}(n/L_2 n) \leq \exp(m_{k,\ell}\}, 1 \leq \ell \leq k+1.$$

Further set

$$I'_{k,\ell} := \{s \geq 1 : n_s \in J'_{k,\ell}\}, I_{k,\ell} := \{s \geq 1 : n_s \in J_{k,\ell}\}, 1 \leq \ell \leq k+1,$$

and let

$$L_k := \{1 \leq \ell \leq k : \sigma_l < \epsilon/60\}, L'_k := \{1, \ldots, k+1\} \setminus L_k, k \geq 1.$$

Noting that we have if $\ell \in L'_k$ and $s \in I_{k,\ell}$,

$$\begin{aligned}\Lambda_s \quad \leq \quad & n_s(\epsilon^2/3600)\{H(\overline{K}(n_s/L_2 n_s)) - H(\exp(m_{k,\ell-1}))\} \\ & + n_s H(\exp(m_{k,\ell-1})),\end{aligned}$$

we therefore see

$$\Lambda_s \leq (e^{-k} + \epsilon^2/3600)\overline{\gamma}^2_{n_s}/2L_2 n_s.$$

Hence we have for large k and $s \in I_{k,\ell}$, $\ell \in L'_k$, that

$$\Lambda_s \leq \epsilon^2\gamma^2_{n_{s-1}}/400 L_2 n_s,$$

which trivially implies,

$$\sum_{k=1}^{\infty}\sum_{\ell \in L'_k}\sum_{s \in I_{k,\ell}} \exp(-\epsilon^2\gamma^2_{n_{s-1}}/324\Lambda_s) < \infty. \tag{2.26}$$

Next using the fact that $\overline{K}(n/L_2 n)/n^{1/3}$ is eventually non-decreasing, we find that for large k, $\# I_{k,\ell} \leq 3(m_{k,\ell} - n_{k,\ell-1})/\log(2)$.

Recalling relation (2.25) and using the fact that $Ln_s \geq 4^{k^4}, s \in I_{k,\ell}, 1 \leq \ell \leq k+1$, we can conclude that for large k,

$$\sum_{\ell \in L_k}\sum_{s \in I_{k,\ell}} \exp(-\epsilon^2\gamma^2_{n_{s-1}}/324\Lambda_s) \leq 3k4^{\epsilon^2 k^4/3600}4^{-\epsilon^2 k^4/3000}/\log(2),$$

whence

$$\sum_{k=1}^{\infty}\sum_{\ell \in L_k}\sum_{s \in I_{k,\ell}} \exp(-\epsilon^2\gamma^2_{n_{s-1}}/324\Lambda_s) < \infty. \tag{2.27}$$

By an obvious modification of the above argument it follows that

$$\sum_{k=1}^{\infty}\sum_{\ell=1}^{k+1}\sum_{s\in I'_{k,\ell}} \exp(-\epsilon^2\gamma^2_{n_{s-1}}/324\Lambda_s) < \infty. \tag{2.28}$$

Combining (2.26)–(2.28), we obtain (2.24), thereby completing the proof of the second part of Theorem 3.

3 Some results for series with independent coefficients

Random series of the form $X = \sum_{k\geq 1}\eta_k e_k$, where $\{\eta_k\}$ are independent real-valued random variables and $\{e_k\} \subset B$, arise quite frequently. For example, every Gaussian vector is of this form, and motivated by the results in Einmahl (1995) for finite series we now turn to the CLIL and the related cluster sets when X is an infinite random series.

To describe our results for random series we need some further notation. Let B have basis $\{u_k\}$ and associated coefficient functionals $\{f_k\}$ such that $f_j(u_k) = \delta_{j,k}$ for $j, k \geq 1$. Set

$$\sigma^2_{n,k} = E(f_k^2(X)I\{|f_k(X)| \leq K_{f_k}(n/L_2n)\}), \tag{3.1}$$

and define

$$\beta_{n,k} = \gamma_n/(2\sigma^2_{n,k}nL_2n)^{1/2}. \tag{3.2}$$

Then $\sigma^2_{n,k} = 0$, implies $\beta_{n,k} = \infty$. Also let $\pi_N(\cdot)$ and $Q_N(\cdot)$ be as in (2.5). For $s > 0$ and $k \geq 1$, define

$$I_k(s) = \{n : \beta_{n,k} \leq s\}. \tag{3.3}$$

Let

$$J(\{s_k\}) = \bigcap_{k=1}^{\infty} I_k(s_k) \tag{3.4}$$

for any sequence $\{s_k\}$ with strictly positive terms. Let $\{t_m : m \geq 2\}$ be an enumeration of the rationals in $(1, \infty)$ and set $t_1 = \infty$. Let Γ_N denote all the sequences of strictly positive integers $\{k_j : j \geq 1\}$ such that $k_j = 1$ for all $j \geq N + 1$. Let $\Gamma = \cup_{N=1}^{\infty}\Gamma_N$, and for $\{k_j\} \in \Gamma$ define

$$\delta(\{k_j\}) = \sup\{\delta > 0 : \sum_{n\in J(\{t_{k_j}\})} n^{-1}\exp\{-\delta^2 L_2 n\} = \infty\}, \tag{3.5}$$

where $\sup \phi = 0$. For $\{k_j\} \in \Gamma$ such that $\delta(\{k_j\}) > 0$, define the ellipsoid

$$E(\{k_j\}) = \{b : \sum_{j=1}^{\infty} f_j^2(b) t_{k_j}^2 / \delta^2(\{k_j\}) \leq 1\}. \tag{3.6}$$

Further let $E(\{k_j\})$ be equal to $\{0\}$ if $\delta(\{k_j\}) = 0$. Since $\{k_j\} \in \Gamma$ implies t_{k_j} is eventually infinite, the corresponding coordinates $f_j(b)$ of b must be 0 for $b \in E(\{k_j\})$. Thus $b = \sum_{j=1}^{\infty} f_j(b) u_j \in E(\{k_j\})$ for $\{k_j\} \in \Gamma$ implies $b = \pi_N(b)$ for some $N \geq 1$.

Theorem 4. *Let X and $\{\gamma_n\}$ be as in Theorem 1, and assume (1.6) and (1.7) hold. Furthermore, assume B has a basis $\{u_k\}$ with associated coefficient functionals $\{f_k\}$ such that $\{f_k(X) : k \geq 1\}$ are independent real-valued random variables. Let $A = C(\{S_n/\gamma_n\})$ with probability 1. Then*

$$A = \bigcap_{N=1}^{\infty} A_N \quad \textit{where} \quad A_N = \pi_N^{-1}(C(\{\pi_N(S_n/\gamma_n)\})) \tag{3.7}$$

decrease in N.

$$A = \{b \in B : J(b, \epsilon) = \infty \quad \forall \epsilon > 0\}, \tag{3.8}$$

where

$$J(b, \epsilon) = \sum_{n=1}^{\infty} \exp\{-Ln - L_2 n \sum_{k=1}^{\infty} \beta_{n,k}^2 (|f_k(b)| - \epsilon)_+^2\}.$$

$$A = cl(\bigcup_{N=1}^{\infty} C(\{\pi_N(S_n/\gamma_n)\})) \quad \textit{with} \quad C(\{\pi_N(S_n/\gamma_n)\}) \tag{3.9}$$

increasing in N.

$$A = cl(\bigcup_{\{k_j\} \in \Gamma} E(\{k_j\})), \tag{3.10}$$

where $\Gamma = \bigcup_{N=1}^{\infty} \Gamma_N$ is as above, and the ellipsoids $E(\{k_j\})$ are given in (3.6). Furthermore, $X \in CLIL$ with respect to $\{\gamma_n\}$ iff for every $\epsilon > 0$ there is an N such that $\overline{\lim}_n \|Q_N(S_n/\gamma_n)\| < \epsilon$ with probability 1.

Remarks. (a) If X is mean-zero, $0 < E\|X\| < \infty$, and $X = \sum_{k \geq 1} \eta_k e_k$, where the $\{\eta_k\}$ are independent real-valued random variables and $\{e_k\}$ is a basic sequence in B, i.e. $\{e_k\}$ is a Schauder basis for the closed subspace $\tilde{B}$ spanned by $\{e_k\}$, then Theorem 4 applies directly if we replace B by $\tilde{B}$.

(b) Also, if X is as above and there exists a sequence $\{f_k\} \in B^*$ such that $f_k(e_j) = \delta_{k,j}$ for $k, j \geq 1$, then

$$\tilde{B} = \{x \in B : \lim_N \sum_{k=1}^{N} f_k(x) e_k \quad \text{exists}\}$$

is a Banach space in the norm

$$|||x||| = \sup_N \|\sum_{k=1}^{N} f_k(x)e_k\|,$$

with Schauder basis $\{e_k\}$ and associated coefficient functionals, the f_k's restricted to $\tilde{B}$, see Proposition 5.3 of Singer (1970). Now $P(X \in \tilde{B}) = 1$, and since $X = \sum_{k\geq 1} \eta_k e_k$ with $\{\eta_k\}$ independent, we know from Theorem 6.11 in Ledoux-Talagrand (1991) that $0 < E\|X\| < \infty \iff 0 < E|||X||| < \infty$. Hence Theorem 4 applies to $(\tilde{B}, |||\cdot|||)$ provided the $\{\gamma_n\}$ are defined in terms of the linear functions on $(\tilde{B}, |||\cdot|||)$, and (1.6) and (1.7) hold for these $\{\gamma_n\}$. Note that from (1.4) and (1.5) the new γ_n are larger. This follows since the class of linear functionals on $(\tilde{B}, |||\cdot|||)$, with norm less than 1, is larger than the corresponding class on $(B, \|\cdot\|)$. Indeed, if $\{e_k\}$ is a basic sequence in B, then the norms $\|\cdot\|$ and $|||\cdot|||$ are equivalent on $\tilde{B}$, and the $\gamma_n's$ defined using these different norms are equivalent, i.e. their ratio is bounded above and below by strictly positive finite constants. Hence (1.6) and (1.7) hold with both versions of the $\gamma_n's$ or neither, i.e. see Lemma 5 of Einmahl (1993).

(c) We conjecture that all the sets described by (3.10) in Theorem 4 actually arise as cluster sets.

Proof of Theorem 4. If $A = C(\{S_n/\gamma_n\})$ with probability 1, then $0 \in A$ by (1.6). Furthermore, if $b \in A$, then $\pi_N(b) \in C(\{\pi_N(S_n/\gamma_n)\})$ for all $N \geq 1$ by continuity of the π_N. Hence $b \in A_N$ for every $N \geq 1$ and $A \subset \bigcap_{N=1}^{\infty} A_N$. In addition, if $b \in A_{N+1}$, then

$$\text{(3.11)} \qquad \begin{aligned} \pi_N(b) = \pi_N(\pi_{N+1}(b)) &\in C(\{\pi_N(\pi_{N+1}(S_n/\gamma_n))\}) \\ &= C(\{\pi_N(S_n/\gamma_k)\}). \end{aligned}$$

Thus $b \in A_N$ and the A_N decrease in N.

Now assume $b \in A_N$ for all $N \geq 1$ and take $\epsilon > 0$. Using Lemma 1 in Einmahl (1995), $b \in A$ provided

$$\text{(3.12)} \qquad \sum_{n=1}^{\infty} \frac{1}{n} P(\|S_n/\gamma_n - b\| < \epsilon) = \infty$$

for each $\epsilon > 0$. Take $N \geq N_1(\epsilon)$ such that $\|Q_N(b)\| < \epsilon/4$. Then independence of the coordinates implies

$$P(\|S_n/\gamma_n - b\| < \epsilon) \geq \alpha\beta,$$

where

$$\alpha = P(\|\pi_N(S_n/\gamma_n - b)\| < \epsilon/2)$$

and

$$\beta = P(\|Q_N(S_n/\gamma_n - b)\| < \epsilon/2).$$

Therefore,

$$P(\|S_n/\gamma_n - b\| < \epsilon)$$
$$\geq P(\|\pi_N(S_n/\gamma_n - b)\| < \epsilon/2)P(\|Q_N(S_n/\gamma_n)\| < \epsilon/4).$$

Since $b \in A_N$ implies $\pi_N(b) \in C(\{\pi_N(S_n/\gamma_n)\})$, Lemma 1 in Einmahl (1995) gives

$$\sum_{n=1}^{\infty} \frac{1}{n} P(\|\pi_N(S_n/\gamma_n) - \pi_N(b)\| < \epsilon/2) = \infty. \tag{3.13}$$

Now $S_n/\gamma_n \xrightarrow{\text{prob}} 0$ implies $Q_N(S_n/\gamma_n) \xrightarrow{\text{prob}} 0$, so the above and (3.13) implies (3.12). Thus $b \in A$ as $\epsilon > 0$ was arbitrary, and (3.7) is proven.

To prove (3.8) recall the notation in (3.1) to (3.6). Applying Proposition 1 in Einmahl (1995), extended to $\pi_N(X)$ with values in $\pi_N(B)$, we have $\pi_N(b) = \sum_{k=1}^{N} f_k(b)u_k \in C(\{\pi_N(S_n/\gamma_n)\}) \iff$ for every $\epsilon > 0$

$$J_N(b, \epsilon) = \infty, \tag{3.14}$$

where

$$J_N(b, \epsilon) = \sum_{n=1}^{\infty} \exp\{-Ln - L_2 n \sum_{k=1}^{N} \beta_{n,k}^2 (|f_k(b)| - \epsilon)_+^2\}.$$

That is, retaining the normalizing constants γ_n obtained from X with values in B, we have Proposition 1 of Einmahl (1995) valid for $\pi_N(X)$ in $\pi_N(B)$, and since all norms are equivalent on $\pi_N(B)$ the result is independent of the norm on $\pi_N(B)$. Thus

$$A_N = \{b \in B : J_N(b, \epsilon) = \infty \quad \forall \epsilon > 0\}, \tag{3.15}$$

and if $b \in A_N$ for all $N \geq 1$, then $\lim_k f_k(b) = 0$ implies b satisfies the divergence criterion in (3.8) for all $\epsilon > 0$. Thus the right-hand side of (3.8) equals $\bigcap_{N=1}^{\infty} A_N$, and this is A by (3.7). Hence (3.8) is proven.

To establish (3.9) observe that $b \in A$ implies $\pi_N(b) \in C(\{\pi_N(S_n/\gamma_n)\})$ by the continuity of π_N. Therefore $b = \lim_N \pi_N(b)$ implies

$$b \in cl(\bigcup_{N=1}^{\infty} C(\{\pi_N(S_n/\gamma_n)\})) \tag{3.16}$$

and

$$(3.17) \qquad A \subset cl(\bigcup_{N=1}^{\infty} C(\{\pi_N (S_n/\gamma_n)\})).$$

On the other hand, if $b \in C(\{\pi_N (S_n/\gamma_n)\})$ for some $N \geq 1$, then $b \in \pi_N(B)$ and $b = \pi_N(b)$. Hence by the extension of Proposition 1 in Einmahl (1995) (3.14) holds for all $\epsilon > 0$. Thus $b = \pi_N(b)$ implies the divergence criterion in (3.8) holds, and hence $b \in A$. Thus

$$\bigcup_{N=1}^{\infty} C(\{\pi_N (S_n/\gamma_n)\}) \subset A,$$

and A closed yields (3.9), since the previous argument also implies the cluster set $C(\{\pi_N (S_n/\gamma_n)\})$ increases in N. Thus (3.9) is proven.

To verify (3.10) observe that if $b \in A$, then by (3.8) and Einmahl (1995), pp. 827–829,

$$(3.18) \qquad \pi_N(b) \in cl(\bigcup_{\{k_j\}\in\Gamma_N} E(\{k_j\}))$$

for all $N \geq 1$, and hence $b \in cl(\bigcup_{\{k_j\}\in\Gamma} E(\{k_j\}))$.

Conversely, recall $b \in E(\{k_j\})$ for $\{k_j\} \in \Gamma$ implies $b = \pi_N(b)$ for some $N \geq 1$ and

$$(3.19) \qquad \sum_{j=1}^{N} f_j^2(b)t_{k_j}^2/\delta^2(\{k_j\}) \leq 1$$

with $t_{k_j} = \infty$ for $j \geq N+1$. Hence if b is also not 0, then (3.19) implies $\delta^2(\{k_j\}) > 0$ and

$$(3.20) \qquad \sum_{j=1}^{N} t_{k_j}^2(|f_j(b)| - \epsilon)_+^2 < \delta^2(\{k_j\})$$

for all $\epsilon > 0$. Thus the definition of $\delta(\{k_j\})$ implies

$$(3.21) \qquad \sum_{n\in J(\{k_j\})} \exp\{-Ln - L_2 n \sum_{j=1}^{N} t_{k_j}^2(|f_j(b)| - \epsilon)_+^2\} = \infty$$

for all $\epsilon > 0$. Since $0 \leq \beta_{n,j} \leq t_{k_j}$ for $j \geq 1$ when $n \in J(\{t_{k_j}\})$, (3.21) implies

$$(3.22) \qquad J_N(b, \epsilon) = \infty$$

for all $\epsilon > 0$. Thus $b = \pi_N(b) \in A$ by (3.8). Of course, $0 \in A$ has already been observed, so $E(\{k_j\}) \subset A$ for all $\{k_j\} \in \Gamma$. Thus $cl(\bigcup_{\{k_j\}\in\Gamma} E(\{k_j\}))$ $\subset A$, and (3.10) is proven.

To complete the proof of Theorem 4 we observe that if for each $\epsilon > 0$ there is an N such that $\overline{\lim}_N \|Q_N(S_n/\gamma_n)\| < \epsilon$ with probability 1, then (1.12) holds and Corollary 1 implies $X \in$ CLIL with respect to $\{\gamma_n\}$. The converse is essentially Proposition 1 with $T = \pi_N$. Hence Theorem 4 is proven.

References

K. Alexander (1989a), Characterization of the cluster set of the LIL sequence in Banach space, *Ann. Probab.* **17**, 737–759.

K. Alexander (1989b), Unusual cluster sets for the LIL sequence in Banach space, *Ann. Probab.* **17**, 1170–1185.

M.M. Day (1973), *Normed Linear Spaces*, Third Edition, Springer-Verlag, Berlin.

U. Einmahl (1993), Toward a general law of the iterated logarithm in Banach space, *Ann. Probab.* **21**, 2012–2045.

U. Einmahl (1995), On the cluster set problem for the generalized law of the iterated logarithm in Euclidean space, *Ann. Probab.* **23**, 817–851.

U. Einmahl and J. Kuelbs (1999), Cluster sets for a generalized law of the iterated logarithm in Banach spaces. *Preprint*

M. Klass, Toward a universal law of the iterated logarithm(I). *Z. Wahrsch. Verw. Gebiete* **36**, 165–178.

J. Kuelbs (1981), When is the cluster set of S_n/a_n empty? *Ann. Probab.* **9**, 377–394.

M. Ledoux and M. Talagrand (1991), *Probability in Banach Spaces*, Springer-Verlag, Berlin.

J. Lindenstrauss and L. Tzafriri (1977), *Classical Banach Spaces I*, Springer-Verlag, Berlin.

I. Singer (1970), *Bases in Banach Spaces I.*, Springer-Verlag, Berlin.

Uwe Einmahl
Departement Wiskunde
Vrije Universiteit Brussel
Pleinlaan 2
1050 Brussels, Belgium
ueinmahl@vub.ac.be

James Kuelbs
Department of Mathematics
University of Wisconsin
Madison, WI 53706
kuelbs@math.wisc.edu

PART 6

Large Deviations

Dominating Points for General Sequences of Random Vectors

A. de Acosta*

1 Introduction

Let E be a separable Banach space, μ a probability measure on E, $\{X_j\}_{j\geq 1}$ an independent and identically distributed [i.i.d.] sequence with common distribution μ, and $S_n = \sum_{j=1}^n X_j$. Cramér's large deviation theorem for $E = \mathbb{R}^d$ and its extension by Donsker-Varadhan [D-V] and Bahadur-Zabell [B-Z] to general E (see also [de A1]) give asymptotic logarithmic bounds for $\{\nu_n(B)\}$, where B is a Borel set in E and $\nu_n = \mathcal{L}(S_n/n)$. In this context, it is of interest to identify subsets of E for which the asymptotic logarithmic bounds can be upgraded to the non-logarithmic level.

The concept of dominating point of an open convex set was introduced by Ney [N1], [N2], for this purpose in the case $E = \mathbb{R}^d$. It is proved in [N1], [N2], that under suitable integrability and smoothness assumptions on

$$\hat{\mu}(\xi) = \int \exp(\langle x, \xi \rangle)\mu(dx),$$

an open convex set D not containing the mean of μ has a unique dominating point. The existence of such a point makes it possible to set up a representation formula for $\nu_n(D)$ which is then used to obtain non-logarithmic bounds.

The result on dominating points and its applications were significantly extended to general E in Einmahl-Kuelbs [E-K] under the integrability assumption

$$\int \exp(t\|x\|)\mu(dx) < \infty$$

for all $t > 0$. In the recent paper [K], Kuelbs refined this result, proving the existence and uniqueness of dominating points in the same framework but under a weaker integrability condition.

For dominating points in the context of Sanov's theorem, see Dinwoodie [D].

---

*This research was partially supported by an NSF grant.

Let $\{\mu_n\}$ be a sequence of probability measures on a topological vector space E. Let

$$\varphi_n(\xi) = \log \int \exp(\langle x, \xi \rangle)\mu_n(dx, \quad \xi \in E^*,$$

where E^* is the dual space of E. Then if

$$\varphi(\xi) = \lim_n n^{-1}\varphi_n(n\xi)$$

exists for all ξ and satisfies a suitable differentiability condition, and $\{\mu_n\}$ is exponentially tight, then $\{\mu_n\}$ satisfies the large deviation principle with rate function

$$\varphi^*(x) = \sup[\langle x, \xi \rangle - \varphi(\xi)], \quad x \in E. \tag{1.1}$$

For a general theorem of this type and further references, see [de A3].

The objectives of the present paper are to prove the existence and uniqueness of a dominating point for (D, φ) (defined in a suitable sense), where D is an open convex set, and to give an initial application to non-logarithmic asymptotic bounds for $\{\mu_n(D)\}$. Theorem 2.3, the main result of the paper, deals with the first objective for a locally convex space E in the general case, and Theorem 3.1 (and Example 3.2) deals with the second. The i.i.d. case for a locally convex space is considered in Corollary 2.6, which in the Banach space situation yields the central result on dominating points in [K], under slightly weaker conditions (see Remark 2.7); Corollary 2.6 therefore also contains as corollaries the results on dominating points in [N1], [N2] and [E-K]. Technically, our proof is in part similar to that in [K] but simpler; we avoid the delicate issues involving strict convexity and supports in [K] by suitable use of certain elementary convex analysis arguments (Lemma 2.4). Theorem 2.3 also contains the result on dominating points for general sequences in $\mathbb{R}^d$ stated in [M-N].

2 Dominating points

Let E be a topological vector space, E^* its dual space. Let $\varphi : E^* \to \overline{\mathbb{R}}$ be a convex function and let $\varphi^* : E \to \overline{\mathbb{R}}$ be given by (1.1).

Definition 2.1 Let D be an open convex subset of E such that $D \cap \operatorname{dom} \varphi^* \neq \phi$. A point $x_0 \in E$ is a *dominating point for* (D, φ) if

(1) $x_0 \in \partial D$,

(2) $\varphi^*(x_0) = \inf_{x \in D} \varphi^*(x)$,

(3) there exists $\xi_0 \in E^*$ such that $D \subset \{x : \langle x, \xi_0 \rangle > \langle x_0, \xi_0 \rangle\}$ and $\varphi^*(x_0) = \langle x_0, \xi_0 \rangle - \varphi(\xi_0)$.

Remarks 2.2. (1) It is always the case that

$$\inf_{x \in D} \varphi^*(x) = \inf_{x \in \overline{D}} \varphi^*(x). \tag{2.1}$$

For, let $x \in \overline{D}, y \in D \cap \operatorname{dom} \varphi^*$. Then for all $\lambda \in (0,1), (1-\lambda)x + \lambda y \in D$, because D is an open convex set (see e.g. [S], p.30). Therefore

$$\inf_{z \in D} \varphi^*(z) \leq \varphi^*((1-\lambda)x + \lambda y) \leq (1-\lambda)\varphi^*(x) + \lambda \varphi^*(y),$$

and letting $\lambda \to 0$ we obtain $\varphi^*(x) \geq \inf_{z \in D} \varphi^*(z)$. (2.1) follows.

(2) If there exists $y_0 \in E$ such that $\varphi^*(y_0) < \varphi^*(x)$ for all $x \in D$, then the condition $x_0 \in \overline{D}$ and (2) of Definition 2.1 imply that $x_0 \in \partial D$. For, suppose $x_0 \in D$. Then for sufficiently small $\lambda > 0$, $(1-\lambda)x_0 + \lambda y_0 \in D$ and

$$\varphi^*((1-\lambda)x_0 + \lambda y_0) \leq (1-\lambda)\varphi^*(x_0) + \lambda \varphi^*(y_0) < \varphi^*(x_0),$$

which is impossible. Therefore $x_0 \in \partial D$.

For $\xi \in E^*$, let $\varphi_\xi : \mathbb{R} \to \overline{\mathbb{R}}$ be defined by

$$\varphi_\xi(t) = \varphi(t\xi).$$

We recall that φ is *E-Gâteaux differentiable at* $\xi \in E^*$ if there exists $\nabla\varphi(\xi) \in E$ such that for all $\eta \in E^*$,

$$\lim_{h \to 0} h^{-1}[\varphi(\xi + h\eta) - \varphi(\xi)] = \langle \nabla\varphi(\xi), \eta \rangle.$$

If $\xi \in E^*$, $t \in \mathbb{R}$ and φ is E-Gâteaux differentiable at $(t\xi)$, then φ_ξ is differentiable at t and

$$\varphi'_\xi(t) = \langle \nabla\varphi(t\xi), \xi \rangle.$$

Let $D_\xi = \operatorname{dom} \varphi_\xi$, $L_a = \{x \in E : \varphi^*(x) \leq a\}, a \geq 0$. We consider the following conditions on φ:

(C1) $\varphi(0) = 0$.

(C2) For all $a \geq 0$, L_a is weakly compact.

(C3) φ is E-Gâteaux differentiable at 0.

(C4) For every $\xi \in E^*$:

(i) for every $t \in D^0_\xi$, φ is E-Gâteaux differentiable at $(t\xi)$; moreover, φ'_ξ is continuous on D^0_ξ.

(ii) $\sup\{\varphi'_\xi(t) : t \in D^0_\xi\} \geq \sup(\text{dom } \varphi^*_\xi)$.

Remarks 2.2. (cont'd) (3) (C1) implies that $\varphi^*(x) \geq 0$ for all $x \in E$.

(4) Since $\varphi'_\xi(t) \in \text{dom}(\partial\varphi^*_\xi) \subset \text{dom } \varphi^*_\xi$ for all $t \in D^0_\xi$ (for example, by Lemma 2.4), it follows that (C4) (ii) is equivalent to

$$\sup\{\varphi'_\xi(t) : t \in D^0_\xi\} = \sup(\text{dom } \varphi^*_\xi).$$

(5) (C4) (ii) is automatically satisfied if φ_ξ is essentially smooth (see [R]).

Theorem 2.3. *Let E be a locally convex space and let D be an open convex subset of E. Let $\varphi : E^* \to \overline{\mathbb{R}}$ be a convex function satisfying (C1)–(C4). Assume*

$$D \cap \textit{dom } \varphi^* \neq \phi \quad \textit{and} \quad a = \inf_{x \in D} \varphi^*(x) > 0.$$

Then there exists a unique dominating point for (D, φ). More precisely,

(a) there exists a unique point satisfying (1) and (2) of Definition 2.1,

(b) there exists $\xi_0 \in E^*$ such that φ is E-Gâteaux differentiable at ξ_0, ξ_0 satisfies (3) of Definition 2.1 and

$$\nabla\varphi(\xi_0) = x_0.$$

We need two lemmas for the proof of Theorem 2.3. In the first place we recall some elementary convex analysis facts, formulated to be suitable for our purposes.

Lemma 2.4. *Let E^* be as in Theorem 2.3, and let $\varphi : E^* \to \overline{\mathbb{R}}$ be a convex function.*

(i) Assume that φ is E-Gâteaux differentiable at ξ. Then

$$\varphi^*(x) = \langle x, \xi\rangle - \varphi(\xi) \quad \textit{if and only if} \quad x = \nabla\varphi(\xi).$$

(ii) Assume that $\varphi(0) = 0$ and φ is E-Gâteaux differentiable at 0. Then

$$\varphi^*(x) = 0 \quad \textit{if and only if} \quad x = \nabla\varphi(0).$$

Proof. (i) The convexity of φ implies that for all $\eta \in E^*$,

$$f(t) = t^{-1}[\varphi(\xi + t(\eta - \xi)) - \varphi(\xi)] \ \ (t > 0)$$

is non-decreasing. Therefore

$$\langle \nabla\varphi(\xi), \eta - \xi\rangle = \lim_{t\downarrow 0} f(t) \le f(1) = \varphi(\eta) - \varphi(\xi),$$

$$\langle \nabla\varphi(\xi), \eta\rangle - \varphi(\eta) \le \langle \nabla\varphi(\xi), \xi\rangle - \varphi(\xi),$$

which implies $\varphi^*(\nabla\varphi(\xi)) = \langle \nabla\varphi(\xi), \xi\rangle - \varphi(\xi)$. Conversely, suppose $\varphi^*(x) = \langle x, \xi\rangle - \varphi(\xi)$. Then for all $\eta \in E^*$, $t > 0$,

$$\langle x, \xi + t\eta\rangle - \varphi(\xi + t\eta) \le \langle x, \xi\rangle - \varphi(\xi),$$

$$\langle x, \eta\rangle \le t^{-1}[\varphi(\xi + t\eta) - \varphi(\xi)].$$

Letting $t \downarrow 0$, it follows that $\langle x, \eta\rangle \le \langle \nabla\varphi(\xi), \eta\rangle$ for all $\eta \in E^*$; therefore $\langle x, \eta\rangle = \langle \nabla\varphi(\xi), \eta\rangle$ for all $\eta \in E^*$, which implies $x = \nabla\varphi(\xi)$. The statement (ii) follows from (i) by taking $\xi = 0$. □

For $\xi \in E^*$, $\xi \ne 0$, $\ell \in \mathbb{R}$, let

$$\begin{aligned} H &= H(\xi, \ell) = \{x \in E : \langle x, \xi\rangle = \ell\}, \\ H_0 &= H_0(\xi, \ell) = \{x \in E : \langle x, \xi\rangle \le \ell\}, \\ H_1 &= H_1(\xi, \ell) = \{x \in E : \langle x, \xi\rangle > \ell\}. \end{aligned}$$

Lemma 2.5. *Let E, φ be as in Theorem 2.3. Let $a > 0$, and let $\xi \in E^*, \ell \in \mathbb{R}$ be such that*

(i) $H \cap L_a \ne \phi$,

(ii) $L_a \subset H_0$,

(iii) $H_1 \cap$ dom $\varphi^* \ne \phi$.

For $t \in D_\xi^0$, let $\rho(t) = \nabla\varphi(t\xi)$. Then there exists $t_0 \in D_\xi^0$, $t_0 > 0$, such that $\rho(t_0) \in H$ and $\varphi^*(\rho(t_0)) = a$.

Proof. (1) We claim that there exists $t_0 \in D_\xi^0, t_0 \ge 0$ such that

$$\varphi_\xi'(t_0) = \ell. \tag{2.2}$$

For, on the one hand by Lemma 2.4 $\rho(0) = \nabla\varphi(0) \in L_a$, and therefore

$$\varphi_\xi'(0) = \langle \rho(0), \xi\rangle \le \ell. \tag{2.3}$$

On the other hand, if $x_1 \in H_1 \cap \text{dom } \varphi^*$, then

$$\ell < \langle x_1, \xi\rangle \in \text{dom } \varphi_\xi^*$$

since $\varphi_\xi^*(\langle x_1, \xi\rangle) \le \varphi^*(x_1) < \infty$. By (C4) (ii) there exists $t_1 \in D_\xi^0$ (necessarily positive, since φ'_ξ is non-decreasing by the convexity of φ_ξ) such that

$$\ell < \varphi'_\xi(t_1). \tag{2.4}$$

By (2.3), (2.4) and (C4)(i), there exists $t_0 \ge 0$ such that (2.2) holds.

(2) Since $\langle \rho(t_0), \xi\rangle = \varphi'_\xi(t_0) = \ell$, it follows that $\rho(t_0) \in H$.

(3) Let $y_0 \in H \cap L_a$. Then by Lemma 2.4

$$\begin{aligned}\varphi^*(\rho(t_0)) &= \langle \nabla\varphi(t_0\xi), t_0\xi\rangle - \varphi(t_0\xi)\\ &= \ell t_0 - \varphi(t_0\xi)\\ &= \langle y_0, t_0\xi\rangle - \varphi(t_0\xi)\\ &\le \varphi^*(y_0) \le a.\end{aligned}$$

On the other hand, if $x_1 \in H_1 \cap \operatorname{dom} \varphi^*$, then for $\lambda \in (0,1)$ we have

$$(1-\lambda)\rho(t_0) + \lambda x_1 \in H_1,$$

$$a < \varphi^*((1-\lambda)\rho(t_0) + \lambda x_1) \le (1-\lambda)\varphi^*(\rho(t_0)) + \lambda\varphi^*(x_1).$$

Letting $\lambda \to 0$ we obtain $\varphi^*(\rho(t_0)) \ge a$. Therefore $\varphi^*(\rho(t_0)) = a$. Finally note that this implies $t_0 > 0$ by Lemma 2.4 (ii). □

Proof of Theorem 2.3. Let $a = \inf_{x\in D} \varphi^*(x)$. Let

$$F = \{x \in E : x \in \partial D \quad \text{and} \quad \varphi^*(x) = a\}.$$

We claim that $F \ne \phi$. For, let $\epsilon > 0$ and define

$$F_\epsilon = L_{a+\epsilon} \cap \overline{D}, \quad F_0 = \bigcap_{\epsilon>0} F_\epsilon.$$

Since $\overline{D}$ is a closed convex set, it is weakly closed. Therefore F_ϵ is non-empty and weakly compact for all $\epsilon > 0$ and $F_0 \ne \phi$. Let $x \in F_0$. Then $\varphi^*(x) \le a$, and, since $x \in \overline{D}$, $\varphi^*(x) \ge a$ by Remark 2.2(1); therefore $\varphi^*(x) = a$. Also, by Remark 2.2(2) with $y_0 = \nabla\varphi(0)$ we have $x \in \partial D$. It follows that $F_0 \subset F$, proving the claim.

Since L_a is a (weakly) closed convex set, D is an open convex set and $L_a \cap D = \phi$, by the Hahn-Banach theorem (see e.g. [S]) there exist $\xi \in E^*, \ell \in \mathbb{R}$ such that, in the notation of Lemma 2.5,

$$L_a \subset H_0 \quad \text{and} \quad D \subset H_1.$$

It easily follows that $F \subset H \cap L_a$.

By Lemma 2.5, there exists $t_0 \in D_\xi^0$ such that $\rho(t_0) \in H$ and $\varphi^*(\rho(t_0)) = a$. Let $\xi_0 = t_0\xi$. We will show that

$$F = \{\nabla\varphi(\xi_0)\}.$$

For, let $x \in F$. Then by Lemma 2.4

$$\begin{aligned}\varphi^*(x) = a &= \varphi^*(\rho(t_0)) \\ &= \langle \rho(t_0), t_0\xi\rangle - \varphi(t_0\xi) \\ &= \ell t_0 - \varphi(t_0\xi) \\ &= \langle x, t_0\xi\rangle - \varphi(t_0\xi),\end{aligned}$$

and again by Lemma 2.4 this implies $x = \rho(t_0) = \nabla\varphi(\xi_0)$. Let $x_0 = \nabla\varphi(\xi_0)$. Then $x_0 \in \partial D$ and $\varphi^*(x_0) = a$. Also, since $\langle x_0, \xi\rangle = \ell$, we have $H_1 = \{x : \langle x, \xi_0\rangle > \langle x_0, \xi_0\rangle\}$ and therefore x_0 is a dominating point for (D, φ). The uniqueness statement follows from the previous argument. □

We obtain the following corollary for the i.i.d. framework.

Corollary 2.6. *Let E be a locally convex space and let $\mathcal{E}$ be a σ-algebra of subsets of E such that every $\xi \in E^*$ is $\mathcal{E}$-measurable. Let μ be a probability measure on E, and for $\xi \in E^*$ let*

$$\hat{\mu}(\xi) = \int_E \exp(\langle x, \xi\rangle)\mu(dx),$$

$\varphi(\xi) = \log\hat{\mu}(\xi)$. Assume that φ satisfies:

(a) For all $a \geq 0$, L_a is weakly compact.

(b) φ is E-Gâteaux differentiable at 0.

(c) For every $\xi \in E^*$:

(i) for every $t \in D_\xi^0$, φ is E-Gâteaux differentiable at $(t\xi)$,

(ii) $\sup\{\varphi'_\xi(t) : t \in D_\xi^0\} \geq \sup(dom\ \varphi_\xi^*)$.

Let D be an open convex set such that

$$D \cap dom\ \varphi^* \neq \phi \quad and \quad a = \inf_{x\in D} \varphi^*(x) > 0.$$

Then (D, φ) has a unique dominating point.

Proof. Since (C1) and the continuity assumption in (C4)(i) are automatic in this case, the statement follows from Theorem 2.3. □

Remark 2.7. Corollary 2.6 implies the main statement of Theorem 1 of [K]. For, assumptions (A–1) and (A–2) of [K] obviously imply (a) and (b).

Assumption (A–3) of [K] is equivalent to (c)(ii). Finally, if $t \in D_\xi^0$ then it follows from (A–2) of [K] using Hölder's inequality that for any $\eta \in E^*$

$$\varphi(t\xi + h\eta) < \infty$$

for sufficiently small $|h|$, and by a standard argument φ is E-Gâteaux differentiable at $(t\xi)$ with

$$\nabla\varphi(t\xi) = (\hat{\mu}(t\xi))^{-1} \int x \exp(\langle x, t\xi\rangle)\mu(dx),$$

which exists in the Bochner sense.

3 Asymptotic non-logarithmic bounds for general sequences

Theorem 3.1. *Let E be a locally convex space and let $\mathcal{E}$ be a σ-algebra of subsets of E such that every $\xi \in E^*$ is $\mathcal{E}$-measurable. Let $\{\mu_n\}$ be a sequence of probability measures on $\mathcal{E}$,*

$$\varphi_n(\xi) = \log \int \exp(\langle x, \xi\rangle)\mu_n(dx), \quad \xi \in E^*.$$

Let $\varphi : E^* \to \overline{\mathbb{R}}$, and assume

(1) For all $\xi \in dom\ \varphi$,

$$\ell(\xi) = \limsup_n |\varphi_n(n\xi) - n\varphi(\xi)| < \infty. \tag{3.1}$$

(2) φ satisfies assumptions (C2)–(C4).

Let D be an open convex set such that $\overline{D} \in \mathcal{E}$,

$$D \cap dom\ \varphi^* \neq \phi \quad and \quad a = \inf_{x \in D} \varphi^*(x) > 0.$$

Let ξ_0 be as in Theorem 2.3. Then

$$\limsup_n e^{na}\mu_n(\overline{D}) \leq e^{\ell(\xi_0)}. \tag{3.2}$$

Proof. Let x_0 be the dominating point for (D, φ) (Theorem 2.3). Using the fact that $\varphi^*(x_0) = \langle x_0, \xi_0\rangle - \varphi(\xi_0)$, we have the representation formula

$$\mu_n(\overline{D}) = e^{-na} \exp[\varphi_n(n\xi_0) - n\varphi(\xi_0)] \int_{\overline{D}} \exp[-n\langle x - x_0, \xi_0\rangle]\mu_{n,\xi_0}(dx),$$

where $d\mu_{n,\xi_0}(x) = \exp[\langle x, n\xi_0\rangle - \varphi_n(n\xi_0)]d\mu_n(x)$. Of course, the important point is that $\langle x - x_0, \xi_0\rangle \geq 0$ for $x \in \overline{D}$. (3.2) follows at once. □

Remark 3.1. The standard large deviation upper bound in this situation is (under suitable assumptions)

$$\limsup_n n^{-1} \log \mu_n(\overline{D}) \leq -a.$$

(3.2) is a stronger statement.

Example 3.2. Let $\{X_j\}_{j\geq 1}$ be an i.i.d. $\mathbb{R}^d$-valued sequence with common distribution μ such that

$$\widehat{\mu}(\alpha) = \int \exp(\langle y, \alpha\rangle) d\mu(y) < \infty$$

for all $\alpha \in \mathbb{R}^d$, and let $S_0 = 0$, $S_n = \sum_{j=1}^n X_j$ for $n \geq 1$. Let $E = C([0,1], \mathbb{R}^d)$, let $\{Z_n\}_{n\geq 1}$ be the E-valued random vectors defined by

$$\begin{aligned} Z_n(t) &= \begin{cases} S_k \text{ if } t = k/n, \text{ k=1, ... ,n,} \\ \text{defined by linear interpolation in } \left[\frac{k-1}{n}, \frac{k}{n}\right] \end{cases} \\ &= S_{[nt]} + (nt - [nt])X_{[nt]+1}, \end{aligned}$$

$(t \in [0,1])$ and let $\mu_n = \mathcal{L}(Z_n/n)$. Then by Mogulskii's theorem [M], the sequence $\{\mu_n\}$ satisfies the large deviation principle with rate function

$$I(f) = \begin{cases} \displaystyle\int_0^1 (\log \widehat{\mu})^*(f'(t))dt & \text{if } f(0) = 0 \text{ and } f \text{ is absolutely continuous,} \\ \infty & \text{otherwise.} \end{cases}$$

Theorem 3.1 applies to this situation. For, let

$$\varphi_n(\nu) = \log \int_E \exp\left[\int_0^1 \langle f(t), d\nu(t)\rangle\right] \mu_n(df)$$

where $\nu \in E^* = M([0,1], \mathbb{R}^d)$, $\varphi(\nu) = \displaystyle\int_0^1 \log \widehat{\mu}(\nu[s,1])ds$. Then by elementary computations

$$\begin{aligned} \varphi_n(n\nu) &= \sum_{j=1}^n \log \widehat{\mu}\Big(\nu\Big[\frac{j}{n}, 1\Big] + \int_{\left[\frac{j-1}{n}, \frac{j}{n}\right)} (nt - [nt]) d\nu(t)\Big), \\ & |\varphi_n(n\nu) - n\varphi(\nu)| \leq C|\nu|([0,1]). \end{aligned}$$

Since $\varphi^* = I$, the level sets of φ^* are strongly compact in E (see [deA2]) and assumption (C2) holds. Also, since dom $\varphi = E^*$ and φ is E-Gâteaux differentiable everywhere on E^*, conditions (C3) and (C4) are easily satisfied. We have thus shown that (1) and (2) of Theorem 3.1 indeed hold in the present situation.

Acknowledgment. The author extends his warmest thanks to the staff of the Department of Mathematics, University of Wisconsin, Madison, where this paper was typed.

References

[deA1] A. de Acosta (1985), On large deviations of sums of independent random vectors. In: *Probability in Banach spaces V*, Lecture Notes in Mathematics 1153, 1–14, Springer-Verlag, Berlin.

[deA2] A. de Acosta (1994), Large deviations for vector-valued Lévy processes, *Stoch. Proc. and Appl.* **51**, 75–115.

[deA3] A. de Acosta (1994), Projective systems in large deviation theory II: some applications. In: *Probability in Banach spaces 9*, Progress in Probability, Vol. 35, 241–250, Birkhäuser, Boston, MA.

[B–Z] R. R. Bahadur and S. Zabell (1979), Large deviations of the sample mean in general vector spaces, *Ann. Probab.* **7**, 587–621.

[D] I. H. Dinwoodie (1992), Mesures dominantes et Théorème de Sanov, *Ann. Inst. Henri Poincaré* **28**, 365–373.

[D–V] M. D. Donsker and S. R. S. Varadhan (1976), Asymptotic evaluation of certain Markov process expectations for large time III, *Comm. Pure Appl. Math.* **29**, 389–461.

[E–K] U. Einmahl and J. Kuelbs (1996), Dominating points and large deviations for random vectors, *Probab. Theory Relat. Fields* **105**, 529–543.

[K] J. Kuelbs, *Large deviation probabilities and dominating points for open convex sets: non-logarithmic behavior*. Preprint.

[M–N] A. Meda and P. Ney (1998), A conditioned law of large numbers for Markov additive chains, *Studia Scient. Math. Hungarica* **34**, 305–316.

[M] A. A. Mogulskii (1976), Large deviations for trajectories of multidimensional random walks, *Th. Probab. Appl.* **21**, 300–315.

[N1] P. Ney (1983), Dominating points and asymptotics of large deviations in $\mathbb{R}^d$, *Ann. Probab.* **11**, 158–167.

[N2] P. Ney (1984), Convexity and large deviations, *Ann. Probab.* **12**, 903–906.

[R] R. T. Rockafellar (1970), *Convex Analysis*, Princeton Univ. Press.

[S] H. H. Schaefer (1970), *Topological vector spaces*, Springer-Verlag, Berlin.

Department of Mathematics
Case Western Reserve University
Cleveland, OH 44106
email: add3@po.cwru.edu

Large Deviations for Local Empirical Measures

Ludovic Menneteau

ABSTRACT We consider the local empirical measure indexed by a function class. Under some regularity assumptions, we give a large deviation principle for this process. A non-standard functional law of the iterated logarithm is inferred from our large deviations principle.

1 Introduction and statement of the main result

For a decade, the notion of local empirical process has been used to study statistics that are functions of the observations in a suitable neighborhood of a point (see e.g., Deheuvels and Mason (1990, 1991, 1994, 1995)). These authors proved functional laws of the iterated logarithm for this process and applied them to obtain consistency results for kernel-type density estimators. In 1997, Einmahl and Mason extended the local empirical process notion to permit them to include kernel regression function estimators and the conditional empirical process within their setup. They established a weak convergence result, a strong invariance principle and a compact law of the iterated logarithm for such a process. In this paper, we consider the same setup as in Einmahl and Mason (1997, 1998) and we are interested in large deviation principles for local empirical measures.

Let $(\Sigma, \mathcal{S})$ be a measure space. Let (ξ_k) be a sequence of independent Σ-valued random variables with common law μ defined on the same probability space $(\Omega, \mathcal{A}, \mathbb{P})$. Choose $\Sigma_0 \in \mathcal{S}$. For each integer $n \geq 1$, consider $V_n : \Sigma \to \Sigma$ an invertible bi-measurable map and set $v_n = \mu\left(V_n\left(\Sigma_0\right)\right)$. Assume that $v_n > 0$ and denote by ν_n the probability measure on $(\Sigma, \mathcal{S})$ such that $\nu_n(B) = v_n^{-1}\mu\left(V_n\left(\Sigma_0 \cap B\right)\right)$.

Let $\mathcal{F}$ be a class of measurable $\mathbb{R}$-valued functions defined on $(\Sigma, \mathcal{S})$ and $(a_n) \subset \mathbb{N}^*$. The local empirical measure indexed by $\mathcal{F}$ is defined by

$$\Theta\left(a_n, .\right) : f \in \mathcal{F} \mapsto \left(nv_n\right)^{-1} \sum_{k=1}^{a_n} f\left(V_n^{-1}\left(\xi_k\right)\right).$$

Consult Einmahl and Mason (1997, 1998) for descriptions of the potential uses of local empirical measures in nonparametric statistics.

Here, we are concerned with large deviation principles for $\left(\Theta\left(a_n, .\right)\right)$. Recall that if E is a topological space, $\mathcal{E}$ a σ-algebra on E, and (ε_n) is a sequence of positive real numbers such that $\varepsilon_n \downarrow 0$, a sequence (X_n) of $(E, \mathcal{E})$-valued random variables defined on a common probability space $(\Omega, \mathcal{A}, \mathbb{P})$ is said to follow the large deviation principle in E with speed (ε_n) and good rate function I whenever, for every $\alpha > 0$, $\{\mathrm{I} \leq \alpha\}$ is compact in E and, for every $A \in \mathcal{E}$,

$$\begin{aligned} - \inf \left\{ \mathrm{I}(x) : x \in \overset{\circ}{A} \right\} &\leq \liminf_{n \to \infty} \varepsilon_n \log \mathbb{P}(X_n \in A) \\ &\leq \limsup_{n \to \infty} \varepsilon_n \log \mathbb{P}(X_n \in A) \\ &\leq - \inf \left\{ \mathrm{I}(x) : x \in \overline{A} \right\}, \end{aligned}$$

where $\overset{\circ}{A}$ (resp. $\overline{A}$) denotes the interior (resp. closure) of A in E. The assumptions and definitions needed to get our main result are listed below.

For the remainder of the paper, we set, for every measure κ on $(\Sigma, \mathcal{S})$ and every κ-integrable function g, $\kappa(g) = \int_\Sigma g(x)\, d\kappa(x)$ and we denote by $L^+(\kappa)$ the space of measurable functions on $(\Sigma, \mathcal{S})$ $\kappa-$almost surely positive. For every class $\mathcal{G}$ of measurable $\mathbb{R}$-valued functions defined on $(\Sigma, \mathcal{S}, \mu)$ and every real valued functional Φ on $\mathcal{G}$, we set

$$\|\Phi\|_{\mathcal{G}} := \sup \{ |\Phi(g)| : g \in \mathcal{G} \},$$

and

$$\operatorname{vect}(\mathcal{G}) := \left\{ \sum_{i=1}^{p} \alpha_i g_i : p \in \mathbb{N}, \alpha_i \in \mathbb{R},\ g_i \in \mathcal{G} \right\}.$$

Let (ε_n) be a sequence of positive real numbers. Assume that the following hold :

(H) $\varepsilon_n \downarrow 0$, $v_n \downarrow 0$ and there exists $b > 0$ such that $n v_n \varepsilon_n \to b$ when $n \uparrow \infty$.

(F.1) $\mathcal{F}$ is a separable class of functions (see e.g. van der Vaart and Wellner (1996) p.115).

(F.2) For every $f \in \mathcal{F}, \operatorname{supp}(f) := \{ s \in \Sigma : f(s) \neq 0 \} \subset \Sigma_0$.

Furthermore, we consider a probability measure ν and we define for every class $\mathcal{G}$ of measurable $\mathbb{R}$-valued functions defined on $(\Sigma, \mathcal{S})$, the metric

$$e_\nu : (f, g) \in \mathcal{G}^2 \mapsto \left[\nu (f - g)^2 \right]^{1/2}.$$

For every $\delta > 0$, we set $\mathcal{F}_\delta = \{ f - g : e_\nu(f, g) < \delta, (f, g) \in \mathcal{F} \times \mathcal{F} \}$.

Our assumptions on ν are :

(F.3) For every $\alpha > 0$, $\nu(\exp(\alpha \|f\|_{\mathcal{F}})) < \infty$.

(F.4) For every $\alpha > 0$, $\limsup_{n \to \infty} \nu_n (\exp(\alpha \|f\|_{\mathcal{F}})) < \infty$.

(F.5) For every $g \in \operatorname{vect}(\mathcal{F})$, $\nu_n(\exp(g)) \to \nu(\exp(g))$ when $n \uparrow \infty$.

(F.6) $(\mathcal{F}, e_\nu)$ is totally bounded.

(F.7) $\lim_{\delta\to 0} \limsup_{n\to\infty} \sup\left\{\nu_n\left(f^2\right) : f \in \mathcal{F}_\delta\right\} = 0.$

In the sequel, we consider a Rademacher sequence (τ_n) independent of (ξ_n) and we set for every class $\mathcal{G}$ of measurable $\mathbb{R}$-valued functions defined on $(\Sigma, \mathcal{S}, \mu)$ and every sequence of integers $(a_n)_{n\in\mathbb{N}}$,

$$T(a_n, \mathcal{G}) = \mathbb{E}\left(\sup\left\{\left|\sum_{k=1}^{a_n} \tau_k g\left(V_n^{-1}(\xi_k)\right)\right| : g \in \mathcal{G}\right\}\right).$$

Our last assumption is :

(F.8) $\limsup_{n\to\infty} (nv_n)^{-1} T(n, \mathcal{F}) = 0.$

Remark. Condition (F.8) is satisfied for a wide set of class $\mathcal{F}$. In particular, following some arguments of the proof of Theorem 2-5-2 of van der Vaart and Wellner (1996), it can be shown that (F.8) holds for Vapnik and Červonenkis classes of functions (see van der Vaart and Wellner (1996) p.141) such that $\limsup_{n\to\infty} \nu_n\left(\|f\|_{\mathcal{F}}^2\right) < \infty$.

Denote by $B(\mathcal{F})$ the set of bounded functions $x : \mathcal{F} \to \mathbb{R}$. Let E_1 be the set $B(\mathcal{F})$ endowed with the topology induced by the semi-norms $\{q_f : f \in \mathcal{F}\}$ where

$$q_f : x \in B(\mathcal{F}) \mapsto |x(f)|,$$

let E_2 be the set $B(\mathcal{F})$ endowed with the uniform topology induced by the norm $\|\,.\,\|_{\mathcal{F}}$, and denote by $\mathcal{E}$ the σ-algebra generated by the closed balls of E_2.

We define the function

$$\begin{aligned} r_b : t \in \mathbb{R} \mapsto\ & b\,(t\log(t) - t + 1) \text{ if } t \geq 0, \\ & +\infty \text{ otherwise,} \end{aligned}$$

where we set $0\log 0 = 0$, and its Fenchel-Legendre transform,

$$r_b^* : s \in \mathbb{R} \mapsto b\left(\exp\left(b^{-1}s\right) - 1\right).$$

Finally, consider the mapping $\mathrm{I}_{a,b}$ such that, for every $\Phi \in B(\mathcal{F})$,

$$\mathrm{I}_{a,b}(\Phi) = \inf\left\{\nu\left(a r_b(\varphi/a)\right) : \varphi \in L^+(\nu), \forall f \in \mathcal{F}, \Phi(f) = \nu(\varphi f)\right\},$$

where $\inf\emptyset = +\infty$.

Our main result is the following theorem :

Theorem 1.1. *Assume that* (H), (F.1) – (F.8) *are satisfied. Then, for every sequence of integers* $(a_n)_{n\in\mathbb{N}}$, *such that* $n^{-1}a_n \to a > 0$, *the sequence* $(\Theta(a_n, .))_{n\in\mathbb{N}}$ *follows the large deviation principle in* E_2 *with speed* (ε_n) *and good rate function* $\mathrm{I}_{a,b}$.

Section 2 is devoted to the proof of the main result and, in Section 3, we derive a functional law of the iterated logarithm from our large deviation principle, which constitutes a "non standard" counterpoint of Einmahl and Mason (1997, 1998) and generalizes the result of Deheuvels and Mason (1995) (see also Theorem 5.7 of Arcones (1997) for the same kind of result in a somewhat different framework).

2 Proof of the main result

The proof of Theorem 1.1 is based on some general principles about large deviations shown, for instance, in Chapter 4 of Dembo and Zeitouni (1991). We first establish that our large deviation principle holds in the weaker topological vector space E_1 and then we control the modulus of continuity of our processes thanks to a Talagrand's inequality which has been already successfully used in the context of large deviations for empirical processes in Wu (1994) and Dembo and Zajic (1997).

For every $f \in \mathcal{F}$, define

$$\pi_f : \Phi \in B(\mathcal{F}) \mapsto \Phi(f).$$

It is easily seen that

$$E_1' := \left\{ \sum_{i=1}^{p} \alpha_i \pi_{f_i} : p \in \mathbb{N}, \alpha_i \in \mathbb{R},\ f_i \in \mathcal{F} \right\},$$

is the topological dual of E_1. In the sequel, $\langle\, . \, , \, . \,\rangle_1$ will denote the duality bracket between E_1 and E_1'.

For $\lambda = \sum_{i=1}^{p} \alpha_i \pi_{f_i} \in E_1'$, we set $g_\lambda = \sum_{i=1}^{p} \alpha_i f_i$. Moreover, we define the functions

$$\Lambda_n : \lambda \in E_1' \mapsto \log \mathbb{E}\left(\exp\left(\langle \Theta_n, \lambda \rangle_1\right)\right),$$

$$\Lambda_{a,b} : \lambda \in E_1' \mapsto ab\nu\left(\exp\left(g_\lambda / b\right) - 1\right),$$

and we consider the Fenchel-Legendre transform of $\Lambda_{a,b}$,

$$\Lambda_{a,b}^* : \Phi \in E_1 \mapsto \sup\left\{\langle \Phi, \lambda \rangle_1 - \Lambda_{a,b}(\lambda) : \lambda \in E_1'\right\}.$$

Finally, for every measurable function f and every $N \in \mathbb{N}^*$, we set

$$\chi_N(f) = (f \wedge N) \vee (-N),$$

$$\chi_N(\mathcal{F}) = \{\chi_N(f) : f \in \mathcal{F}\},$$

and

$$\overline{\chi_N}(\mathcal{F}) = \{f - \chi_N(f) : f \in \mathcal{F}\}.$$

Lemma 2.1. *For every $\alpha > 0$, $\{\mathrm{I}_{a,b} \leq \alpha\}$ is compact in E_2.*

Proof. It is enough to prove the result for $a = 1$. First we show that for every $\alpha \in \mathbb{R}$, $\{\mathrm{I}_{1,b} \leq \alpha\}$ is totally bounded in E_2. By the Ascoli Theorem we have to show that for every $\alpha \in \mathbb{R}$, $\{\mathrm{I}_{1,b} \leq \alpha\}$ is bounded in E_2 and equicontinuous for e_ν.

For every $\Phi \in \{\mathrm{I}_{1,b} \leq \alpha\}$ and every $\varepsilon > 0$, there exists $\varphi \in L^+(\nu)$ with

$$\int r_b(\varphi)\, d\nu < \alpha + \varepsilon$$

such that,

$$\Phi : f \in \mathcal{F} \mapsto \nu(\varphi f).$$

We first assert that for every $c > 0$ and every measurable function f we have,

$$\int \varphi |f|\, d\nu \leq c\left(\int r_b^*\left(c^{-1}|f|\right) d\nu + \alpha + \varepsilon\right) \tag{2.1}$$

Indeed, since r_b^* is the Fenchel-Legendre transform of r_b we have, for all real numbers s and t,

$$st \leq r_b^*(s) + r_b(t).$$

Hence, for every measurable function f and every $c > 0$,

$$\begin{aligned}\int \varphi |f|\, d\nu &= c\int \varphi \left|c^{-1} f\right| d\nu \\ &\leq c\left(\int r_b^*\left(c^{-1}|f|\right) d\nu + \alpha + \varepsilon\right).\end{aligned}$$

Therefore, for $c = 1$, since ε is arbitrary,

$$\sup\{\|\Phi\|_{\mathcal{F}} : \Phi \in \{\mathrm{I}_{1,b} \leq \alpha\}\} \leq \int r_b^*(\|f\|_{\mathcal{F}})\, d\nu + \alpha \tag{2.2}$$

i.e., by (F.3), $\{\mathrm{I}_{1,b} \leq \alpha\}$ is bounded.

To show the equicontinuity of $\{\mathrm{I}_{1,b} \leq \alpha\}$, that is

$$\limsup_{\delta \to 0} \sup\left\{\|\Phi\|_{\mathcal{F}_\delta} : \Phi \in \{\mathrm{I}_{1,b} \leq \alpha\}\right\} = 0, \tag{2.3}$$

we prove that for any $\eta > 0$, there exists $A_\eta > 0$ such that for every $\alpha > 0$ and every $\delta > 0$,

$$\sup\left\{\|\Phi\|_{\mathcal{F}_\delta} : \Phi \in \{\mathrm{I}_{1,b} \leq \alpha\}\right\} \leq \eta(\alpha + 1) + A_\eta \delta. \tag{2.4}$$

First note that by (2.1) and since ε is arbitrary, we have, for every $c > 0$,

$$\begin{aligned}
&\sup\left\{\|\Phi\|_{\mathcal{F}_\delta} : \Phi \in \{\mathrm{I}_{1,b} \le \alpha\}\right\} \\
&\le c\alpha + c\sup_{f\in\mathcal{F}_\delta} \nu\left(r_b^*\left(c^{-1}|f|\right)\right) \\
&\le c\alpha + cr_b^*\left(c^{-1}N\right) \sup_{f\in\mathcal{F}_\delta} \nu\left(|f| \le N\right) + c\sup_{f\in\mathcal{F}_\delta} \nu\left(r_b^*\left(c^{-1}|f|\right)\mathbf{1}_{\{|f|>N\}}\right) \\
&\le c\alpha + cN^{-1}r_b^*\left(c^{-1}N\right)\delta + c\nu\left(r_b^*\left(2c^{-1}\|f\|_{\mathcal{F}}\right)\mathbf{1}_{\{2\|f\|_{\mathcal{F}}>N\}}\right),
\end{aligned} \tag{2.5}$$

where we have used the Cauchy-Schwarz inequality and the fact that $\|f\|_{\mathcal{F}_\delta} \le 2\|f\|_{\mathcal{F}}$. Fix $\eta > 0$. Since by (F.3) and dominated convergence, for every $c > 0$,

$$\limsup_{N\to\infty} \nu\left(r_b^*\left(2c^{-1}\|f\|_{\mathcal{F}}\right)\mathbf{1}_{\{2\|f\|_{\mathcal{F}}>N\}}\right) = 0,$$

there exists N_η such that

$$\nu\left(r_b^*\left(2\eta^{-1}\|f\|_{\mathcal{F}}\right)\mathbf{1}_{\{2\|f\|_{\mathcal{F}}>N_\eta\}}\right) \le 1.$$

Hence, using (2.5), we obtain (2.4) with $A_\eta = \eta N_\eta^{-1}r_b^*\left(\eta^{-1}N_\eta\right)$, and for every $\eta > 0$,

$$\limsup_{\delta\to 0} \sup\left\{\|\Phi\|_{\mathcal{F}_\delta} : \Phi \in \{\mathrm{I}_{1,b} \le \alpha\}\right\} \le \eta(\alpha+1),$$

which gives (2.3).

Now we show that for every $\alpha \in \mathbb{R}$, $\{\mathrm{I}_{1,b} \le \alpha\}$ is closed in E_2. Set $(\Phi_n) \subset \{\mathrm{I}_{1,b} \le \alpha\}$ and $\Psi \in E_2$ such that $\Phi_n \to \Psi$ in E_2. For every $\varepsilon > 0$ set

$$A_{\alpha,\varepsilon} = \left\{\varphi \in L^+(\nu) : \nu\left(r_b(\varphi)\right) < \alpha + \varepsilon\right\}.$$

There exists $(\varphi_n) \subset A_{\alpha,\varepsilon}$ such that for each n,

$$\Phi_n : f \in \mathcal{F} \mapsto \nu\left(\varphi_n f\right).$$

Denote by $L^1(\nu)$ (resp. $L^\infty(\nu)$) the set of $\nu-$integrable (resp. $\nu-$bounded) functions defined on $(\Sigma, \mathcal{S})$ and by $\sigma\left(L^1(\nu),\ L^\infty(\nu)\right)$ the weak topology induced by $L^\infty(\nu)$ in $L^1(\nu)$. Since, for every $t \in \mathbb{R}$,

$$|t| \le b^{-1}r_b(t) + (e-1),$$

we have $A_{\alpha,\varepsilon} \subset L^1(\nu)$. Furthermore by a little algebra, we get that

$$\limsup_{u\to\infty} \sup\left\{\int_{\{|\varphi|>u\}} |\varphi|\, d\nu : \varphi \in A_{\alpha,\varepsilon}\right\} = 0.$$

That is $A_{\alpha,\varepsilon}$ is uniformly integrable and so, by the Dunford-Pettis Theorem (see e.g. Theorem C.7 of Dembo and Zeitouni (1991)), it is $\sigma\left(L^1(\nu), L^\infty(\nu)\right)$ sequentially compact. Therefore, there is a subsequence $(\varphi_{n'})$ of (φ_n) and $\varphi_0 \in L^1(\nu)$ such that

$$\varphi_{n'} \to \varphi_0 \text{ in } \sigma\left(L^1(\nu), L^\infty(\nu)\right). \tag{2.6}$$

To show that $\varphi_0 \in A_{\alpha,\varepsilon}$, we prove that $A_{\alpha,\varepsilon}$ is $\sigma\left(L^1(\nu), L^\infty(\nu)\right)$-closed. Since $A_{\alpha,\varepsilon}$ is convex it is enough to prove that it is closed in $L^1(\nu)$. For that, take $(\theta_n) \subset A_{\alpha,\varepsilon}$ such that $\theta_n \to \theta$ in $L^1(\nu)$. Then, there is a subsequence $(\theta_{n'})$ of (θ_n) such that $\theta_{n'} \to \theta$ ν−a.s. Therefore, by Fatou's lemma,

$$\begin{aligned}\nu(r_b(\theta)) &\le \liminf_{n'\to\infty} \nu(r_b(\theta_{n'})) \\ &< \alpha + \varepsilon,\end{aligned}$$

and since obviously, $\theta \ge 0$, we get that $\theta \in A_{\alpha,\varepsilon}$.

Set

$$\Phi_0 : f \in \mathcal{F} \mapsto \nu(\varphi_0 f).$$

The last step is to prove that $\Phi_0 = \Psi$. Consider the sequence $(\varphi_{n'})$ defined in (2.6) and

$$\Phi_{n'} : f \in \mathcal{F} \mapsto \nu(\varphi_{n'} f).$$

For every $N > 0$, $f \in \mathcal{F}$ and $\eta > 0$, we have,

$$\begin{aligned}&|\Phi_0 - \Psi|(f) \\ &\le |\nu((\varphi_0 - \varphi_{n'})\chi_N(f))| \\ &+ |\nu((\varphi_0 - \varphi_{n'})(f - \chi_N(f)))| + \|\Phi_{n'} - \Psi\|_{\mathcal{F}} \\ &\le |\nu((\varphi_0 - \varphi_{n'})\chi_N(f))| + \|\Phi_{n'} - \Psi\|_{\mathcal{F}} \\ &+ 2\sup\left\{\|\Phi\|_{\overline{\chi_N}(\mathcal{F})} : \Phi \in \{\mathrm{I}_{1,b} \le \alpha\}\right\}.\end{aligned}$$

By (2.1) and dominated convergence, we may choose $N(\eta)$ such that

$$\sup\left\{\|\Phi\|_{\overline{\chi_{N(\eta)}}(\mathcal{F})} : \Phi \in \{\mathrm{I}_{1,b} \le \alpha\}\right\} \le \eta,$$

and since $\chi_{N(\eta)}(f) \in L^\infty(\nu)$,

$$\begin{aligned}&|\Phi_0 - \Psi|(f) \\ &\le 2\eta + \limsup_{n'\to\infty}\left(\left|\nu\left((\varphi_0 - \varphi_{n'})\chi_{N(\eta)}(f)\right)\right| + \|\Phi_{n'} - \Psi\|_{\mathcal{F}}\right) \\ &\le 2\eta,\end{aligned}$$

which gives the result since η is arbitrary. □

In the next lemma, we establish the exponential tightness (see e.g. p. 8 of Dembo and Zeitouni (1991)) of $(\Theta(a_n, .))_{n\geq 1}$ in the topological space E_1 and with speed (ε_n).

Lemma 2.2. *Under assumptions* (F.1), (F.2), (F.4) *and* (H), *there exists a family* $(K_D)_{D>0}$, *of compacts sets of* E_1 *such that each* $K_D \in \mathcal{E}$ *and,*

$$\lim_{D\to\infty} \limsup_{n\to\infty} \varepsilon_n \log \mathbb{P}(\Theta(a_n, .) \notin K_D) = -\infty.$$

Proof. For $D > 0$, set $K_D = \{\Phi \in B(\mathcal{F}) : \|\Phi\|_{\mathcal{F}} \leq D\}$. By Tychonov's theorem, K_D is compact in E_1. Moreover, note that

$$\mu\left(\exp \left\|fV_n^{-1}\right\|_{\mathcal{F}}\right) = v_n \nu_n\left(\exp \|f\|_{\mathcal{F}}\right) + 1 - v_n.$$

We may choose $c > 0$ such that, for n large enough,

$$\nu_n\left(\exp \|f\|_{\mathcal{F}}\right) < c.$$

Hence, by the Chebyshev exponential inequality,

$$\begin{aligned}
&\limsup_{n\to\infty} \varepsilon_n \log \mathbb{P}\left(\|\Theta(a_n, .)\|_{\mathcal{F}} > D\right) \\
&\leq \limsup_{n\to\infty} \left(-Dnv_n\varepsilon_n\right) \\
&+ \limsup_{n\to\infty} \varepsilon_n a_n \log\left(v_n\nu_n\left(\exp\|f\|_{\mathcal{F}}\right) + 1 - v_n\right) \\
&\leq -Db + ab(c-1),
\end{aligned}$$

which gives the intended result. □

Proposition 2.3. *Under assumptions* (F.1) – (F.5) *and* (H), $(\Theta(a_n, .))_{n\in\mathbb{N}}$ *follows the large deviation principle in* E_1 *with speed* (ε_n) *and good rate function* $\mathrm{I}_{a,b}$.

Proof. First, we prove that the large deviation principle holds with the rate function $\Lambda^*_{a,b}$. Because $\Lambda_{a,b}$ is a finite valued and Gateaux differentiable function (see e.g. p. 142 of Dembo and Zeitouni (1991)) and since by Lemma 2.2, $(\Theta(a_n, .))$ is (ε_n)-exponentially tight in E_1, it is enough by Corollary 4-5-27 of Dembo and Zeitouni (1991) to prove that for every $\lambda \in E_1'$,

$$\lim_{n\to\infty} \varepsilon_n \Lambda_n\left(\varepsilon_n^{-1}\lambda\right) = \Lambda_{a,b}(\lambda).$$

Set for convenience, $b_n = \varepsilon_n n\, v_n$. For $\lambda = \sum_{i=1}^{p} \alpha_i \pi_{f_i} \in E_1'$,

$$\begin{aligned}
\varepsilon_n \Lambda_n\left(\varepsilon_n^{-1}\lambda\right) &= \varepsilon_n \log \mathbb{E}\left(\exp\left(b_n^{-1}\sum_{k=1}^{a_n} g_\lambda\left(V_n^{-1}(\xi_k)\right)\right)\right) \\
&= a_n\varepsilon_n \log \mathbb{E}\left(\exp\left(b_n^{-1} g_\lambda\left(V_n^{-1}(\xi_1)\right)\right)\right),
\end{aligned}$$

and, by (F.2),

$$\mathbb{E}\left(\exp\left(b_n^{-1}g_\lambda\left(V_n^{-1}(\xi_1)\right)\right)\right) = v_n\nu_n\left(e^{b_n^{-1}g_\lambda}\right) + 1 - v_n.$$

Furthermore, by the Mean Value Theorem, there exists a function β_n with range in $\left[b_n^{-1}\wedge b^{-1}, b_n^{-1}\vee b^{-1}\right]$ such that

$$e^{b_n^{-1}g_\lambda} - e^{b^{-1}g_\lambda} = \left(b_n^{-1} - b^{-1}\right)g_\lambda e^{\beta_n g_\lambda}.$$

Hence, for n large enough,

$$\begin{aligned}
&\left|\nu_n\left(e^{b_n^{-1}g_\lambda}\right) - \nu\left(e^{b^{-1}g_\lambda}\right)\right| \\
&\leq \left|b_n^{-1} - b^{-1}\right|\nu_n\left(\left|g_\lambda\right|e^{\beta_n\left|g_\lambda\right|}\right) + \left|\left(\nu_n - \nu\right)\left(e^{b^{-1}g_\lambda}\right)\right| \\
&\leq \left|b_n^{-1} - b^{-1}\right|\nu_n\left(e^{\left(\beta_n+1\right)\left|g_\lambda\right|}\right) + \left|\left(\nu_n - \nu\right)\left(e^{b^{-1}g_\lambda}\right)\right| \\
&\leq \left|b_n^{-1} - b^{-1}\right|\nu_n\left(e^{2\left(b^{-1}+1\right)g_\lambda} + e^{-2\left(b^{-1}+1\right)g_\lambda}\right) + \left|\left(\nu_n - \nu\right)\left(e^{b^{-1}g_\lambda}\right)\right|.
\end{aligned}$$

Therefore, by (H), (F.4) and (F.5),

$$\nu_n\left(e^{b_n^{-1}g_\lambda}\right) = \nu\left(e^{b^{-1}g_\lambda}\right) + o(1),$$

and,

$$\begin{aligned}
\varepsilon_n\Lambda_n\left(\varepsilon_n^{-1}\lambda\right) &= a_n\varepsilon_n\log\left(1 - v_n + v_n\nu\left(e^{b^{-1}g_\lambda}\right) + o\left(v_n\right)\right) \\
&= a_n\varepsilon_n v_n\left(\nu\left(e^{b^{-1}g_\lambda}\right) - 1 + o(1)\right) \\
&= \Lambda_{a,b}(\lambda) + o(1),
\end{aligned}$$

which ends the first part of the proof.

Now, it remains to prove that $\Lambda_{a,b}^* = \mathrm{I}_{a,b}$. Of course, it is enough to show the result for $a = 1$. By the Duality Lemma (see e.g. Lemma 4-5-8 of Dembo and Zeitouni (1991)), we have only to show that $\mathrm{I}_{1,b}$ is a proper convex function lower semicontinuous (in E_1) and that $\mathrm{I}_{1,b}^* = \Lambda_{1,b}$, where

$$\mathrm{I}_{1,b}^* : \lambda \in E_1' \mapsto \sup\left\{\langle\Phi,\lambda\rangle_1 - \mathrm{I}_{1,b}(\Phi) : \Phi \in E_1\right\}.$$

The fact that $\mathrm{I}_{1,b}$ is a proper convex function is obvious. Moreover, by Lemma 2.1, $\mathrm{I}_{1,b}$ is lower semicontinuous in E_2. Now, for every $\lambda \in E_1'$,

$$\begin{aligned}
\mathrm{I}_{1,b}^*(\lambda) &= \sup\left\{\langle\Phi,\lambda\rangle_1 - \mathrm{I}_{1,b}(\Phi) : \Phi \in E_1\right\} \\
&= \sup\left\{\int \varphi\left(g_\lambda - b\log(\varphi) + b\right)d\nu : \varphi \in L^+(\nu)\right\} + b.
\end{aligned}$$

Set $\varphi_\lambda = \exp\left(b^{-1} g_\lambda\right)$, then,

$$\mathrm{I}^*_{1,b}(\lambda) \geq \int \varphi_\lambda \left(g_\lambda - b \log\left(\varphi_\lambda\right) + b\right) d\nu + b = \Lambda_{1,b}(\lambda).$$

Furthermore, since as can be easily seen,

$$\sup\left\{\varphi\left(g_\lambda - b\log(\varphi) + b\right) : \varphi \in L^+(\nu)\right\} = b\exp\left(b^{-1} g_\lambda\right),$$

we obtain

$$\mathrm{I}^*_{1,b}(\lambda) \leq \int b \exp\left(b^{-1} g_\lambda\right) d\nu + b = \Lambda_{1,b}(\lambda).$$

□

To extend our large deviation principle for $(\Theta(a_n, .))$ in the weak topological space E_1 to the stronger topological space E_2, we have to control the modulus of continuity of $\Theta(a_n, .)$. For that, we use an inequality of Talagrand (1996), in a form obtained by Ledoux (1996). Since this inequality holds only for bounded classes of functions, we will need a truncation argument.

For every $\delta > 0$ and $N \in \mathbb{N}^*$, we set

$$\mathcal{F}_{N,\delta} = \left\{\chi_N(f) - \chi_N(g) : (f,g) \in \mathcal{F} \times \mathcal{F}, e_\nu(f,g) < \delta\right\}.$$

Moreover, some new parameters have to be introduced. For every $N \in \mathbb{N}^*$ and $\delta > 0$, we set,

$$\sigma_n^2\left(\mathcal{F}_{N,\delta}\right) = \sup\left\{\int_\Sigma \left(g\left(V_n^{-1}\right) - \mu\left(g\left(V_n^{-1}\right)\right)\right)^2 d\mu : g \in \mathcal{F}_{N,\delta}\right\}, \tag{2.7}$$

and

(2.8)

$$\overline{T}\left(a_n, \mathcal{F}_{N,\delta}\right) = \mathbb{E}\left(\sup\left\{\left|\sum_{k=1}^{a_n}\left(g\left(V_n^{-1}(\xi_k)\right) - \mu\left(g\left(V_n^{-1}\right)\right)\right)\right| : g \in \mathcal{F}_{N,\delta}\right\}\right).$$

Lemma 2.4. *(i) Assume that* (F.7) *holds, then, for every $N \in \mathbb{N}^*$,*

$$\lim_{\delta\to 0} \limsup_{n\to\infty} v_n^{-1} \sigma_n^2\left(\mathcal{F}_{N,\delta}\right) = 0. \tag{2.9}$$

(ii) Assume that (F.8) *holds, then, for every $N \in \mathbb{N}^*$*

$$\lim_{\delta\to 0} \limsup_{n\to\infty} \left(n v_n\right)^{-1} \overline{T}\left(a_n, \mathcal{F}_{N,\delta}\right) = 0. \tag{2.10}$$

Proof. (i) Straightforward.

(ii) By Lemma 2-3-6 of van der Vaart and Wellner (1996),

$$\overline{T}\left(a_n, \mathcal{F}_{N,\delta}\right) \leq 2T\left(a_n, \mathcal{F}_{N,\delta}\right).$$

Moreover, for n large enough, by the Lévy inequality (see Ledoux and Talagrand (1991) p. 47) and an integration by parts,

$$\begin{aligned} T\left(a_n, \mathcal{F}_{N,\delta}\right) &\leq \left\lfloor n^{-1} a_n \right\rfloor T\left(n, \mathcal{F}_{N,\delta}\right) + 2T\left(n, \mathcal{F}_{N,\delta}\right) \\ &\leq (a+3)\, T\left(n, \mathcal{F}_{N,\delta}\right) \leq 2\,(a+3)\, T\left(n, \chi_N\left(\mathcal{F}\right)\right). \end{aligned}$$

And, since Theorem 4-12 of Ledoux and Talagrand (1991) implies

$$T\left(n, \chi_N\left(\mathcal{F}\right)\right) \leq 2T\left(n, \mathcal{F}\right),$$

we get that, for n large enough,

$$\overline{T}\left(a_n, \mathcal{F}_{N,\delta}\right) \leq 8\,(a+3)\, T\left(n, \mathcal{F}\right).$$

Therefore, (F.8) implies the intended result. □

For every $m \in \mathbb{N}^*$, by (F.6), we could define a map

$$w_m : \mathcal{F} \to \mathcal{F}, \tag{2.11}$$

such that

$$w_m\left(\mathcal{F}\right) = \left\{w_m\left(f\right) : f \in \mathcal{F}\right\} \tag{2.12}$$

is a finite set and

$$\sup\left\{e_\nu\left(f, w_m\left(f\right)\right) : f \in \mathcal{F}\right\} < m^{-1}.$$

For every $m \in \mathbb{N}^*$, define

$$u_m : E_1 \to E_2$$

such that

$$u_m\left(\Phi\right) : f \in \mathcal{F} \mapsto \Phi\left(w_m\left(f\right)\right). \tag{2.13}$$

Lemma 2.5. *(i) Assume that* (F.4) *and* (H) *hold, then, for all* η,

$$\limsup_{N\to\infty}\ \limsup_{n\to\infty} \varepsilon_n \log \mathbb{P}\left(\left\|\Theta\left(a_n, \,.\,\right)\right\|_{\overline{\chi_N}(\mathcal{F})} \geq \eta\right) = -\infty.$$

(ii) For all $\eta > 0$,

$$\lim_{\delta\to 0}\ \limsup_{n\to\infty} \varepsilon_n \log \mathbb{P}\left(\left\|\Theta\left(a_n, \,.\,\right)\right\|_{\mathcal{F}_\delta} \geq \eta\right) = -\infty. \tag{2.14}$$

In particular, for all $\eta > 0$,

$$\lim_{m\to\infty}\ \limsup_{n\to\infty} \varepsilon_n \log \mathbb{P}\left(\left\|u_m\left(\Theta\left(a_n, \,.\,\right)\right) - \Theta\left(a_n, \,.\,\right)\right\|_{\mathcal{F}} \geq \eta\right) = -\infty. \tag{2.15}$$

Proof. (i) For every $\alpha > 0$,

$$\begin{aligned}&\exp\left(\alpha \|f\|_{\overline{\mathcal{X}_N}(\mathcal{F})}\right) - 1\\ &\leq \exp\left(\alpha \|f\|_{\mathcal{F}}\right) \mathbf{1}\left(\|f\|_{\mathcal{F}} > N\right)\\ &\leq \exp\left(\left(\alpha + 1\right) \|f\|_{\mathcal{F}}\right) \exp\left(-N\right).\end{aligned}$$

Hence, by (F.4), there exists $c(\alpha) > 0$ such that, for every n large enough and every $N \in \mathbb{N}$,

$$\nu_n\left(\exp\left(\alpha \|f\|_{\overline{\mathcal{X}_N}(\mathcal{F})}\right) - 1\right) \leq c(\alpha) \exp\left(-N\right). \tag{2.16}$$

Moreover, by the Chebyshev exponential inequality,

$$\begin{aligned}&\mathbb{P}\left(\|\Theta\left(a_n, \,.\,\right)\|_{\overline{\mathcal{X}_N}(\mathcal{F})} \geq \eta\right)\\ &\leq \exp\left(-\alpha\eta n v_n\right) \left(\mathbb{E}\left(\exp\left(\alpha \left\|f\left(V_n^{-1}\left(\xi_1\right)\right)\right\|_{\overline{\mathcal{X}_N}(\mathcal{F})}\right)\right)\right)^{a_n}.\end{aligned}$$

Therefore, by (2.16) and (H),

$$\begin{aligned}&\limsup_{n\to\infty} \varepsilon_n \log \mathbb{P}\left(\|\Theta\left(a_n, \,.\,\right)\|_{\overline{\mathcal{X}_N}(\mathcal{F})} \geq \eta\right)\\ &\leq -\alpha\eta b + \limsup_{n\to\infty} \varepsilon_n a_n \log\left(1 + v_n \nu_n\left(\exp\left(\alpha \|f\|_{\overline{\mathcal{X}_N}(\mathcal{F})}\right) - 1\right)\right)\\ &\leq -\alpha\eta b + abc(\alpha) \exp\left(-N\right).\end{aligned}$$

And, we obtain for all $\alpha > 0$,

$$\limsup_{N\to\infty} \limsup_{n\to\infty} \varepsilon_n \log \mathbb{P}\left(\|\Theta\left(a_n, \,.\,\right)\|_{\overline{\mathcal{X}_N}(\mathcal{F})} \geq \eta\right) \leq -\alpha\eta b.$$

(ii) By the triangle inequality, for every $N \in \mathbb{N}^*$,

$$\|\Theta\left(a_n, \,.\,\right)\|_{\mathcal{F}_\delta} \leq 2 \|\Theta\left(a_n, \,.\,\right)\|_{\overline{\mathcal{X}_N}(\mathcal{F})} + \|\Theta\left(a_n, \,.\,\right)\|_{\mathcal{F}_{N,\delta}}.$$

Hence, for every $N \in \mathbb{N}^*$,

$$\begin{aligned}&\lim_{\delta\to 0} \limsup_{n\to\infty} \varepsilon_n \log \mathbb{P}\left(\|\Theta\left(a_n, \,.\,\right)\|_{\mathcal{F}_\delta} \geq \eta\right)\\ &\leq \max\left(\begin{array}{c}\lim_{\delta\to 0} \limsup_{n\to\infty} \varepsilon_n \log \mathbb{P}\left(\|\Theta\left(a_n, \,.\,\right)\|_{\mathcal{F}_{N,\delta}} \geq \eta/2\right),\\ \limsup_{n\to\infty} \varepsilon_n \log \mathbb{P}\left(\|\Theta\left(a_n, \,.\,\right)\|_{\overline{\mathcal{X}_N}(\mathcal{F})} \geq \eta/4\right)\end{array}\right).\end{aligned}$$

Therefore, by (i), it is enough to show that for every $N \in \mathbb{N}^*$, and $\eta > 0$,

$$\lim_{\delta\to 0} \limsup_{n\to\infty} \varepsilon_n \log \mathbb{P}\left(\|\Theta\left(a_n, \,.\,\right)\|_{\mathcal{F}_{N,\delta}} \geq \eta\right) = -\infty. \tag{2.17}$$

To this end, set

$$\widetilde{\Theta}(a_n, .) : f \mapsto (nv_n)^{-1} \sum_{k=1}^{a_n} \left(f\left(V_n^{-1}(\xi_k)\right) - \mu\left(f\left(V_n^{-1}\right)\right)\right).$$

By the triangle inequality,

$$\begin{aligned} &\|\Theta(a_n, .)\|_{\mathcal{F}_{N,\delta}} \\ &\leq \left\|\widetilde{\Theta}(a_n, .)\right\|_{\mathcal{F}_{N,\delta}} + n^{-1} a_n \sup\left\{|\nu_n(f)| : f \in \mathcal{F}_{N,\delta}\right\}. \end{aligned}$$

Now, by the Cauchy-Schwartz inequality, and (F.7),

$$\lim_{\delta\to 0} \limsup_{n\to\infty} n^{-1} a_n \sup\left\{|\nu_n(f)| : f \in \mathcal{F}_{N,\delta}\right\} = 0.$$

Hence, to prove (2.17), we only have to show that for all $\eta > 0$,

$$\lim_{\delta\to 0} \limsup_{n\to\infty} \varepsilon_n \log \mathbb{P}\left(\left\|\widetilde{\Theta}(a_n, .)\right\|_{\mathcal{F}_{N,\delta}} \geq \eta\right) = -\infty.$$

Now, by Corollary 2.6 of Ledoux (1996) and since for every $f \in \mathcal{F}_{N,\delta}$, $|f| \leq 2N$, we get that there exists a constant K such that,

$$\begin{aligned} &\mathbb{P}\left(\left\|\widetilde{\Theta}(a_n, .)\right\|_{\mathcal{F}_{N,\delta}} \geq \eta\right) \\ &\leq \mathbb{P}\left(\left\|\sum_{k=1}^{a_n} \left(f\left(V_n^{-1}(\xi_k)\right) - \mu\left(f\left(V_n^{-1}\right)\right)\right)\right\|_{\mathcal{F}_{N,\delta}} \geq \overline{T}\left(a_n, \mathcal{F}_{N,\delta}\right) + \frac{\eta}{2} n v_n\right) \\ &\leq 3 \exp\left(-\frac{\eta n v_n}{4KN} \log\left(1 + N\eta\alpha_{N,n}(\delta)\right)\right), \end{aligned}$$

where,

$$\alpha_{N,n}(\delta) = \left(n^{-1} a_n v_n^{-1} \sigma_n^2\left(\mathcal{F}_{N,\delta}\right) + 2N (nv_n)^{-1} \overline{T}\left(a_n, \mathcal{F}_{N,\delta}\right)\right)^{-1}.$$

Therefore, by Lemma 2.4,

$$\begin{aligned} &\lim_{\delta\to 0} \limsup_{n\to\infty} \varepsilon_n \log \mathbb{P}\left(\left\|\widetilde{\Theta}(a_n, .)\right\|_{\mathcal{F}_{N,\delta}} \geq \eta\right) \\ &\leq -\frac{\eta b}{4KN} \log\left(1 + N\eta \lim_{\delta\to 0} \liminf_{n\to\infty} \alpha_{N,n}(\delta)\right) \\ &= -\infty. \end{aligned}$$

□

Lemma 2.6. *(i) For every $m \in \mathbb{N}^*$, $u_m : E_1 \to E_2$ is continuous.*
(ii) For every $a > 0$, $\lim_{m\to\infty} \sup\left\{\|u_m(\Phi) - \Phi\|_{\mathcal{F}} : \Phi \in \{\mathrm{I}_{a,b} \leq \alpha\}\right\} = 0.$

Proof. (i) Straightforward.
(ii) Obvious since $\|u_m(\Phi) - \Phi\|_{\mathcal{F}} \le \|\Phi\|_{\mathcal{F}_{m-1}}$ and $\{I_{a,b} \le \alpha\}$ is equicontinuous. □

Proof of Theorem 1.1. Using Proposition 2.3, (2.15) and Lemma 2.6 it is readily seen that the intended result is a simple consequence of Theorem 4-2-23 of Dembo and Zeitouni (1991). □

3 The law of the iterated logarithm

In this part, we prove a functional law of the iterated logarithm for the processes $(\Theta(a_n, .))$ when (a_n) is a sequence of positive integers such that $n^{-1}a_n \to a > 0$. We refer to O'Brien and Vervaat (1991) for a study of the links between large deviation principles and laws of the iterated logarithm.

For convenience, we set $K_{a,b} := \{I_{a,b} \le 1\}$, and B denotes the open unit ball of E_2.

Lemma 3.1. *(i) For every $\eta > 0$,*

$$\inf_{(K_{a,b}+\eta B)^c} I_{a,b} > 1. \tag{3.1}$$

(ii) There exists $M > 0$ such that

$$\inf_{(MB)^c} I_{a,b} \ge 3. \tag{3.2}$$

(iii) For every $\eta > 0$, there exists $\beta_\eta > 0$ such that for every $\beta \in (0, \beta_\eta)$,

$$\inf_{(\eta B)^c} I_{\beta,b} \ge 3. \tag{3.3}$$

Proof. (i) and (ii) are straightforward.
(iii) Set $\eta > 0$. It is enough to show that, for every $\beta > 0$ small enough, $\{I_{\beta,b} \le 3\} \subset \eta B$. Note that $\{I_{\beta,b} \le 3\} = \beta\left\{I_{1,b} \le 3\beta^{-1}\right\}$ and take $\Phi \in \{I_{\beta,b} \le 3\}$, by (2.1), since $\beta^{-1}\Phi \in \left\{I_{1,b} \le 3\beta^{-1}\right\}$,

$$\left\|\beta^{-1}\Phi\right\|_{\mathcal{F}} \le \frac{\eta}{12}\left(6\beta^{-1} + \nu\left(r_b^*\left(12\eta^{-1}\|f\|_{\mathcal{F}}\right)\right)\right).$$

Hence,

$$\begin{aligned}\|\Phi\|_{\mathcal{F}} &\le \frac{\eta}{12}\left(6 + \beta\nu\left(r_b^*\left(12\eta^{-1}\|f\|_{\mathcal{F}}\right)\right)\right)\\ &\le \eta\end{aligned}$$

for β small enough. □

For every $q > 1$ and $p \in \mathbb{N}^*$, we set

$$n_p = n_p(q) = \lfloor q^p \rfloor .$$

Furthermore, in order to state our law of the iterated logarithm, we introduce the following assumptions :

(F.9) For each $n \geq 1$ and $l \geq n$, $f\left(V_n^{-1}\right)$ and $f\left(V_l^{-1} \circ V_n\right) \in \mathcal{F}$.

(F.10)

$$\lim_{q \downarrow 1} \limsup_{p \to \infty} \sup_{f \in \mathcal{F}} \max_{n_p+1 \leq l \leq n_{p+1}} e_\nu\left(f_{l,p}, f\right) = 0$$

where $f_{l,p} = f\left(V_l^{-1} \circ V_{n_p}\right) \in \mathcal{F}$.

Theorem 3.2. *Assume that* (F.1)–(F.10) *and* (H) *with* $\varepsilon_n = (\log\log(n))^{-1}$ *hold, then, for every sequence of integers* (a_n), *such that* $n^{-1}a_n \to a > 0$, $(\Theta(a_n, .))_{n\geq 1}$ *is almost surely relatively compact in* E_2 *with set of limit points equal to* $K_{a,b}$.

Proof. In the following, $C(.)$ will denote a constant which depends only on its argument and may change from line to line. By our assumptions and Theorem 1.1, $(\Theta(a_n, .))_{n\geq 1}$ follows the large deviation principle in E_2 with speed (ε_n) and good rate function $\mathrm{I}_{a,b}$.

Step 1. $\limsup_{n\to\infty} \inf_{\Phi \in K_{a,b}} \|\Theta(a_n, .) - \Phi\|_{\mathcal{F}} = 0$ a.s.
It is enough to show that for every $\eta > 0$ and every $q > 1$,

$$\limsup_{p\to\infty} \inf_{\Phi\in K_{a,b}} \left\|\Theta\left(a_{n_p}, .\right) - \Phi\right\|_{\mathcal{F}} \leq \eta \text{ a.s.} \tag{3.4}$$

and for every $\eta > 0$, there exists $q > 1$ such that

$$\limsup_{p\to\infty} \max_{n_p+1\leq l\leq n_{p+1}} \left\|\Theta(a_l, .) - \Theta\left(a_{n_p}, .\right)\right\|_{\mathcal{F}} \leq 3\eta \text{ a.s.} \tag{3.5}$$

To prove (3.4), note that, by the large deviation principle for $(\Theta(a_n, .))_{n\geq 1}$ and by (3.1), there exists $\alpha \in \left(1, \inf_{(K_{a,b}+\eta B)^c} \mathrm{I}_{a,b}\right)$ such that, for every large enough p,

$$\begin{aligned}\mathbb{P}\left(\inf_{\Phi\in K_{a,b}} \left\|\Theta\left(a_{n_p}, .\right) - \Phi\right\|_{\mathcal{F}} \geq \eta\right) &= \mathbb{P}\left(\Theta\left(a_{n_p}, .\right) \in (K_{a,b} + \eta B)^c\right) \\ &\leq \exp\left(-\alpha\varepsilon_{n_p}\right) \\ &\leq C(\alpha, q)\, p^{-\alpha},\end{aligned}$$

which is the general term of a convergent series. Therefore the result holds by the Borel-Cantelli lemma.

To prove (3.5), note that, by the triangle inequality,

$$\max_{n_p+1\le l\le n_{p+1}} \left\|\Theta\left(a_l,\,.\,\right)-\Theta\left(a_{n_p},\,.\right)\right\|_{\mathcal{F}} \le \Delta_{1,p}\left(q\right)+\Delta_{2,p}\left(q\right)+\Delta_{3,p}\left(q\right),$$

where

$$\Delta_{1,p}\left(q\right):=\left(n_p v_{n_p}\right)^{-1}\max_{n_p+1\le l\le n_{p+1}}\left\|\sum_{k=1}^{a_l} f_{l,p}\left(V_{n_p}^{-1}\left(\xi_k\right)\right)-f\left(V_{n_p}^{-1}\left(\xi_k\right)\right)\right\|_{\mathcal{F}},$$

$$\Delta_{2,p}\left(q\right):=\left(n_p v_{n_p}\right)^{-1}\max_{n_p+1\le l\le n_{p+1}}\left\|\sum_{k=a_{n_p}+1}^{a_l} f\left(V_{n_p}^{-1}\left(\xi_k\right)\right)\right\|_{\mathcal{F}},$$

$$\Delta_{3,p}\left(q\right):=\left(1-\frac{n_p v_{n_p}}{n_{p+1}v_{n_{p+1}}}\right)\left\|\Theta\left(a_{n_p},\,.\right)\right\|_{\mathcal{F}}.$$

By (2.14), we may choose $\delta>0$, such that, for p large enough,

$$\mathbb{P}\left(\left\|\Theta\left(a_{n_p},\,.\right)\right\|_{\mathcal{F}_\delta}\ge\frac{\eta}{384}\right)\le\exp\left(-2\varepsilon_{n_p}\right)\le C\left(q\right)p^{-2},$$

and, by (F.9) and (F.10), we have, for all q close enough to 1 and p large enough,

$$\left\{f_{l,p}-f: f\in\mathcal{F}, n_p+1\le l\le n_{p+1}\right\}\subset\mathcal{F}_\delta.$$

Hence, for such a q and for p large enough,

$$\mathbb{P}\left(\Delta_{1,p}\left(q\right)\ge\eta\right)\le\mathbb{P}\left(\left(n_p v_{n_p}\right)^{-1}\max_{n_p+1\le l\le n_{p+1}}\left\|\sum_{k=1}^{a_l} f\left(V_{n_p}^{-1}\left(\xi_k\right)\right)\right\|_{\mathcal{F}_\delta}\ge\eta\right)$$

and, by the Lévy inequality for asymmetric random vectors (see e.g. Corollary 1.12 of Giné (1996)), we get that, for all q close enough to 1 and p large enough,

$$\begin{aligned}
&\mathbb{P}\left(\Delta_{1,p}\left(q\right)\ge\eta\right)\\
&\le 30\mathbb{P}\left(\left(n_p v_{n_p}\right)^{-1}\left\|\sum_{k=1}^{a_{n_{p+1}}} f\left(V_{n_p}^{-1}\left(\xi_k\right)\right)\right\|_{\mathcal{F}_\delta}\ge\frac{\eta}{192}\right)\\
&\le 30\mathbb{P}\left(\left\|\Theta\left(a_{n_p},\,.\right)\right\|_{\mathcal{F}_\delta}\ge\frac{\eta}{384}\right)\\
&\quad+30\mathbb{P}\left(\left(n_p v_{n_p}\right)^{-1}\left\|\sum_{k=a_{n_p}+1}^{a_{n_{p+1}}} f\left(V_{n_p}^{-1}\left(\xi_k\right)\right)\right\|_{\mathcal{F}_\delta}\ge\frac{\eta}{384}\right)\\
&\le 30C\left(q\right)p^{-2}+30\mathbb{P}\left(\Delta_{4,p}\left(q\right)\ge\frac{\eta}{384}\right), \qquad (3.6)
\end{aligned}$$

where

$$\Delta_{4,p}(q) = \left(n_p v_{n_p}\right)^{-1} \left\| \sum_{k=a_{n_p}+1}^{a_{n_{p+1}}} f\left(V_{n_p}^{-1}(\xi_k)\right) \right\|_{\mathcal{F}}$$
$$=_d \left\| \Theta\left(a_{4,n_p}, \,.\right) \right\|_{\mathcal{F}},$$

with

$$a_{4,n} = \lfloor a(q-1)n \rfloor \quad \text{for } n \notin \{n_p : p \geq 1\}$$
$$= a_{n_{p+1}} - a_{n_p} - 1 \text{ for } n = n_p.$$

Now, by (3.3), we can take q close enough to 1 such that,

$$\inf_{\left(\frac{\eta}{364}B\right)^c} \mathrm{I}_{a(q-1),b} > 3. \tag{3.7}$$

And, since $n^{-1}a_{4,n} \to a(q-1)$, the large deviation principle for $(\Theta(a_{4,n}, \,.))$ and (3.7) give

$$\mathbb{P}\left(\Delta_{4,p}(q) \geq \frac{\eta}{364}\right) \leq C(q)\, p^{-2} \text{ eventually.} \tag{3.8}$$

Moreover, by the Lévy inequality for asymmetric r.v. and by (3.8), we get

$$\mathbb{P}(\Delta_{2,p}(q) \geq \eta) \leq 30\mathbb{P}\left(\Delta_{4,p}(q) \geq \frac{\eta}{192}\right)$$
$$\leq C(q)\, p^{-2} \text{ eventually.} \tag{3.9}$$

Finally, use (3.2) to choose M such that

$$\inf_{(\eta MB)^c} \mathrm{I}_{a,b} > 3.$$

Then, since it is easily seen that

$$\frac{n_p v_{n_p}}{n_{p+1} v_{n_{p+1}}} \to 1$$

when $p \uparrow \infty$, we obtain by the large deviation principle for $(\Theta(a_n, \,.))$, that

$$\limsup_{p\to\infty} \varepsilon_{n_p} \log \mathbb{P}(\Delta_{3,p}(q) \geq \eta) \leq \limsup_{p\to\infty} \varepsilon_{n_p} \log \mathbb{P}\left(\Theta\left(a_{n_p}, \,.\right) \notin \eta MB\right)$$
$$\leq -3.$$

Hence, eventually,

$$\mathbb{P}(\Delta_{3,p}(q) \geq \eta) \leq C(q)\, p^{-2}. \tag{3.10}$$

Now, (3.5) is an easy consequence of (3.6), (3.8), (3.9), (3.10), and of the Borel-Cantelli Lemma.

Step 2. For every $\Phi \in K_{a,b}$, $\liminf_{n\to\infty} \|\Theta(a_n, .) - \Phi\|_{\mathcal{F}} = 0$ a.s.
By a classical diagonal argument, we may assume that

$$\mathrm{I}_{a,b}(\Phi) < 1. \tag{3.11}$$

Furthermore, it is enough to show that for every $\eta > 0$, there exists $q > 1$, such that

$$\liminf_{p\to\infty} \left\|\Theta\left(a_{n_{p+1}}, .\right) - \Phi\right\|_{\mathcal{F}} \leq 3\eta \text{ a.s.} \tag{3.12}$$

By the triangle inequality,

$$\liminf_{p\to\infty} \left\|\Theta\left(a_{n_{p+1}}, .\right) - \Phi\right\|_{\mathcal{F}} \leq \limsup_{p\to\infty} \Delta_{5,p}(q) + \liminf_{p\to\infty} \Delta_{6,p}(q, \Phi), \tag{3.13}$$

where

$$\Delta_{5,p}(q) := \left\|\Theta\left(a_{5,n_{p+1}}, .\right)\right\|_{\mathcal{F}},$$

with

$$\begin{aligned} a_{5,n} &= \lfloor aq^{-1}n \rfloor \quad \text{for } n \notin \{n_p : p \geq 1\} \\ &= a_{n_p} \text{ for } n = n_{p+1}, \end{aligned}$$

and

$$\Delta_{6,p}(q, \Phi) := \left\| \left(n_{p+1} v_{n_{p+1}}\right)^{-1} \sum_{k=a_{n_p}+1}^{a_{n_{p+1}}} f\left(V_{n_p}^{-1}(\xi_k)\right) - \Phi \right\|_{\mathcal{F}}.$$

Now, by (3.3), for q large enough,

$$\inf_{(\eta B)^c} \mathrm{I}_{aq^{-1},b} > 3. \tag{3.14}$$

And, since $n^{-1}a_{5,n} \to aq^{-1}$, the large deviation principle for $(\Theta(a_{5,n}, .))$ and (3.14) give,

$$\mathbb{P}\left(\Delta_{5,p}(q) \geq \eta\right) \leq C(q)\, p^{-2} \text{ eventually,} \tag{3.15}$$

and the Borel-Cantelli Lemma leads to

$$\limsup_{p\to\infty} \Delta_{5,p}(q) < \eta \text{ a.s.} \tag{3.16}$$

Furthermore, note that

$$\begin{aligned} &\mathbb{P}\left(\Delta_{6,p}\left(q\right) < 2\eta\right) \\ (3.17) \qquad &\geq \mathbb{P}\left(\left\|\Theta\left(a_{n_{p+1}}, .\right) - \Phi\right\|_{\mathcal{F}} < \eta\right) - \mathbb{P}\left(\Delta_{5,p}\left(q\right) \geq \eta\right). \end{aligned}$$

And, by the large deviation principle for $(\Theta(a_n, .))$ and (3.11),

$$\begin{aligned} \limsup_{p\to\infty} \varepsilon_{n_{p+1}} \log \mathbb{P}\left(\left\|\Theta\left(a_{n_{p+1}}, .\right) - \Phi\right\|_{\mathcal{F}} < \eta\right) &\geq -\inf_{\Phi+\eta B} \mathrm{I}_{a,b} \\ &\geq -\mathrm{I}_{a,b}\left(\Phi\right) \\ &> -1. \end{aligned}$$

Hence, we eventually have,

$$(3.18) \qquad \mathbb{P}\left(\left\|\Theta\left(a_{n_{p+1}}, .\right) - \Phi\right\|_{\mathcal{F}} < \eta\right) \geq C\left(q\right)\left(p+1\right)^{-1},$$

which is the general term of a divergent series.

Therefore, (3.15), (3.17) and (3.18), give, for q large enough,

$$\sum_{p=1}^{\infty} \mathbb{P}\left(\Delta_{6,p}\left(q\right) < 2\eta\right) = \infty,$$

and since $\left(\Delta_{6,p}\left(q\right)\right)_{p\geq 1}$ are independent, the second part of the Borel Cantelli Lemma leads to

$$\liminf_{p\to\infty} \Delta_{6,p}\left(q\right) < 2\eta \text{ a.s.},$$

which, combined with (3.16) and (3.13), give (3.12). This completes the proof. □

References

M. Arcones (1997), The law of the iterated logarithm for a triangular array of empirical processes, *Electronic Journal of Probability* **2**, 1–39.

P. Deheuvels and D. M. Mason (1990), Nonstandard functional laws of the iterated logarithm for tail empirical and quantile processes, *Annals of Probability* **18**, 1693–1722.

P. Deheuvels and D. M. Mason (1991), A tail empirical approach to some nonstandard laws of the iterated logarithm, *Journal of Theoretical Probability* **4**, 53–85.

P. Deheuvels and D. M. Mason (1994), Functional laws of the iterated logarithm for local empirical processes indexed by sets, *Annals of Probability* **22**, 1619–1661.

P. Deheuvels and D. M. Mason (1995), Nonstandard local empirical processes indexed by sets, *Journal of Statistical Planning and Inference* **45**, 91–112.

A. Dembo and T. Zajic (1997), Uniform moderate and large deviations for functional empirical processes, *Stochastic processes and their applications* **67**, 195–211.

A. Dembo and O. Zeitouni (1991), *Large Deviations Techniques and Applications*, Jones and Bartlett.

U. Einmahl and D. M. Mason (1997), Gaussian approximation of local empirical processes indexed by functions, *Probability Theory and Related Fields* **107**, 283–301.

U. Einmahl and D. M. Mason (1998), Strong Approximations to the Local Empirical Process, In: *in High Dimensional Probability*, E. Eberlein, M. Hahn, and M. Talagrand, eds., Birkhäuser, Progress in Probability, 43, 75–92.

E. Giné (1996), Lectures on some aspects of the bootstrap, *Ecole d'été de probabilité de Saint Flour*, Lecture Note in Math. 1665, Springer.

M. Ledoux (1996), On Talagrand's deviation inequalities for product measures, *ESAIM* **1**, 63–87.

M. Ledoux and M. Talagrand (1991), *Probability in Banach Spaces*, Springer-Verlag, Berlin.

G.L. O'Brien and V. Vervaat (1991), Capacities, large deviations and loglog laws, in: *Stable Processes and Related Topics*, Birkhäuser, Progress in Probability, 25, 43–83.

M. Talagrand (1996), New concentration inequalities in product spaces, *Invent. Math.* **126**, 505–563.

A. W. van der Vaart and J.A. Wellner (1996), *Weak Convergence and Empirical Processes with Applications to Statistics*, Springer-Verlag, Berlin.

L. Wu (1994), Large deviations, moderate deviations and L.I.L. for empirical processes, *Annals of Probability* **22**, 17–27.

Université Paris VI and CREST
56-58 Cours de Vincennes
75012 PARIS
e-mail: `mennet@ensae.fr`

PART 7

Sums of Independent Random Variables in High Dimensions

An Example Concerning Tightness of Sums of B-Valued Random Variables

Richard C. Bradley*

1 Introduction

Suppose B is a separable real Banach space.

In the development of limit theory for a given sequence $X_1, X_2, X_3, \ldots$ of (say) independent B-valued random variables, an important role is played by the behavior of the real-valued random variables $f(X_1), f(X_2), f(X_3), \ldots$, where f is an arbitrary real bounded linear functional on B. Such linear functionals have played a key role in the development (for B-valued random variables) of the central limit theorem (see e.g. [1] or [10, section 3]), the laws of the iterated logarithm (see e.g. [8]), the almost sure invariance principles (see e.g. [6, Theorem 3]), convergence of types (see e.g. [12, section 2.7]), and the Ito-Nisio theorem involving convergence of infinite sums (see e.g. [1, p. 105, Theorem 2.10]).

There has been an extensive development of central limit theory for weakly dependent sequences of B-valued random variables. Here too, linear functionals have played a key role. In this theory, the absolute regularity condition (defined below) seems to be a natural condition of weak dependence. See e.g. the almost sure invariance principle proved by Dehling and Philipp [6, Theorem 4], involving strictly stationary, absolutely regular sequences of B-valued random variables. In a different vein, Jim Kuelbs [13] posed the problem of to what extent the Ito-Nisio theorem can be extended to absolutely regular sequences of B-valued random variables, with the absolute regularity condition formulated in a manner suitable to nonstationary sequences. (This problem will not be addressed here.)

For a given sequence $X := (X_k, \ k \in \mathbb{Z})$ of B-valued random variables, define for each $n \geq 1$ the partial sum (a B-valued random variable)

$$S_n := X_1 + X_2 + \ldots + X_n.$$

If $(X_k, \ k \in \mathbb{Z})$ is a strictly stationary sequence of *real*-valued random variables such that $EX_0 = 0$, $EX_0^2 < \infty$, and $EX_0X_n \to 0$ as $n \to \infty$, then

*This research was partially supported by an NSF grant DMS–9703712.

by a result of Leonov [15] (see e.g. [11, Theorem 18.2.2]), either $ES_n^2 \to \infty$ as $n \to \infty$ or ES_n^2 is bounded. However, as the construction of Herrndorf [9] shows, even in the case where $ES_n^2 \to \infty$ as $n \to \infty$, the partial sums $(S_1, S_2, S_3, \dots)$ can still satisfy tightness, and can fail to be asymptotically normally distributed. For his example, Herrndorf verified the property of strong mixing (defined below), but in fact by a similar argument his example satisfies the (stronger) property of absolute regularity.

For strictly stationary sequences of B-valued random variables, the questions arise as to (1) what kinds of similar "bad behavior" might be possible for the partial sums, and (2) to what extent one can infer such "bad behavior" from the real bounded linear functionals of the partial sums. For strictly stationary, absolutely regular sequences of B-valued random variables, it was shown by the author [5] (see Theorem 1.4 below) that the property of tightness of the partial sums follows from tightness of the bounded linear functionals of those sums, and the same is true for "shift-tightness." Here we shall show with an example that this fails to hold if the assumption of absolute regularity is replaced by the (weaker) assumption of a trivial double tail σ-field.

We shall start by directly repeating, in an abridged form, some relevant material from the beginning of the paper [5].

Notation 1.1. Suppose B is a separable real Banach space, with norm $\|.\|$.

Of course B is a metric space with the usual metric $d(x, y) := \|x - y\|$. We shall let $\mathcal{B}$ denote the "standard σ–field" on B: the σ–field generated by the open balls (in that metric).

Throughout this paper, we shall be working on a probability space $(\Omega, \mathcal{F}, P)$. If $(X_i, i \in I)$ is a family of B–valued random variables (where I is a nonempty index set), then $\sigma(X_i, i \in I)$ denotes the σ–field ($\subset \mathcal{F}$) generated by these random variables — the smallest σ–field on Ω which contains all of the events $\{X_i \in A\}$, $i \in I$, $A \in \mathcal{B}$. Of course, whenever $B = \mathbb{R}$ (e.g. when "real–valued" random variables are being discussed), the usual norm $\|x\| = |x|$ (the absolute value) is used, and $\mathcal{B}$ is the usual Borel σ–field on $\mathbb{R}$.

Definition 1.2. Suppose $Y := (Y_1, Y_2, Y_3, \dots)$ is a sequence of random variables taking their values in a separable real Banach space B.

The sequence Y satisfies "tightness" (in B) if for every $\varepsilon \in (0, 1)$, there exists a compact set $C \subset B$ such that for all $n \geq 1$, $P(Y_n \in C) \geq 1 - \varepsilon$.

The sequence Y satisfies "shift–tightness" (in B) if there exist elements $b_1, b_2, b_3, \dots \in B$ such that the random sequence $(Y_1 - b_1,\ Y_2 - b_2,\ Y_3 - b_3,\ \dots)$ satisfies tightness (in B).

The sequence Y satisfies "complete dissipation" (in B) if for every $r > 0$, one has that

$$\kappa(r, n) := \sup_{b \in B} P(\|Y_n - b\| \leq r) \longrightarrow 0 \text{ as } n \to \infty.$$

This quantity $\kappa(r,n)$ is a well known "concentration function" for the random variable Y_n. For sums of independent random variables, it has been studied in numerous papers, including the works of Esseen [7] and Larin [14]. Yoshihara [20] studied the concentration function for sums from random sequences satisfying the "absolute regularity" condition (see Definition 1.3 below) under a certain assumption on the mixing rate.

The notion of "shift–tightness" has played a role in various papers, such as that of Schmidt [18].

Definition 1.3. Suppose $X := (X_k, k \in \mathbb{Z})$ is a strictly stationary sequence of random variables taking their values in a separable real Banach space B. For $-\infty \le J \le L \le \infty$, define the σ–field

$$\mathcal{F}_J^L := \sigma(X_k,\ J \le k \le L\ (k \in \mathbb{Z})).$$

For σ–fields $\mathcal{G}, \mathcal{H} \subset \mathcal{F}$, define the measures of dependence

$$\alpha(\mathcal{G},\mathcal{H}) := \sup_{G \in \mathcal{G}, H \in \mathcal{H}} |P(G \cap H) - P(G)P(H)|, \quad \text{and} \tag{1.1}$$

$$\beta(\mathcal{G},\mathcal{H}) := \sup \frac{1}{2} \sum_{i=1}^{I} \sum_{j=1}^{J} |P(G_i \cap H_j) - P(G_i)P(H_j)| \tag{1.2}$$

where this latter supremum is taken over all pairs of finite partitions $\{G_1, \dots, G_I\}$ and $\{H_1, \dots, H_J\}$ of Ω such that $G_i \in \mathcal{G}$ for all i and $H_j \in \mathcal{H}$ for all j.

For each $n = 1, 2, 3, \dots$, define for the given strictly stationary sequence X the dependence coefficients

$$\alpha(n) = \alpha(X,n) := \alpha(\mathcal{F}_{-\infty}^0, \mathcal{F}_n^\infty), \quad \text{and}$$
$$\beta(n) = \beta(X,n) := \beta(\mathcal{F}_{-\infty}^0, \mathcal{F}_n^\infty).$$

The given strictly stationary sequence X is "strongly mixing" [16] if $\alpha(n) \to 0$ as $n \to \infty$, and it is "absolutely regular" [19] if $\beta(n) \to 0$ as $n \to \infty$. From (1.1) and (1.2), one trivially has that absolute regularity implies strong mixing.

Theorem 1.4 ([5, Theorem 1.4]). *Suppose B is a separable real Banach space. Suppose $X := (X_k, k \in \mathbb{Z})$ is a strictly stationary, absolutely regular sequence of random variables taking their values in B. Then the following four statements hold:*

(1) The sequence $(S_1, S_2, S_3, \dots)$ of (B–valued) partial sums satisfies either shift–tightness or complete dissipation (in B).

(2) Suppose there exists a real bounded linear functional f on B such that the sequence $(f(S_n), n = 1, 2, 3, \dots)$ of (real–valued) random variables fails

to satisfy shift–tightness (in $\mathbb{R}$*). Then the sequence* $(S_1, S_2, S_3, \dots)$ *satisfies complete dissipation (in* B*).*

(3) Suppose that for every real bounded linear functional f *on* B*, the sequence* $(f(S_n), n = 1, 2, 3, \dots)$ *satisfies shift–tightness (in* $\mathbb{R}$*). Then the sequence* $(S_1, S_2, S_3, \dots)$ *satisfies shift–tightness (in* B*).*

(4) Suppose that for every real bounded linear functional f *on* B*, the sequence* $(f(S_n), n = 1, 2, 3, \dots)$ *satisfies tightness (in* $\mathbb{R}$*). Then the sequence* $(S_1, S_2, S_3, \dots)$ *satisfies tightness (in* B*), and there exists a strictly stationary, absolutely regular sequence* $Z := (Z_k, k \in \mathbb{Z})$ *of random variables taking their values in* B*, such that for all* $k \in \mathbb{Z}$*,* $X_k = Z_k - Z_{k-1}$ *a.s.*

Obviously statements (2) and (3) together imply statement (1).

In statement (4), in the case $B = \mathbb{R}^d$ for some positive integer d, the random sequence Z comes from a result of Klaus Schmidt [17, Theorem 11.8], and for general (separable real) Banach spaces B it came from an extension of Schmidt's result by the author [4]. (The references [4, 17] did not deal with the absolute regularity condition.)

For strictly stationary sequences of real–valued random variables, statement (1) was proved by the author [3, Theorem 1] under the strong mixing condition (which is weaker than absolute regularity). There is no obvious way to extend it to the case of B–valued random variables without strengthening the dependence assumption to absolute regularity (as in the theorem here). It is an open question whether Theorem 1.4 still holds under strong mixing.

This completes the repetition of relevant material from [5]. Now let us turn our attention to the example, described in Theorem 1.5 below. First some more terminology is needed.

Recall our probability space $(\Omega, \mathcal{F}, P)$. A σ–field $\mathcal{A} \subset \mathcal{F}$ is said to be "trivial" if $P(A) = 0$ or 1 for every $A \in \mathcal{A}$.

If $X := (X_k, k \in \mathbb{Z})$ is a strictly stationary sequence of random variables taking their values in a separable real Banach space, the "double tail σ–field of X" is the σ–field $\mathcal{D}_X$ defined by

$$\mathcal{D}_X := \bigcap_{n=1}^{\infty} \left(\mathcal{F}_{-\infty}^{-n} \vee \mathcal{F}_n^{\infty} \right).$$

If X is absolutely regular, then its double tail σ–field $\mathcal{D}_X$ is trivial. This is well known. (See e.g. [2, Theorem 2.2 (the case $p = 1$)].)

Let c_0 denote the (separable real) Banach space of sequences $x := (x_1, x_2, x_3, \dots)$ of real numbers such that $\lim_{n \to \infty} x_n = 0$, with the usual sup norm $\|x\| := \sup_{n \geq 1} |x_n|$.

Here is our main result:

Theorem 1.5. *There exists a strictly stationary sequence $X := (X_k, k \in \mathbb{Z})$ of random variables taking their values in the (separable real) Banach space c_0, such that the following three statements hold:*

(1) The double tail σ–field $\mathcal{D}_X$ is trivial.

(2) For each $n \geq 1$, $\|S_n\| \leq 1$ a.s.

(3) For every sequence $a_1, a_2, a_3, \ldots$ of elements of c_0, and every compact set $C \subset c_0$, one has that $P(S_n - a_n \in C) \to 0$ as $n \to \infty$.

Remark 1.6. Theorem 1.5 will be proved in section 2. Several observations on this theorem will be made here.

The proof of property (3) (see eq. (2.23)) will actually yield much stronger information: For some constant $\lambda \in (0,1)$ (namely $\lambda := (31/32)^{1/8}$), one has that $P(S_n - a_n \in C) = O(\lambda^n)$ as $n \to \infty$, and hence also by a simple argument, $P(S_m - a_m \in C$ for some $m \geq n) = O(\lambda^n)$ as $n \to \infty$, and in particular,

$$P(S_n - a_n \in C \text{ for infinitely many } n) = 0.$$

The multiplicative constants implied in the two "O's" do not depend on the elements $a_1, a_2, a_3, \ldots$ of c_0, but (unavoidably) they do depend on the compact set C.

The partial sums are confined to the unit ball in c_0 (see property (2)); but in a certain sense, as $n \to \infty$, those partial sums "spill sideways into more and more dimensions," escaping from every compact set. In particular, those partial sums do not satisfy tightness, or even shift-tightness.

Of course, for every real bounded linear functional f on c_0, the sequence of real–valued random variables $f(S_n), n = 1, 2, 3, \ldots$ is bounded (a.s.) and therefore satisfies tightness. (One has that $|f(S_n)| \leq A \cdot \|S_n\| \leq A$ a.s., where A is the norm of f.) Consequently, in Theorem 1.4, statements (1), (3), and (4) become false if the assumption of absolute regularity is omitted or replaced by the (weaker) assumption of a trivial double tail σ–field.

The example constructed for Theorem 1.5 does not appear to be strongly mixing. It leaves open the question (alluded to above) whether Theorem 1.4 still holds under strong mixing.

2 Proof of Theorem 1.5

Let $\mathbb{N}$ denote the set of all positive integers. Let $V_k^{(m)}, m \in \mathbb{N}, k \in \mathbb{Z}$, be independent random variables that take only the values 0 and 1, such that for each $m \in \mathbb{N}$ and each $k \in \mathbb{Z}$,

$$(2.1) \qquad P(V_k^{(m)} = 0) = 1 - 1/(8m) \quad \text{and} \quad P(V_k^{(m)} = 1) = 1/(8m).$$

For each $m \in \mathbb{N}$ and each $k \in \mathbb{Z}$, define the event

(2.2)
$$D_k^{(m)} := \{V_k^{(m)} = 1 \quad \text{and} \quad V_{k+1}^{(m)} = V_{k+2}^{(m)} = V_{k+3}^{(m)} = \ldots = V_{k+4m-1}^{(m)} = 0\}.$$

The following fact is a trivial but useful consequence of (2.2):

Lemma 2.1. *If $m \in \mathbb{N}$ and $i, j \in \mathbb{Z}$ are such that $1 \le |i-j| \le 4m-1$, then the event $D_i^{(m)} \cap D_j^{(m)}$ is empty.*

For each $m \in \mathbb{N}$ and each $\omega \in \Omega$, define the set

$$Q^{(m)}(\omega) := \{k \in \mathbb{Z} : \ \omega \in D_k^{(m)}\}. \tag{2.3}$$

By Lemma 2.1, the elements of $Q^{(m)}(\omega)$ are separated from each other by at least $4m$.

Let the indicator function of a given event D be denoted $I(D)$.

For each $m \in \mathbb{N}$ and each $k \in \mathbb{Z}$, define the random variable $X_k^{(m)}$ as follows: For each $\omega \in \Omega$,

$$X_k^{(m)}(\omega) := \frac{1}{8m}\left[\sum_{j=0}^{2m-1} I(D_{k-j}^{(m)})(\omega) - \sum_{j=2m}^{4m-1} I(D_{k-j}^{(m)})(\omega)\right]. \tag{2.4}$$

For a given $m \in \mathbb{N}$, $k \in \mathbb{Z}$, and $\omega \in \Omega$, by Lemma 2.1, at most one term $I(D_{k-j}^{(m)})(\omega)$, $0 \le j \le 4m-1$, can be 1 (instead of 0); and hence by (2.3) and (2.4),

(2.5)
$$X_k^{(m)}(\omega) = \begin{cases} 1/(8m) & \text{if } k-j \in Q^{(m)}(\omega) \text{ for some } j \in \{0, 1, \ldots, 2m-1\}, \\ -1/(8m) & \text{if } k-j \in Q^{(m)}(\omega) \text{ for some } \\ & \qquad j \in \{2m, 2m+1, \ldots, 4m-1\}, \\ 0 & \text{otherwise.} \end{cases}$$

By (2.5), for each $k \in \mathbb{Z}$ and each $\omega \in \Omega$, $X_k^{(m)}(\omega) \to 0$ as $m \to \infty$. For each $k \in \mathbb{Z}$, define the c_0–valued random variable X_k as follows: For each $\omega \in \Omega$,

$$X_k(\omega) := (X_k^{(1)}(\omega), X_k^{(2)}(\omega), X_k^{(3)}(\omega), \ldots). \tag{2.6}$$

Our remaining task is to verify that this random sequence $X := (X_k, k \in \mathbb{Z})$ is strictly stationary and satisfies properties (1), (2), and (3) in Theorem 1.5.

Proof that X is strictly stationary. Refer to (2.1). Note that for each $m \in \mathbb{N}$, $V^{(m)} := (V_k^{(m)}, k \in \mathbb{Z})$ is a sequence of independent, identically distributed random variables; and these sequences $V^{(m)}, m \in \mathbb{N}$, are independent of each other. From (2.2), it follows that for each $m \in \mathbb{N}$, the random sequence $\eta^{(m)} := (I(D_k^{(m)}), k \in \mathbb{Z})$ is strictly stationary; and these sequences $\eta^{(m)}, m \in \mathbb{N}$, are independent of each other. From (2.4), it follows that for each $m \in \mathbb{N}$, the random sequence $X^{(m)} := (X_k^{(m)}, k \in \mathbb{Z})$ is strictly stationary; and these sequences $X^{(m)}, m \in \mathbb{N}$, are independent of each other. Hence by (2.6) and an elementary measure–theoretic argument, the random sequence X $(= (X_k, k \in \mathbb{Z}))$ is strictly stationary.

Proof of property (1). For each positive integer L, define the σ–field

$$\mathcal{G}_L := \sigma(V_k^{(m)}, (m,k) \in \mathbb{N} \times \mathbb{Z} \text{ such that } \max\{m, |k|\} = L). \tag{2.7}$$

Since the random variables $V_k^{(m)}$ are independent, these σ–fields $\mathcal{G}_1, \mathcal{G}_2, \mathcal{G}_3, \ldots$ are independent. Hence their "tail σ–field" $\mathcal{T} := \bigcap_{L=1}^{\infty}(\bigvee_{\ell=L}^{\infty} \mathcal{G}_\ell)$ is trivial. This is really an application of the Kolmogorov zero–one law. To prove property (1), it suffices to prove that $\mathcal{D}_X \subset \mathcal{T}$.

Let $L \geq 1$ be arbitrary but fixed. It suffices to prove that $\mathcal{D}_X \subset \bigvee_{\ell=L}^{\infty} \mathcal{G}_\ell$.

For each $m \in \mathbb{N}$ and each $k \in \mathbb{Z}$, $\sigma(X_k^{(m)}) \subset \sigma(V_j^{(m)},\ k - 4m + 1 \leq j \leq k + 4m - 1)$ by (2.2) and (2.4). Hence by (2.6),

$$\begin{aligned} \mathcal{D}_X &\subset \sigma(X_k, |k| \geq 5L) = \sigma(X_k^{(m)}, m \geq 1, |k| \geq 5L) \\ &\subset \sigma(V_k^{(m)}, m \geq 1, |k| \geq 5L - 4m). \end{aligned} \tag{2.8}$$

Now if $m \geq 1$ and $|k| \geq 5L - 4m$, then $\max\{m, |k|\} \geq L$. (Either $m \geq L$, or else $1 \leq m < L$ and $|k| \geq 5L - 4m \geq L$.) Hence by (2.7) and (2.8), $\mathcal{D}_X \subset \bigvee_{\ell=L}^{\infty} \mathcal{G}_\ell$, which is the desired equation. This completes the proof of property (1).

For each $m \in \mathbb{N}$ and each $n \in \mathbb{N}$, define the partial sum

$$S_n^{(m)} := X_1^{(m)} + X_2^{(m)} + \cdots + X_n^{(m)}.$$

Then by (2.6), for each $n \geq 1$,

$$S_n = (S_n^{(1)}, S_n^{(2)}, S_n^{(3)}, \ldots). \tag{2.9}$$

In order to prove properties (2) and (3) (in Theorem 1.5), we first need to prove the following lemma:

Lemma 2.2. *Suppose $m \geq 1$. Then for each $n \geq 1$, one has that*

$$|S_n^{(m)}| \leq 1 \quad a.s.; \tag{2.10}$$

and for each $n \geq 8m$, one has that

$$P(S_n^{(m)} \geq 1/8) \geq 1/32 \quad \text{and} \quad P(S_n^{(m)} \leq -1/8) \geq 1/32. \tag{2.11}$$

Proof. Let m be an arbitrary fixed positive integer.

By (2.5), for every $n \in \{1, 2, \ldots, 8m\}$ and every $\omega \in \Omega$, one has that $|S_n^{(m)}(\omega)| \leq \sum_{k=1}^{n} |X_k^{(m)}(\omega)| \leq 1$.

Now let n be an arbitrary fixed integer such that

$$n \geq 8m. \tag{2.12}$$

To complete the proof of Lemma 2.2 (both parts), it suffices to prove for this n that (2.10) and (2.11) hold.

First some comments on notation will be given. For pairs of integers $i \leq j$, let the set $\{k \in \mathbb{Z} : i \leq k \leq j\}$ be denoted by $\mathbb{Z}(i, j)$. An "empty sum" $\sum_{i \in \emptyset} c_i$ is interpreted as 0. Since m is fixed, the set $Q^{(m)}(\omega)$ in (2.3) will (for a given $\omega \in \Omega$) be written simply as $Q(\omega)$, for typographical convenience.

Suppose $\omega \in \Omega$. Then (for our given m and n) one has that

$$S_n^{(m)}(\omega) = \sum_{h \in Q(\omega) \cap \mathbb{Z}(2-4m, n)} \quad \sum_{k \in \mathbb{Z}(h, h+4m-1) \cap \mathbb{Z}(1,n)} X_k^{(m)}(\omega). \tag{2.13}$$

Let us verify this. Keep in mind that (obviously) the right side of (2.13) does not count any k outside of the set $\mathbb{Z}(1, n)$. Since elements of $Q(\omega)$ are separated by at least $4m$ (by Lemma 2.1), the sets $\mathbb{Z}(h, h + 4m - 1), h \in Q(\omega)$, are (pairwise) disjoint. Hence, no element $k (\in \mathbb{Z}(1, n))$ is counted more than once in the right side of (2.13). Also, by (2.5), for each $k \in \mathbb{Z}(1, n)$ such that $X_k^{(m)}(\omega) \neq 0$, there exists (a unique) $h \in Q(\omega) \cap \mathbb{Z}(2-4m, n)$ such that $k \in \mathbb{Z}(h, h + 4m - 1)$. Hence the right side of (2.13) counts (exactly once) every $k \in \mathbb{Z}(1, n)$ such that $X_k^{(m)}(\omega) \neq 0$. Hence for any "uncounted" elements $k \in \mathbb{Z}(1, n)$ one has that $X_k^{(m)}(\omega) = 0$. Hence the right side of (2.13) equals $\sum_{k=1}^{n} X_k^{(m)}(\omega)$, or $S_n^{(m)}(\omega)$. This completes the verification of (2.13).

Now let us simplify (2.13) a little. For any $h \in Q(\omega) \cap \mathbb{Z}(1, n - 4m + 1)$ (if such an h exists), one has by (2.5) that

$$\begin{aligned} \textstyle\sum_{k \in \mathbb{Z}(h, h+4m-1) \cap \mathbb{Z}(1,n)} X_k^{(m)}(\omega) &= \textstyle\sum_{k=h}^{h+4m-1} X_k^{(m)}(\omega) \\ &= \textstyle\sum_{k=h}^{h+2m-1} \frac{1}{8m} + \sum_{k=h+2m}^{h+4m-1} \left(-\frac{1}{8m}\right) = 0. \end{aligned}$$

Hence by (2.13),

$$S_n^{(m)}(\omega) = \zeta_1(\omega) + \zeta_2(\omega), \tag{2.14}$$

where

$$\zeta_1(\omega) := \sum_{h \in Q(\omega) \cap \mathbb{Z}(2-4m, 0)} \quad \sum_{k \in \mathbb{Z}(h, h+4m-1) \cap \mathbb{Z}(1,n)} X_k^{(m)}(\omega) \tag{2.15}$$

and

$$(2.16) \qquad \zeta_2(\omega) := \sum_{h\in Q(\omega)\cap\mathbb{Z}(n-4m+2,n)} \quad \sum_{k\in\mathbb{Z}(h,h+4m-1)\cap\mathbb{Z}(1,n)} X_k^{(m)}(\omega).$$

Since the elements of $Q(\omega)$ are separated by at least $4m$ (by Lemma 2.1), there can be at most one element $h \in Q(\omega) \cap \mathbb{Z}(2-4m, 0)$. Hence at most $4m$ elements k are counted in the right side of (2.15). Hence by (2.15) and (2.5), $|\zeta_1(\omega)| \le 4m \cdot 1/(8m) = 1/2$. By an analogous argument using (2.16), $|\zeta_2(\omega)| \le 1/2$. Hence $|S_n^{(m)}(\omega)| \le 1$ by (2.14). Since $\omega \in \Omega$ was arbitrary, the proof of (2.10) is complete.

Our task now is to prove (2.11) (for our given m and n such that (2.12) holds).

For pairs of integers $i \le j$, define the event $F(i,j) := \bigcup_{k=i}^{j} D_k^{(m)}$. (Since m is fixed, the dependence of $F(i,j)$ on m can be suppressed.) By (2.1) and (2.2), for each $k \in \mathbb{Z}$,

$$\begin{aligned}(8m)^{-1} &\ge P(D_k^{(m)}) = (8m)^{-1}[1-1/(8m)]^{4m-1} \\ &\ge (8m)^{-1}[1-(4m-1)/(8m)] \ge (8m)^{-1}[1-1/2] = (16m)^{-1}.\end{aligned}$$

Hence

$$(2.17) \qquad P(F(i,j)) \le (j-i+1)\cdot(8m)^{-1} \quad \text{if} \quad i \le j;$$

and also by Lemma 2.1,

$$(2.18) \qquad P(F(i,j)) \ge (j-i+1)\cdot(16m)^{-1} \quad \text{if} \quad i \le j \le i+4m-1.$$

Our next task is to show that

$$(2.19) \qquad \begin{aligned}&F(1-3m, 1-2m) \subset \{\zeta_1 \le -1/8\};\\ &F(n-2m+1, n-m+1) \subset \{\zeta_2 \ge 1/8\};\\ &[F(2-4m,0)]^c \subset \{\zeta_1 = 0\}; \text{ and } [F(n-4m+2,n)]^c \subset \{\zeta_2 = 0\}.\end{aligned}$$

To prove the first equation in (2.19), suppose $\omega \in F(1-3m, 1-2m)$. Our task is to show that $\zeta_1(\omega) \le -1/8$. Let $h \in \mathbb{Z}(1-3m, 1-2m)$ be such that $\omega \in D_h^{(m)}$. Then by (2.3) and Lemma 2.1, $Q(\omega) \cap \mathbb{Z}(2-4m, 0) = \{h\}$. Also, $\mathbb{Z}(h, h+4m-1) \cap \mathbb{Z}(1,n) = \mathbb{Z}(1, h+4m-1)$ by (2.12). Hence $\zeta_1(\omega) = \sum_{k=1}^{h+4m-1} X_k^{(m)}(\omega)$ by (2.15). By (2.5), $X_k^{(m)}(\omega) = -(8m)^{-1}$ for each $k \in \mathbb{Z}(h+2m, h+4m-1)$, and in particular, for each $k \in \mathbb{Z}(1, h+4m-1)$. Hence $\zeta_1(\omega) = \sum_{k=1}^{h+4m-1}(-(8m)^{-1})$. Now $h+4m-1 \ge m$. Hence $\zeta_1(\omega) \le -1/8$, which is the desired inequality. Thus the first equation in (2.19) holds.

The second equation in (2.19) holds by a similar argument: Suppose $\omega \in F(n-2m+1, n-m+1)$. Then $\omega \in D_h^{(m)}$ for some $h \in \mathbb{Z}(n-2m+1, n-m+1)$; and $Q(\omega) \cap \mathbb{Z}(n-4m+2, n) = \{h\}$. Also, $\mathbb{Z}(h, h+4m-1) \cap \mathbb{Z}(1, n) = \mathbb{Z}(h, n)$ by (2.12), and $\zeta_2(\omega) = \sum_{k=h}^{n} X_k^{(m)}(\omega)$ by (2.16). By (2.5), $X_k^{(m)}(\omega) = 1/(8m)$ for each $k \in \mathbb{Z}(h, h+2m-1)$, and in particular, for each $k \in \mathbb{Z}(h, n)$. Hence $\zeta_2(\omega) = \sum_{k=h}^{n} 1/(8m) \geq \sum_{k=n-m+1}^{n} 1/(8m) = 1/8$, which gives the second equation in (2.19).

The third equation in (2.19) holds by (2.15) and (2.3), since the set $Q(\omega) \cap \mathbb{Z}(2-4m, 0)$ is empty for any $\omega \in [F(2-4m, 0)]^c$. Similarly the fourth equation in (2.19) holds by (2.16) and (2.3).

Now we are ready to verify the two inequalities in (2.11). By (2.2), the events $F(1-3m, 1-2m)$ and $[F(n-4m+2, n)]^c$ are respectively members of the σ–fields $\sigma(V_k^{(m)}, k \leq 2m)$ and $\sigma(V_k^{(m)}, k \geq n-4m+2)$. By (2.12), those two σ–fields are independent (since the random variables $V_k^{(m)}, k \in \mathbb{Z}$, are independent). Hence by (2.14), (2.19) (first and fourth equations), (2.18), and (2.17), one has that

(2.20)
$$\begin{aligned} P(S_n^{(m)} \leq -1/8) &\geq P(\zeta_1 \leq -1/8 \text{ and } \zeta_2 = 0) \\ &\geq P(F(1-3m, 1-2m) \cap [F(n-4m+2, n)]^c) \\ &= P(F(1-3m, 1-2m)) \cdot P([F(n-4m+2, n)]^c) \\ &\geq [m/(16m)] \cdot [1 - 4m/(8m)] = 1/32. \end{aligned}$$

Similarly, the events $[F(2-4m, 0)]^c$ and $F(n-2m+1, n-m+1)$ are members of the independent σ–fields $\sigma(V_k^{(m)}, k \leq 4m-1)$ and $\sigma(V_k^{(m)}, k \geq n-2m+1)$ (see (2.2) and (2.12)), and

(2.21)
$$\begin{aligned} P(S_n^{(m)} \geq 1/8) &\geq P(\zeta_1 = 0 \text{ and } \zeta_2 \geq 1/8) \\ &\geq P([F(2-4m, 0)]^c \cap F(n-2m+1, n-m+1)) \\ &\geq [1 - 4m/(8m)] \cdot m/(16m) = 1/32. \end{aligned}$$

From (2.20) and (2.21), one has both inequalities in (2.11). This completes the proof of Lemma 2.2.

We return to the main argument for Theorem 1.5.

Proof of property (2). Apply (2.9) and (2.10) (in Lemma 2.2).

Proof of property (3). Suppose $a_1, a_2, a_3, \dots$ are each an element of c_0, and C is a compact subset of c_0. Our task is to verify that $P(S_n - a_n \in C) \to 0$ as $n \to \infty$.

Since C is compact, there exists a finite subset $C_1 \subset C$ such that for every $x \in C$, there exists $y \in C_1$ such that $\|x - y\| \leq 1/20$. Let M be a positive integer such that for each element $y := (y_1, y_2, y_3, \dots) \in C_1$, one has that

$|y_m| \le 1/20$ for all $m \ge M$. Then for every $x := (x_1, x_2, x_3, \dots) \in C$, one has that $|x_m| \le 1/10$ for all $m \ge M$.

For each $n \ge 1$, represent the element $a_n := (a_n^{(1)}, a_n^{(2)}, a_n^{(3)}, \dots)$.

Suppose m and n are positive integers such that $m \ge M$ and $n \ge 8m$. By (2.11) in Lemma 2.2, one has that if $a_n^{(m)} \ge 0$ then

$$P(|S_n^{(m)} - a_n^{(m)}| > 1/10) \ge P(S_n^{(m)} \le -1/8) \ge 1/32,$$

and if instead $a_n^{(m)} < 0$ then

$$P(|S_n^{(m)} - a_n^{(m)}| > 1/10) \ge P(S_n^{(m)} \ge 1/8) \ge 1/32.$$

Thus for all (m, n) such that $m \ge M$ and $n \ge 8m$,

$$P(|S_n^{(m)} - a_n^{(m)}| > 1/10) \ge 1/32. \tag{2.22}$$

For a given $n \ge 1$, the random variables $S_n^{(m)}, m \in \mathbb{N}$, are independent (since they are respectively measurable with respect to the independent σ–fields $\sigma(V_k^{(m)}, k \in \mathbb{Z}),\ m \in \mathbb{N}$). Hence for each $n \ge 8M$, by (2.22), one has that

(2.23)

$$\begin{aligned} P(S_n - a_n \in C) &\le P(|S_n^{(m)} - a_n^{(m)}| \le 1/10 \quad \text{for all } m \ge M) \\ &\le P(|S_n^{(m)} - a_n^{(m)}| \le 1/10 \quad \text{for all } m = M, M+1, \dots, [n/8]) \\ &= \textstyle\prod_{m=M}^{[n/8]} P(|S_n^{(m)} - a_n^{(m)}| \le 1/10) \\ &\le \textstyle\prod_{m=M}^{[n/8]} (31/32) = (31/32)^{[n/8]-M+1}. \end{aligned}$$

Hence $P(S_n - a_n \in C) \to 0$ as $n \to \infty$. This completes the proof of property (3), and of Theorem 1.5.

Acknowledgement. The author thanks the numerous people at the Second International Conference on High-Dimensional Probability who made various comments and suggestions relevant to this paper.

References

[1] A. Araujo and E. Giné, *The Central Limit Theorem for Real and Banach Valued Random Variables,* (Wiley, New York, 1980).

[2] H. Berbee, Periodicity and absolute regularity, *Israel J. Math.* **55** (1986), 289–304.

[3] R.C. Bradley, On the dissipation of partial sums from a stationary strongly mixing sequence, *Stochastic Process. Appl.* **54** (1994), 281–290.

[4] R.C. Bradley, A "coboundary" theorem for sums of random variables taking their values in a Banach space, *Pacific J. Math.* **178** (1997), 201–224.

[5] R.C. Bradley, On tightness of partial sums from strictly stationary, absolutely regular sequences of B-valued random variables, *J. Math. Anal. Appl.* **240** (1999), 128–144.

[6] H. Dehling and W. Philipp, Almost sure invariance principles for weakly dependent vector-valued random variables, *Ann. Probab.* **10** (1982), 689–701.

[7] C.–G. Esseen, On the concentration function of a sum of independent random variables, *Z. Wahrsch. verw. Gebiete* **9** (1968), 290–308.

[8] V. Goodman, J. Kuelbs, and J. Zinn, Some results on the LIL in Banach space with applications to weighted empirical processes, *Ann. Probab.* **9** (1981), 713-752.

[9] N. Herrndorf, Stationary strongly mixing sequences not satisfying the central limit theorem, *Ann. Probab.* **11** (1983), 809–813.

[10] J. Hoffmann-Jørgensen and G. Pisier, The law of large numbers and the central limit theorem in Banach spaces, *Ann. Probab.* **4** (1976), 587–599.

[11] I.A. Ibragimov and Yu.V. Linnik, *Independent and Stationary Sequences of Random Variables,* (Wolters–Noordhoff, Groningen, 1971).

[12] Z.J. Jurek and J.D. Mason, *Operator-Limit Distributions in Probability Theory,* (Wiley, New York, 1993).

[13] J. Kuelbs, private communication, (1999).

[14] Yu.V. Larin, On concentration of distributions of sums of independent random vectors on bounded sets, *Theor. Probab. Appl.* **38** (1993), 743–751.

[15] V.P. Leonov, On the dispersion of time averages of a stationary random process, *Theor. Probab. Appl.* **6** (1961), 93–101.

[16] M. Rosenblatt, A central limit theorem and a strong mixing condition, *Proc. Nat. Acad. Sci. USA* **42** (1956), 43–47.

[17] K. Schmidt, *Cocycles on Ergodic Transformation Groups,* (Macmillan India, Delhi, 1977).

[18] K. Schmidt, On recurrence, *Z. Wahrsch. verw. Gebiete* **68** (1984), 75–95.

[19] V.A. Volkonskii and Yu.A. Rozanov, Some limit theorems for random functions I., *Theory Probab. Appl.* **4** (1959), 178–197.

[20] K. Yoshihara, Some estimates of the concentration function of sums of absolutely regular sequences, *Yokohama Math. J.* **27** (1979), 57–61. (Correction: *Yokohama Math. J.* **28** (1980), 169.)

Department of Mathematics
Indiana University
Bloomington, IN 47405
e-mail: bradleyr@indiana.edu

Images and Level Sets of Additive Random Walks*

Davar Khoshnevisan and Yimin Xiao

1 Introduction

We define $\mathbb{N}_0 = \{0, 1, \dots\}$ and say that a stochastic process $X = \{X(n); n \in \mathbb{N}_0^N\}$ is an N-parameter, $\mathbb{Z}^d$-valued, *additive random walk*, if there are N independent random walks $X_1, \dots, X_N$ on $\mathbb{Z}^d$, such that

$$X(n) = X_1(n_1) + \cdots + X_N(n_N), \qquad n \in \mathbb{N}_0^N.$$

Here, n_j denotes the jth coordinate of $n \in \mathbb{N}_0^N$ and we are following the standard convention of starting our (ordinary) random walks at the origin. That is, $X_j(0) = 0$ for all $j = 1, \dots, N$. From now on, N will always denote the temporal dimension and d the spatial one, in accordance with the conventional language of stochastic processes.

The continuous time analogue of such random walks plays an important rôle in the analysis of Lévy sheets and, in particular, the Brownian sheet; cf. Ref.'s [4, 5, 6, 7, 12, 13, 14, 17, 18] to cite only some of the references. These additive random walks also occur naturally in the theory of intersections of Lévy processes; cf. [9, 10, 16] and in the applications of probability theory in constructing exceptional sets and in harmonic analysis; cf. [11] and its extensive list of references.

In this article, we study the distributional properties of the level sets of X by analyzing the 'size' of the random image $X(E)$ where $E \subset \mathbb{N}_0^N$ is any nonrandom set. We will also show how such general results characterize 'recurrence' and 'transience' of X. As such, the results of the present paper are closely related to the companion paper [15]. In order to present the main result of this paper, we define the function $\Phi : \mathbb{N}_0^N \times \mathbb{N}_0^N \mapsto [0, 1]$ as

$$\Phi(n, m) = \mathbb{P}\{X(n) = X(m)\}, \qquad n, m \in \mathbb{N}_0^N, \tag{1.1}$$

and refer to this Φ as the *gauge function* of the additive random walk X. For all probability measures μ on $\mathbb{N}_0^N$, we can define the Φ-*energy* of μ as

$$\mathcal{E}_\Phi(\mu) = \iint \Phi(x, y)\, \mu(dx)\, \mu(dy). \tag{1.2}$$

---

*This research was partially supported by an NSF grant and a grant from NATO.

Then, the Φ-*capacity* $\mathcal{C}_\Phi(E)$ of any $E \subset \mathbb{N}_0^N$ can be defined by the principle of minimum energy:

$$\mathcal{C}_\Phi(E) = \left[\inf \left\{ \mathcal{E}_\Phi(\mu) : \ \mu \in \mathcal{P}(E) \text{ and } \mathcal{E}_\Phi(\mu) > 0 \right\} \right]^{-1}, \tag{1.3}$$

where $\mathcal{P}(E)$ denotes the collection of all probability measures on E, $\inf \varnothing = +\infty$ and $1/\infty = 0$.

The first main result of this article is the following theorem that estimates the size of the image of E under the random map X.

Theorem 1.1. *Let X be an N-parameter, d-dimensional additive random walk with gauge function Φ. For all $E \subset \mathbb{N}_0^N$,*

$$\mathcal{C}_\Phi(E) \leqslant \mathbb{E}\{\#X(E)\} \leqslant 16^N \mathcal{C}_\Phi(E), \tag{1.4}$$

where $\#X(E)$ denotes the cardinality of $X(E)$.

Remark 1.1. When $N = 1$ and X is an ordinary random walk on $\mathbb{Z}^d$, constant 16 in (1.4) can be replaced by 2. cf. Remark 4.1 in §4. □

Next, we illustrate the power of such a general estimate by studying the level sets of a large class of additive random walks.

We say that our additive random walk X is *symmetric*, if for each $n \in \mathbb{N}_0^N$, $-X(n)$ has the distribution as $X(n)$. Clearly the symmetry of the additive random walk X implies that for each $j = 1, \ldots, N$, the classical random walk X_j is also symmetric.

Note that in case X is symmetric, $\Phi(n, m)$ is a function of $n - m \in \mathbb{Z}^N$. That is, we can define $\Psi : \mathbb{Z}^N \mapsto [0, 1]$ in a unique and well-defined manner by

$$\Psi(n - m) = \Phi(n, m), \tag{1.5}$$

where $n, m \in \mathbb{N}_0^N$. Let $\overline{n}$ denote the point in $\mathbb{N}_0^N$ whose jth coordinate is $|n_j|$. Then, we point out that for all $n, m \in \mathbb{N}_0^N$,

$$\Phi(n, m) = \Psi(n - m) = \mathbb{P}\{X(\overline{n - m}) = 0\}.$$

We also write $\mathcal{C}_\Psi(E)$ and $\mathcal{E}_\Psi(\mu)$ for $\mathcal{C}_\Phi(E)$ and $\mathcal{E}_\Phi(\mu)$, in the symmetric case, and also refer to Ψ as the (symmetrized) *gauge function* of X.

In order to describe a natural partial order on $\mathbb{Z}^N$, we define the partial order '$\preccurlyeq$' on $\mathbb{Z}^N$ by: for all $m, n \in \mathbb{Z}^N$, $m \preccurlyeq n$ if and only if $m_j \leqslant n_j$ for all $j = 1, \ldots, N$. This partial order induces a minimum (maximum, resp.) operator $\curlywedge$ ($\curlyvee$, resp.): for all $m, n \in \mathbb{Z}^N$, $m \curlywedge n$ ($m \curlyvee n$, resp.) denotes the point in $\mathbb{Z}^N$ whose jth coordinate is $\min\{m_j, n_j\}$ ($\max\{n_j, m_j\}$, resp.) for all $j = 1, \ldots, N$.

We say that a $\mathbb{Z}^d$–valued random variable Y is *κ-weakly unimodal* for a positive constant $\kappa \geqslant 1$, when

$$\sup_{a\in\mathbb{Z}^d} \mathbb{P}\{Y = a\} \leqslant \kappa \mathbb{P}\{Y = 0\}.$$

Throughout the remainder of the Introduction, we will assume the existence of a fixed constant $\kappa \geqslant 1$ such that whenever $m, n \in \mathbb{N}_0^N$ with $\mathbb{P}\{X(m) = 0, X(n) = 0\} > 0$, $X(m \curlywedge n)$ is κ-weakly unimodal. When this is so, we say that the N-parameter random walk $X := \{X(n); n \in \mathbb{N}_0^N\}$ is *weakly unimodal.* We observe that the above definition of weak unimodality of X implies that whenever $m \in \mathbb{N}_0^N$ with $\mathbb{P}\{X(m) = 0\} > 0$, $X(m)$ is weakly unimodal.

For all $E \subset \mathbb{N}_0^N$, we write δ_E (Δ_E, resp.) for any choice of $n \in \mathbb{N}_0^N$ with the following property: $\mathbb{P}\{X(n) = 0\} > 0$ and $n \preccurlyeq m$ ($n \succcurlyeq m$, resp.) for all $m \in E$. We note that δ_E always exists in $\mathbb{N}_0^N$, and Δ_E also exists as long as E is bounded and X symmetric. If $\mathbb{P}\{X(n) = 0\} = 0$ for all $n \succcurlyeq E$ (i.e. $n \succcurlyeq m$ for all $m \in E$) or E is infinite, we define $\Delta_E = \infty$. Of course, it is *not* necessarily true that δ_E or Δ_E are themselves in E.

Theorem 1.2. *Let X be a symmetric, weakly unimodal, N-parameter additive random walk with symmetrized gauge function Ψ. Then, for all $E \subset \mathbb{N}_0^N$,*

$$\begin{aligned} \frac{\Psi^2(\Delta_E)}{\kappa^3 \Psi(\delta_E)} \cdot \mathcal{C}_\Psi(E) \;&\leqslant\; \mathbb{P}\big\{0 \in X(E)\big\} \\ &\leqslant \sup_{a\in\mathbb{Z}^d} \mathbb{P}\big\{a \in X(E)\big\} \\ &\leqslant \kappa 16^N \Psi(\delta_E) \cdot \mathcal{C}_\Psi(E), \end{aligned}$$

where $\Psi(\infty) = 0$ and $0/0 = 0$.

Amongst other things, the above estimates the probability that the zero set of X intersects a given set E and this estimate is essentially sharp, as long as $\Psi(\delta_E)$ and $\Psi(\Delta_E)$ are of the same order of magnitude.

Next, let us denote the level set of X (at 0) by $\mathcal{L}$. That is,

$$\mathcal{L} = \Big\{n \in \mathbb{N}_0^N : \; X(n) = 0\Big\}. \tag{1.6}$$

When $N = 1$, a classical problem of probability has been to establish when $\mathcal{L}$ is an infinite set. Equivalently, the question is: "when is 0 *recurrent*?" We shall continue using this notation for all $N \geqslant 1$. Our next theorem gives a necessary and sufficient condition for 0 to be recurrent for the additive random walk X. Needless to say, one needs to better understand the notion of recurrence in order to work in the multiparameter setting. For instance, the reader is reminded that notions of '*stopping times*', '*first time to* $\cdots$', etc. no longer even make sense. As such, other ideas are needed to prove the following.

Theorem 1.3. *Suppose X is a symmetric, weakly unimodal, N-parameter additive random walk on $\mathbb{Z}^d$ and $X(0) = 0$, the following are equivalent:*

1. *0 is recurrent with positive probability;*
2. *0 is recurrent with probability one; and*
3. $\sum_{n \in \mathbb{N}_0^N} \mathbb{P}\{X(n) = 0\} = \infty$.

It follows from Theorem 1.3 that if $\sum_{n \in \mathbb{N}_0^N} \mathbb{P}\{X(n) = 0\} = \infty$, then $\lim_{n\to\infty} \#(\mathcal{L} \cap [0,n]^N) = \infty$, almost surely, where $\#$ denotes cardinality. Due to its close connections to the various fractal dimensions of $\mathcal{L}$, the rate at which $\#(\mathcal{L} \cap [0,n]^N) \to \infty$ is a matter of some importance; see [1, 2] for results about discrete fractals determined by classical random walks. Here, we define the following two indices:

$$\begin{aligned}
\overline{\gamma} &= \inf\Big\{\beta > 0 : \limsup_{n\to\infty} n^{-\beta} \sum_{k\in[0,n]^N} \Psi(k) < \infty\Big\}, \text{ and} \\
\underline{\gamma} &= \sup\Big\{\beta > 0 : \liminf_{n\to\infty} n^{-\beta} \sum_{k\in[0,n]^N} \Psi(k) > 0\Big\}.
\end{aligned}$$

It is easy to verify that $0 \leqslant \underline{\gamma} \leqslant \overline{\gamma} \leqslant N$. Whenever $\underline{\gamma} = \overline{\gamma}$, we write their common value as γ. Later on, we will use the following elementary fact: for any $\varepsilon > 0$, there exists a positive and finite constant K such that

$$K^{-1} n^{\underline{\gamma}-\varepsilon} \leqslant \sum_{k\in[0,n]^N} \Psi(k) \leqslant K n^{\overline{\gamma}+\varepsilon}, \tag{1.7}$$

for all n large enough.

In the following, we denote $\mathcal{L}(n) = \#(\mathcal{L} \cap [0,n]^N)$.

Theorem 1.4. *Suppose X is a symmetric, weakly unimodal, N-parameter additive random walk on $\mathbb{Z}^d$. Then, almost surely,*

$$\underline{\gamma} \leqslant \limsup_{n\to\infty} \frac{\log \mathcal{L}(n)}{\log n} \leqslant \overline{\gamma}. \tag{1.8}$$

In particular, under the conditions of Theorem 1.4, whenever $\underline{\gamma} = \overline{\gamma}$,

$$\limsup_{n\to\infty} \frac{\log \mathcal{L}(n)}{\log n} = \gamma,$$

almost surely.

The outline of the remainder of this article is as follows. In Section 2, we establish a series of Markov properties for additive random walks; this is an important step in our proof of Theorem 1.1. In Section 3, we prove some technical lemmas and give some examples of weakly unimodal additive random walks. Theorems 1.1 and 1.2 are proved in Sections 4 Section 5, respectively. Finally, in Section 6, we prove Theorems 1.3 and 1.4.

2 Markov properties

The additive random walks of §1 satisfy 2^N '*Markov properties*' that are immediately relevant to the results of this paper. However, to get these Markov properties. we will need to use non-probability measures. To do this concisely, we work on the canonical probability space $(\Omega, \mathcal{F})$, where $\Omega = (\mathbb{Z}^d)^{\mathbb{N}_0} \oplus \cdots \oplus (\mathbb{Z}^d)^{\mathbb{N}_0}$ is the collection of all functions $\omega : \mathbb{N}_0^N \mapsto \mathbb{Z}^d$ of form $\omega(n) = \omega_1(n_1) + \cdots + \omega_N(n_N)$, where $\omega_j \in (\mathbb{Z}^d)^{\mathbb{N}_0}$. Moreover, $\mathcal{F}$ denotes the Borel field based on the product topology. The processes $X, X_1, \ldots, X_N$ are then described in the usual way as follows: for all $\omega \in \Omega$, all $n \in \mathbb{N}_0^N$ and all $j = 1, \ldots, N$, $X(n)(\omega) = \omega(n)$, $X_j(n_j) = \omega_j(n_j)$. Let $\mathbb{P}$ denote the probability measure—derived from Kolmogorov's existence theorem—that makes this coordinatewise definition of X a version of the additive random walk we need, and define for all $A \in \mathcal{F}$,

$$\mathbb{Q}(A) = \sum_{a \in \mathbb{Z}^d} \mathbb{P}\{\omega \in \Omega : \ \omega + a \in A\}, \tag{2.1}$$

where, as usual, $(\omega + a)(n) = \omega(n) + a$ $(n \in \mathbb{N}_0^N)$. The measure $\mathbb{Q}$ is σ-finite but *not* finite. In fact, for all $n \in \mathbb{N}_0^N$ and $b \in \mathbb{Z}^d$, $\mathbb{Q}\{X(n) = b\} = 1$. We denote by $\mathbb{E}_{\mathbb{Q}}$ the abstract integral (or 'expectation') with respect to the infinite measure $\mathbb{Q}$. That is, for all $Z : \Omega \mapsto \mathbb{R}_+$, $\mathbb{E}_{\mathbb{Q}}\{Z\} = \int_\Omega Z(\omega)\, \mathbb{Q}(d\omega)$, and this uniquely defines $\mathbb{E}_{\mathbb{Q}}\{Z\}$, at least whenever $\mathbb{E}_{\mathbb{Q}}\{|Z|\} < \infty$. There are 2^N standard partial orders on $\mathbb{N}_0^N$, all of which we will need. A good way to keep track of them is as follows: let Π denote the collection of all subsets of $\{1, \ldots, N\}$. For all $A \in \Pi$, we then define a partial order $\preccurlyeq_{(A)}$ on $\mathbb{N}_0^N$ by declaring $n \preccurlyeq_{(A)} m$ whenever for all $j \in A$, $n_j \leqslant m_j$ and for all $j \notin A$, $n_j \geqslant m_j$. We will not distinguish between $n \preccurlyeq_{(A)} m$ and $m \succcurlyeq_{(A)} n$. It is important to notice that the partial order "$\preccurlyeq_{(\{1,\ldots,N\})}$" is the same as "$\preccurlyeq$" of the Introduction.

For each $A \in \Pi$ and for all $n \in \mathbb{N}_0^N$, we define $\mathcal{F}^A(n)$ to be the σ-field generated by the collection $\{X(m);\ m \preccurlyeq_{(A)} n\}$. Clearly, $\mathcal{F}^A$ is a filtration in the partial order $\preccurlyeq_{(A)}$. That is, whenever $n \preccurlyeq_{(A)} m$, $\mathcal{F}^A(n) \subset \mathcal{F}^A(m)$. We will also need a kind of '*reversed filtration*' $\mathcal{G}^A$ for each $A \in \Pi$: for all $n \in \mathbb{N}_0^N$, $\mathcal{G}^A(n)$ denotes the σ-field generated by $\{X(m);\ m \succcurlyeq_{(A)} n\}$.

Proposition 2.1 (Markov Properties). *For every $A \in \Pi$ and all $n \in \mathbb{N}_0^N$, $\mathcal{G}^A(n)$ is conditionally independent of $\mathcal{F}^A(n)$, given $X(n)$, under the measure $\mathbb{Q}$.*

Remark 2.1. When $N \geq 2$, this is not true under $\mathbb{P}$.

In order to explain Proposition 2.1, let us fix $A \in \Pi$ and suppose $p, q \in \mathbb{N}_0$ and $k_{-p}, \dots, k_q \in \mathbb{N}_0^N$ satisfy the following: for all $1 \leqslant i \leqslant p$ and $1 \leqslant j \leqslant q$, $k_{-i} \preccurlyeq_{(A)} k_0 \preccurlyeq_{(A)} k_j$. Then, Proposition 2.1 states that for all $f_\ell : \mathbb{Z}^d \mapsto \mathbb{R}_+$ $(-p \leqslant \ell \leqslant q)$,

$$\begin{aligned} &\mathbb{E}_{\mathbb{Q}}\Big\{ \prod_{\ell=-p}^{q} f_\ell(X(k_\ell)) \,\Big|\, X(k_0) \Big\} \\ &= f_0(X(k_0)) \times \mathbb{E}_{\mathbb{Q}}\Big\{ \prod_{\ell=-p}^{-1} f_\ell(X(k_\ell)) \,\Big|\, X(k_0) \Big\} \\ &\qquad \times \mathbb{E}_{\mathbb{Q}}\Big\{ \prod_{\ell=1}^{q} f_\ell(X(k_\ell)) \,\Big|\, X(k_0) \Big\}, \end{aligned} \tag{2.2}$$

$\mathbb{Q}$-almost surely.

Proof. Using the notation used in Eq. (2.2),

$$\begin{aligned} &\mathbb{E}_{\mathbb{Q}}\Big\{ \prod_{\ell=-p}^{q} f_\ell(X(k_\ell)) \Big\} \\ &= \sum_{a \in \mathbb{Z}^d} \mathbb{E}\Big\{ \prod_{\ell=-p}^{q} f_\ell(X(k_\ell) + a) \Big\} \\ &= \sum_{a \in \mathbb{Z}^d} f_0(a) \cdot \mathbb{E}\Big\{ \prod_{\substack{-p \leqslant \ell \leqslant q \\ \ell \neq 0}} f_\ell(X(k_\ell) - X(k_0) + a)) \Big\}, \end{aligned}$$

as long as any of the latter sums is finite. We now use the easily verifiable fact that the following collections of random variables are independent from one another:

(i) $\{X(k_{-p}) - X(k_0), \dots, X(k_{-1}) - X(k_0)\}$; and

(ii) $\{X(k_1) - X(k_0), \dots, X(k_q) - X(k_0)\}$.

By the monotone convergence theorem, Eq. (2.2) holds for all $f_{-p}, \dots, f_q$ and the proposition follows. □

Indeed, the above argument has the following immediate but important consequence.

Corollary 2.1. *For all $A \in \Pi$, every $n, m \in \mathbb{N}_0^N$ with $n \preccurlyeq_{(A)} m$, and for all $f : \mathbb{Z}^d \mapsto \mathbb{R}_+$,*

$$\mathbb{E}_{\mathbb{Q}}\Big\{f \circ X(m) \;\Big|\; \mathcal{F}^A(n)\Big\} = \mathbb{E}\Big\{f\big[X(m) - X(n) + \beta\big]\Big\},$$

$\mathbb{Q}$-almost surely on $\{X(n) = \beta\}$.

3 Preliminaries on unimodal walks

In this section, we present a sufficient condition for an additive random walk to be weakly unimodal. We also prove a technical result that will be useful in our proof of Theorem 1.2.

It is easy to verify directly that for $N = 1$, any simple (nearest neighbor with equal probabilities) random walk on $\mathbb{Z}^d$ is weakly unimodal with $\kappa = 1$. This can also be derived from the following lemma.

Lemma 3.1. *Let Y be a random variable on $\mathbb{Z}^d$. If $\mathbb{E}\{e^{i\xi\cdot Y}\} \geqslant 0$,*

$$\sup_{a \in \mathbb{Z}^d} \mathbb{P}\{Y = a\} \leqslant \mathbb{P}\{Y = 0\}.$$

Proof. By Fourier's inversion formula, for all $g \in \mathbb{Z}^d$,

$$\mathbb{P}\{Y = g\} = (2\pi)^{-d} \int_{[-\pi,\pi]^d} e^{-ig\cdot\xi}\, \mathbb{E}\{e^{i\xi\cdot Y}\}\, d\xi,$$

and the result follows immediately. □

Example. There are many random variables in $\mathbb{Z}$ that have a positive characteristic function. Here is one construction: consider a sequence $\{p_k\}$ of non-negative numbers such that $p_0 + \sum_{k=1}^{\infty} p_k = 1$ and $\sum_{k=1}^{\infty} p_k \leqslant p_0$ and define Y to be a random variable with

$$\mathbb{P}\{Y = 0\} = p_0, \;\; \mathbb{P}\{Y = k\} = \mathbb{P}\{Y = -k\} = p_k/2 \;\; (k \geqslant 1).$$

Then, it is clear that $\mathbb{E}\{e^{i\xi\cdot Y}\} \geqslant 0$. Now, consider $X_1, \cdots, X_N$—N independent random walks on $\mathbb{Z}^d$—and denote the characteristic function of $X_j(1)$ by Ψ_j ($j = 1, \cdots, N$), respectively. Then, for any $n \in \mathbb{N}_0^N$, $\mathbb{E}[\exp\{i\xi \cdot X(n)\}] = \prod_{j=1}^{N} [\Psi_j(\xi)]^{n_j}$. There are many possible choices for N and Ψ_j ($j = 1, \ldots, N$) such that the above characteristic function is non-negative for all $n \in \mathbb{N}_0^N$, or more generally $\mathbb{E}[\exp\{i\xi \cdot X(n)\}] \geqslant 0$ for all $n \in \mathbb{N}_0^N$ with $\mathbb{P}\{X(n) = 0\} > 0$. It follows from Lemma 3.1 that for such choices the corresponding additive random walks are weakly unimodal with $\kappa = 1$. In particular, this is the case for the simple random walk in $\mathbb{Z}^d$. □

Next, we use the order structure on $\mathbb{N}_0^N$ described in the Introduction: for all $m, n \in \mathbb{N}_0^N$, $m \preccurlyeq n$ if and only if $m_\ell \leqslant n_\ell$, for all $\ell = 1, \ldots, N$. We also recall the definitions of $m \curlywedge n$ and Ψ.

Lemma 3.2 (Weak Monotonicity). *Let X be an N-parameter weakly unimodal additive random walk. For $m, n \in \mathbb{N}_0^N$, if $m \preccurlyeq n$ and $\mathbb{P}\{X(m) = 0\} > 0$, then $\Psi(n) \leqslant \kappa\Psi(m)$.*

Proof. Since $m \preccurlyeq n$, $X(m)$ and $X(n) - X(m)$ are independent. Hence,

$$\begin{aligned}\Psi(n) &= \mathbb{P}\{X(m) + (X(n) - X(m)) = 0\} \\ &\leqslant \sup_{a \in \mathbb{Z}^d} \mathbb{P}\big\{X(m) + a = 0\big\} \leqslant \kappa\Psi(m),\end{aligned}$$

by weak unimodality. This proves the lemma. □

Lemma 3.3. *Let X be a symmetric and weakly unimodal N-parameter additive random walk. For all $m, n \in \mathbb{N}_0^N$,*

$$\mathbb{P}\{X(m) = 0, X(n) = 0\} \leqslant \kappa\Psi(m \curlywedge n)\Psi(m - n).$$

Proof. If $\mathbb{P}\{X(m) = 0, X(n) = 0\} = 0$, then there is nothing to prove. So, we may assume that this probability is positive. Recall that $n \curlywedge m$ denotes the point in $\mathbb{N}_0^N$ whose jth coordinate is $n_j \wedge m_j$ (the minimum). For all $n, m \in \mathbb{N}_0^N$, let $Z_1 = X(n) - X(n \curlywedge m)$ and $Z_2 = X(m) - X(n \curlywedge m)$. Then,

$$\begin{aligned}&\mathbb{P}\big\{X(n) = 0\ ,\ X(m) = 0\big\} \\ &\quad = \mathbb{P}\big\{X(n \curlywedge m) + Z_1 = 0\ ,\ X(n \curlywedge m) + Z_2 = 0\big\} \\ &\quad \leqslant \mathbb{P}\big\{X(n \curlywedge m) + Z_2 = 0\ ,\ X(n) - X(m) = 0\big\}.\end{aligned} \tag{3.1}$$

Since $X(n \curlywedge m)$, Z_1 and Z_2 are mutually independent, the above is bounded above by

$$\begin{aligned}&\sup_{a \in \mathbb{Z}^d} \mathbb{P}\big\{X(n \curlywedge m) = a\big\} \cdot \mathbb{P}\big\{X(n) - X(m) = 0\big\} \\ &\quad \leqslant \kappa\mathbb{P}\big\{X(n \curlywedge m) = 0\big\} \cdot \mathbb{P}\big\{X(n) - X(m) = 0\big\} \\ &\quad = \kappa\Psi(n \curlywedge m) \cdot \mathbb{P}\big\{X(n) - X(m) = 0\big\}.\end{aligned}$$

We have used weak unimodality in the above. By symmetry and independence, one can also directly check that $\mathbb{P}\big\{X(n) - X(m) = 0\big\} = \Psi(n - m)$, and our proof is complete. □

4 Proof of Theorem 1.1

We begin by stating a multiparameter martingale inequality. When the underlying measure is a probability measure, this is due to R. Cairoli; cf. [3, 18]. When the underlying measure is our infinite but σ-finite measure $\mathbb{Q}$, a proof can be set forth by following through the standard proof of the probability case carefully. In a one-parameter, discrete setting, the details can be found in [8, Eq. (40.2), p. 34]. One generalizes this development to our present multiparameter setting by applying one-parameter arguments one parameter at a time; this idea is due to R. Cairoli and is well explained in [18].

Lemma 4.1 (Cairoli's (2,2) Maximal Inequality). *For $A \in \Pi$ and for any bounded random variable Y, define the N-parameter $\mathcal{F}^A$–martingale $M(m) = \mathbb{E}_{\mathbb{Q}}\{Y \mid \mathcal{F}^A(m)\}$, $(m \in \mathbb{N}_0^N)$. Then,*

$$\mathbb{E}_{\mathbb{Q}}\Big\{ \sup_{m \in \mathbb{N}_0^N} M(m)^2 \Big\} \leqslant 4^N \mathbb{E}_{\mathbb{Q}}\{Y^2\}.$$

Define

$$J_\mu(a) = \sum_n \mathbf{1}_{\{X(n)=a\}}\ \mu(\{n\}), \qquad a \in \mathbb{Z}^d,\ \mu \in \mathcal{P}(\mathbb{N}_0^N). \tag{4.1}$$

As an immediate consequence of Corollary 2.1, we obtain

Lemma 4.2. *For all $a \in \mathbb{Z}^d$ and $\mu \in \mathcal{P}(\mathbb{N}_0^N)$,*

$$\begin{aligned} \mathbb{E}_{\mathbb{Q}}\big\{J_\mu(a)\big\} &= 1, \\ \mathbb{E}_{\mathbb{Q}}\big\{|J_\mu(a)|^2\big\} &= \mathcal{E}_\Phi(\mu). \end{aligned}$$

We are ready to verify the easier half of Theorem 1.1.

Proof of Theorem 1.1: Lower Bound. By the Paley–Zygmund inequality, for all $a \in \mathbb{Z}^d$ and all $\mu \in \mathcal{P}(E)$,

$$\mathbb{Q}\{J_\mu(a) > 0\} \geqslant \frac{\Big[\mathbb{E}_{\mathbb{Q}}\big\{J_\mu(a)\big\}\Big]^2}{\mathbb{E}_{\mathbb{Q}}\big\{|J_\mu(a)|^2\big\}};$$

see [11] for a proof in the probability case. The standard argument for this works to prove the above formulation for $\mathbb{Q}$, as well. Applying Lemma 4.2 and optimizing over $\mu \in \mathcal{P}(E)$, we can deduce that

$$\mathbb{Q}\big\{\exists n \in E :\ X(n) = a\big\} \geqslant \mathcal{C}_\Phi(E).$$

On the other hand,

$$\begin{aligned} \mathbb{Q}\big\{\exists n \in E :\ X(n) = a\big\} &= \sum_{x \in \mathbb{Z}^d} \mathbb{P}\big\{\exists n \in E :\ X(n) + x = a\big\} \\ &= \mathbb{E}\big\{\#X(E)\big\}. \end{aligned} \tag{4.2}$$

This completes our proof of the lower bound in Theorem 1.1. □

In order to prove the upper bound, we need one technical estimate that involves martingale estimates. Recalling Eq. (4.1), we define for all $\mu \in \mathcal{P}(E)$, $A \in \Pi$ and for every $a \in \mathbb{Z}^d$,

$$M_\mu^{a,A}(n) = \mathbb{E}_{\mathbb{Q}}\{J_\mu(a) \mid \mathcal{F}^A(n)\}, \qquad n \in \mathbb{N}_0^N. \tag{4.3}$$

Lemma 4.3. *For all $\mu \in \mathcal{P}(E)$, $A \in \Pi$, $a \in \mathbb{Z}^d$ and for all $n \in \mathbb{N}_0^N$,*

$$M_\mu^{a,A}(n) \geqslant \sum_{\ell \succcurlyeq_{(A)} n} \Phi(\ell, n)\ \mu(\{\ell\}) \cdot \mathbf{1}_{\{X(n)=a\}},$$

$\mathbb{Q}$-*almost surely.*

Proof. Combining Eq.'s (4.1) and (4.3), we see that

$$M_\mu^{a,A}(n) \geqslant \sum_{\ell \succcurlyeq_{(A)} n} \mathbb{Q}\{X(\ell) = a \mid \mathcal{F}^A(n)\}\ \mu(\{\ell\}) \cdot \mathbf{1}_{\{X(n)=a\}},$$

$\mathbb{Q}$-almost surely. By Corollary 2.1, $\mathbb{Q}$-almost surely on $\{X(n) = a\}$,

$$\mathbb{Q}\{X(\ell) = a \mid \mathcal{F}^A(n)\} = \Phi(\ell, n).$$

This completes our proof. □

We can now proceed with our

Proof of Theorem 1.1: Upper Bound. By standard Choquet capacity arguments, we can assume that E is a finite set. Let τ denote any measurable selection of $n \in E$, such that $X(\tau) = a$. If $a \notin X(E)$, we *define* $\tau \equiv +\infty$. Squaring the expression in Lemma 4.3 and taking suprema over all $n \in \mathbb{N}_0$, we see that for all $a \in \mathbb{Z}^d$, $A \in \Pi$ and for all $\mu \in \mathcal{P}(E)$,

$$\sup_{n \in \mathbb{N}_0^N} \left|M_\mu^{a,A}(n)\right|^2 \geqslant \Big[\sum_{\ell \succcurlyeq_{(A)} \tau} \Phi(\ell, \tau)\ \mu(\{\ell\})\Big]^2 \cdot \mathbf{1}_{\{\exists n \in E: X(n)=a\}},$$

$\mathbb{Q}$-almost surely. Since $\sum_{A\in\Pi} |x_A|^2 \geqslant 2^{-N} \left|\sum_{A\in\Pi} x_A\right|^2$ for every sequence $\{x_A; A \in \Pi\}$, we can add over all $A \in \Pi$ and see that for any $\mu \in \mathcal{P}(E)$,

$$\sum_{A\in\Pi} \sup_{n\in\mathbb{N}_0^N} \left|M_\mu^{a,A}(n)\right|^2 \geqslant 2^{-N} \Big[\sum_{\ell} \Phi(\ell, \tau)\ \mu(\{\ell\})\Big]^2 \cdot \mathbf{1}_{\{\exists n \in E: X(n)=a\}}, \tag{4.4}$$

$\mathbb{Q}$-almost surely. Now, we choose a special μ by defining

$$\mu(\{\bullet\}) = \mathbb{Q}\{\tau = \bullet \mid \tau \neq +\infty\}.$$

Since E is assumed to be finite, by Eq. (4.2),

$$1 \leq \mathbb{Q}\{\tau \neq \infty\} = \mathbb{E}\{\#X(E)\} \leqslant \#E < \infty.$$

Thus, $\mu \in \mathcal{P}(E)$ and by Eq. (4.4), applied to this specific choice of μ,

$$\begin{aligned}\sum_{A\in\Pi} \mathbb{E}_{\mathbb{Q}}\Big\{ \sup_{n\in\mathbb{N}_0^N} \big|M_\mu^{a,A}(n)\big|^2\Big\} \\ \geqslant 2^{-N} \sum_n \Big[\sum_\ell \Phi(\ell,n)\ \mu(\{\ell\})\Big]^2 \mu(\{n\}) \cdot \mathbb{Q}\{\exists n \in E:\ X(n)=a\} \\ \geqslant 2^{-N}\Big[\mathcal{E}_\Phi(\mu)\Big]^2 \cdot \mathbb{E}\big\{\#X(E)\big\}.\end{aligned}$$

We have used Eq. (4.2) in the last step, together with the Cauchy–Schwarz inequality. On the other hand, by Lemmas 4.1 and 4.2, the left-hand side is bounded above by

$$4^N \sum_{A\in\Pi} \mathcal{E}_\Phi(\mu) = 8^N \mathcal{E}_\Phi(\mu).$$

We have shown the existence of some $\mu \in \mathcal{P}(E)$, such that

$$\mathbb{E}\big\{\#X(E)\big\} \cdot \Big[\mathcal{E}_\Phi(\mu)\Big]^2 \leqslant 16^N \mathcal{E}_\Phi(\mu). \tag{4.5}$$

Since $\Phi(n,n) = 1$, $\mathcal{E}_\Phi(\mu) \geqslant \sum_{n\in E} \big|\mu(\{n\})\big|^2 > 0$. Thus, we can deduce the requisite upper bound by dividing each side of (4.5) by $\mathcal{E}_\Phi(\mu)$. This completes our proof. □

Remark 4.1. As mentioned after the statement of Theorem 1.1, in the $N = 1$ case, constant 16 in (1.4) can be improved to 2. We now give the requisite argument. The special feature of this 1-parameter setting is that the parameter set $\mathbb{N}_0$ is *totally ordered* by the linear order $\preccurlyeq$ which is now the same as the usual $\leqslant$. Thus, we need consider only the latter partial order. We recall that for all $m \in \mathbb{N}_0$, $\mathcal{F}_m$ is the σ–algebra generated by $X(n)$, for all $0 \leqslant n \leqslant m$. For each $\mu \in \mathcal{P}(\mathbb{N}_0)$, consider the martingale $M = \{M_\mu^a(n); n \in \mathbb{N}_0\}$ defined by

$$M_\mu^a(n) = \mathbb{E}_{\mathbb{Q}}\Big\{J_\mu(a)\Big|\ \mathcal{F}_n\Big\}, \qquad n \in \mathbb{N}_0.$$

Similar to Lemma 4.3, we obtain that for all $\mu \in \mathcal{P}(\mathbb{N}_0)$,

$$M_\mu^a(n) \geqslant \sum_{\ell \geqslant n} \Phi(\ell,n)\ \mu(\{\ell\}) \cdot \mathbf{1}_{\{X(n)=a\}}. \tag{4.6}$$

This holds $\mathbb{Q}$-almost surely for all $n \in \mathbb{N}_0$. Define τ to be the first $n \in \mathbb{N}_0$ such that $X(n) = a$, with the usual stipulation that $\inf \varnothing = +\infty$. Then, τ is a stopping time with respect to the filtration $\mathcal{F} = \big\{\mathcal{F}_n;\ n \in \mathbb{N}_0\big\}$. By the optional stopping theorem for $\mathbb{E}_{\mathbb{Q}}$, which is standard to verify, we have

$$\begin{aligned}\mathbb{E}_{\mathbb{Q}}\big\{M_\mu^a(\tau)\mathbf{1}_{(\tau<+\infty)}\big\} &\leqslant \lim_{k\to\infty} \mathbb{E}_{\mathbb{Q}}\big\{M_\mu^a(\tau \wedge k)\big\} \\ &= \mathbb{E}_{\mathbb{Q}}\big\{M_0^a\big\} = 1.\end{aligned}$$

Applying (4.6) to $n = \tau$ and taking expectations with respect to $\mathbb{Q}$, we obtain

$$1 \geqslant \mathbb{E}_{\mathbb{Q}}\Big\{ \sum_{n \geqslant \tau} \Phi(n, \tau)\ \mu(\{\ell\})\ \cdot \mathbf{1}_{\{\tau<+\infty\}} \Big\}.$$

As this holds for all $\mu \in \mathcal{P}(\mathbb{N}_0)$, we can apply it to

$$\mu(\bullet) = \mathbb{Q}\{\tau \in \bullet \mid \tau < +\infty\}$$

and see that

$$\mathbb{Q}\{\tau < +\infty\} \leqslant \Big[\sum_{\ell \geqslant m} \Phi(\ell, m)\ \mu(\{\ell\})\ \mu(\{m\}) \Big]^{-1}.$$

Since the above double sum is clearly bounded below by $\frac{1}{2}\mathcal{E}_\Phi(\mu)$, we obtain the upper bound of Theorem 1.1 with the better constant of 2, in case $N = 1$. This proves the assertion of Remark 1.1. □

5 Proof of Theorem 1.2

We are ready to prove Theorem 1.2. In contrast to our presentation of §4, this time, we begin with the harder half, since we have already done most of the preliminary work.

Proof of Theorem 1.2: Upper Bound. The upper bound follows readily from the upper bound given in Theorem 1.1. Indeed, for any $a \in \mathbb{Z}^d$,

$$\begin{aligned}
&\mathbb{P}\{\exists n \in E:\ X(n) = a\} \\
&\quad = \sum_{x \in \mathbb{Z}^d} \mathbb{P}\{X(\delta_E) = x\} \cdot \mathbb{P}\{\exists n \in E:\ X(n) - X(\delta_E) = a - x\} \\
&\quad \leqslant \Psi(\delta_E)\kappa \cdot \sum_{x \in \mathbb{Z}^d} \mathbb{P}\{\exists n \in E:\ X(n) - X(\delta_E) = a - x\}.
\end{aligned}$$

We have used weak unimodality in the penultimate line. By Theorem 1.1,

$$\mathbb{P}\{\exists n \in E:\ X(n) = a\} \leqslant 16^N \kappa \Psi(\delta_E) \cdot \mathcal{C}_\Psi(E \ominus \delta_E),$$

where $E \ominus x = \{n - x:\ n \in E\}$. It is clear from Eq. (1.5), that $F \mapsto \mathcal{C}_\psi(F)$ is translation invariant. Thus, $\mathcal{C}_\Psi(E \ominus \delta_E) = \mathcal{C}_\Psi(E)$ and the upper bound is proved. □

The 'lower bound version' of the above argument is not sufficiently sharp. Instead, we adapt the arguments that lead to the lower bound in Theorem 1.1.

Proof of Theorem 1.2: Lower Bound. If $\Psi(\Delta_E) = 0$, there is nothing to prove. Hence, we may assume, with no loss in generality, that $\Psi(\Delta_E) > 0$. Using the notation Eq. (4.1) with $\mu \in \mathcal{P}(E)$, let $J_\mu = J_\mu(0)$ for brevity. Then, by weak unimodality, weak monotonicity (Lemma 3.2) and after a few lines of calculations, we obtain the following.

$$\begin{aligned} \mathbb{E}\{J_\mu\} &\geqslant \frac{1}{\kappa}\Psi(\Delta_E) \\ \mathbb{E}\{|J_\mu|^2\} &\leqslant \kappa\Psi(\delta_E) \cdot \mathcal{E}_\Psi(\mu). \end{aligned}$$

By the Paley–Zygmund inequality and after optimizing over all $\mu \in \mathcal{P}(E)$, we obtain

$$\mathbb{P}\{\exists n \in E : \ X(n) = 0\} \geqslant \frac{\Psi^2(\Delta_E)}{\kappa^3\Psi(\delta_E)} \cdot \mathcal{C}_\Psi(E).$$

This completes our argument. □

6 Proof of Theorems 1.3 and 1.4

In this section, we prove Theorem 1.3 and 1.4.

Proof of Theorem 1.3. Since the event $\{\omega : \ \#\mathcal{L} = +\infty\}$ is a tail event for all of $X_1, \ldots, X_N$. The first two assertions of Theorem 1.3 are equivalent. By Fubini's theorem, if assertion 3 does not hold, then $\mathbb{E}\{\#\mathcal{L}\} < \infty$, which implies that 0 is a.s. not recurrent. It suffices to prove the converse. That is,

$$\mathbb{E}\{\#\mathcal{L}\} = +\infty \implies \mathbb{P}\{\#\mathcal{L} = +\infty\} > 0. \tag{6.1}$$

In order to prove this, we first note that by Lemma 3.3,

$$\begin{aligned} \mathbb{E}\{|\mathcal{L}(n)|^2\} &= \sum\sum_{i,j \in [0,n]^N} \mathbb{P}\{X(i) = X(j) = 0\} \\ &\leqslant \kappa \sum\sum_{i,j \in [0,n]^N} \Psi(i \curlywedge j)\ \Psi(i - j) \\ &= \kappa \sum\sum_{i,j \in [0,n]^N} \Psi(i \curlywedge j)\ \Psi(i \curlyvee j - i \curlywedge j). \end{aligned}$$

Thus, we can rearrange the sum to see that $\mathbb{E}\{|\mathcal{L}(n)|^2\}$ is bounded above by

$$\kappa \sum_{\substack{p,q\in[0,n]^N:\\ p\succcurlyeq q}} \Psi(q)\ \Psi(p-q)\cdot \#\big\{i,j\in[0,n]^N : p=i\curlyvee j, q=i\curlywedge j\big\}$$
$$\leqslant 2^N\kappa \sum_{\substack{p,q\in[0,n]^N:\\ p\succcurlyeq q}} \Psi(q)\ \Psi(p-q)$$
$$\leqslant 2^N\kappa\Big\{\sum_{p\in[0,n]^N}\Psi(p)\Big\}^2.$$

This proves that

$$\mathbb{E}\{|\mathcal{L}(n)|^2\}\leqslant \kappa 2^N[\mathbb{E}\{\mathcal{L}(n)\}]^2.$$

Consequently, it follows from the Paley–Zygmund inequality that

$$\begin{aligned}\mathbb{P}\Big\{\mathcal{L}(n)\geqslant \frac{1}{2}\mathbb{E}\big[\mathcal{L}(n)\big]\Big\} &\geqslant \frac{1}{4}\frac{(\mathbb{E}\{\mathcal{L}(n)\})^2}{\mathbb{E}\{|\mathcal{L}(n)|^2\}}\\ &\geqslant \frac{1}{\kappa 2^{N+3}};\end{aligned} \tag{6.2}$$

see [11, p.8]. On the other hand, whenever $\mathbb{E}\{\#\mathcal{L}\}=+\infty$, $\lim_n \mathbb{E}\{\mathcal{L}(n)\}=\infty$. Since $\mathcal{L}(n)\leqslant \#\mathcal{L}$,

$$\mathbb{P}\{\mathcal{L}=+\infty\}\geqslant \liminf_{n\to\infty}\mathbb{P}\Big\{\mathcal{L}(n)\geqslant\frac{1}{2}\mathbb{E}\big[\mathcal{L}(n)\big]\Big\}\geqslant \kappa^{-1}2^{-N-3}.$$

We have verified Eq. (6.1). □

Proof of Theorem 1.4. We start with our proof of the upper bound in (1.8). For any $\varepsilon>0$, by Eq. (1.7),

$$\mathbb{E}\{\mathcal{L}(n)\}=\sum_{k\in[0,n]^N}\Phi(k)\leqslant Kn^{-(\overline{\gamma}+\varepsilon)}\quad\text{for all } n\in\mathbb{N}_0^N. \tag{6.3}$$

For every integer $\ell\geqslant 1$, we take $n_\ell=2^\ell$. Then, by (6.3) and Chebyshev's inequality we obtain

$$\mathbb{P}\Big\{\mathcal{L}(n_\ell)>2^{-\ell(\overline{\gamma}+2\varepsilon)}\Big\}\leqslant 2^{-\varepsilon\ell}. \tag{6.4}$$

The Borel-Cantelli lemma implies that almost surely for all integers ℓ that are large enough,

$$\mathcal{L}(n_\ell)\leqslant 2^{-\ell(\overline{\gamma}+2\varepsilon)}. \tag{6.5}$$

If n is large and if ℓ satisfies $2^\ell \leqslant n < 2^{\ell+1}$, Eq. (6.5) implies that

$$\mathcal{L}(n) \leqslant \mathcal{L}(n_{\ell+1}) \leqslant 2^{\overline{\gamma}+2\varepsilon}\, n^{\overline{\gamma}+2\varepsilon}.$$

Therefore,

$$\limsup_{n\to\infty} \frac{\log \mathcal{L}(n)}{\log n} \leqslant \overline{\gamma} + 2\varepsilon. \qquad \text{a.s.}$$

Since $\varepsilon > 0$ is arbitrary, this proves the upper bound in (1.8). In order to prove the corresponding lower bound, we note that by Eq. (6.2),

$$\mathbb{P}\Big\{\mathcal{L}(n) \geqslant \frac{1}{2}\mathbb{E}\big[\mathcal{L}(n)\big] \text{ infinitely often }\Big\} \geqslant \frac{1}{\kappa 2^{N+3}}. \tag{6.6}$$

By Eq. (1.7), for any $\varepsilon > 0$ and all $n \geqslant 1$, there exists a finite and positive constant K, such that $\mathbb{E}[\mathcal{L}(n)] \geqslant K^{-1} n^{\underline{\gamma}-\varepsilon}$ for all n large enough. This and Eq. (6.6), together imply that

$$\limsup_{n\to\infty} \frac{\log \mathcal{L}(n)}{\log n} \geqslant \underline{\gamma} - \varepsilon,$$

with positive probability. We obtain the asserted lower bound in Theorem 1.4 by making an appeal to the 0-1 law of Hewitt and Savage. □

Let us conclude this section by studying a class of additive random walks that have nice, regular behavior. Namely, we assume that $X_1, \ldots, X_N$ are all independent random walks on $\mathbb{Z}^d$ and they all have the same law. Furthermore, we assume the existence of $\alpha > 0$, such that for all $n \geqslant 1$,

$$\mathbb{P}\big\{X_1(n) = 0\big\} \asymp n^{-\alpha}, \tag{6.7}$$

where $a(x) \asymp b(x)$ means there exists a finite $C > 1$, such that for all x under consideration,

$$C^{-1} \leqslant \frac{a(x)}{b(x)} \leqslant C.$$

We note that, in this case, for all $n \in \mathbb{Z}^N$,

$$\Psi(n) \asymp \{|n| + 1\}^{-\alpha}, \tag{6.8}$$

where $|\bullet|$ denotes any of the Euclidean norms on $\mathbb{R}^N$. It is easy to verify that, in the present setting, $\underline{\gamma} = \overline{\gamma} = (N - \alpha)^+$. It follows from Theorems 1.3 and 1.4 that X is recurrent if and only if $N > \alpha$. Moreover, with probability one,

$$\limsup_{n\to\infty} \frac{\log \mathcal{L}(n)}{\log n} = (N - \alpha)^+.$$

References

[1] M. Barlow and S. J. Taylor (1989), Fractal dimensions of sets in discrete spaces, *J. Phys.* **A 22**, 2621–2626.

[2] M. Barlow and S. J. Taylor (1992), Defining fractal subsets of $\mathbb{Z}^d$, *Proc. London Math. Soc.* **64**(3), 125–152.

[3] R. Cairoli (1968), Une inegalité pour martingales à indices multiples et ses applications, *Séminaire de Probabilités IV*, 1–27.

[4] R. C. Dalang and T. S. Mountford (1997), Points of increase of the Brownian sheet, *Prob. Th. Rel. Fields*, **108**(1), 1–27.

[5] R. C. Dalang and T. S. Mountford (1996), Nondifferentiability of curves on the Brownian sheet, *Ann. Probab.*, **24**(1), 182–195.

[6] R. C. Dalang and J. B. Walsh (1993), Geography of the level sets of the Brownian sheet, *Prob. Th. Rel. Fields*, **96**(2), 153–176.

[7] R. C. Dalang and J. B. Walsh (1993), The structure of a Brownian bubble, *Prob. Th. Rel. Fields*, **96**(4), 475–501.

[8] C. Dellacherie and P.-A. Meyer (1982), *Probabilities and Potential, B.* Theory of Martingales. North–Holland Publishing Co., Amsterdam. (J. P. Wilson, Trans.)

[9] S. N. Evans (1987), Multiple points in the sample paths of a Lévy process, *Prob. Th. Rel. Fields*, **76**(3), 359–367.

[10] P. J. Fitzsimmons and T. S. Salisbury (1989), Capacity and energy for multiparameter processes, *Ann. Inst. Henri Poincaré: Prob. Stat.*, **25**(3), 325–350.

[11] J.-P. Kahane (1985), *Some Random Series of Functions*, Cambridge Studies in Advanced Mathematics, Cambridge, U.K.

[12] W. S. Kendall (1980), Contours of Brownian process with several–dimensional times, *Z. Wahr. Verw. Geb.* **52**, 267–276.

[13] D. Khoshnevisan (1999), Brownian sheet images and Bessel-Riesz capacity, *Trans. Amer. Math. Soc.*, **351**(7), 2607–2622.

[14] D. Khoshnevisan and Z. Shi (1999), Brownian sheet and capacity, *Ann. of Probab.*, **27**(3), 1135–1159.

[15] D. Khoshnevisan and Y. Xiao (1999), Level sets of additive Lévy processes. *Preprint.*

[16] J.-F. LeGall, J. Rosen and N.-R. Shieh (1989), Multiple points of Lévy processes, *Ann. of Probab.*, **17**(2), 503–515.

[17] T. S. Mountford (1993), Estimates of the Hausdorff dimension of the boundary of positive Brownian sheet components, *Séminaire de Probabilités, XXVII*, 233–255, Lecture Notes in Math., #1557, Springer, Berlin.

[18] J. B. Walsh (1986), *Martingales with a Multidimensional Parameter and Stochastic Integrals in the Plane*, Lecture Notes in Math. #1215, Springer, Berlin, Ed's: G. del Piño and R. Robodello.

Davar Khoshnevisan
Department of Mathematics
University of Utah
155 S 1400 E, JWB 233
Salt Lake City, UT 84112-0090
URL: `http://www.math.utah.edu/~davar`
E-mail: `davar@math.utah.edu`

Yimin Xiao
Department of Mathematics
University of Utah
155 S 1400 E, JWB 233
Salt Lake City, UT 84112-0090
E-mail: `xiao@math.utah.edu`
and
Microsoft Corporation
One Microsoft Way
Redmond, WA 98052

PART 8

Random Vectors and Processes

Lee-Yang Models, Selfdecomposability and Negative-Definite Functions

Joël De Coninck* and Zbigniew J. Jurek‡

ABSTRACT A class $\mathcal{L}$ of Ising models is introduced via Lévy class L characteristic functions. The critical temperature for these new models is associated with the weak law of large numbers, and it is proved that the critical exponent δ is greater than or equal to 1. New inequalities for the Ursell functions are proposed via the Schoenberg Theorem. Moreover, with the functions u_0 and u_1 one associates some Fourier transforms as functions of the external field.

1 Introduction

Statistical mechanics refers to the microscopic description of thermodynamics. The challenging difficulty is that it deals with correlated variables. In the theory of critical phenomena, the problem becomes even worse in the sense that the elementary variables are then strongly correlated.

Due to the Lee-Yang property, it has been shown rather recently that there exists a deep relation between correlated models in statistical mechanics and limit theorems for sums of *independent* random variables in probability theory; cf. De Coninck (1984). This intriguing property opens new perspectives into more detailed descriptions of free energies for models which are still unsolved, such as the three dimensional Ising model.

For definiteness, let us consider a lattice Z^d, at each vertex $j = (j_1, ..., j_d)$ we assign a spin variable S_j with an *a priori* free spin probability distribution ρ. The Van Hove sequence of blocks $(\Lambda_n)_{n\geq 1}$ will be defined by

$$\Lambda_n = \{j \in Z^d : 0 \leq |j_k| < n \ for \ k = 1, 2, ..., d\}.$$

The energy of a given block configuration is given by some Hamiltonian function $H_\Lambda(\sigma_1, \sigma_2, ..., \sigma_{|\Lambda|})$ and the associated joint probability distribution is given by the Gibbs measure. Then the moment generating function

*This work, in part, was completed at the Research Center for Molecular Modelling, University of Mons, Mons, Belgium (October 1996– January 1997) with the support of Fondes National de la Recherche Scientifique (FNRS).

‡The second author was also supported by Grants No. 2 P03 A02914 and A01408 from KBN, Warsaw, Poland.

of the total magnetization $M_{\Lambda_n} = \sum_{i\in\Lambda_n} S_i$, can be viewed as the inverse of a selfdecomposable characteristic function $\varphi_{\widetilde{M}_{\Lambda_n}}(h)$. Thus with the sum of *dependent* variables, M_{Λ_n}, we associate a selfdecomposable random variable $\widetilde{M}_{\Lambda_n}$, which is a limit of a normalized *independent* summand. The random integral representation of selfdecomposable distribution is the most crucial property for this note; cf. Appendix (A1).

The structure of the paper is as follows: in Section 2, we introduce the class $\mathcal{L}$ of Ising models, prove the existence of the so-called first Ursell function u_1 and, among others, its new property: $u_1(2t) \leq 2u_1(t)$ for $t > 0$. [Functions satisfying such inequalities are called O-regularly varying (cf. Bingham,Goldie & Tuegels (1987)); see also Δ_2-conditions in the theory of Orlicz spaces (cf. Krasnoselski & Rutickii (1961)).] Moreover, we show that $\exp[-tu_1(t)]$ is an infinitely divisible characteristic function. Using Schoenberg's Theorem, new inequalities for u_1 are given. In Section 3, new definitions of critical temperatures are introduced and related to the classical weak law of large numbers. It is proved that for class $\mathcal{L}$ Ising models the critical exponent δ is greater than or equal to 1. Under an additional assumption, concavity of u_1 is established.

2 Class $\mathcal{L}$ of Ising models

Let us consider the lattice $\mathbf{Z}^d$ with a spin *rv* S_i, and the spin probability distribution ρ (independent of vertices i). For a given finite block $\Lambda \subset \mathbf{Z}^d$ we define the Hamiltonian

$$H_\Lambda(\sigma_1, \sigma_2, ..., \sigma_{|\Lambda|}) := -\sum_{\substack{i,j\in\Lambda \\ i\neq j}} J_{ij}\sigma_i\sigma_j - h\sum_{i\in\Lambda}\sigma_i, \tag{2.1}$$

where $J_{ij} \geq 0$ is the interaction between sites i and j, $|\Lambda|$ is the number of sites in the block Λ, σ_i are the values of spin $rv's$ S_i and h represents the external field. For a fixed finite Λ_n, we assume that the random vector $(S_1, S_2, ..., S_{|\Lambda|})$ has a joint probability distribution given by the Gibbs measure μ_{Λ_n}, i.e.,

$$d\mu_\Lambda := Z_\Lambda(\beta, \beta h)^{-1} \exp[-\beta H_\Lambda(\sigma_1, ..., \sigma_{|\Lambda|})] d\rho(\sigma_1)...d\rho(\sigma_{|\Lambda|}), \tag{2.2}$$

where $Z_\Lambda(\beta, \beta h)$ is the *partition function* normalizing the Gibbs state and $0 < \beta$ is the inverse temperature parameter. The aim of this paper is to study the free energy of these models. We assume that the limit

$$g(\beta, h) := -\beta^{-1} \lim_{\Lambda\uparrow\mathbf{Z}^d} |\Lambda|^{-1} \log Z_\Lambda(\beta, h) \quad \textit{exists}, \tag{2.3}$$

when Λ approach $\mathbf{Z}^d$ (in van Howe's sense), and for each β, the function

$$\varphi(\beta, h) \equiv \varphi(h) := \exp[\beta(g(\beta, h) - g(\beta, 0))] \tag{2.4}$$

is a symmetric selfdecomposable characteristic function. We simply write $\varphi \in L$ and L is the class of all selfdecomposable distributions. We call such statistical models *the class $\mathcal{L}$ Ising models* and briefly write $g \in \mathcal{L}$. Let us recall that the class L (or selfdecomposable) probabilities are characterized by the condition:

$$(2.5) \quad \varphi \in L \quad \text{iff} \quad \forall (0 < c < 1)\ h \mapsto \varphi(h)/\varphi(ch) \quad \textit{is a char. function.}$$

Thus, in terms of the free energy function g, (2.5) means equivalently that

$$(2.6) \quad g \in \mathcal{L} \quad \text{iff} \quad \forall (0 < c < 1) h \mapsto \exp[\beta(g(\beta,h) - g(\beta,ch))] \quad \textit{is a char. function.}$$

How large the Ising class $\mathcal{L}$ is follows from the following comments.

Remarks 2.1. (a) If for a fixed $\Lambda_n \subset \mathbf{Z}^d$,

$$(2.7) \qquad [Z_{\Lambda_n}(\beta,h)/Z_{\Lambda_n}(\beta,0)]^{-1/|\Lambda_n|} := \varphi_{\Lambda_n}(h) \in L, \quad h \in \mathbf{R},$$

and the limit (2.3) exists, then $g \in \mathcal{L}$.

(b) Statistical models from De Coninck (1984) belong to the class $\mathcal{L}$.

(c) Random variables considered in Newman (1975) and examples 2.1 and 2.3 in Newman (1974) lead to the $\mathcal{L}$ models.

Let us recall that the Ursell functions are defined as follows

$$(2.8) \quad u_0(h) := -\beta(g(\beta,h) - g(\beta,0)); \quad u_n(h) := -\beta\partial^n_{\beta h} g(\beta,h), \quad n \geq 1,$$

for a fixed temperature β^{-1}. Now we have

Theorem 2.1. *For Ising $\mathcal{L}$ models (i.e. (2.3) and (2.4)hold) the Ursell function u_1 exists on $\mathbf{R} \setminus \{0\}$, is increasing, odd and continous. Also it satisfies the inequality $u_1(2h) \leq 2u_1(h)$ for $h > 0$ and $\exp[-hu_1(h)]$, $h \in \mathbf{R}$ is a characteristic function of an infinitely divisible distribution with finite logarithmic moment.*

Proof. By Hölder's inequality, one gets that the partition function: $h \mapsto Z_{\Lambda_n}(\beta,h)$ is log-convex and therefore, by (2.3), functions $-\beta\, g(\beta,\cdot)$ are convex. Hence so is u_0. Consequently it is differentiable (except possibly at countable many points) with non-decreasing derivative; cf. Krasnoselskii & Rutickii (1961), p.5. However, (2.4) *gives much more.* Namely,

$$(2.9) \quad -u_0(h) = \beta(g(\beta,h) - g(\beta,0)) = \log\varphi(h) = \int_0^h \log\psi(r)/r dr,$$

for $h \neq 0$; cf. Appendix. This with (2.8) gives

$$(2.10) \quad u_1(h) = -\log\psi(h)/h = \sigma^2 h + \frac{1}{h}\int_{\mathbf{R}\setminus\{0\}} (1 - \cos(hx))N(dx),$$

where $h \neq 0$, $\sigma^2 \geq 0$ (Gaussian part of ψ) and N is a Lévy spectral measure of the infinitely divisible (background) *rv* ψ, with finite logarithmic moment. From elementary inequality $(1 - \cos 2\alpha) \leq 4(1 - \cos \alpha)$, we get the inequality for u_1, which completes the proof.

Corollary 2.1. (a) *Class L characteristic functions corresponding to class $\mathcal{L}$ Ising models are non-degenerate and log-concave.*
(b) *For class $\mathcal{L}$ Ising models and their Ursell functions u_0, u_1, the functions $-u_0(h)$ and $-hu_1(h)$ are logarithms of infinitely divisible characteristic functions.*
(c) *For class $\mathcal{L}$ Ising models one has, for $h \neq 0$,*

$$(2.11) \quad u_1(h) = \frac{1}{h} \int_{\mathbf{R}\setminus\{0\}} (1-\cos(hx))N(dx) = \int_0^\infty N[|u| > y] \sin(hy)dy,$$

where N is σ-finite measure on $\mathbf{R} \setminus \{0\}$ such that

$$\int_{\mathbf{R}\setminus\{0\}} (|x|^{-1} \wedge |x|^2)N(dx) < \infty.$$

Proof. (a) In (2.4), $\psi \equiv 1$ corresponds to Ising models with a constant free energy per site g. Log-concavity follows from the proof of Theorem 2.1.
(c) Since u_1 (magnetization) must be a bounded function of h, from (2.10) we conclude $\sigma = 0$. The second equality follows from Fubini's Theorem. The integrability condition is a combination of facts that Lévy spectral measure N integrates x^2 on $(|x| \leq 1)$ and $\log |x|$ on $(|x| > 1)$.

A function $f : \mathbf{R}^d \to \mathbf{C}$ is said to be *negative-definite* if *for any integer $m \geq 1$, any complex numbers $z_1, ..., z_m$ and any $t_1, ..., t_m$ from $\mathbf{R}^d$*

$$\sum_{j=1}^m \sum_{k=1}^m z_j\overline{z}_k(f(t_j - t_k) - f(t_j) - f(-t_k)) \geq 0.$$

Using Schoenberg's Theorem, cf. Cuppens(1975), pp.80-82, we conclude that the Ising free energy can be viewed as a negative definite function. Applying it for the class $\mathcal{L}$ models, we get new inequalities for Ursell functions:

Corollary 2.2. *Let u_0 denote the difference of free energies $g(0) - g(h)$ where $g \in \mathcal{L}$, then for each $m \geq 1$ and each complex number $z_1, z_2, ..., z_m$ and each $h_1, h_2, ..., h_m \in \mathbf{R}$,*

$$\sum_{1\leq j,k\leq m} z_j\overline{z}_k[u_0(h_j - h_k) - u_0(h_j) - u_0(h_k)] \leq 0.$$

Corollary 2.3. *If u_1 is the first derivative with respect to the external field of the free energy $g \in \mathcal{L}$, then for each $m \geq 1$ and each complex number $z_1, z_2, ..., z_m$ and each $h_1, h_2, ..., h_m \in \mathbf{R}$,*

$$\sum_{1\leq j,k\leq m} z_j\overline{z}_k[(h_j - h_k)u_1(h_j - h_k) - h_ju_1(h_j) - h_ku_1(h_k)] \leq 0.$$

3 Critical temperature and exponent: Concavity

Of course, the background driving *rv* ψ or its Lévy spectral measure N depend on the temperature β^{-1}, although so far it was not explicitly written. In particular, we have for fixed β, the magnetization $u_1(\beta, h) := u_1(h)$ and

(3.1) $m_0(\beta) \equiv m_0 := \lim_{h\to 0+} u_1(\beta, h)$ *is the spontaneous magnetization.*

It exists since u_1 is (strictly) monotone on $(0, \infty)$, cf. Theorem 2.1. Moreover, $\lim_{h\to 0-} u_1(\beta, h) = -m_0$ and therefore no spontenous magnetization, at given temperature, means $m_0(\beta) = 0$.

Proposition 3.1. *For class $\mathcal{L}$ Ising models, with the background driving rv ψ_β, no spontanous magnetization i.e., $m_0(\beta) = 0$, is equivalent to the condition that $\psi'_\beta(0)$ exists (then is necessarily to equal 0).*

Proof. From (2.10) and the equality

$$-u_1(\beta, h) = \frac{\log \psi_\beta(h)}{h} = \frac{\log \psi_\beta(h) - \log \psi_\beta(0)}{\psi_\beta(h) - \psi_\beta(0)} \cdot \frac{\psi_\beta(h) - \psi_\beta(0)}{h},$$

we conclude that $m_0(\beta) = 0$ which means $\psi'_\beta(0)$ exists. Since ψ_β is a real characteristic function, then $\psi'_\beta(0) = 0$, which concludes the proof.

Let us recall that the usual definition of the *critical temperature* β_c is as follows

$$\beta_c := \sup\{\beta > 0 : \lim_{h\to 0+} u_1(\beta, h) = 0\}. \tag{3.2}$$

Here are other equivalent forms.

Corollary 3.1. *For class $\mathcal{L}$ Ising models*

$$\begin{aligned}\beta_c &= \sup\{\beta : \psi'_\beta(0) \text{ exists}\} \\ &= \sup\left\{\beta : \lim_{h\to 0+} h^{-1} \int_{\mathbf{R}\setminus\{0\}} [1 - \cos(hx)] N_\beta(dx) = 0\right\} \\ &= \sup\left\{\beta : \lim_{h\to 0+} \int_0^\infty N_\beta[|u| > y] \sin(hy) dy = 0\right\}.\end{aligned}$$

where ψ_β is the background driving infinitely divisible rv and N_β is its Lévy spectral measure. [*Since ψ_β is symmetric, $\psi'_\beta(0)$ is zero.*]

Proof. It is a consequence of (3.2), (2.11) and Proposition 3.1.

Remark 3.1. Let Y_β be the background driving *rv's* corresponding to the characteristic functions ψ_β, $\beta > 0$. Then it means that one has a process

Y_β, $\beta > 0$ (the temperature is the time parameter) such that for $\beta < \beta_c$, Y_β satisfies WLLN (weak law of large numbers) but for $\beta > \beta_c$, Y_β does not satisfy WLLN; cf. for instance W.Feller (1971).

Corollary 3.2. *For class $\mathcal{L}$ Ising models,*

(i) $$u_2(\beta,0) = \lim_{t\to 0+} \frac{1-\psi_\beta(t)}{t^2} = \psi''_\beta(0) > 0 \quad \textit{for} \quad \beta < \beta_c,$$

(ii) $$u_2(\beta,0) = +\infty \quad \textit{for} \quad \beta > \beta_c.$$

Proof. From (2.10), the definition of the derivative and Corollary 3.1, one gets

$$\begin{aligned} u_2(\beta,0) &:= \lim_{h\to 0} \frac{u_1(\beta,h)-u_1(\beta,0)}{h} = -\lim_{h\to 0} \frac{\log\psi_\beta(h)}{h^2} \\ &= \lim_{h\to 0} \Big[\frac{\log\psi_\beta(h)-\log\psi_\beta(0)}{\psi_\beta(h)-\psi_\beta(0)}\Big]\Big[\frac{1-\psi_\beta(h)}{h^2}\Big] \\ &= \lim_{h\to 0} \frac{1-\psi_\beta(h)}{h^2} > 0 \end{aligned}$$

because limit zero is equivalent to degenerate ψ (cf. Corollary 2.1 (a)). For $\beta > \beta_c$, $\lim_{h\to 0+} u_1(\beta,h) = m_0(\beta) > 0$ and therefore $u_2(\beta,0) = +\infty$, which completes the proof.

Critical behaviours of thermodynamic functions are those observed at the critical temperature β_c. It is expected, cf. Baxter(1982), p. 28, that

(3.3) $$u_1(\beta_c,h) \sim h^{1/\delta} \quad \text{as} \quad h\to 0,$$

where the power-law exponent δ is known as a *critical exponent* of the magnetization function.

Corollary 3.3. *For all class $\mathcal{L}$ Ising models, $\delta \geq 1$.*

Proof. Since by Theorem 2.1. $u_1(\beta_c, 2h) \leq 2u_1(\beta_c,h)$, $h > 0$, then (3.3) holds only if $\delta \geq 1$, which gives the proof.

This extends, to the class $\mathcal{L}$ models, Sokal's result on the critical exponent which was established on the basis of concavity of $u_1(t)$, i.e., on the Griffiths-Hurst-Shermann inequality; cf. Griffiths, Hurst,Shermann (1970) and Sokal (1981).

For the rest of this section let us assume that the variance $u_2(0) \equiv u_2(\beta,0)$ exists (so $\beta < \beta_c$). Then (2.11) and the equality in the proof of Corollary 3.2. give us that

(3.4) $$u_2(0) = \psi''_\beta(0) = \int_0^\infty N[|x| > y]y\,dy.$$

Hence Y_β (the background *rv* corresponding to ψ_β; cf. Appendix, (A2)) has a finite second moment ($\mathbf{E}Y_\beta^2 < \infty \Longleftrightarrow \varphi''_\beta(0)$ exists). Consequently, u_1 is differentiable and

$$(3.5) \qquad u_2(h) = \int_0^\infty N[|x| > y] y \cos(hy) dy \quad \textit{exists and is positive;}$$

cf. Theorem 2.1. [The second Ursell function $u_2(h)$ is the susceptibility function in physical terms.] We have the following partial result.

Proposition 3.2. *If* $\mathbf{E}|Y_\beta|^3 < \infty$, *where* Y_β, *with Lévy spectral measure* N_β, *is the background rv corresponding to an Ising* $\mathcal{L}$ *model and* $y \mapsto N_\beta[|x| \geq y]y^2$ *is decreasing on* $(0, \infty)$, *then* u_1 *is a concave function on* $(0, \infty)$.

Proof. From the existence of a third moment, we have that $y^2 N_\beta[|x| \geq y]$ is integrable on $(0, \infty)$. Consequently, u_3 exists and

$$(3.6) \qquad u_3(h) = -\int_0^\infty y^2 N[|x| \geq y] \sin(hy)\, dy, \qquad h \in \mathbf{R},$$

by (3.5). Let $l(y)$ be any decreasing function on $(0, \infty)$ and $\alpha > 0$. Then

$$\begin{aligned}\int_0^\infty l(y) \sin(\alpha y)\, dy &= \sum_{k=0}^{\infty} \int_{\alpha^{-1}(2\pi k, 2\pi(k+1)]} l(y) \sin(\alpha y)\, dy \\ &= \alpha^{-1} \sum_k \int_{(0,2\pi]} l(\alpha^{-1}(x + 2k\pi)) \sin x\, dx\end{aligned}$$

and

$$\begin{aligned}\int_{(0,2\pi]} l(\alpha^{-1}(x + 2k\pi) \sin x\, dx &= \int_{(0,\pi]} \cdots + \int_{(\pi,2\pi]} \cdots \\ &= \int_{(0,\pi]} [l(\alpha^{-1}(x + 2k\pi)) - l(\alpha^{-1}(x + (2k+1)\pi)] \sin x\, dx > 0.\end{aligned}$$

Consequently, $u_3 < 0$ and the proof is completed.

Remark 3.2. The argument used in Proposition 3.2 gives an alternative proof that $u_1 > 0$ on $(0, \infty)$; cf. (2.11).

Corollary 3.4. *For the class* $\mathcal{L}$ *Ising models with* $\varphi \in L$ *and* $\varphi''(0)$ *finite (in (2.4)) Ursell function* u_2 *is the Fourier transform of a positive density (not necessarily a probability density).*

Proof. Since $\varphi''(0)$ exists iff $\psi''(0)$ exists (ψ is the background *rv* for φ), and N is the Lévy spectral measure of ψ, therefore by (3.5)

$$u_2(h) = \int_0^\infty N[|x| \geq y] y \cos(hy)\, dy = \int_{-\infty}^\infty e^{itx} g(x)\, dx$$

where $g(x) := \frac{1}{2}|x|N[|s| \geq |x|]$, $x \in \mathbf{R}$, is the density of u_2. Of course, $u_2(h)/u_2(0)$ is a characteristic function of a probability density and the proof is complete.

4 Appendix

For more details on class L distributions see Loeve (1963), p. 322 or Jurek & Mason (1993), p. 177. The most crucial characterization of selfdecomposability needed here is the *random integral representation*,i.e.,

$$\text{(A1).} \quad \varphi \in L \quad \text{iff} \quad \varphi \textit{ is char. f. of a random integral } \int_{(0,\infty)} e^{-t} dY(t),$$

for uniquely determined Lévy process Y satisfying $\mathbf{E}\log(1+|Y(1)|) < \infty$. [We refer to Y as the **B**ackground **D**riving **L**évy **P**rocess, in short: BDLP, of φ.]

Equivalently, in terms of characteristic functions, the above reads

$$\text{(A2).} \quad \varphi \in L \quad \text{iff} \quad \log\varphi(t) = \int_0^1 \log\psi(st)s^{-1}ds = \int_0^t \log\psi(r)r^{-1}dr,$$

where ψ is the characteristic function of $Y(1)$ in (A1). We refer to ψ as the *background driving char. function* (rv) of φ from L.

(A3). Besides stable laws, L contains among others: gamma random variables, (in particular, χ^2-distributions), Student's t-distribution, log-normal, Fisher's F-distributions, generalized inverse Gaussian, hyperbolic distributions, hyperbolic sine and cosine, logarithms of gamma, etc. (cf. Jurek (1997)).

References

R.J. Baxter (1982). *Exactly Solved Models in Statistical Mechanics*, Academic Press, New York.

N.H. Bingham, C.M. Goldie, J.L. Teugels (1987). *Regular Variation*, Cambridge University Press, Cambridge, U.K.

J. De Coninck (1984). Infinitely divisible distributions functions of class L and the Lee-Yang Theorem, *Commun. Math. Phys.* **96**, 373–385.

R. Cuppens (1975). *Decomposition of Multivariate Probabilities*, Academic Press, New York.

W. Feller (1971). *An Introduction to Probability Theory and its Applications, vol. II*, (2nd edition), J. Wiley, New York.

R.B. Griffiths, C.A. Hurst and S. Sherman (1970). Concavity of magnetization of an Ising ferromagnet in a positive external field, *J. Math. Phys.* **11**, 790–798.

Z.J. Jurek (1996). Series of independent exponential random variables. In: *Proc 7th Japan-Russia Symposium on Probab. Theory and Math. Statistics*, S. Watanabe, M. Fukushima, Yu.V. Prohorov, A.N. Shiryaev, eds., World Scientific, Singapore, pp. 174–182.

Z.J. Jurek (1997). Selfdecomposability: an exception or a rule? *Annales Univ. M. Curie-Sklodowska, Lublin-Polonia* **51** 1,10, Section A, 93–107.

Z.J. Jurek, J.D. Mason (1993). *Operator-Limit Distributions in Probability Theory*, J. Wiley & Sons, New York.

M.A. Krasnoselskii, Ya. B. Rutickii (1961). *Convex Functions and Orlicz Spaces*, P. Noordhoff, Groningen.

M. Loeve (1963). *Probability Theory*, D. van Nostrand Co., Princeton, New Jersey.

Ch. M. Newman (1974). Zeros of the partition function for generalized Ising systems, *Commun. Pure and Appl. Math.* **27**, 143–159.

Ch. M. Newman (1975). Inequalities for Ising models and field theories which obey the Lee-Yang theorem, *Commun. Math. Phys.* **41**, 1–9.

A. Sokal (1981). More inequalities for critical exponents, *J. Stat. Phys.* **25**, 25–50.

Joël De Connick
Faculté des Sciences
Université de Mons-Hainaut
B-7000 Mons, Belgium
joel@gibbs.umh.ac.be

Zbigniew J. Jurek
Institute of Mathematics
University of Wrocław
50-384 Wrocław, Poland
zjjurek@math.uni.wroc.pl

When an Isotropic Random Vector is Pseudo-Isotropic

Jolanta K. Misiewicz and Krzysztof Tabisz

ABSTRACT A random vector $(X_1, \dots X_n)$ is called *isotropic* if there exists a quasi-norm q on $\mathbb{R}^n$ such that the density level curves for this vector are of the form $\{x : q(x) = const\}$. A random vector $(X_1, \dots X_n)$ is called pseudo-isotropic if the level curves for its characteristic function are of the form $\{x : c(x) = r\}$, $r \geq 0$, for some quasi-norm c on $\mathbb{R}^n$. We discuss in the paper random vectors having both properties. We show for example that an isotropic random vector which is symmetric and stable must be sub-Gaussian.

1 Isotropic random vectors

We define the random vector U_q on the unit sphere $S_q = \{x \in \mathbb{R}^n : q(x) = 1\}$ having the distribution concentrated on S and defined by the differential form

$$\omega(dx) = c(n,q) \sum_{j=1}^{\infty} (-1)^{j+1} x_j dx_1 \dots dx_{j-1} dx_{j+1} \dots dx_n,$$

where c(n,q) is the normalizing constant.

Definition 1. *A random vector $X = (X_1, \dots, X_n)$ is isotropic if there exists a quasi-norm q on $\mathbb{R}^n$ and a nonnegative random variable Θ independent of U_q on S_q such that*

$$X \stackrel{d}{=} U_q \cdot \Theta.$$

Notice that the above definition coincides with the natural definition of ℓ_p-isotropic random vectors when q is given by the ℓ_p-norm $\|\cdot\|_p$ on $\mathbb{R}^n$. Indeed, if X is ℓ_p-isotropic, then its density is of the form $f(\|x\|_p)$ for some measurable and integrable function f on $\mathbb{R}$. Writing now the vector X in the form

$$X = \left(\frac{X_1}{\|X\|_p}, \dots, \frac{X_n}{\|X\|_p} \right) \cdot \|X\|_p,$$

we can easily see that $U = X/\|X\|_p$ and $\Theta = \|X\|_p$ are independent and the density of the variable Θ is equal to $c(n, \|\cdot\|_p)^{-1} f(r)$. The normalizing constant $c(n, \ell_p)$ can be easily calculated in this case:

$$1 = \int_{S_p^n} \cdots \int \omega(dx) = nc(n, \ell_p) \int_{B_p^n} \cdots \int dx_1 \dots dx_n = \frac{2^n \Gamma(1/p)^n c(n, \ell_p)}{p^{n-1} \Gamma(n/p)},$$

where B_p^n is the unit ball, and S_p^n is the unit sphere in the quasi-norm $\| \, \|_p$. For an isotropic random vector $X \in \mathbb{R}^n$ it is not true that its projections into lower dimensional subspaces are also isotropic. However, if the quasi-norm q is of the form

$$q(\xi)^p = \sum_{k=1}^{n} \alpha_k |\xi_k|^p, \qquad \xi \in \mathbb{R}^n,$$

for some positive constants $\alpha_1, \ldots, \alpha_n$, then for every choice of coordinates $k_1, \ldots, k_m$, $m < n$ the random vector $(X_{k_1}, \ldots, X_{k_m})$ is isotropic with the function q_1 given by

$$q_1(\xi)^p = \sum_{i=1}^{m} \alpha_{k_i} |\xi_{k_i}|^p, \qquad \xi \in \mathbb{R}^m.$$

The characteristic function for the isotropic random vector $X = U\Theta$, with the quasi-norm q, U and Θ independent and $\mathbf{P}\{\Theta \geq 0\} = 1$ can be expressed as

$$\mathbf{E} \exp\{i < \xi, X >\} = \int_0^\infty I_q(\xi r) \lambda(dr), \qquad \xi \in \mathbb{R}^n,$$

where λ is the distribution of Θ and I_q is the characteristic function of the random vector U:

$$I_q(\xi) = \int \underset{S}{\cdots} \int \cos < \xi, u > \omega(du).$$

Only in a very few special cases can something more be said about the function $I_q(\xi)$. The best known example is the characteristic function of the rotationally invariant random vector U_2^n concentrated on the Euclidean sphere $S_2^n \subset \mathbb{R}^n$. In this case we know that $I_2(\xi) = \Omega_n(\|\xi\|_2)$ and

$$\begin{aligned}\Omega_n(\|\xi\|_2) \overset{def}{=} \ & \mathbf{E} \exp\left\{ i \sum_{k=1}^{n} \xi_k U_k \right\} = \Gamma\left(\frac{n}{2}\right) \left(\frac{2}{\|\xi\|_2}\right)^{n/2-1} J_{\frac{n-2}{2}}(\|\xi\|_2) \\ = \ & \frac{2\Gamma(n/2)}{\sqrt{\pi}\Gamma(n-1/2)} \int_0^{\pi/2} \cos(\|\xi\|_2 \sin\varphi) \cos^{n-2}(\varphi) d\varphi,\end{aligned}$$

where J_ν denotes the Bessel function of the first kind (for details see I. J. Schoenberg (1938)). As we can see here, the characteristic function of any rotationally invariant random vector (i.e. isotropic with the function q defined by the Euclidean norm) $X = U_2^n \Theta$ depends only on $\|\xi\|_2$, so it is also rotationally invariant. This property seems to be exceptional, i.e. characteristic functions of isotropic random vectors would rather have unbounded level curves, except in the Euclidean case.

The next well known example is connected with ℓ_1-isotropic random vectors (for details see S. Cambanis, R. Keener, and G. Simons). For $n = 2$, and $|a| \neq |b|$ we have

$$I_1((a,b) = \frac{a \sin a - b \sin b}{a^2 - b^2}.$$

For $n = 3$ and $|a| \neq |b| \neq |c|$,

$$I_1(a,b,c) = 8\left[\frac{a^2 \cos a}{(a^2-b^2)(c^2-a^2)} + \frac{b^2 \cos b}{(a^2-b^2)(b^2-c^2)} + \frac{c^2 \cos c}{(b^2-c^2)(c^2-a^2)}\right].$$

If $p = \infty$ and $p = 3$, we obtain (for details see. J.K. Misiewicz (1989))

$$I_\infty(a,b,c) = \tfrac{2}{abc}\left[(-a+b+c)\cos(-a+b+c) + (a-b+c)\cos(a-b+c)\right.$$
$$\left. + (a+b-c)\cos(a+b-c) - (a+b+c)\cos(a+b+c)\right],$$

when $abc(a+b+c)(a+b-c)(a-b+c)(-a+b+c) \neq 0$.

Finally let us present here for the first time the function $I_{1/2}(a,b)$:

$$I_{1/2}(a,b) = \tfrac{2ab\left(b \sin a + a \sin b\right)}{(a^2-b^2)^2} + \tfrac{\sqrt{2\pi}}{4(a^2-b^2)^{5/2}} \times$$
$$\left\{\left[C\left(\tfrac{b}{a-b}\right) - C\left(\tfrac{a}{a-b}\right)\right](a+b)^{5/2}\left[(a-b)\sin\left(\tfrac{ab}{a-b}\right) + 2ab\cos\left(\tfrac{ab}{a-b}\right)\right]\right.$$
$$+\left[S\left(\tfrac{b}{a-b}\right) - S\left(\tfrac{a}{a-b}\right)\right](a+b)^{5/2}\left[2ab\sin\left(\tfrac{ab}{a-b}\right) - (a-b)\cos\left(\tfrac{ab}{a-b}\right)\right]$$
$$+\left[C\left(\tfrac{b}{a+b}\right) + C\left(\tfrac{a}{a+b}\right)\right](a-b)^{5/2}\left[(a+b)\sin\left(\tfrac{ab}{a+b}\right) + 2ab\cos\left(\tfrac{ab}{a+b}\right)\right]$$
$$\left.+\left[S\left(\tfrac{b}{a+b}\right) + S\left(\tfrac{a}{a+b}\right)\right](a-b)^{5/2}\left[2ab\sin\left(\tfrac{ab}{a+b}\right) + (a+b)\cos\left(\tfrac{ab}{a+b}\right)\right]\right\},$$

where $S(r)$ and $C(r)$ are the Fresnel Sine and Cosine functions.

Let us recall here that a random vector $X \in \mathbb{R}^n$ is called sub-Gaussian if there exists a Gaussian random vector $Y \in \mathbb{R}^n$ and a positive random variable Θ such that X and $Y\sqrt{\Theta}$ have the same distribution.

Lemma 1. *Assume that a symmetric random vector (R,S) has compact support in $\mathbb{R}^2$, and assume that a symmetric α-stable random vector $X = (X_1, X_2)$ has the same distribution as $(R,S)\Theta$ for some nonnegative random variable Θ independent of (R,S). Then X is sub-Gaussian.*

Proof. Let R_1 be a random variable with the same distribution as R independent of R, S and Θ. We will consider the random vector $Y = X|R_1|$. Since X is symmetric α-stable, the characteristic function of Y can be written in the following way:

$$\varphi(a,b) = \mathbf{E}\exp\{i(aX_1 + bX_2)|R_1|\} = \mathbf{E}\exp\{-c(a,b)^\alpha |R|^\alpha\},$$

where the quasi-norm c is defined as the L_α norm appearing in the characteristic function of X: $\mathbf{E}\exp\{i(aX_1+bX_2)\} = \exp\{-c(a,b)^\alpha\}$. On the other hand we have:

$$\varphi(a,b) = \mathbf{E}\exp\{i(aR+bS)Q\},$$

where $Q = \Theta\cdot|R_1|$. Both formulas coincide for all $a, b \in \mathbb{R}^2$. For $b=0$ we obtain

$$\mathbf{E}\exp\{iaX_1|R|\} = \frac{1}{2}\mathbf{E}\exp\{ia|R|Q\} + \frac{1}{2}\mathbf{E}\exp\{-ia|R|Q\},$$

which means that the random variable $\widetilde{Q}$ defined as $\widetilde{Q} = Q$ with probability $1/2$ and $\widetilde{Q} = -Q$ with probability $1/2$ has the same distribution as X_1, i.e. symmetric α-stable with the characteristic function $\exp\{-|a|^\alpha c(1,0)^\alpha\}$. Since the random vector (R,S) is symmetric, it is easy to see that $(R,S)Q$ has the same distribution as $(R,S)\widetilde{Q}$. Finally we obtain that for every fixed $k \in \mathbb{R}$

$$\mathbf{E}\exp\{-c(a,ka)^\alpha|R|^\alpha\} = \mathbf{E}\exp\Big\{i(aR+kaS)\widetilde{Q}\Big\} = \mathbf{E}\exp\{-|aR+kaS|^\alpha\}.$$

The above equality we treat as the equality of Laplace transforms of two nonnegative random variables at the point $|a|^\alpha$, which implies that for every $k\in\mathbb{R}$ the random variables $c(1,k)^\alpha|R|^\alpha$ and $|R+kS|^\alpha$ have the same distribution. Since the random vector (R,S) has compact support, all its weak moments are finite and

$$c(a,b)^2\mathbf{E}R^2 = a^2\mathbf{E}\Big(c(1,k)^\alpha|R|^\alpha\Big)^{2/\alpha} = a^2\mathbf{E}\Big(|R+kS|^\alpha\Big)^{2/\alpha} = \mathbf{E}\Big|aR+bS\Big|^2.$$

This shows that the function $c(a,b)$ is an L_2-norm on $\mathbb{R}^2$. Now it is easy to see that X has the same distribution as $\Gamma\Theta_{\alpha/2}$, where $\Theta_{\alpha/2}$ is $\alpha/2$-stable positive random variable with the Laplace transform $\exp\{-2t^{\alpha/2}\}$ independent of the symmetric Gaussian random vector Γ with the covariance matrix $(\mathbf{E}(R^iS^j))_{i+j=2}$, which ends the proof. □

As an easy consequence of the Lemma we have

Theorem 2. *Assume that a symmetric α-stable random vector $X\in\mathbb{R}^2$ is isotropic with the quasi-norm $q:\mathbb{R}^2\mapsto[0,\infty)$. Then X is sub-Gaussian and the quasi-norm q is a norm defined by an inner product in $\mathbb{R}^2$.*

2 Pseudo-isotropic random vectors

We give here only some necessary information about psudo-isotropic random vectors. Much more can be found in J.K. Misiewicz (1996).

Definition 3. *A symmetric random vector $X = (X_1, \dots, X_n)$ is pseudo-isotropic if for every $\xi \in \mathbb{R}^n$ there exists a non-negative constant $c(\xi)$ such that*

$$< \xi, X > \stackrel{d}{=} c(\xi) X_1.$$

The characteristic function of the pseudo-isotropic random vector X at the point $t\xi$ can be written as

$$\mathbf{E} \exp\{i < t\xi, X >\} = \mathbf{E} \exp\{ic(t\xi)X_1\} \stackrel{def}{=} \varphi(|t| c(\xi)),$$

which means that all one-dimensional projections $< \xi, X >$ of the random vector X have the same distributions up to a scale parameter $c(\xi)$. It is easy to see that all rotationally invariant random vectors, symmetric Gaussian random vectors and symmetric stable random vectors are pseudo-isotropic. Without loss of generality we will assume that $c(\xi) = 0$ if and only if $\xi = 0$.

There are still open questions in this area: we do not know the full characterization of the functions c for which there exists a non-trivial function φ such that $\varphi(c(\xi))$ is a characteristic function on $\mathbb{R}^n$. Also the full characterization of pseudo-isotropic random vectors is unknown.

Fixing the function $c : \mathbb{R}^n \mapsto [0, \infty)$ by $\mathcal{M}(c, n)$ we will denote the set of all distributions of pseudo-isotropic random vectors with the characteristic function of the form $\varphi(c(\xi))$ for some function φ. It is known (see J. K. Misiewicz (1996)) that the set $\mathcal{M}(c, n)$ is weakly closed, convex and is equal to the weak closure of all linear combinations of its extreme points. However, the full characterization of extreme points fot $\mathcal{M}(c, n)$ in finite dimensional space $\mathbb{R}^n$, exists only in only two cases:

if $c(\xi)^2 =< \xi, \Sigma\xi >$ for some positive definite matrix $\Sigma = A^* A$, then $\{T_a \mu : a \geq 0\}$ is the set of extreme points for $\mathcal{M}(c, n)$, where $\mu = \mathcal{L}(AU_2^n)$ and $T_a \mu(A) = \mu(A/a)$ if $a > 0$ and $T_0 \mu = \delta_0$;

if $c(\xi) = \sum_{k=1}^n |\xi_k|$ then $\{T_a \nu : a \geq 0\}$ is the set of extreme points for $\mathcal{M}(c, n)$, where

$$\nu = \mathcal{L}\left(\frac{U_1}{D_1}, \dots, \frac{U_n}{D_n}\right),$$

vector $U_2^n = (U_1, \dots, U_n)$ has uniform distribution on the unit sphere S_2^n, $D = (D_1, \dots, D_n)$ has Dirichlet distribution with parameters $(1/2, \dots, 1/2)$. This representation follows from the very suprising formula given by Cambanis, Keener and Simons (1983):

$$\int_0^{\pi/2} f\left(\frac{a^2}{\sin^2 x} + \frac{b^2}{\cos^2 x}\right) dx = \int_0^{\pi/2} f\left(\frac{(|a| + |b|)^2}{\sin^2 x}\right) dx,$$

which holds for every measurable function f for which one of these integrals exists. Rewriting this equality in the language of Dirichlet distributions we have

$$\int_0^1 f\left(\frac{a^2}{t^2} + \frac{b^2}{(1-t^2)}\right) (1-t^2)^{-1/2} dt = \int_0^1 f\left(\frac{(|a| + |b|)^2}{t^2}\right) (1-t^2)^{-1/2} dt.$$

We can give now some more information about the set of extreme points of the set $\mathcal{M}(c,n)$:

Theorem 4. *Assume that the class $\mathcal{M}(c,n)$ contains a measure $\mu \neq \delta_0$ such that $\int |x|^\varepsilon \mu(dx) < \infty$ for some $\varepsilon > 0$. If the function c is not a norm defined by some inner product on $\mathbb{R}^n$, then there exist at least two extreme point of $\mathcal{M}(c,n)$ having different density level curves, or there exists at least one extreme point of $\mathcal{M}(c,n)$ which is not isotropic.*

Proof. Assume that every extreme point of $\mathcal{M}(c,n)$ is isotropic with the fixed function $q : \mathbb{R}^n \mapsto [0,\infty)$. It was proven by J. Misiewicz in 1996 that under our assumptions for every $r \in (0, \varepsilon \wedge 2)$, the set $\mathcal{M}(c,n)$ contains a symmetric r stable measure γ_r with the characteristic function

$$\widehat{\gamma_r}(\xi) = \exp\left\{ - \int | < \xi, x > |^r \mu(dx) \right\} = \exp\left\{ -c(\xi)^r \int_{-\infty}^{\infty} |t|^r \mu_1(dt) \right\},$$

where μ_1 is the distribution of X_1 if $\mathcal{L}((X_1,\dots,X_n)) = \mu$. This means that for every such r the measure γ_r is a weak limit of convex linear combinations of the extreme points of $\mathcal{M}(c,n)$, i.e.

$$\gamma_r \Leftarrow \sum_{k=1}^{n} \alpha_{k,n}\nu_{k,n}, \qquad \sum_{k=1}^{n} \alpha_{k,n} = 1,\ \alpha_{k,n} > 0,$$

where $\nu_{k,n}$ is an extreme point of $\mathcal{M}(c,n)$ for every k and n. If all measures $\nu_{k,n}$ have the same level curves of the form $\{x : q(x) = const\}$, then also γ_r would have the same level curves. This however is impossible in view of Theorem 2 and the assumption that c is not a norm defined by some inner product in $\mathbb{R}^n$. □

Our definitions of isotropic and pseudo-isotropic random vectors can be extended to vectors taking values in infinite dimensional spaces. In the simplest case we can extend it to sequences of random variables:

A sequence of random variables $X_1, X_2, \dots$ is *isotropic* if for every $n \in \mathbb{N}$ the random vector $(X_1,\dots,X_n)$ is isotropic;

A sequence of random variables $X_1, X_2, \dots$ is *pseudo-isotropic* if for every $n \in \mathbb{N}$ the random vector $(X_1,\dots,X_n)$ is pseudo-isotropic.

By $\mathbb{R}^{(\infty)}$ we will denote the set of all sequences $\xi \in \mathbb{R}^{\mathbb{N}}$ for which the number of these $k \in \mathbb{N}$ that $\xi_k \neq 0$ is finite. We will consider the space $\mathbb{R}^{(\infty)}$ equipped with the quasi-norm c defined by the pseudo-isotropic sequence of random variables $X_1, X_2, \dots$ in such a way that for every $\xi \in \mathbb{R}^{(\infty)}$, $c(\xi)$ is the same constant for which $\sum_{k=1}^{\infty} \xi_k X_k \stackrel{def}{=} c(\xi) X_1$. The definition is correct, because in this case only a finite number of components are different from 0.

Theorem 5. *In both of the following cases none of the extreme points of $\mathcal{M}(c)$ is isotropic:*

1) $c(\xi) = \sum_{k=1}^{n} |\xi_k|$, $n \geq 2$;
2) the space $(\mathbb{R}^{(\infty)}, c)$ *contains* ℓ_α *(at least uniformly) for some* $\alpha < 2$ *and* α *is the maximal such* β *that* $(\mathbb{R}^{(\infty)}, c)$ *embeds isometrically into some* L_β*-space.*

Proof. It is enough to notice that in both cases there exists a measure ν_0 such that all the extreme points of $\mathcal{M}(c)$ are of the form $T_a\nu_0$, $a \geq 0$. In case 1) this follows from Cambanis, Keener and Simons' result. In case 2) it was proven by Misiewicz in 1996 that ν_0 is the distribution of a sequence of symmetric α-stable variables. Thus either all the extreme points are isotropic or none are, and in view of the previous theorem the second possibility holds. □

It is not true that convex linear combinations of non-isotropic measures are non-isotropic, thus the general question is still open: is it true that a measure μ which is both isotropic and pseudo-isotropic must be sub-Gaussian?

Since Cambanis, Keener and Simons published their formula in 1983 many mathematicians have tried to obtain a similar result replacing the ℓ_1 norm with the ℓ_p-norm for at least one $p \in (0,1) \cup (1,2)$. We propose here studying the following function for $p > 0$:

$$F_p(a,b) = \int_0^1 f\left(\frac{a^2}{t^2} + \frac{b^2}{(1-t^p)^{2/p}}\right)(1-t^p)^{1/p-1}dt.$$

Using Mathematica, the function $F_2(a,b)$ has level curves similar to the following for many different continuous functions:

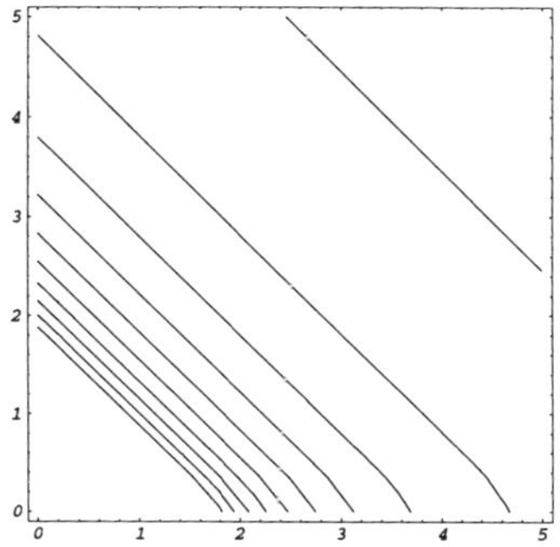

$p = 2.0$

Although the calculations are not very precise, the obtained level curves are pretty close to the real $\{(a,b) : |a| + |b| = constant\}$. We give here several examples of the maps obtained for different parameters $p > 0$. The level curves of $F_p(a,b)$ define some isotropic convex sets for $p \geq 1$, and some isotropic concave sets for $p < 1$.

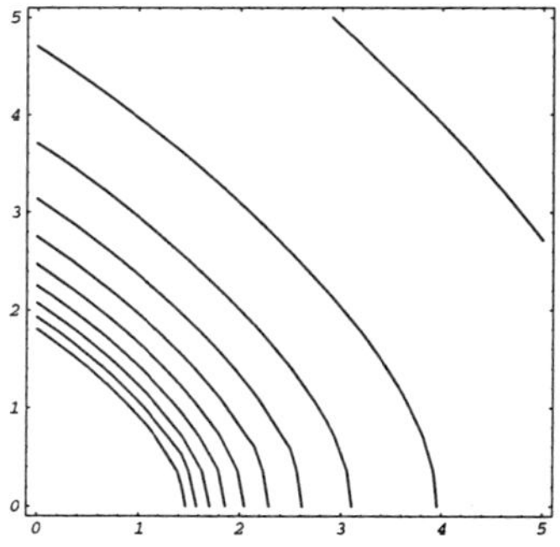

$p = 5.0$

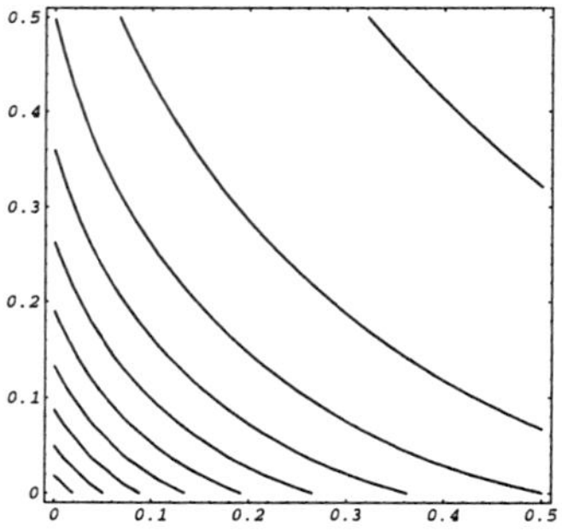

$p = 0.5$

References

S. Cambanis, R. Keener, and G. Simons (1983). On α-symmetric distributions. *J. Multivariate Anal.* **13**, 213–233.

W. Feller (1966). *An Introduction to Probability Theory and its Application*, volume II. Wiley, New York.

R.D. Gupta, J.K. Misiewicz, D.St.P. Richards (1996). Infinite sequences with sign–symmetric Liouville type distributions. *Probab. and Math. Statist.* **16**, 29–44.

J.K. Misiewicz (1989). Positive definite functions on l^∞. *Statist. Probab. Letters* **8**, 255–260.

J.K. Misiewicz (1996). Sub-stable and pseudo-isotropic processes. Connections with the geometry of sub-spaces of L_α-spaces. *Dissertationes Mathematicae* **CCCLVIII**.

D.St.P. Richards (1985). Positive definite symmetric functions on finite dimensional spaces. *Statist. Probab. Letters* **3**, 325–329.

D.St.P. Richards (1986). Positive definite symmetric functions on finite dimensional spaces. I. Application of the Radon transform. *J. Multivariate Anal.* **19**, 280–298.

I. J. Schoenberg (1938). Metric spaces and completely monotonic functions. *Ann. Math.* **38**, 811–841.

Jolanta K. Misiewicz
Instytut Matematyki
Politechnika Zielonogórska
ul. Podgórna 50
65-246 Zielona Góra, Poland
jolanta@axel.im.pz.zgora.pl

Krzysztof Tabisz
Instytut Matematyczny
Uniwersytet Wrocławski
Plac Grunwaldzki 2/4
50-384 Wrocław, Poland
tabisz@math.uni.wroc.pl

Support Fragmentation for Multiplicative Cascade Measures

Mina Ossiander

ABSTRACT The family of h-cascades arises naturally in the study of multiplicative cascades. We show that the limiting h-cascades deriving from a single collection of cascade generators have disjoint supports with probability 1 and are consequently a.s. mutually singular. This result relies upon the derivation of a fragmentation level for the individual multiplicative h-cascade measures using large deviation type techniques.

1 Introduction

The history of multiplicative random cascades goes back to Kolmogorov (1941), who proposed an early version as a model of energy dissipation in turbulence. They have since then been explored as models for highly intermittent random phenomena by a diverse group of researchers; c.f. Frisch (1995). Random cascade measures can be viewed as random measures on the unit interval. In this setting they arise as the a.s. weak limit of a sequence of measures that are absolutely continuous with respect to Lebesgue measure. Paradoxically, cascade measures are themselves generally singular with respect to Lebesgue measure. Indeed their support sets are typically highly intermittent, with Hausdorff dimension less than 1. The goal of this paper is to show that a family of multiplicative cascades, constructed quite naturally from a single family of cascade generators, admits a family of support sets that are disjoint with probability 1. As a consequence, this family of cascade measures forms, with probability 1, a collection of mutually singular random measures.

Let $b \geq 2$ be a natural number and let

$$\mathbf{T} = \{0, 1, 2, \ldots, b-1\}^{\mathbf{N}} \tag{1.1}$$

denote an infinite-dimensional product space. $\mathbf{T}$ can be thought of as a b-ary tree with the elements $t = (t_1, t_2, t_3, \ldots)$ representing paths through the tree. The space $\mathbf{T}$ can be endowed with the metric

$$\rho(s,t) = b^{-|s \wedge t|}, s, t \in \mathbf{T}, \tag{1.2}$$

where $|s \wedge t| = \inf\{n \geq 0 : s_{n+1} \neq t_{n+1}\}, s = (s_1, s_2, \dots), t = (t_1, t_2, \dots) \in \mathbf{T}$. For $t \in \mathbf{T}$, let $t|n = (t_1, t_2, \dots, t_n)$. That is, $t|n$ can be thought of as the truncation of the path t through the tree $\mathbf{T}$ at the node that is distance n from the root. Denote the closed ball of radius b^{-n} containing t by $\Delta_n(t)$ and let λ represent normalized Haar measure on $\mathbf{T}$. Note that

$$\Delta_n(t) = \{s \in \mathbf{T} : s|n = t|n\}, \tag{1.3}$$

with

$$\lambda(\Delta_n(t)) = b^{-n}. \tag{1.4}$$

Let $\mathcal{B}(\mathbf{T})$ denote the Borel σ-field on $\mathbf{T}$ generated by $\{\Delta_n(t) : n \geq 1, t \in \mathbf{T}\}$. If $\mathbf{T}$ is identified with the unit interval $[0,1]$, then λ can be thought of as Lebesgue measure on $([0,1], \mathcal{B}([0,1]))$.

Let W be a non-negative random variable with finite mean equal to 1, and define the structure function (a modified cumulant generating function) associated with W to be

$$\chi_b(h) = \log_b E(W^h \mathbf{1}_{[W>0]}) - (h-1) \tag{1.5}$$

for all h in $\mathbf{R}$ with $E(W^h \mathbf{1}_{[W>0]}) < \infty$. (Here we understand $0/0 = 0$.) Notice that $\chi_b(1) = 0$. It is easy to verify that χ_b is defined on an interval (possibly infinite) containing $[0,1]$, and on the interior of this interval is twice differentiable with non-negative second derivative. Indeed, $\chi_b''(h) \equiv 0$ corresponds to $P(W = p^{-1}) = 1 - P(W = 0) = p$ for some $p \in (0,1]$. This particular cascade model is sometimes referred to as the beta model.

Define on a probability space $(\Omega, \mathcal{F}, P)$ a denumerable collection $\{W_v : v = t|n, n \geq 1, t \in \mathbf{T}\}$ of i.i.d. copies of the random variable W, indexed by the nodes of the tree $\mathbf{T}$. Let $\{\mathcal{F}_n, n \geq 1\}$ denote the filtration defined by

$$\mathcal{F}_n = \sigma\{W_v : |v| \leq n\} \text{ for } n \geq 1, \tag{1.6}$$

where $|v| = n$ if $v = t|n$ for some $t \in \mathbf{T}$. For $n \geq 1$, define a sequence $\{\lambda_n : n \geq 1\}$ of random measures on $(\mathbf{T}, \mathcal{B}(\mathbf{T}))$ via

$$\frac{d\lambda_n}{d\lambda}(t) = \prod_{i=1}^{n} W_{t|i}, \qquad t \in \mathbf{T}. \tag{1.7}$$

An appeal to $\mathbf{T}$-martingale theory (see Kahane and Peyrière (1976)) shows that there exists a random measure λ_∞ on $(\mathbf{T}, \mathcal{B}(\mathbf{T}))$ with λ_n converging vaguely to λ_∞ a.s. P. The measure λ_∞ is known as a multiplicative cascade measure. This a.s. convergence allows the following representation of λ_∞ on the balls $\Delta_n(t)$ as a product of $\lambda_n(\Delta_n(t))$ and a copy, $Z(t|n)$, of $\lambda_\infty(\mathbf{T})$ which is independent of $\mathcal{F}_n$:

$$\lambda_\infty(\Delta_n(t)) = Z(t|n)\lambda_n(\Delta_n(t)). \tag{1.8}$$

The random measure $\lambda_\infty(\mathbf{T})$ has positive mass with positive probability (i.e. λ_n converges weakly to λ_∞) if and only if $EW \log_b W < 1$. This is stated in terms of the structure function as part of the following well-known theorem. Parts (a) and (b) of this theorem are due to Kahane and Peyrière (1976). Part (c) can be found in Kahane (1987).

Theorem 1. STRUCTURE THEOREM

(a) (Nondegeneracy) $E\lambda_\infty(\mathbf{T}) > 0$ *iff* $\chi_b'(1-) < 0$.

(b) (Convergence of moments) $E\lambda_\infty^h(\mathbf{T}) < \infty$ *for* $0 \le h \le 1$ *and if* $h_c := \sup\{h \ge 1 : \chi_b(h) \le 0\} > 1$ *then* $E\lambda_\infty^h(\mathbf{T}) < \infty$ *for* $1 < h < h_c$.

(c) (Support dimension) If $\lambda_\infty(\mathbf{T}) > 0$, *then* λ_∞ *is a.s. supported on a set with Hausdorff dimension* $-\chi_b'(1)$.

Note that there is a close connection between cascades and branching processes. For example, the beta cascade model mentioned above can be identified with a Galton-Walton branching process with each individual independently giving birth to a binomial (b, p) number of offspring. With this identification, degeneracy of λ_∞ ($P(\lambda_\infty(\mathbf{T}) = 0) = 1$) corresponds to the branching process dying out with probability 1. Since $\chi_b'(1) = \log_b(1/bp)$, it is easy to see that the cascade measure λ_∞ is positive with positive probability exactly when the expected number of offspring in the corresponding Galton-Watson branching process is greater than 1.

The Rényi exponent associated with a realization of the multiplicative cascade measure λ_∞ is defined to be

$$\tau(h) = \lim_{n\to\infty} \frac{\ln \sum_{|t|=n} \lambda_\infty^h(\Delta_n(t))}{\ln(b^{-n})}. \tag{1.9}$$

This exponent is of interest to physicists (c.f. Molchan (1996)) and also arises in the problem of estimating an unknown structure function from observation of a single realization of λ_∞ on the balls $\Delta_n(t)$. (See Troutman and Vecchia (1999) or Ossiander and Waymire (2000) for details on the latter.) The decomposition (1.8) gives

$$\lambda_\infty^h(\Delta_n(t)) = Z^h(t|n) \prod_{1\le i\le n} W_{t|i}^h \quad b^{-nh}. \tag{1.10}$$

This suggests thinking of $\sum_{|t|=n} \lambda_\infty^h(\Delta_n(t))$ as an integral with respect to a measure formed from the suitably renormalized collection $\{W_v^h : v = t|n, n \ge 1, t \in \mathbf{T}\}$. This in turn suggests that the multiplicative cascade measure λ_∞ is just one among a natural family of multiplicative cascades indexed by h. In the following section we give a careful definition of the family of h-cascades and state the main support theorem. In Section 3 we give fragmentation theorems for the h-cascades along with the proof of the support theorem. We conclude with some remarks in Section 4.

2 The h-cascades and the support theorem

The family of multiplicative h-cascade measures is constructed as follows. Fix h with $E(W^h \mathbf{1}_{[W>0]})$ finite. Define $\lambda_n(h, \cdot)$ via

$$\text{(2.1)} \qquad \frac{d\lambda_n}{d\lambda}(h,t) = \frac{\prod_{i=1}^n W_{t|i}^h \mathbf{1}_{[W_{t|i}>0]}}{E^n W^h \mathbf{1}_{[W>0]}}, \qquad t \in \mathbf{T}.$$

Notice that these sequences of measures are formed by letting the random functions

$$\text{(2.2)} \qquad \frac{W_v^h \mathbf{1}_{[W_v>0]}}{EW^h \mathbf{1}_{[W>0]}}$$

take on the role of the W_v's in (1.7) above. In particular, the sequence of random measures $\lambda_n(\cdot)$ defined in (1.7) becomes the sequence $\lambda_n(1, \cdot)$ in (2.1). The Kahane-Peyrière $\mathbf{T}$-martingale theory and the structure theorem used together provide a fairly complete picture of vague convergence for each fixed h. Indeed, for h fixed, $\lambda_n(h, \cdot)$ converges vaguely to a multiplicative cascade measure $\lambda_\infty(h, \cdot)$ a.s. P. Using the analog to the decomposition given in (1.8) above, we have the following iterative decomposition of each $\lambda_\infty(h, \cdot)$ measure on the balls $\Delta_n(t)$:

$$\text{(2.3)}\ \lambda_\infty(h, \mathbf{T}) = \sum_{|t|=n} \lambda_\infty(h, \Delta_n(t)) = \sum_{|t|=n} Z(h, t|n) \lambda_n(h, \Delta_n(t)),$$

where for each fixed $n \geq 1$, $\{Z(h, t|n) : t \in \mathbf{T}\}$ is a collection of i.i.d. copies of $\lambda_\infty(h, \mathbf{T})$. The structure function associated with $\frac{W^h \mathbf{1}_{[W>0]}}{EW^h \mathbf{1}_{[W>0]}}$ is defined to be

$$\text{(2.4)} \qquad \chi_{b,h}(r) = \log_b E\Big(\frac{W^h}{E(W^h \mathbf{1}_{[W>0]})}\Big)^r - (r-1).$$

Some simplification gives

$$\text{(2.5)} \qquad \chi_{b,h}(r) = \chi_b(hr) - r\chi_b(h)$$

for both h and hr in the domain of χ_b.

Corollary 2. STRUCTURE THEOREM FOR h-CASCADES

(a) (Nondegeneracy) $E\lambda_\infty(h, \mathbf{T}) > 0$ *iff* $\chi'_{b,h}(1-) = h\chi'_b(h-) - \chi_b(h) < 0$.

(b) (Convergence of moments) $E\lambda^r_\infty(h, \mathbf{T}) < \infty$ *for* $0 \leq r \leq 1$ *and if* $r_c(h) := \sup\{r \geq 1 : \chi_{b,h}(r) \leq 0\} > 1$ *then* $E\lambda^r_\infty(h, \mathbf{T}) < \infty$ *for* $1 < r < r_c(h)$.

(c) (Support dimension) If $\lambda_\infty(h, \mathbf{T}) > 0$, *then* $\lambda_\infty(h, \cdot)$ *is a.s. supported on a set with Hausdorff dimension* $-\chi'_{b,h}(1) = \chi_b(h) - h\chi'_b(h)$.

It is easy to check that for h in the interior of the domain of $\chi_b(h)$, $\chi'_{b,h}(1)$ decreases in h for $h > 0$ and increases in h for $h < 0$. This gives a clear picture of the dimension of the support of these measures. It is at a maximum at $h = 0$ (corresponding to a beta model with $p = P(W > 0)$) and decreases to $-\chi'_{b,H_c^{\pm}}(1) = \chi_b(H_c^{\pm}) - H_c^{\pm}\chi'_b(H_c^{\pm})$ at $h = H_c^{\pm}$. (These values are typically 0.) The same calculation also gives a critical interval in h on which the $\lambda_\infty(h,\cdot)$ cascades live with positive probability.

Corollary 3. CRITICAL INTERVAL

$$P(\lambda_\infty(h, \mathbf{T}) > 0) > 0 \quad \textit{for} \quad h \in (H_c^-, H_c^+) \tag{2.6}$$

where $H_c^- = \inf\{h : h\chi'_b(h) - \chi_b(h) < 0\}$ *and* $H_c^+ = \sup\{h : h\chi'_b(h) - \chi_b(h) < 0\}$. *In addition,*

$$P(\lambda_\infty(h, \mathbf{T}) = 0) = 1 \quad \textit{for} \quad h > H_c^+ \quad \textit{if} \quad EW^h \mathbf{1}_{[W>0]} < \infty, \tag{2.7}$$

and

$$P(\lambda_\infty(h, \mathbf{T}) = 0) = 1 \quad \textit{for} \quad h < H_c^- \quad \textit{if} \quad EW^h \mathbf{1}_{[W>0]} < \infty. \tag{2.8}$$

The values $H_c^{\pm}$ also appear as critical values, or points of phase transition, in the estimation of Rényi exponents; see Molchan (1996) and Ossiander and Waymire (2000). The usual assumption of $\chi'_b(1-) < 0$ gives $H_c^- \leq 0$ and $H_c^+ \geq 1$. If $\chi'_{b,H_c^{\pm}}(1) = 0$, as is typically the case, then $P(\lambda_\infty(H_c^{\pm}, \mathbf{T}) = 0) = 1$.

Notice that not only are the $\lambda_n(h,\cdot)$'s absolutely continuous with respect to λ, they are all mutually absolutely continuous a.s. P as well. The following result shows that, with the exception of the beta model, the $\lambda_\infty(h,\cdot)$'s are all mutually singular a.s. P.

Theorem 4. DISJOINT SUPPORTS FOR h-CASCADES *If $\chi_b(h)$ is non-linear, then for any $h, r \in (H_c^-, H_c^+)$ with $h \neq r$, the supports of the random cascade measures $\lambda_\infty(h,\cdot)$ and $\lambda_\infty(r,\cdot)$ are disjoint a.s. P. Consequently $\lambda_\infty(h,\cdot)$ and $\lambda_\infty(r,\cdot)$ are mutually singular a.s. P.*

Remark: If $\chi_b(h)$ is linear (the beta model) then for fixed n, the $\lambda_n(h,\cdot)$ measures are all identical. Thus for the beta model the $\lambda_\infty(h,\cdot)$ measures are all identical as well.

3 Fragmentation of the h-cascades and the proof of Theorem 4

The following fragmentation theorem results from exploitation of a large deviation-type principle governing the fluctuation of the $\lambda_n(h, \Delta_n(t))$'s.

This same principle also underlies the success of size-bias methods for multiplicative cascades; see for example Waymire and Williams (1996). Seen here, it makes evident that the transform $\prod_{i=1}^n W_{t|i} \mapsto \prod_{i=1}^n W_{t|i}^h$ brings out different levels of the fluctuation behavior of $\prod_{i=1}^n W_{t|i}$.

Theorem 5. THE BASIC FRAGMENTATION THEOREM *If for each $n \geq 1$, $\{X_n(t) : |t| = n\}$ is a collection of independent random variables, independent of $\mathcal{F}_n$, with $\sup_{n,t} E|X_n(t)| = c$ for some $c < \infty$, then, for any $\epsilon > 0$, the following holds both a.s. P and in L_1:*

$$\sum_{|t|=n} X_n(t)\lambda_n(\Delta_n(t))\mathbf{1}_{[\lambda_n(\Delta_n(t))<b^{n(\chi_b'(1-)-\epsilon)}]} \to 0. \tag{3.1}$$

If, in addition, for some $\delta > 0$, $EW^{1+\delta} < \infty$ then, for any $\epsilon > 0$, the following holds both a.s. P and in L_1:

$$\sum_{|t|=n} X_n(t)\lambda_n(\Delta_n(t))\mathbf{1}_{[\lambda_n(\Delta_n(t))>b^{n(\chi_b'(1)+\epsilon)}]} \to 0. \tag{3.2}$$

The value $\chi_b'(1-)$ can be thought of as a fragmentation level for the sequence of cascade measures $\lambda_n(\cdot)$. The fragmentation levels for the $\lambda_n(h, \cdot)$ measures are necessarily given in terms of the $\chi_{b,h}'(1-)$ values. These levels are specified in Corollary 6, which follows directly from Theorem 5 after using the obvious scaling argument.

Corollary 6. FRAGMENTATION OF h-CASCADES *If for each $n \geq 1$, $\{X_n(t) : |t| = n\}$ is a collection of independent random variables, independent of $\mathcal{F}_n$, with $\sup_{n,t} E|X_n(t)| < \infty$, then, for any h with $EW^h\mathbf{1}_{[W>0]} < \infty$ and any $\epsilon > 0$, the following holds both a.s. P and in L_1:*

$$\sum_{|t|=n} X_n(t)\lambda_n(h, \Delta_n(t))\mathbf{1}_{[\lambda_n(h,\Delta_n(t))<b^{n(h\chi_b'(h-)-\chi_b(h)-\epsilon)}]} \to 0. \tag{3.3}$$

If, in addition, for some $\delta > 0$, $EW^{h(1+\delta)}\mathbf{1}_{[W>0]} < \infty$, then, for any $\epsilon > 0$, the following holds both a.s. P and in L_1:

$$\sum_{|t|=n} X_n(t)\lambda_n(h, \Delta_n(t))\mathbf{1}_{[\lambda_n(h,\Delta_n(t))>b^{n(h\chi_b'(h)-\chi_b(h)+\epsilon)}]} \to 0. \tag{3.4}$$

We now give the proof of Theorem 5 followed by the proof of Theorem 4.

Proof (Theorem 5). First consider the case $P(W = p^{-1}) = 1 - P(W = 0) = p$. Then $\chi_b(h)$ is linear with slope $\chi_b'(h) = \ln_b(1/bp)$ and for each $t \in \mathbf{T}$, $\lambda_n(\Delta_n(t))$ is either 0 or $(pb)^{-n} = b^{n\chi_b'(1)}$, so that $\mathbf{1}_{[\lambda_n(\Delta_n(t))>b^{n(\chi_b'(1)+\epsilon)}]} \equiv 0$. Likewise $\lambda_n(\Delta_n(t))\mathbf{1}_{[\lambda_n(\Delta_n(t))<b^{n(\chi_b'(1)-\epsilon)}]} \equiv 0$. This gives (3.1) and (3.2) trivially.

For the remainder of the proof, assume that χ_b is non-linear. The mean-value theorem will manifest itself throughout via

$$\chi_b(h) = (h-1)\frac{(\chi_b(h) - \chi_b(1))}{h-1} = (h-1)\chi_b'(\xi) \tag{3.5}$$

for some ξ between h and 1. Recall that $\chi_b''(h) > 0$, so $\chi_b'(1-) > \chi_b'(0)$.

To verify (3.1), fix $\epsilon \in (0, \chi_b'(1-) - \chi_b'(0))$ and choose $\gamma \in (0,1)$ to satisfy $\chi_b'(1-\gamma) = \chi_b'(1-) - \epsilon$. In addition, take $\xi \in (1-\gamma, 1)$ to satisfy $\chi_b'(\xi) = -\chi_b(1-\gamma)/\gamma$. Then, taking $t_0 \in \mathbf{T}$ fixed,

$$\begin{aligned}
E|\sum_{|t|=n} & X_n(t)\lambda_n(\Delta_n t)\mathbf{1}_{[\lambda_n(\Delta_n(t)) < b^{n(\chi_b'(1-)-\epsilon)}]}| \\
&\le \sup_t E|X_n(t)| b^n E\lambda_n(\Delta_n(t_0))\mathbf{1}_{[\lambda_n(\Delta_n(t_0)) < b^{n(\chi_b'(1-)-\epsilon)}]} \\
&\le cE\prod_{i=1}^n W_{t_0|i}\mathbf{1}_{[\prod_{i=1}^n W_{t_0|i} < b^{n(\chi_b'(1-)+1-\epsilon)}]} \\
&\le cE\prod_{i=1}^n W_{t_0|i}^{1-\gamma} b^{n\gamma(\chi_b'(1-)+1-\epsilon)} \\
&= cb^{n(\chi_b(1-\gamma)+\gamma\chi_b'(1-\gamma))} \\
&= cb^{n\gamma(\chi_b'(1-\gamma)-\chi_b'(\xi))}.
\end{aligned} \tag{3.6}$$

Since $\chi_b''(h) > 0$ and χ_b non-linear, $\chi_b'(1-\gamma) < \chi_b'(\xi)$. Thus this expectation is decreasing exponentially and (3.1) holds both a.s. P and in L_1.

To verify (3.2), fix $\epsilon > 0$ sufficiently small to have $\chi_b'(1) + \epsilon = \chi_b'(1+\gamma)$ for some $\gamma < \delta$. Notice that for some $\xi \in (1, 1+\gamma)$,

$$\chi_b(1+\gamma) - \gamma\chi_b'(1+\gamma) = \gamma(\chi_b'(\xi) - \chi_b'(1+\gamma)) < 0. \tag{3.7}$$

Then

$$\begin{aligned}
E|\sum_{|t|=n} & X_n(t)\lambda_n(\Delta_n(t))\mathbf{1}_{[\lambda_n(\Delta_n(t)) > b^{n(\chi_b'(1)+\epsilon)}]}| \\
&\le \sum_{|t|=n} E|X_n(t)|\lambda_n(\Delta_n(t))\mathbf{1}_{[\lambda_n(\Delta_n(t)) > b^{n(\chi_b'(1)+\epsilon)}]} \\
&\le \sum_{|t|=n} E|X_n(t)|\lambda_n^{1+\gamma}(\Delta_n(t)) b^{-n\gamma(\chi_b'(1)+\epsilon)} \\
&\le \sup_t E|X_n(t)| E\prod_{i=1}^n W_{t_0|i}^{1+\gamma} b^{-n\gamma(1+\chi_b'(1+\gamma))} \\
&\le cb^{n(\chi_b(1+\gamma)-\gamma\chi_b'(1+\gamma))}.
\end{aligned} \tag{3.8}$$

This is decreasing exponentially, so (3.2) holds.

Proof (Theorem 4). We assume that $\chi_b'(h)$ is not constant. By use of an obvious scaling argument, it suffices to show that the theorem holds in each of the following two scenarios. First we assume that $H_c^+ > 1$ and show that the supports of $\lambda_\infty(1,\cdot)$ and $\lambda_\infty(h,\cdot)$ are disjoint for fixed $h \in (1, H_c^+)$. We then consider the case $H_c^- < 0$ and $H_c^+ \geq 1$ and show that the supports of $\lambda_\infty(1,\cdot)$ and $\lambda_\infty(h,\cdot)$ are disjoint for fixed $h \in (H_c^-, 0)$.

Case (1): Suppose that $H_c^+ > 1$ and fix $h \in (1, H_c^+)$. Note that for any real numbers x and y, and subset A of $\mathbf{R}^2$,

$$\begin{aligned} |x-y| &= (x-y)(\mathbf{1}_{[x>y]} - \mathbf{1}_{[y>x]}) \\ &\geq (x-y)(\mathbf{1}_A(x,y) - \mathbf{1}_{A^c}(x,y)) \\ &= x(1-2\mathbf{1}_{A^c}(x,y)) + y(1-2\mathbf{1}_A(x,y)). \end{aligned} \tag{3.9}$$

For $|t| = n$, set $A_n(t) = [\lambda_n(1,\Delta_n(t)) > \lambda_n(h,\Delta_n(t))]$. After some manipulation, we can rewrite

$$A_n(t) = [\lambda_n(h,\Delta_n(t)) < b^{n\chi_b(h)/(h-1)}] \tag{3.10}$$

and

$$A_n^c(t) = [\lambda_n(1,\Delta_n(t)) \geq b^{n\chi_b(h)/(h-1)}]. \tag{3.11}$$

Notice that for some $\xi \in (1,h)$, $\chi_b(h)/(h-1) = \chi_b'(\xi) \in (\chi_b'(1), \chi_b'(h))$. Using the cascade decomposition given in (2.3) for $\lambda_\infty(r,\cdot)$ with both $r = 1$ and $r = h$, we have

$$\lambda_\infty(r,\mathbf{T}) = \sum_{|t|=n} Z(r,t|n)\lambda_n(r,\Delta_n(t)) \tag{3.12}$$

where for each fixed n, $\{Z(r,t|n) : t \in \mathbf{T}\}$ is a collection of i.i.d. copies of the non-negative r. v. $\lambda_\infty(r,\mathbf{T})$, independent of $\mathcal{F}_n$. Using (b) from both Theorem 1 and Corollary 2 we see that $EZ(r,t|n) = 1$ for all $t \in \mathbf{T}$ for $r = 1, h$ as well as $EZ^{1+\delta}(1,t|n) = c_\delta < \infty$ for some $\delta > 0$. Take $X_n(t) = Z(1,t|n)$, and apply (3.2) of Theorem 5, to see that

$$\begin{aligned} &\sum_{|t|=n} \lambda_\infty(1,\Delta_n(t))\mathbf{1}_{A_n^c(t)} \\ &= \sum_{|t|=n} Z(1,t|n)\lambda_n(1,\Delta_n(t))\mathbf{1}_{A_n^c(t)} \to 0 \end{aligned} \tag{3.13}$$

a.s. P. Similiarly, take $r = h$ in (3.12), $X_n(t) = Z(h,t|n)$, and appeal to (3.3) of Corollary 6 to see that

$$\begin{aligned} &\sum_{|t|=n} \lambda_\infty(h,\Delta_n(t))\mathbf{1}_{A_n(t)} \\ &= \sum_{|t|=n} Z(h,t|n)\lambda_n(h,\Delta_n(t))\mathbf{1}_{A_n(t)} \to 0 \end{aligned} \tag{3.14}$$

a.s. P. Here one should note that the appropriate fragmentation level associated with $\lambda_n(h,\cdot)$ is given by $h\chi_b'(h) - \chi_b(h)$. It is easy to check that $\chi_b(h)/(h-1) < \chi_b'(h)$ is equivalent to $\chi_b(h)/(h-1) < h\chi_b'(h) - \chi_b(h)$.

Let ν denote the non-negative measure on $(\mathbf{T}, \mathcal{B}(\mathbf{T}))$ given by the total variation of the signed measure $\lambda_\infty(1,\cdot) - \lambda_\infty(h,\cdot)$. Clearly

$$\lambda_\infty(1,\mathbf{T}) + \lambda_\infty(h,\mathbf{T}) \geq \nu(\mathbf{T}). \tag{3.15}$$

In addition, for each $n \geq 1$,

$$\begin{aligned} \nu(\mathbf{T}) &\geq \sum_{|t|=n} |\lambda_\infty(1,\Delta_n(t)) - \lambda_\infty(h,\Delta_n(t))| \\ &\geq \sum_{|t|=n} (\lambda_\infty(1,\Delta_n(t)) - \lambda_\infty(h,\Delta_n(t)))(\mathbf{1}_{A_n(t)} - \mathbf{1}_{A_n^c(t)}) \\ &= \lambda_\infty(1,\mathbf{T}) - 2\sum_{|t|=n} \lambda_\infty(1,\Delta_n(t))\mathbf{1}_{A_n^c(t)} \\ &\quad +\lambda_\infty(h,\mathbf{T}) - 2\sum_{|t|=n} \lambda_\infty(h,\Delta_n(t))\mathbf{1}_{A_n(t)}. \end{aligned} \tag{3.16}$$

Let $n \to \infty$ to see that $\nu(\mathbf{T}) = \lambda_\infty(1,\mathbf{T}) + \lambda_\infty(h,\mathbf{T})$ a.s. P. Thus the supports of $\lambda_\infty(1,\cdot)$ and $\lambda_\infty(h,\cdot)$ are disjoint a.s. P. This completes the proof of case (1).

Case (2): Suppose that $H_c^- < 0$ and fix $h \in (H_c^-, 0)$. Again take $A_n(t) = [\lambda_n(1,\Delta_n(t)) > \lambda_n(h,\Delta_n(t))]$. Then rewrite

$$A_n(t) = [\lambda_n(1,\Delta_n(t)) < b^{n\chi_b(h)/(h-1)}] \tag{3.17}$$

and

$$A_n^c(t) = [\lambda_n(h,\Delta_n(t)) \geq b^{n\chi_b(h)/(h-1)}]. \tag{3.18}$$

After noting that $\chi_b'(h) < \chi_b(h)/(h-1) < \chi_b'(1)$, the proof follows exactly as case (1) with the roles of $A_n(t)$ and $A_n^c(t)$ reversed.

4 Remarks

A more rudimentary form of the fragmentation argument seen in the proof of Theorem 5 is used to derive a.s. convergence of the Rényi exponent along with a central limit theorem in Ossiander and Waymire (2000). These same techniques are further developed in Ossiander (2000) to derive strong laws for random functions of the form $\int X_n(t)d\lambda_n(t)$. These strong laws then give fluctuation levels of both $\lambda_n(h,\cdot)$ and $\lambda_\infty(h,\cdot)$ on small sets as well as sets of varying Hausdorff dimension.

References

Frisch, U., *Turbulence: The Legacy of A.N. Kolmogorov*, (Cambridge University Press, NY, 1995).

Kahane, J.P., Multiplications aléatoires et dimensions de Hausdorff, *Ann. Inst. Poincaré* **23** (1987), 289–296.

Kahane, J.P. and J. Peyrière, Sur certaines martingales de B. Mandelbrot, *Adv. in Math.* **22** (1976), 131–145.

Kolmogorov, A.N., The local structure of turbulence in incompressible viscous fluid for very large Reynolds number, *Dokl. Akad. Nauk SSSR*, **30** (1941), 9–13.

Molchan, G.M., Scaling exponents and multifractal dimensions for independent random cascades, *Comm. Math. Phys.*, **179** (1996), 681–702.

Ossiander M., Strong laws and fluctuation levels of multiplicative cascades, 2000, *preprint.*

Ossiander, M. and E.C. Waymire, Statistical estimation for multiplicative cascades, to appear in *Ann. Stat.*, (2000).

Troutman, B. and A.V. Vecchia, Estimation of Rényi exponents in random cascades, *Bernoulli*, **5** (1999), 191–207.

Waymire, E. and S.C. Williams, A cascade decomposition theory with applications to Markov and exchangeable cascades, *Trans. Amer. Math. Soc.*, **348**:2 (1996), 585–632.

Department of Mathematics
Oregon State University
Corvallis, OR 97331
e-mail: `ossiand@math.orst.edu`

On Simulating Fractional Brownian Motion

Jerzy Szulga and Fred Molz

ABSTRACT We discuss how a computer simulation affects the properties of random trajectories, like stationarity or self-similarity, focusing on the Weierstrass-Mandelbrot approximation of the fractional Brownian motion.

1 Introduction

We address the mathematical side of a hydrological fractal model from Molz, Szulga, and Liu (1997). In the spirit of Mandelbrot (1983), the fractional Brownian motion was approximated by a certain random Fourier series, the Weierstrass-Mandelbrot process, in the manner suggested in Berry and Lewis (1980) and formalized in Szulga and Molz (2000). The scope of that variation of the functional central limit theorem was extended recently in Pipiras and Taqqu (1998).

To represent hydrologic phenomena we consider second order complex valued stochastic processes $W(t) = X(t)+iY(t)$. For simplicity, we assume that $\mathsf{E}W(t) = 0$, without loss of generality. The *mixed covariance* and the *autocovariance* are defined by the formulas

$$C_{W,W'}(s,t) \stackrel{\text{def}}{=} \mathsf{E}W(s)\overline{W'(t)}, \qquad C = C_W = C_{W,W}.$$

A process W has stationary increments (in the wide sense) if

$$\mathsf{E}\Big(W(t+h) - W(t)\Big)\Big(\overline{W(t+k) - W(t)}\Big) = \mathsf{E}W(h)\overline{W(k)}$$

(for details, cf., e.g., Yaglom (1987)). Then, $W(t)$ admits a *spectral representation* as a stochastic integral with respect to a unique (a.s.) orthogonally scattered random measure $Z(dw) = Z_1(dw) + iZ_2(dw)$,

$$\begin{aligned} W(t) = & \int_{-\infty}^{\infty} \Big(1 - e^{itw}\Big)\, Z(dw) \\ = & \int_{-\infty}^{\infty} (1-\cos tw)\, Z_1(dw) + \int_{-\infty}^{\infty} \sin tw\, Z_2(dw) \\ & +i\left(-\int_{-\infty}^{\infty} \sin tw\, Z_1(dw) + \int_{-\infty}^{\infty} (1-\cos tw)\, Z_2(dw)\right) \end{aligned}$$

(the "integrals" with respect to Z_1 and Z_2 are merely symbols barring additional properties of the process). We call $\Phi(dw) = \mathsf{E}|Z(dw)|^2$ the *spectral measure*, its density (if it exists) — the *spectral density*, its support S — the *spectrum*, etc. The stochastic integral is well defined if

$$D(t) \stackrel{\text{def}}{=} D_W(t) \stackrel{\text{def}}{=} \mathsf{E}|W(t)|^2 = 2\int_{-\infty}^{\infty} \Big(1 - \cos(wt)\Big)\,\Phi(dw) < \infty, \tag{1.1}$$

for every $t > 0$. This happens if and only if Φ is a *Lévy measure*, i.e., $\int_{\mathbb{R}} (1 \wedge w^2)\Phi(dw) < \infty$. Then,

$$\begin{aligned} \Re C(s,t) &= \int_{-\infty}^{\infty} \Big(1 - \cos sw - \cos tw + \cos\,(s-t)w\Big)\,\Phi(dw) \\ &= \frac{1}{2}\Big(D(s) + D(t) - D(s-t)\Big). \end{aligned} \tag{1.2}$$

The *first structure function* $D(t)$ determines only the even modification of the Lévy measure Φ, $\Phi_e(A) = (\Phi(A) + \Phi(-A))/2$, $A \subset [0,\infty)$. The *second structure function* is any function $I(t)$ such that

$$\begin{aligned} \Im C(s,t) &= \int_{-\infty}^{\infty} \Big(\sin\,(s-t)w - \sin sw + \sin tw\Big)\,\Phi(dw) \\ &= \frac{1}{2}\Big(I(s-t) - I(s) + I(t)\Big). \end{aligned} \tag{1.3}$$

If $I(t)$ are $I_1(t)$ are two continuous functions of this category, then the choice of $t = 2s$ leads to the equation $I(2s) - I_1(2s) = 2\big(J(s) - I_1(s)\big)$. Whence, $I(s) - I_1(s) = cs$, for some constant c. E.g., one may choose

$$I(t) = 2\int_{-\infty}^{\infty} \Big(\sin tw - tw\mathbf{1}\{|w| \le a\}\Big)\,\Phi(dw). \tag{1.4}$$

Stronger integrability conditions at 0 or infinity may allow one to choose $a = 0$ or $a = \infty$, respectively. $I(t)$ determines uniquely only the odd modification of the Lévy measure, $\Phi_o(A) = (\Phi(A) - \Phi(-A))/2$ on $[0,\infty)$.

The spectral formula for a *fractional Brownian motion*, FBM, is

$$W(t) = W_H(t) = \alpha_H \int_S \Big(1 - e^{itw}\Big)\,|w|^{-H-1/2}\,(B_1 + iB_2)\,(dw), \tag{1.5}$$

where $0 < H < 1$, and $S \subset \mathbb{R}$ is a Borel set. B_1 and B_2 are real Brownian motions, so we may disregard the warning following the spectral representation. From now on, we choose $S = [0,\infty)$, and call the FBM an *MBL-process* (for "Mandelbrot-Berry-Lewis"). The scaling constant α_H can be set as $2^{-1/2}$. The parameters of FBM can be augmented by $H = 0$ and $H = 1$, but these lead to trivial processes, $W_0(t) = X_0$ and $W_1(t) = tX_1$.

We check directly that

(1.6)

$$D_H(t) \quad = r_H \; |t|^{2H}, \qquad r_H = \int_0^\infty \frac{1-\cos w}{|w|^{2H+1}} dw;$$

$$I_H(t) \quad = i_H t^{\langle 2H \rangle}, \qquad i_H = \begin{cases} \displaystyle\int_0^\infty \frac{\sin w}{|w|^{2H+1}} dw, & \text{if } H < 1/2, \\ \displaystyle\int_0^\infty \frac{\sin w - w}{|w|^{2H+1}} dw, & \text{if } H > 1/2, \end{cases}$$

where $t^{\langle a \rangle} \stackrel{\text{def}}{=} \text{sgn}(t) \cdot |t|^a$, for real t and $a \neq 0$. The remaining function is different, $I_{1/2}(t) = -t \ln |t|$. We choose it according to (1.4). Put

$$J(t) = \int_0^\infty \Big(\sin wt - wt \mathbf{1}\{0 \le w \le a\} \Big) \frac{dw}{w^2}.$$

The function $g(t)$ below is constant, $g(t) = c$, where

$$g(t) = \frac{J(t)}{t} + \ln t = \int_0^{at} (\sin w - w) \frac{dw}{w^2} + \int_{at}^\infty \sin w \frac{dw}{w^2} + \ln t,$$

since $dg/dt = 0$. The choice of a that makes $c = 0$ leads to $I_{1/2}$.

2 The proximity of WM and FBM structures

For an increasing sequence $\boldsymbol{s} = \{s_n\}$, put $I_n = (s_n, s_{n+1}]$, $d_n^2 = \mathsf{E}|Z(I_n)|^2$, and $\zeta_n = d_n^{-1} Z(I_n)$. The process with discrete spectrum

(2.1)
$$W_{\boldsymbol{s}}(t) = \sum_{n \in \mathbb{Z}} (1 - e^{is_n t}) d_n \zeta_n$$

satisfies the formula

$$\mathsf{E}|W(t) - W_{\boldsymbol{s}}(t)|^2 = 2 \sum_n \int_{s_n}^{s_{n+1}} \Big(1 - \cos(w - s_n)t\Big) \Phi(dw)$$

$$= \quad 2 \int_{\mathbb{R}} \sum_n \mathbb{I}_n(w) \Big(1 - \cos(w - s_n)t\Big) \, \Phi(dw).$$

Let the spectrum depend on a parameter p such that $s_{n+1} - s_n \to 0$. Then $h(w) = \sum_n \mathbb{I}_n(w)\Big(1 - \cos(w - s_n)t\Big) \to 0$ a.e. Since $h(w) \le 2(1 \wedge wt)^2$, which is Φ-integrable, then $\mathsf{E}|W(t) - W_{\boldsymbol{s}}(t)|^2 \to 0$, by the dominated convergence theorem. When $W(t)$ is a Gaussian process, so is $W_{\boldsymbol{s}}$. Then, the convergence in variance implies the convergence of finite dimensional distributions.

The covariance function remains identical but the probability distribution may change when ζ_n's are orthonormal random variables unrelated to Z. Yet, for some non-Gaussian processes, called quasi-Gaussian in Szulga and Molz (2000), the covariance determines the limits. E.g., it suffices to consider independent ζ_n's with orthogonal real and imaginary parts.

The difference of the characteristic functions

$$\left| \mathsf{E} e^{i \sum_k \left(a_k X(t_k) + b_k Y(t_k) \right)} - \mathsf{E} e^{i \sum_k \left(a_k X_r(t_k) + b_k Y_r(t_k) \right)} \right| \tag{2.2}$$

measures well the proximity of distributions of two processes $W_r = X_r + iY_r$ and $W = X + iY$. For quasi-Gaussian processes, this difference can be estimated by the difference of covariance functions, whence the convergence of structure functions will lead to the convergence of probability distributions.

The WM-process (for "*Weierstrass-Mandelbrot*") involves the geometric spectrum $s_n = r^n$ and Steinhaus variables,

$$W_r(t) = (r-1)^{1/2} \sum_{n \in \mathbb{Z}} (1 - e^{ir^n t}) r^{-nH} e^{2\pi i U_n}, \tag{2.3}$$

where $r > 1$, $0 < H < 1$, and U_n's are standard i.i.d. variables.

The normalizing factor $(r-1)^{1/2} \sim \ln^{1/2} r$ appears also in the covariance function of the original process (2.3). As shown in Szulga and Molz (2000), the difference (2.2) is majorized by $\sup_{s,t \le T} |C_r(s,t) - C(s,t)|$, uniformly with respect to $t_k \in [0,T]$ and bounded coefficients (a_k, b_k), and converges to 0, as $r \to 1$ (even for quasi-Gaussian processes). We wish now to find numerical estimates in the specific approximation of a fractional Brownian motion. To this end, it suffices to estimate the deviation of structure functions $D_r(t) - D(t)$ and $I_r(t) - I(t)$.

The natural partition of the H-range leads to the three types of random behavior, *anti-persistent* ($H < 1/2$), *persistent* ($H > 1/2$), and *chaotic* ($H = 1/2$). The estimates deteriorate at the extremes, i.e., for H close to 0 or 1. A singularity at $H = 1/2$ shows a phase transition, pertinent to passing through the threshold value $H = 1/2$. With some effort, the estimates can be smoothened in the vicinity of the threshold point.

It is convenient to replace the normalizing factor $(r-1)$ by its approximation $p = \ln r$ in the structure function of the normalized WM-process

$$t^{-2H} \mathsf{E}|W_r(t)|^2/2 = (r-1) \sum_n (1 - \cos r^n t) r^{-2Hn} t^{-H},$$

and, then, rewrite the series as follows,

$$p \sum_n \left(1 - \cos e^{np+a} \right) e^{-2H(np+a)}, \tag{2.4}$$

where $a = \ln t$. By the same token, we rewrite $I(t)$ as

$$
\text{(2.5)} \qquad
\begin{aligned}
&p\sum_n \left(\sin e^{np+a} \right) e^{-2H(np+a)}, && \text{if } H < 1/2, \text{ or} \\
&p\sum_n \left(\sin e^{np+a} - e^{np+a} \right) e^{-2H(np+a)}, && \text{if } H > 1/2.
\end{aligned}
$$

The matching quantities for the limit FBM are constants.

The functions and the series fit the following pattern of Szulga and Molz (2000). If a measurable function f on the real line is dominated by an integrable and eventually monotone function (i.e., $g(x)$ and $g(-x)$ are decreasing and $|f(\pm x)| \leq g(x)$, when $x \geq N$, for some N) then the 'averages'

$$\Phi_p(a) = p\sum_n f(np + a)$$

are well defined as periodic functions of period $p > 0$. That is, their properties are determined over the domain $-p \leq a \leq 0$. Further,

$$\text{(2.6)} \qquad \int_{\mathbb{R}} f(w)dw = \sum_{n\in\mathbb{Z}} \int_0^p f(np + a + x)dx = \frac{1}{p}\int_0^p \Phi_p(a + x)dx$$

suggests that $\Phi_p(a) \approx \langle f \rangle \stackrel{\text{def}}{=} \int_{\mathbb{R}} f$, when $p \approx 0$. The relation is false, in general, even when f is continuous. On the other hand, if f is differentiable and f' is dominated by an integrable and eventually monotone function, $|f'(x)| \leq h(x)$, then the mean value theorem and (2.6) imply that

$$\text{(2.7)} \quad |\Phi_p(a) - \langle f \rangle| \leq \int_0^p \sum_n \Big| f(np + a + x) - f(np + a) \Big| \, dx \leq C \langle h \rangle \, p,$$

where C is a constant that may further depend on the dominant h.

In the case when bounds of f' do not decrease but can be controlled by a non-rapidly increasing function, an optimization algorithm may still produce a reasonable estimate. This happens for the WM-process. The following summary relies on a routine but tedious calculation.

Summary. *The proximity between the MBL-process $W(t)$ and the WM-process $W_r(t)$, for $p = \ln r \to 0$, can be measured as the discrepancy between the covariance functions, uniformly on $[0, T]$. An upper estimate of the discrepancy is proportional to $T^{2H} \cdot G(H, p)$. The function $G(H, p)$ behaves as follows,*

$$G(H, p) = \begin{cases} \dfrac{p}{(1 - H)(H - 0.5)} & \textit{if } H > 1/2, \\ \dfrac{p}{1 - H} - p \ln p & \textit{if } H \geq 1/2, \\ \dfrac{p^{2H}}{H} + \dfrac{c_{H,p} p^{2H}}{H - 0.5} & \textit{if } H < 1/2, \end{cases}$$

where $c_{H,p}$ *is a stabilizing factor that causes the second term to approach* $-p\ln p$, *for* $H \to 0.5$.

3 Simulation of WM and FBM

The normalized Weierstrass-Mandelbrot processes $W_r(t) = \sqrt{\ln r}M_r(t)$, have often been used to simulate trajectories of fractional Brownian motions. In generating a trajectory of a WM-process, a double-infinite random Fourier series is represented by a finite sum, of course. The programming is casual, given the right size of the curtate series. The variance of the remainder will control the error of approximation. The search for the appropriate cut-off is the only nontrivial issue and we will discuss it here.

If a domain $[0,T]$ is partitioned into k subintervals, and K is the size of the curtate series, then the simulation embraces a time-space matrix of size $K \times k$. While the path-resolution variable k may be chosen at will, the spatial variable K is constrained by the permissible error. Quick crude estimates force K to be very large, and consequently, lead to a huge data matrix with billions of entries. These estimates worsen further for extreme values of the parameters T, r, and H, and thus yield serious running time and memory problems.

Therefore, the assessment of errors must include evidence that these monstrous matrices appear for a reason and not because of a flaw of the estimate. This concern adds to the burden of computation.

3.1 Truncation effects

Define a scaled process $aW\cdot b$ by putting $aW\cdot b(t) \overset{\text{def}}{=} aW(bt)$. W is *invariant under scaling*, if W and $aW\cdot b$ can be identified, for some parameters a and b. Note a direct consequence of the definition:

"Theorem". *If* W *and* $aW\cdot b$ *are identifiable, for a single pair* (a,b) *with* $a>0$, *then* W *and* $c^{-H}W\cdot c$ *are identifiable, for every factor of the form* $c=b^m$, *where* $m\in\mathbb{Z}$ *and* $H=-\ln b/\ln a$.

W is called *self-similar*, if for every b there is a constant a such that W and $aW\cdot b$ are identifiable.

Self-similarity, stationarity, the shape of trajectories, etc., make but a sample of properties affected by the unavoidable truncation of a double-infinite series. The first reason to consider asymmetric cutoffs lies in the asymmetry of the spectral function itself. Further, each of the sub-series implicit in (2.3), one for positive and one for non-positive integers n, plays a different role in behavior of the process $\sqrt{\ln r}W_r(t)$.

For positive integers M, N, consider

$$\begin{aligned} W_r^-(M,t) &= \sqrt{\ln r} \sum_{n=-M}^{0} (1 - e^{ir^n t}) r^{-Hn} e^{i\phi_n}; \quad W_r^-(t) \stackrel{\text{def}}{=} W_r^-(\infty, t), \\ W_r^+(N,t) &= \sqrt{\ln r} \sum_{n=1}^{N} (1 - e^{ir^n t}) r^{-Hn} e^{i\phi_n}; \quad W_r^+(t) \stackrel{\text{def}}{=} W_r^+(\infty, t). \end{aligned}$$

The sum $W_r(M,N,t) = W_r^-(M,t) + W_r^+(M,t)$ of size $K = M + N + 1$ is supposed to approximate the true process $W_r(t)$. In addition to the parameters T, H, and r, the cut-off values M and N may depend on an acceptable predetermined error ε. The behavioral discrepancy of two parts will become more visible when the limit for $r \to 1$ is simulated, especially for long time intervals.

Although the WM process is not stationary and almost all its trajectories are nowhere differentiable, any curtate finite sum will be a random translation of a stationary (cf. the spectral representation in Yaglom (1987)) and smooth process with the divergent random shift:

$$\begin{aligned} W_r(M,N,t) &= \theta_{M,N} - \sum_{-M \le n \le N} e^{ir^n t} r^{-Hn} e^{i\phi_n}, \\ \theta_{M,N} &= \sum_{-M \le n \le N} r^{-Hn} e^{i\phi_n}. \end{aligned}$$

The negative part W_r^- is solely responsible for the shift $\theta_{M,N}$ escaping to infinity. Nevertheless, a single simulated trajectory will appear before an experienced eye as a typical path of a stationary process. Yet, this is but an illusion, caused by forcing an undetermined expression '$\infty - \infty$' to become determined in a way that is not allowed mathematically. On the other hand, the negative part $W_r^-(t)$ is differentiable everywhere while the positive part is differentiable nowhere. The preceding discussion has indicated two cut-off errors:

$$\begin{aligned} \varepsilon^- = \varepsilon^-(r, M, T) &\stackrel{\text{def}}{=} \sup_{t \le T} Var\Big(W_r^-(t) - W_r^-(M,t)\Big), \\ \varepsilon^+ = \varepsilon^+(r, N, T) &\stackrel{\text{def}}{=} \sup_{t \le T} Var\Big(W_r^+(t) - W_r^+(N,t)\Big). \end{aligned}$$

Theorem 1. *Let $T > 0$, $0 < H < 1$, and M, N be nonnegative integers. There exist constants $A \le 1$ and $B \le 2$ such that*

$$\begin{aligned} \varepsilon^- &\le A T^2 \frac{r^{-2(1-H)M}}{2(1-H)}, \\ \varepsilon^+ &\le B \frac{r^{-2HN}}{2H}. \end{aligned} \tag{3.1}$$

Assume that r^N and r^M are large and $r-1$ is small. More precisely, let $r^N T > \pi/2$, $r^M > T$, and $r < \sqrt{2}$. Then, there exist positive constants A_0, B_0 such that

$$(3.2) \qquad \begin{aligned} \varepsilon^- &\geq A_0\, T^2\, \frac{r^{-2(1-H)M}}{2(1-H)}, \\ \varepsilon^+ &\geq B_0\, \frac{r^{-2HN}}{2H}. \end{aligned}$$

Proof. *Upper estimates*: Note that $\ln r/(1-r^{-a}) \leq r^a/a$, for $a > 0$. Applying the inequality with $a = 2H$:

$$\varepsilon^+ = 2\ln r \sup_{t\leq T} \sum_{n=N+1}^{\infty} (1-\cos r^n t) r^{-2Hn} \leq 2\ln r \sum_{n=N+1}^{\infty} r^{-2Hn} \leq \frac{r^{-2HN}}{H}.$$

Similarly, since $2(1-\cos(r^{-n}t)) \leq r^{-2n}t^2$, the inequality with $a = 2(1-H)$ leads to the estimate

$$\varepsilon_- \leq T^2 \frac{r^{-2(1-H)M}}{2(1-H)}.$$

Lower estimates: To find a lower bound of ε_+ we lessen the L_∞-norm to the L^1-norm:

$$\sup_{t\leq T} |F(t)| \geq \frac{1}{T}\int_0^T |F(t)|dt.$$

Hence,

$$\begin{aligned} \varepsilon^+ &\geq 2\ln r \sum_{n>N} \frac{1}{T}\int_0^T (1-\cos r^n t)\, dt\; r^{-2Hn} \\ &= 2\ln r \sum_{n>N} \left(1 - \frac{\sin r^n T}{r^n T}\right) r^{-2Hn} \overset{\text{def}}{=} Q. \end{aligned}$$

Since $r^N T$ is large, i.e., $x = r^N T > \pi/2$, then $1 - \dfrac{\sin x}{x} \geq 1 - \dfrac{2}{\pi}$, which, in turn, implies that

$$\varepsilon_+ \geq Q \geq (1-2/\pi) 2 \ln r \frac{r^{-2H(N+1)}}{1-r^{-2H}}.$$

The assumption $r^{2H} < r^2 < 2$, leads to the inequality

$$\frac{\ln r}{1-r^{-2H}} \geq \ln 2\, \frac{r^{2H}}{2H},$$

and to the second lower estimate (3.2) with $B_0 = 2(1-2/\pi)\ln 2$.

Now, let us estimate ε_- from below. Suppose that $r^M > T$ and $r \le 2$. For brevity, denote $V = Var\,(W_r^-(t) - W_r^-(M,t))$. Then,

$$V = 2t^{2H-1}\frac{\ln r}{r-1}\sum_{n=M+1}^{\infty}(1-\cos r^{-n}t)(r^{-n}t)^{1-2H}(r^{n+1}-r^n).$$

Since $r^n > r^M > T > t$, then $u = r^{-n}t < 1$. Thus, the function $f(u) = (1-\cos u)u^{1-2H}$ increases, when $u \in (0,1)$. Equivalently, the function $f_1(x) = (1-\cos(t/x))(t/x)^{1-2H}$ decreases, if $t \le x < \infty$. Considering consecutive intervals $[r^n, r^{n+1})$, the quantity A is but an integral sum over the semi-line $[r^M, \infty)$. The restriction $r < \sqrt{2} < 2$ implies $2\ln r/(r-1) \ge 2\ln 2$. Hence V is bounded from below by

$$\begin{aligned}&2\ln 2\, t^{2H-1}\int_{r^{M+1}}^{\infty}(1-\cos(t/x))(t/x)^{1-2H}\,dx\\ &= 2\ln 2\, t^{2H}\int_0^{tr^{-M-1}}(1-\cos v)v^{-1-2H}\,dv.\end{aligned}$$

Both terms increase in t. Hence, taking the supremum over $t \in [0,T]$,

$$\varepsilon_- \ge 2\ln 2\, T^{2H}\int_0^{Tr^{-M-1}}(1-\cos v)v^{-1-2H}\,dv.$$

Since $Tr^{-M-1} < Tr^{-M} < 1$, and for $v \in (0,1)$, $1-\cos v \ge \cos 1\, v^2/2$, then

$$\varepsilon_- \ge \cos 1\ \ln 2\ T^{2H}\int_0^{Tr^{-M-1}} v^{1-2H}\,dv = T^{2H}\frac{T^{2-2H}r^{-(2-2H)(M+1)}}{2-2H}.$$

The first lower estimate in (3.2) follows, with $A_0 = (\cos 1\ \ln 2)/2$, since $r^{-(2-2H)} \ge 1/2$.

3.2 Remarks and corollaries

1. The averaging method, used in the lower estimate of ε_+, works equally well in obtaining the lower estimate of ε_-. However, the integral sum method that was applied to ε^- cannot be adjusted to obtain the lower estimate of ε_+. Further, this method could be used to derive the integral limit of D_r^-. It is unsuitable for D_r^+ (again, superscripts mark the suitable sub-series), because jammed oscillations defy the use of monotonicity.

2. Truncation of a random Fourier series, like the WM process, destroys the self-similarity, even the one restricted to integer powers of r. The 'positive' part is insensitive to re-scaling. In particular, in 'up-scaling' it violates the spirit of self-similarity but even more so in 'down-scaling' (while passing from the unit to an infinitesimal interval). The 'negative' part, in turn,

follows the line of increase of T^2 instead of T^{2H}, related to the increase of the structure function of the original process.

3. For one fixed T, both cut-off values M and N can be made dependent on T so that the errors would be proportional to T^{2H}.

4. The dependence on the exponent H also differentiates both parts. For H close to 0, N must be increased multifold to adjust for the effect of the multiplier $1/H$. By the same token, when H is near 1, M needs to overcome the presence of the factor $1/(1-H)$.

Let ε be an acceptable error, either absolute (i.e., independent of T) or relative (i.e., placed against T^{2H}). Minimizing routinely $M+N$, subject to the constraint $\varepsilon_- + \varepsilon_+ \leq \varepsilon$, we arrive at the estimates

$$N \geq \frac{-\ln H(1-H)\varepsilon}{2H\ln r} \quad \text{and} \quad M \geq \frac{-\ln H(1-H)\varepsilon) + 2\ln T - \ln 2}{2(1-H)\ln r}.$$

5. The direct simulation of the FBM that uses the linear covariance operator applied to a randomly generated standard Gaussian random vector takes a fraction of the time and effort needed to generate the WM process. In the spirit of the classical Central Limit Theorem, the FBM-approximation of the WM-process rather than the WM-approximation of the FBM-process seems more fitting if not only more efficient. Yet, the increase of the time resolution (i.e., the number of temporal nodes) affects the simulation of the WM-process linearly. Roughly speaking, a greater precision requires just a proportionally longer simulation. In contrast, the same increase may have a detrimental effect on the covariance-based procedure because the method relies on the Cholesky factorization of a huge matrix and exhausts a machine's memory. Thus, it is more expensive and less efficient. Similar controversies arise in simulating one of many integral representations of the FBM (see, e.g. Mandelbrot and Van Ness (1968)), for a number of non-spectral formulas). The bottom line is: the WM-method surpasses the direct method when one desires high resolution paths of the fractal Brownian motion.

References

Berry, M.V. and Lewis Z.V., On the Weiertrass-Mandelbrot fractal function. *Proc. R. Soc. Lond. A*, **370** (1980), 459-484.

Mandelbrot, B., *The Fractal Geometry of Nature*, (Freeman and Co., NY, 1983).

Mandelbrot, B. and J. Van Ness, Fractional Brownian motion, fractional noises and applications. *SIAM Review*, **10**, (1968), 422–437.

Molz, F.J., Szulga, J. and Liu, H.H., Fractional Brownian Motion and Fractional Gaussian Noise in Subsurface Hydrology: A Review of Fundamental Properties. *Water Resources Research* **33**:10, (1997), 2273–2286.

Pipiras, V. and Taqqu, M.S., Convergence of the Weierstrass-Mandelbrot process to Fractional Brownian Motion, *preprint*, (1998).

Szulga, J. and Molz, F.J., The Weierstrass-Mandelbrot process revisited, to appear in *J. Stat. Physics*, (2000).

Yaglom, A.M., *Correlation Theory of Stationary and Related Random Functions*, (Springer-Verlag, NY, 1987).

Jerzy Szulga
Department of Mathematics
Auburn University
Auburn, AL 36849
szulgje@mail.auburn.edu

Fred Molz
ESE Department
Clemson University
Clemson, SC 29634
fredi@ese.clemson.edu

PART 9

Function Estimation

On Robust Recursive Nonparametric Curve Estimation

Eduard Belitser and Sara van de Geer

ABSTRACT Suppose we observe $X_k = \theta(x_k) + \xi_k$. The function $\theta : [0,1] \to \mathbb{R}$, is assumed to belong a priori to a given nonparametric smoothness class, the ξ_k's are independent identically distributed random variables with zero medians. The only prior information about the distribution of the noise is that it belongs to a rather wide class. The assumptions describing this class include cases in which no moments of the noises exist, so that linear estimation methods (for example, kernel methods) can not be applied. We propose a robust estimator based on a stochastic approximation procedure and derive its rate of convergence, as the frequency of observations n tends to infinity, in almost sure as well as in mean square sense, uniformly over the smoothness class. Finally, we discuss a multivariate formulation of the problem, a robust nonparametric M-estimator (the least deviations estimator), the so called penalized estimator, and the case when the noises are not necessarily identically distributed.

1 Introduction

Suppose we observe deterministic signal values on a discrete grid of a closed bounded interval $I \subset \mathbb{R}$, contaminated with random noises, i.e.

$$X_k = \theta(x_k) + \xi_k, \qquad k = 0, 1, \dots, n , \tag{1.1}$$

where $\{x_k\} = \{x_{k,n}\}$ is a design to be specified later, $\theta(x) \in \Theta_\beta(H, L)$, $x \in I$, with nonparametric class $\Theta_\beta(H, L)$ which we define below, and where the ξ_k's are independent identically distributed random variables. Without loss of generality we assume the interval I to be unit $I = [0, 1]$.

The distribution of ξ_k is absolutely continuous with density f_ξ, which is assumed to be unknown. We will need the following conditions on f_ξ: for positive constants δ, p, L_ξ,

A1. $F_\xi(0) = \int_{-\infty}^{0} f_\xi(u)du = 1/2$, i.e. the distribution of ξ_k has zero median;

A2. $\inf_{|x|\le\delta} f_\xi(x) \ge p$;

A3. The functions $f_\xi(u)$ satisfy the Lipschitz condition:

$$|f_\xi(u_1) - f_\xi(u_2)| \le L_\xi |u_1 - u_2|, \qquad u_1, u_2 \in \mathbb{R} .$$

The design $\{x_k\}$ is assumed to be almost equidistant. To be precise, it satisfies the following conditions:

D1. $0 = x_0 \le x_1 \le \ldots \le x_n = 1$;

D2. $|x_l - x_m| \le D|l - m|/n$ for all $0 \le l, m \le n$ and some fixed positive constant D.

One could consider a general compact set $T \subset \mathbb{R}$ instead of the unit interval. We can also consider an unbounded set which can be represented as a countable union of disjoint compact sets. In the latter case the parameter n would play the role of density of observations, number of observations per unit interval in $\mathbb{R}$. Other possible generalizations of the model can be found in the remarks below. Since these kind of generalizations do not contribute much to the essence of the estimation problem, we will not pursue this pseudo-generality.

The unknown function $\theta(x)$ on the interval $[0,1]$ belongs to the Lipschitz function class $\Theta_\beta = \Theta_\beta(H, L)$ with the smoothness β. For some positive H, L and $0 < \beta \le 1$ define

$$\Theta_\beta(H,L) = \Big\{\theta : \ |\theta(0)| \le H\,,\ |\theta(u) - \theta(v)| \le L|u - v|^\beta,\ u, v \in [0,1]\Big\} .$$

We wish to estimate a function value $\theta(x)$, $x \in I$. An estimator $\hat{\theta}_n = \hat{\theta}_n(X_0, \ldots, X_n) = \hat{\theta}_n(x_0, \ldots, x_n, X_0, \ldots, X_n)$ is an arbitrary measurable function of the observations.

The integer parameter n is the frequency of observations, i.e. a number of observations per unit interval. We study the estimation problem in asymptotic setup when this parameter tends to infinity. Note that in fact we study the sequence of models

$$X_{k,n} = \theta(x_{k,n}) + \xi_{k,n}, \qquad k = 0, 1, \ldots, n .$$

To avoid typographical excess, we omit the subscript n in our notations, for instance $X_k = X_{k,n}$, $\xi_k = \xi_{k,n}$ etc.

Note that the model (1.1) is in general not a nonparametric regression since the conditional expectation may not exist under conditions (A1)–(A3). This implies also that, in general, linear methods (for instance, kernel methods) can not be applied in the situations when we measure the quality of estimator $\hat{\theta}_n(x)$ by a risk function of the form $E|\hat{\theta}_n(x) - \theta(x)|^\kappa$, $\kappa > 0$. Notice that the function value $\theta(x_k)$ has a meaning of conditional median, so we have to come up with some sort of nonparametric robust estimator.

Related nonparametric estimation problems have been studied by Korostelev (1987), Truong (1989), Belitser and Korostelev (1992).

The aim of this paper is to estimate the function θ. We propose an estimator based on a stochastic approximation procedure and derive its rate of convergence, as the density of observations n tends to infinity, in almost sure as well as in mean square sense, uniformly over the smoothness class and the class of all distributions of noise. A robust nonparametric M-estimator, the least deviations estimator, and the so called penalized estimator are discussed.

2 A recursive estimator

First we introduce some notation. Let $I\{A\}$ denote the indicator function of the set A. Introduce the signum function $\operatorname{sign}(u) = I\{u \geq 0\} - I\{u < 0\}$. Define further for some fixed positive h (for example $h = 1$)

$$S(u,v) = \begin{cases} \operatorname{sign}(u-v), & |v| \leq H+L+h\ , \\ -v, & |v| > H+L+h\ , \end{cases}$$

$$C_0 = 2/\min\{2p, 2p\delta/(2(H+L)+h), 1/2\} \tag{2.1}$$

and the sequence $\gamma_n = n^{-2\beta/(2\beta+1)} \log n$, where the constants β, H and L appear in the definition of the class Θ_β. For the sake of brevity, denote also $\theta_k = \theta(x_k)$ and by c and C generic constants which are in general different in different expressions.

The following recursive formula gives an estimator for the function value θ_k:

$$\hat{\theta}_{k+1} = \hat{\theta}_k + \gamma_n S(X_k, \hat{\theta}_k), \qquad k = 0, 1, \ldots, n-1\,, \tag{2.2}$$

with the initial value $\hat{\theta}_0 = 0$.

If we interpret index k as time-moment, then the above estimating procedure is a filtering algorithm. Filtering algorithms are most appropriate in situations when observations appear successively so that at a fixed time-moment k only those observations are accessible which have been obtained during the preceding time interval, i.e. only X_i's with $i \leq k$. Then an estimator of function value $\theta(x_k)$ is an arbitrary measurable function of the observations accessible by the time-moment k.

Clearly, the estimator $\hat{\theta}_k$ can not be consistent for all values θ_k, $0 \leq k \leq n$, since the true initial value θ_0 need not be zero. Nevertheless, this estimator is consistent beginning with a certain moment. Moreover, the following theorem specifies also the rate of convergence.

Theorem 1. *Suppose conditions (A1) and (A2) are fulfilled. Let* $K_n = \{k \in \mathbb{N} : C_0 n^{2\beta/(2\beta+1)} \leq k \leq n\}$, *where* C_0 *is defined by (2.1). Then, for* *some positive constant* C_1*, any fixed* $\alpha, \epsilon > 0$ *and any sequence* $\{k_n\}$ *such that* $k_n \in K_n$*, the relations*

$$\sum_{n=1}^{\infty} P_\theta\Big\{n^{\beta/(2\beta+1)}(\log n)^{-(3/2+\alpha)}|\hat\theta_{k_n} - \theta_{k_n}| > \epsilon\Big\} \quad < \quad \infty\,, \tag{2.3}$$

$$\limsup_{n\to\infty} \max_{k\in K_n} \frac{n^{2\beta/(2\beta+1)}}{(\log n)^2} E_\theta(\hat\theta_k - \theta_k)^2 \quad \le \quad C_1 \tag{2.4}$$

hold uniformly over $\theta \in \Theta_\beta$*.*

Notice that, for any $0 < \epsilon < 1$, $\{k \in \mathbb{N} : \; \epsilon \le k/n \le 1\} \subset K_n$ for sufficiently large n.

Let $\hat\theta_n(u)$ be a piecewise constant continuation of $\hat\theta_k = \hat\theta(x_k)$, $k = 0, 1, \ldots, n$, i.e. $\hat\theta_n(u) = \hat\theta_k$ for $x_k \le u < x_{k+1}$, $i = 0, 1, \ldots, n-1$, and $\hat\theta_n(1) = \hat\theta_n$. Observe that $\hat\theta_n(u)$ and $\hat\theta_n$ denote different objects.

The following corollary follows immediately from Theorem 1 and the Lipschitz condition on functions from the class Θ_β.

Corollary 1. *Suppose conditions (A1) and (A2) are fulfilled. Let* $T_n = \{x : \; C_0 n^{-1/(2\beta+1)} \le x \le 1\}$*. Then, for some positive constant* C_1*, any fixed* $\alpha, \epsilon > 0$ *and any fixed* $x \in (0, 1]$ *the relations*

$$\sum_{n=1}^{\infty} P_\theta\Big\{n^{\beta/(2\beta+1)}(\log n)^{-(3/2+\alpha)}|\hat\theta_n(x) - \theta(x)| > \epsilon\Big\} < \infty\,,$$

$$\limsup_{n\to\infty} \max_{u\in T_n} \frac{n^{2\beta/(2\beta+1)}}{(\log n)^2} E_\theta(\hat\theta_n(u) - \theta(u))^2 \le C_1$$

hold uniformly in $\theta \in \Theta_\beta$*.*

Remarks.

1. From the first assertion of the corollary and the Borel-Cantelli lemma it follows that

$$\lim_{n\to\infty} \frac{n^{\beta/(2\beta+1)}}{(\log n)^{3/2+\alpha}} |\hat\theta_n(x) - \theta(x)| = 0 \quad \text{almost surely,}$$

   for any fixed $x \in (0, 1]$ and any $\theta \in \Theta_\beta$.

2. Analyzing the proof of the theorem, one can see that the choice γ_n comes essentially from the balancing of the terms in the upper bound for the risk of the estimator and the size of the set K_n. Note also that taking $\gamma_n = C_n n^{-2\beta/(2\beta+1)} \log n$ leads to

$$K_n = \{i : \; C_0 C_n^{-1} n^{2\beta/(2\beta+1)} \le i \le n\}\,.$$

   So, we can enlarge the set K_n by taking a sequence $C_n \to \infty$, but then the rate of convergence of the estimator becomes worse with the factor C_n^{-1}.

3. As one can see, we need to know a constant in the algorithm which must be bigger than $H+L$ (we took $H+L+h$). Although this seems to be a restrictive feature of the algorithm, we can assume this without loss of generality, because we can prove the result with a sequence H_n converging to infinity sufficiently slowly instead of constant $H+L+h$. This would make the proof of the result lengthier. In practice one can use simply $S(u,v) = \text{sign}(u-v)$.

4. It is easy to see that $E\{\hat{\theta}_{k+1} | X_0, X_1, \dots, X_{k-1}\} = \hat{\theta}_k + \gamma_n(1 - 2F_\epsilon(\hat{\theta}_k - \theta_k))I\{|\hat{\theta}_k| \le H + L + h\} - \gamma_n\hat{\theta}_k I\{|\hat{\theta}_k| > H + L + h\}$. In view of (A1), this expresses, in a way, the requirement for the algorithm (2.2) to update the estimator $\hat{\theta}_k$ correctly in the sense of shifting the estimator in the right direction, towards θ_k.

5. If we give up the requirement for the estimator to be recursive, we can get an estimator for the set $\bar{K}_n = \{k \in \mathbb{N} : 0 \le k \le n - C_0 n^{2\beta/(2\beta+1)}\}$ by using the algorithm in the opposite direction:

$$\hat{\theta}_k = \hat{\theta}_{k+1} + \gamma_n S(X_{k+1}, \hat{\theta}_{k+1}), \qquad k = n-1, \dots, 0,$$

with the initial value $\hat{\theta}_n = 0$. In so doing, we can provide an estimator with the same properties as in Theorem 1 with the set $\{0, \dots, n\}$ instead of K_n.

6. Since the unknown signal $\theta(t)$ satisfies the Lipschitz condition, it is natural to take a continuous continuation of $\hat{\theta}_k = \hat{\theta}(x_k)$, $k = 0, 1, \dots, n$. Obviously, the corollary is also true for piecewise linear continuation: for $x_k \le u < x_{k+1}$,

$$\hat{\theta}_n(u) = \frac{(\hat{\theta}_{k+1} - \hat{\theta}_k)u}{x_{k+1} - x_k} + \frac{\hat{\theta}_k x_{k+1} - \hat{\theta}_{k+1} x_k}{x_{k+1} - x_k},$$

$k = 0, \dots, n-1$; and $\hat{\theta}_n(1) = \hat{\theta}_n$.

7. The results still hold if we allow ξ_k's to have different distributions, all satisfying the conditions (A1) and (A2). Moreover, if we define the class $\mathcal{P}_\xi = \mathcal{P}_\xi(\delta, p, M)$ of densities satisfying (A1), (A2) and the condition $\sup_{u\in\mathbb{R}} |f(u)| \le M$ for some $M > 0$, then all the results hold uniformly over all joint distributions of $(\xi_0, \dots, \xi_n)$, such that ξ_k's are independent and the distribution of each ξ_k is from $\mathcal{P}_\xi$.

The proof of Theorem 1 is based on the following lemma. Denote

$$g(u, \theta_k) = E_\theta S(X_k, u).$$

Lemma 1. *Suppose conditions (A1) and (A2) are fulfilled. For any $u \in \mathbb{R}$, $n \in \mathbb{N}$, $k \in \mathbb{N}$, $0 \le k \le n$, the representation $g(u, \theta_k) = -G(u, \theta_k)(u - \theta_k)$ holds uniformly over $\theta \in \Theta_\beta$ for some function $G(u, \theta_k)$ such that $G_1 \le G(u, \theta_k) \le G_2$ with some positive constants G_1 and G_2.*

Proof. Consider first the case $|u| \leq H + L + h$. Then we compute

$$g(u, \theta_k) = \int \operatorname{sign}(x - u) f_\epsilon(x - \theta_k) dx = 1 - 2F_\epsilon(u - \theta_k) \,.$$

This implies that

$$G(u, \theta_k) = -g(u, \theta_k)/(u - \theta_k) = (2F_\epsilon(u - \theta_k) - 1)/(u - \theta_k) \,.$$

According to the assumption (A1), distribution F_ϵ has zero median. Therefore $G(u, \theta_k)$ is always nonnegative. Denote $v = u - \theta_k$. If $|v| \leq \delta$, with constant δ from (A2), we can expand $F_\epsilon(v)$ around zero in a Taylor series up to the first order term. According to (A1), we get $F_\epsilon(v) = 1/2 + f_\epsilon(v^*)v$, which implies $G(u, \theta_k) = 2f_\epsilon(v^*)$, with $|v^*| \leq |v| \leq \delta$. Using the assumption (A2), we have that $0 < 2p \leq G(u, \theta_k) \leq 2 \sup_{|v| \leq \delta} f_\epsilon(v) < \infty$.

If $|u - \theta_k| > \delta$, then obviously

$$G(u, \theta_k) = (2F_\epsilon(u - \theta_k) - 1)/(u - \theta_k) \leq 1/\delta \,.$$

On the other hand,

$$\begin{aligned} G(u, \theta_k) &\geq (2(H + L) + h)^{-1} \min\{2F_\epsilon(\delta) - 1, 1 - 2F_\epsilon(-\delta)\} \\ &\geq 2p\delta/(2(H + L) + h) > 0 \end{aligned}$$

since

$$|u - \theta_k| \leq |u| + |\theta_k| \leq 2(H + L) + h \,,$$

which is in turn due to $|\theta_k| \leq \sup_{u \in [0,1]} |\theta(u)| \leq H + L$ for all k.

Now let $|u| > H + L + h$. In this case $g(u, \theta_k) = -u$, so $G(u, \theta_k) = u/(u - \theta_k)$. Recall also that $|\theta_k| \leq H + L$ for all k. Since $|u - \theta_k| \geq |u| - |\theta_k| \geq h$,

$$G(u, \theta_k) = u/(u - \theta_k) \leq 1 + |\theta_k|/|u - \theta_k| \leq 1 + (H + L)/h \,.$$

On the other hand, if θ_k and u have the same sign, then certainly $G(u, \theta_k) = u/(u - \theta_k) \geq 1$. If θ_k and u have opposite signs, then $|u - \theta_k| \geq H + L + h + |\theta_k|$ and therefore

$$\begin{aligned} G(u, \theta_k) &= u/(u - \theta_k) \geq 1 - |\theta_k|/|u - \theta_k| \\ &\geq 1 - |\theta_k|/(H + L + h + |\theta_k|) \\ &= (H + L + h)/(H + L + h + |\theta_k|) \\ &\geq (H + L + h)/(2(H + L) + h) \geq 1/2 \,. \end{aligned}$$

We can take $G_1 = \min\{2p, 2p\delta/(2(H + L) + h), 1/2\}$ and

$$G_2 = \max\left\{2 \sup_{|v| \leq \delta} f_\epsilon(v), 1/\delta, 1 + (H + L)/h\right\} .$$

The lemma is proved. $\square$

Proof of Theorem 1. Establish the following conventions: $\sum_{i=m+1}^{m} b_i = 0$, $\prod_{i=m+1}^{m} b_i = 1$ for any sequence $\{b_i\}$ and let c and C denote generic constants which can be different in different expressions. Define the differences $\delta_k = \delta_{k,n} = \hat{\theta}_k - \theta_k$ and $\Delta\theta_k = \theta_k - \theta_{k+1}$. Denote also $M(X_k, \hat{\theta}_k, \theta_k) = S(X_k, \hat{\theta}_k) - g(\hat{\theta}_k, \theta_k)$. Using the relation $g(\hat{\theta}_k, \theta_k) = -G(\hat{\theta}_k, \theta_k)\delta_k$ from Lemma 1, we represent the relation (2.2) in new terms as follows:

$$\begin{aligned}\delta_k &= \delta_{k-1} + \gamma_n(M(X_{k-1}, \hat{\theta}_{k-1}, \theta_{k-1}) + g(\hat{\theta}_{k-1}, \theta_{k-1})) + \Delta\theta_{k-1} \\ &= \delta_{k-1}(1 - \gamma_n G(\hat{\theta}_{k-1}, \theta_{k-1})) + \gamma_n M(X_{k-1}, \hat{\theta}_{k-1}, \theta_{k-1}) + \Delta\theta_{k-1},\end{aligned}$$

$k = 1, \ldots, n$. For any k_0, $0 \le k_0 \le k$, by iterating the above relation, we obtain that

$$\begin{aligned}\delta_k &= \delta_{k_0} \prod_{i=k_0}^{k-1} [1 - \gamma_n G(\hat{\theta}_i, \theta_i)] \\ &\quad + \sum_{i=k_0}^{k-1} \left\{ [\Delta\theta_i + \gamma_n M(X_i, \hat{\theta}_i, \theta_i)] \prod_{j=i+1}^{k-1} [1 - \gamma_n G(\hat{\theta}_j, \theta_j)] \right\}.\end{aligned} \tag{2.5}$$

Now we denote $A_i = \sum_{j=k_0}^{i} a_i$, with $a_i = \Delta\theta_j + \gamma_n M(X_j, \hat{\theta}_j, \theta_j)$, and apply Abel's transformation for series to the second term:

$$\begin{aligned}&\sum_{i=k_0}^{k-1} a_i \prod_{j=i+1}^{k-1} [1 - \gamma_n G(\hat{\theta}_j, \theta_j)] \\ &= A_{k-1} - \sum_{i=k_0}^{k-2} A_i \gamma_n G(\hat{\theta}_{i+1}, \theta_{i+1}) \prod_{j=i+2}^{k-1} [1 - \gamma_n G(\hat{\theta}_j, \theta_j)].\end{aligned} \tag{2.6}$$

From (2.5) and (2.6) it follows that

$$\begin{aligned}\delta_k &= \delta_{k_0} \prod_{i=k_0}^{k-1} [1 - \gamma_n G(\hat{\theta}_i, \theta_i)] + A_{k-1} \\ &\quad - \sum_{i=k_0}^{k-2} A_i \gamma_n G(\hat{\theta}_{i+1}, \theta_{i+1}) \prod_{j=i+2}^{k-1} [1 - \gamma_n G(\hat{\theta}_j, \theta_j)].\end{aligned}$$

Note that the relation (2.6) holds for any sequence $\{a_i\}$. So, by taking in particular $a_{k_0} = 1$ and $a_i = 0$ for $i > k_0$, we derive that

$$\sum_{i=k_0}^{k-2} \gamma_n G(\hat{\theta}_{i+1}, \theta_{i+1}) \prod_{j=i+2}^{k-1} [1 - \gamma_n G(\hat{\theta}_j, \theta_j)] = 1 - \prod_{i=k_0}^{k-1} [1 - \gamma_n G(\hat{\theta}_i, \theta_i)] \le 1$$

for sufficiently large n, by Lemma 1.

By analyzing the procedure (2.2) one can see that

$$|\hat{\theta}_k| \leq H + L + h + b, \tag{2.7}$$

with $b = \sup_{u \geq 1} u^{-2\beta/(2\beta+1)} \log u$.

From the definition of the class Θ_β, it follows that $|\theta_k| \leq H+L$ uniformly in k and $\theta \in \Theta_\beta$. Thus $|\delta_k|$ is bounded uniformly in k and $\theta \in \Theta_\beta$ as well. Combining the last three relations, we obtain that

$$\begin{aligned} |\delta_k| &\leq |\delta_{k_0}|(1-\gamma_n G_1)^{k-k_0} + 2 \max_{k_0 \leq i \leq k} |A_i| \\ &\leq c(1-\gamma_n G_1)^{k-k_0} + 2 \max_{k_0 \leq i \leq k} |A_i| \end{aligned} \tag{2.8}$$

for sufficiently large n, uniformly in $\theta \in \Theta_\beta$. Recalling the definition of A_i, we get, uniformly in $\theta \in \Theta_\beta$,

$$\begin{aligned} &\delta_k^2 \leq 2 \max_{k_0 \leq i \leq k} \delta_{k_0}^2 (1-\gamma_n G_1)^{2(k-k_0)} \\ &+ c \max_{k_0 \leq i \leq k} \Big\{ \sum_{j=k_0}^{i} \Delta\theta_j \Big\}^2 + C\gamma_n^2 \max_{k_0 \leq i \leq k} \Big\{ \sum_{j=k_0}^{i} M(X_j, \hat{\theta}_j, \theta_j) \Big\}^2 . \end{aligned} \tag{2.9}$$

Choose the level of backward recursion k_0 as follows: $k_0 = k_0(k) = k_0(k,n) = \lfloor k - C_0 \gamma_n^{-1} \log n \rfloor$. So,

$$C_0 \gamma_n^{-1} \log n \leq k - k_0 \leq C_0 \gamma_n^{-1} \log n + 1$$

and therefore $k_0(k)$ are properly defined for those k which satisfy

$$k \in \{ i \in \mathbb{N} : \ C_0 \gamma_n^{-1} \log n \leq i \leq n \} = J_n .$$

Now it is easy to derive the following bound for the first term in the right-hand side of (2.9): uniformly in $\theta \in \Theta_\beta$,

$$(1-\gamma_n G_1)^{k-k_0} \leq C(1-\gamma_n G_1)^{C_0 \gamma_n^{-1} \log n} \leq C e^{-2 \log n} = C n^{-2} \tag{2.10}$$

for sufficiently large n because the constant C_0 is chosen in such a way that $C_0 G_1 \geq 2$.

The second term in the right-hand side of (2.9) is bounded as follows:

$$\max_{k_0 \leq i \leq k} \Big\{ \sum_{j=k_0}^{i} \Delta\theta_j \Big\}^2 \leq \max_{k_0 \leq i \leq k} L^2 (x_i - x_{k_0})^{2\beta} \leq c(k-k_0)^{2\beta}/n^{2\beta}$$

uniformly in $\theta \in \Theta_\beta$.

To evaluate the last term from the right-hand side of (2.9), note first that the sequence $\big\{ M(X_j, \hat{\theta}_j, \theta_j) \big\}_{j=k_0}^{k}$ is a martingale difference relative to the natural filtration $\{\mathcal{A}_j\}_{j=k_0}^k$, i.e. $\mathcal{A}_j = \sigma(X_0, X_1, \dots, X_j)$ is the σ-algebra generated by $X_0, X_1, \dots, X_j$. Indeed, by the definition of $M(X_j, \hat{\theta}_j, \theta_j)$ and (2.7), one can easily see that $|M(X_j, \hat{\theta}_j, \theta_j)| \le 2\max\{1, H+L+h+b\}$ almost surely and

$$E\{M(X_{j+1}, \hat{\theta}_{j+1}, \theta_{j+1}) \,|\, \mathcal{A}_j\} = g(\hat{\theta}_j, \theta_j) - g(\hat{\theta}_j, \theta_j) = 0\,, \quad j = 1, \dots, n-1\,.$$

Therefore $\{\sum_{j=k_0}^m M(X_j, \hat{\theta}_j, \theta_j)\}_{m=k_0}^k$ is a martingale. Applying Doob's maximal inequality for submartingales, we obtain that

$$E\Big\{\max_{k_0 \le i \le k}\Big[\sum_{j=k_0}^{i} M(X_j, \hat{\theta}_j, \theta_j)\Big]^2\Big\} \le 4E\Big\{\sum_{j=k_0}^{k} M(X_j, \hat{\theta}_j, \theta_j)\Big\}^2 = C(k-k_0)$$

uniformly in $\theta \in \Theta_\beta$ and over $k \in J_n$.

As $k - k_0 \le C\gamma_n^{-1}\log n$ and $\gamma_n = n^{-2\beta/(2\beta+1)}\log n$, collecting all the relations beginning with (2.9) yields that, for sufficiently large n,

$$E\delta_k^2 \le Cn^{-2} + c(k-k_0)^{2\beta}n^{-2\beta} + C\gamma_n^2(k-k_0) = O\Big(n^{-2\beta/(2\beta+1)}(\log n)^2\Big)$$

uniformly in $\theta \in \Theta_\beta$ and over $k \in J_n$, which proves the relation (2.4) of the theorem.

Denote for brevity $\phi_n = n^{\beta/(2\beta+1)}(\log n)^{-(3/2+\alpha)}$. From (2.8) it follows that, for sufficiently large n and any k such that $k \in J_n$,

$$\begin{aligned} P\Big\{\phi_n|\delta_k| > \epsilon\Big\} \;\le\; & P\Big\{C\phi_n\Big[(1-\gamma_n G_1)^{k-k_0} + \max_{k_0\le i\le k}\Big|\sum_{j=k_0}^{i}\Delta\theta_j\Big|\Big] > \epsilon/2\Big\} \\ & + P\Big\{c\phi_n\gamma_n \max_{k_0\le i\le k}\Big|\sum_{j=k_0}^{i} M(X_j, \hat{\theta}_j, \theta_j)\Big| > \epsilon/2\Big\}. \end{aligned}$$

The first term in the right-hand side of the last inequality is zero for sufficiently large n because, by (2.10) and the definition of the class Θ_β,

$$\begin{aligned} \phi_n\Big[(1-\gamma_n G_1)^{k-k_0} + \max_{k_0\le i\le k}\Big|\sum_{j=k_0}^{i}\Delta\theta_j\Big|\Big] \;&\le\; \frac{C\phi_n}{n^2} + \frac{L\phi_n(k-k_0)^\beta}{n^\beta} \\ &\le\; \frac{c}{(\log n)^{3/2+\alpha}} \end{aligned}$$

uniformly in $\theta \in \Theta_\beta$ and over $k \in J_n$.

To evaluate the second term, we apply the Azuma-Hoeffding inequality [see, for example, Williams (1991); the inequality can also be derived from a general result in de la Peña and Giné (1999)] for a martingale $\{M_l\}_{l=1}^m$ whose increments $d_l = M_l - M_{l-1}$ (with $M_0 = 0$) are bounded in absolute value by c_l: $P\{\sup_{1\le l\le m} M_l \ge x\} \le \exp\{-x^2/(2\sum_{l=1}^m c_l^2)\}$. We know already that $\{\sum_{j=k_0}^{m} M(X_j, \hat{\theta}_j, \theta_j)\}_{m=k_0}^{k}$ is a martingale with bounded increments. Applying the Azuma-Hoeffding inequality yields

$$P\Big\{c\phi_n\gamma_n \max_{k_0 \le i \le k} \Big|\sum_{j=k_0}^{i} M(X_j, \hat{\theta}_j, \theta_j)\Big| > \epsilon/2\Big\}$$

$$\le \quad 2\exp\Big\{-\frac{Cn^{2\beta/(2\beta+1)}(\log n)^{1+2\alpha}}{k-k_0}\Big\} \le 2\exp\{-c(\log n)^{1+2\alpha}\} \le Cn^{-2}$$

uniformly in $\theta \in \Theta_\beta$ and over $k \in K_n$. Since the above estimates are uniform in $\theta \in \Theta_\beta$ and over $k \in K_n$ (constants do not depend on k), we conclude that for any sequence $\{k_n\}$ such that $k_n \in K_n$

$$\sum_{n=1}^{\infty} P\{\phi_n|\delta_{k_n}| > \epsilon\} < \infty$$

uniformly in $\theta \in \Theta_\beta$. The relation (2.3) follows and the theorem is proved. □

3 Multivariate case

Now we discuss shortly the multivariate case. Suppose we observe deterministic signal values on a discrete grid of the unit cube $I_d = [0,1]^d$ and the unknown function $\theta(\mathbf{x})$ on cube $[0,1]^d$ belongs to the multivariate Lipschitz function class $\Theta_\beta = \Theta_\beta(H, L, d)$: for some positive H and L,

$$\Theta_\beta(H,L,d) = \Big\{\theta : \sup_{\mathbf{u}\in[0,1]^d} |\theta(\mathbf{u})| \le H;$$
$$|\theta(\mathbf{u}_1) - \theta(\mathbf{u}_2)| \le L\|\mathbf{u}_1 - \mathbf{u}_2\|^\beta, \ \mathbf{u}_1, \mathbf{u}_2 \in [0,1]^d\Big\},$$

where $0 < \beta \le 1$.

The design $\{\mathbf{x}_k\}$ is assumed to be "dense" and "uniform" enough. To be precise, instead of the conditions D1 and D2, it should satisfy the following condition: for some fixed positive constant A_d

D3. $\|\mathbf{x}_m - \mathbf{x}_l\| \le A_d(|m-l|/n)^{1/d}$ for any $0 \le m, l \le n$.

One could consider a general compact set $\mathcal{T} \subset \mathbb{R}^d$ instead of the unit cube. We can also consider an unbounded set which can be represented as a countable union of disjoint compact sets. In this case parameter n would play a role of density of observations, number of observations per unit volume in $\mathbb{R}^d$.

Minor modifications in the proofs are needed for the multivariate case, in fact, only in the place where we evaluate the quantity

$$\max_{k_0 \le i \le k} \Big\{\sum_{j=k_0}^{i} \Delta\theta_j\Big\}^2.$$

In the multivariate case we bound this by (D3) as follows:

$$\max_{k_0 \le i \le k} \Big\{ \sum_{j=k_0}^{i} \Delta\theta_j \Big\}^2 \le \max_{k_0 \le i \le k} L^2 \|\mathbf{x}_i - \mathbf{x}_{k_0}\|^{2\beta} \le c((k-k_0)/n)^{2\beta/d}.$$

Thus, the same kind of results hold for the recursive estimator $\hat{\theta}$ defined by (2.2) with $\gamma'_n = n^{-2\beta/(2\beta+d)} \log n$, $K'_n = \{k \in \mathbb{N} : C_0 n^{2\beta/(2\beta+d)} \le k \le n\}$. This optimal choice of step size γ comes from balancing of corresponding terms, similarly to the one-dimensional case. As to the estimation of $\theta(\mathbf{x})$, note that for any fixed $\mathbf{x} \in [0,1]^d \setminus \mathbf{0}$ there exists a sequence $\{k_n\}$ such that $\|\mathbf{x}_{k_n} - \mathbf{x}\| \le Cn^{-1/d}$ and $k_n \in K'_n$ for sufficiently large n. Then we take $\hat{\theta}_n(\mathbf{x}) = \hat{\theta}_{k_n}$ as estimator of $\theta(\mathbf{x})$. Notice that the approximation error $|\theta(\mathbf{x}_{k_n}) - \theta(\mathbf{x})| \le cn^{-\beta/d}$ is of a smaller order compared to the convergence rate $n^{-\beta/(2\beta+d)}$.

Results analogous to Theorem 1 and Corollary 1 hold. The difference is that now we have the sets K'_n, $[0,1]^d \setminus \mathbf{0}$, the expressions $\beta/(2\beta+d)$ and $2\beta/(2\beta+d)$ instead of K_n, $(0,1]$, $\beta/(2\beta+1)$ and $2\beta/(2\beta+1)$ respectively.

4 Nonparametric M- and penalized estimators

Suppose now that the unknown function $\theta(x)$ on the interval $[0,1]$ belongs to the Hölder function class $\mathcal{H}_\alpha = \mathcal{H}_\alpha(H,L)$ of the smoothness α: for some positive H and L,

$$\mathcal{H}_\alpha(H,L) = \Big\{ \theta : |\theta^{(m)}(0)| \le H\,,\ m = 0,\dots,r\,; \\ |\theta^{(r)}(u) - \theta^{(r)}(v)| \le L|u-v|^{\beta_0},\ u,v \in [0,1] \Big\},$$

where $r \in \mathbb{N}$, $0 < \beta_0 \le 1$, $\alpha = r + \beta_0$. Here $\theta^{(r)}(u)$ denotes the r-th derivative of $\theta(u)$. Notice that for $0 < \alpha \le 1$ $\mathcal{H}_\alpha = \Theta_\alpha$.

The estimator $\hat{\theta}$ we studied in the second section can be regarded as a modified nonparametric recursive version of the least absolute deviations estimator, a certain M-estimator with a particular choice of a criterion function. We give here some heuristic arguments. Neglecting for the moment the trimming, we have that $\hat{\theta}_{k+1} = \hat{\theta}_k - \gamma_n \nabla J_k(\hat{\theta}_k)$, with $\nabla J_k(u) = \partial |X_k - u|/\partial u$ for $u \ne X_k$ and zero otherwise; compare this also with the gradient method for determining the extreme points of functions.

Let us consider now this M-estimator itself. This is an estimator which minimizes $\sum_{k=0}^{n} |X_k - \theta_k|$ over $\theta \in \mathcal{H}_\alpha$. We denote it by $\tilde{\theta}$. The following proposition characterizes the performance of this estimator, this time not pointwisely but in L_2-type sense (in fact, a discrete approximation of the L_2-risk, based on the function values from the observation model $\{\theta(x_k)\}$).

Proposition 1. *Suppose the conditions (A1) and (A2) are met. Then, uniformly over $\theta \in \mathcal{H}_\alpha$, for some positive constant c and all $T \geq c$,*

$$P\Big\{\sum_{k=0}^{n}(\tilde{\theta}_k - \theta_k)^2 \geq Tn^{1/(2\alpha+1)}\Big\} \leq c\exp\big\{-n^{1/(2\alpha+1)}T^2/c\big\}.$$

The proof of the proposition follows from Theorem 12.3 in van de Geer (2000) and is based on the fact that the ϵ-entropy of the set $\mathcal{H}_\alpha$ is bounded by $A\epsilon^{-1/\alpha}$ for all $\epsilon > 0$, with constant A depending on α, H and L, appearing in the definition of $\mathcal{H}_\alpha$. The above assertion says essentially that the convergence rate is in a certain sense $n^{\alpha/(2\alpha+1)}$; this rate is shown by Stone (1982) to be minimax over the class $\mathcal{H}_\alpha$ in case of the pointwise risk.

The case of the Lipschitz class ($\mathcal{H}_\alpha$ with $\alpha = 1$) can also be handled by the so called penalized estimator. Introduce the total variation of function θ, $\theta_k = \theta(x_k)$, $0 = x_0 \leq \ldots \leq x_n = 1$,

$$TV(\theta) = \sum_{k=1}^{n}|\theta_k - \theta_{k-1}|.$$

Let $\bar{\theta}$ be an estimator (called penalized estimator), which minimizes

$$\sum_{k=0}^{n}|X_k - \theta_k| + Cn^{1/3}TV(\theta),$$

where C is any positive constant.

Proposition 2. *Suppose the conditions (A1) and (A2) are met. Let M be a fixed positive constant. Then, uniformly over $\theta \in \{\theta : TV(\theta) \leq M\}$, for some positive constant c and all $T \geq c$*

$$P\Big\{\sum_{k=0}^{n}(\bar{\theta}_k - \theta_k)^2 \geq Tn^{1/3}\Big\} \leq c\exp\big\{-n^{1/3}T^2/c\big\}.$$

This proposition follows from the Corollary 4.4 in van de Geer (1999).

References

E.N. Belitser, A.P. Korostelev (1992), Pseudovalues and minimax filtering algorithms for the nonparametric median, *Adv. in Sov. Math.* **12**, 115–124.

V.H. de la Peña, E. Giné (1999), *Decoupling. From dependence to independence. Randomly stopped processes. U-statistics and processes. Martingales and beyond*, Probability and its Applications. Springer-Verlag, New York.

A.P. Korostelev (1987), *Asymptotic minimax filtering of nonparametric signals*, Technical report, Institute of System Studies, Moscow.

C.J. Stone (1982), *Optimal global rates of convergence for nonparametric regression*, Ann Statist 10, 1040-1053.

Y.K. Truong (1989), Asymptotic properties of kernel estimators based on local medians, *Ann. Statist.* **18**, 606–617.

S. van de Geer (2000), *Empirical Processes in M-estimation*, Cambridge University Press, Cambridge, UK.

S. van de Geer (1999), *M-estimation using penalties or sieves*, Technical Report MI 31-99, University of Leiden, to appear in JSPI.

D. Williams (1991), *Probability with Martingales*, Cambridge University Press, Cambridge, UK.

Eduard Belitser and Sara van de Geer
Mathematical Institute
Leiden University
2333 CA Leiden
Niels Bohrweg 1
The Netherlands
e-mail:belitser@math.leidenuniv.nl
e-mail: geer@math.leidenuniv.nl

Variable Kernel Estimates: on the Impossibility of Tuning the Parameters

Luc Devroye* and Gábor Lugosi**

ABSTRACT For the standard kernel density estimate, it is known that one can tune the bandwidth such that the expected L1 error is within a constant factor of the optimal L1 error (obtained when one is allowed to choose the bandwidth with knowledge of the density). In this paper, we pose the same problem for variable bandwidth kernel estimates where the bandwidths are allowed to depend upon the location. We show in particular that for positive kernels on the real line, for any data-based bandwidth, there exists a density for which the ratio of expected L1 error over optimal L1 error tends to infinity. Thus, the problem of tuning the variable bandwidth in an optimal manner is "too hard". Moreover, from the class of counterexamples exhibited in the paper, it appears that placing conditions on the densities (monotonicity, convexity, smoothness) does not help.

1 Introduction

We are given an i.i.d. sample $X_1, \ldots, X_n$ drawn from an unknown density f on $\mathbb{R}$. A density estimate $f_n(x) = f_n(x, X_1, \ldots, X_n)$ is a real-valued measurable function of its arguments. One of the most popular estimates is

$$f_{nh}(x) = \frac{1}{n} \sum_{i=1}^{n} K_h(x - X_i),$$

where $K : \mathbb{R} \to \mathbb{R}$ is a fixed kernel with $\int K = 1$, $K_h(x) = (1/h^d)K(x/h)$, and $h > 0$ is the smoothing factor (Akaike, 1954; Parzen, 1962; Rosenblatt, 1956). Many data-dependent choices for h have been proposed in the literature. The question of whether h can be tuned in an optimal manner has been answered in the affirmative by Devroye and Lugosi (1996, 1997),

*Research was supported by NSERC Grant A3456 and by FCAR Grant 90-ER-0291.
**Research was supported by DIGES Grant PB96-0300.
Keywords and phrases: Density estimation, variable kernel estimate, convergence, smoothing factor, minimax lower bounds, asymptotic optimality.

where data-dependent smoothing factors H are introduced for which

$$\sup_f \limsup_{n\to\infty} \frac{\mathbf{E}\int |f_{nH}-f|}{\inf_h \mathbf{E}\int |f_{nh}-f|} \le 3 ,$$

whenever the kernel K is nonnegative, Lipschitz, and of a compact support.

The variable kernel estimate (Breiman, Meisel and Purcell, 1977; Raatgever and Duin, 1978; Habbema, Hermans and Remme, 1978) allows h to depend upon either i or x. In the BMP (Breiman-Meisel-Purcell) form

$$f_n(x) = \frac{1}{n}\sum_{i=1}^{n} K_{h_i}(x-X_i),$$

f_n is still a density. Various ways of letting the data pick the h_i's were proposed by Breiman, Meisel and Purcell (1977) and others later. As each h_i should be tuned to what is happening near X_i, the behavior of the BMP estimate should not be radically different from that of the raw variable kernel estimate in which $h = h(x)$. In a data-based form, we have $H = H(x; X_1, \ldots, X_n)$. It is this form that will be studied in this paper.

We note here some general references on variable kernel estimates, such as Devroye (1985), Devroye and Penrod (1986), Jones (1990), Terrell and Scott (1992), Hall (1992), and Marron, Hall and Hu (1995), who all deal with convergence issues. A particularly influential paper was that of Abramson (1982), who showed how variable bandwidths with positive kernels can nevertheless induce convergence rates usually attainable with fixed bandwidths and fourth order kernels (see also Hall and Marron, 1988). This sets variable bandwidth kernels apart from fixed bandwidth kernels. Particular variable bandwidths include the cross-validation methods of Hall and Marron (1988), Hall and Schucany (1989) and Mielniczuk, Sarda and Vieu (1989), Sheather's solve-the-equation bandwidth (1983, 1986) (Thombs and Sheather, 1992), Sain's (1994) and Sain and Scott's bootstrap bandwidth (1996), and Hazelton's (1996) and Farmen's (1996) smoothed bootstrap bandwidths. None of the cited papers addresses the question posed here. Our work was inspired by interesting observations by David Scott at presentations in Louvain La Neuve and Montreal in 1997 which are summarized in Sain and Scott (1997).

To introduce our main result, we fix the notation as follows: the raw variable kernel estimate is

$$f_{n,h(x)}(x) = \frac{1}{n}\sum_{i=1}^{n} K_{h(x)}(x-X_i),$$

and the data-based variable kernel estimate is

$$f_{n,H(x)}(x) = \frac{1}{n}\sum_{i=1}^{n} K_{H(x)}(x-X_i),$$

where it is understood that $H(x) = H(x; X_1, \ldots, X_n)$. Let $\mathcal{F}_B$ be the class of nondecreasing, convex-shaped densities f on $[t, t+s]$ with

$$s \sup_{(t,t+s)} f(x) \leq B,$$

where $t \in \mathbb{R}$ and $s > 0$ are arbitrary. (Note that t is just a translation constant and that if $f \in \mathcal{F}_B$, then so are all rescaled and translated versions of f.) One of the results we show is the following: if $K \geq 0$ is a symmetric square-integrable kernel on $[-1, 1]$, then there exists a positive constant C not depending upon n such that

$$\inf_{H:\mathbb{R}^{n+1}\to(0,\infty)} \sup_{f\in\mathcal{F}_B} \frac{\mathbf{E}\int |f_{n,H(x)}(x) - f(x)|dx}{\inf_{h:\mathbb{R}\to(0,\infty)} \mathbf{E}\int |f_{n,h(x)}(x) - f(x)|dx} \geq Cn^{\frac{1}{10}}.$$

So, even with the knowledge that $f \in \mathcal{F}_B$, one cannot efficiently design a variable bandwidth. Generalizations and extensions of this result round out the paper.

It is generally accepted that some versions of variable bandwidths outperform fixed bandwidth estimates, and our results do not contradict this. However, we only say that the class of variable bandwidth kernel estimates is too large to be optimized. Devroye, Lugosi and Udina (1998) basically describe how far one can go in the optimization: they partition the real line into k intervals and use a different (fixed) bandwidth on each interval. Simultaneous optimization of the k intervals and the k bandwidths is possible in the sense that there exists a data-based choice of a piecewise constant function $H(x)$ with k pieces such that for all densities

$$\mathbf{E}\int |f_{n,H(x)}(x) - f(x)|dx \leq c_0 \inf \mathbf{E}\int |f_{n,h(x)}(x) - f(x)|dx + c_1\sqrt{\frac{k\log n}{n}},$$

where c_0 and c_1 are absolute constants, and the infimum is taken over all variable kernel estimates where $h(x)$ is piecewise constant with k pieces. However, as shown in this paper, optimization becomes impossible when $k = n$. As a happy by-product, any heuristic for variable bandwidths is suboptimal, and therefore, no claim of superiority can be made for any method.

2 The main result

We state our main result for the simplest possible kernels and will generalize later. No attempt is made to optimize the constants in the bounds. We introduce a shape parameter for nonnegative symmetric kernels K:

$$\rho \stackrel{\text{def}}{=} \text{support}(K) \times \int K^2 .$$

Note that by the Cauchy-Schwarz inequality, $\rho \geq 1$, and equality is reached for the uniform kernel. In any case, ρ is a scale-invariant parameter that will appear in our bounds.

Theorem 1. *Let K be a symmetric nonnegative square-integrable kernel with shape parameter ρ. Then, for $n \geq 288$,*

$$\inf_{H:\mathbb{R}^{n+1}\to(0,\infty)} \sup_{f\in\mathcal{F}_{7/3}} \frac{\mathbf{E}\int |f_{n,H(x)}(x) - f(x)|dx}{\inf_{h:\mathbb{R}\to(0,\infty)} \mathbf{E}\int |f_{n,h(x)}(x) - f(x)|dx} \geq \frac{n^{\frac{1}{10}}}{41\sqrt{\rho}}.$$

The choice of the constant $B = 7/3$ is irrelevant, and is motivated by convenience. To prove Theorem 1, we combine two results, an estimate of the L1 error for raw variable kernel estimates inspired by recent work of Sain and Scott (1997), and a minimax lower bound for $\mathcal{F}_B$. Sain and Scott remarked that for most locations x, the kernel estimate could be made unbiased by taking $h(x)$ fixed and positive (not depending upon n). A similar but less explicit remark may also be found in Hazelton (1996, p. 223). This prompted us to consider $f \in \mathcal{F}_B$, where every x has this property. In particular, the rate of convergence could be $O(1/\sqrt{n})$ if we knew f and thus the optimal map $h : \mathbb{R} \to (0, \infty)$.

Lemma 1. *Let K be a symmetric nonnegative square-integrable kernel on the real line with shape parameter ρ. Then*

$$\sup_{f\in\mathcal{F}_B} \inf_{h:\mathbb{R}\to(0,\infty)} \mathbf{E} \int |f_{n,h(x)}(x) - f(x)|dx \leq \sqrt{\frac{4B\rho}{n}}.$$

Proof. Fix $f \in \mathcal{F}_B$. By a linear transformation, we may and will assume that f is supported on $[0,1]$ and bounded by B. Let the support of K be $[-c,c]$. Note that we may set $h(x) = |x|/c$ for $x < 0$ and $h(x) = (x-1)/c$ for $x > 1$, so that since f has support on $[0,1]$ and K on $[-c,c]$, we have $f_{n,h(x)} \equiv f(x) \equiv 0$ outside $[0,1]$. Thus, pick $x \in (0,1)$. We use the notation $*$ for the convolution operator, and drop the argument of $h(x)$. Note that as long as $h = h(x)$ is less than $\min(x, 1-x)/c$,

$$\begin{aligned}
\mathbf{E}f_{n,h}(x) &= f * K_h(x) \\
&= (1/h)\int f(y)K((x-y)/h)dy \\
&= \int f(x+hw)K(w)dw \\
&= \mathbf{E}f(x+hW) \quad \text{(where } W \text{ has density } K) \\
&\geq f(x+h\mathbf{E}\{W\}) \quad \text{(by Jensen's inequality)} \\
&= f(x).
\end{aligned}$$

Also, it is trivial to check that $\lim_{h\to\infty} f*K_h(x) = 0$. Thus, since $f*K_h(x)$, considered as a function of h, is continuous in h, we see that necessarily

$f(x) = f * K_h(x)$ for some $h > 0$, and, in fact, there exists at least one such h with $h \geq \min(x, 1-x)/c$. For $x \in (0,1)$, we define

$$h(x) = \sup\{z > 0 : f * K_z(x) = f(x)\} ,$$

and note for further use that $h(x) \geq \min(x, 1-x)/c$. From here on, h denotes this choice. Then, by routine calculations,

$$\begin{aligned}
\mathbf{E}\{|f_{n,h}(x) - f(x)|\} &= \mathbf{E}\{|f_{n,h}(x) - \mathbf{E}f_{n,h}(x)|\} \\
&\leq \sqrt{\mathbf{E}\{(f_{n,h}(x) - \mathbf{E}f_{n,h}(x))^2\}} \\
&= \sqrt{\frac{\mathrm{Var}\{K_h(x - X_1)\}}{n}} \\
&\leq \sqrt{\frac{\mathbf{E}\{(K_h(x - X_1))^2\}}{n}} \\
&= \sqrt{\frac{\int K^2((x-y)/h)h^{-2}f(y)dy}{n}} \\
&\leq \sqrt{\frac{B\int K^2}{nh}} .
\end{aligned}$$

This is true for all $x \in (0,1)$, and integration yields

$$\begin{aligned}
\mathbf{E}\left\{\int |f_{n,h}(x) - f(x)|dx\right\} &\leq \sqrt{\frac{B\int K^2}{n}} \int_0^1 \frac{1}{\sqrt{h(x)}}dx \\
&\leq 2\sqrt{\frac{Bc\int K^2}{n}} \int_0^{1/2} \frac{1}{\sqrt{x}}dx \\
&= \sqrt{\frac{4B\rho}{n}}
\end{aligned}$$

as required. □

It is of course impossible to estimate all f's in $\mathcal{F}_B$ at the rate $1/\sqrt{n}$ with any density estimate, let alone the data-based variable kernel density estimate. To support this, we merely require a minimax lower bound. For general results in density estimation, this was done by Birgé (1985, 1986) and Devroye (1987), and details for monotone densities are worked out in Birgé (1987a, 1987b) and Yang (1996). For the family $\mathcal{F}_B$, we merely require a good non-asymptotic bound with the right dependence on n but not B, so a bound is derived here which is optimal in n but suboptimal in B.

Lemma 2. *For $n \geq 288$, we have*

$$\inf_{f_n} \sup_{f \in \mathcal{F}_{7/3}} \mathbf{E} \int |f_n - f| \geq \frac{1}{37.37192\ldots n^{2/5}},$$

where the infimum is over all density estimates.

Proof. Let k be a suitable positive integer to be determined further on. We will construct a family of 2^k densities contained in $\mathcal{F}_B$, where $B = 7/3$. The interval $[0,1]$ is partitioned into k intervals denoted by $A_1, \ldots, A_k$. Thus, $A_i = [(i-1)/k, i/k)$. On $[(i-1)/k, i/k]$, we consider two piecewise linear functions f_i and g_i with the following properties:

A. $f_i = g_i$ at the ends of the interval, and $f_{i-1}((i-1)/k) = f_i((i-1)/k)$;

B. $\int_{A_i} f_i = \int_{A_i} g_i$;

C.

$$\sup_{x\in((i-2)/k,(i-1)/k)} \max(f'_{i-1}(x), g'_{i-1}(x)) \leq \inf_{x\in((i-1)/k,i/k)} \min(f'_i(x), g'_i(x));$$

D. $\sup_{x\in A_{i-1}} \max(f_{i-1}(x), g_{i-1}(x)) \leq \inf_{x\in A_i} \min(f_i(x), g_i(x))$;

E. $f_1(0) = g_1(0) = A > 0$, where $A = 1/3$;

F. $f_k(1) = g_k(1) \leq 7/3$;

G. $\sum_{i=1}^k \int_{A_i} f_i = 1$.

If we piece together a function f by choosing either f_i or g_i on A_i, then we obtain a bona fide density (by B, D, E and G). Furthermore, f is increasing (by D), continuous on (0,1) (by A) and convex (by A and C). There are 2^k such possible functions, and we may parametrize the family by a bit vector $b = (b_1, \ldots, b_k)$, where $b_i = 1$ if the corresponding f picks f_i on A_i, and $b_i = 0$ otherwise. We will denote this f by f_b and apologize for using the same notation as in f_i.

Define $d_i = 4ai$ for some constant a. We will pick the f_i's such that on A_i, both f'_i and g'_i are between d_{i-1} and d_i. This will then insure nondecreasing derivatives, and thus convexity for f. The positivity is also insured. On A_i, we set

$$f'_i = \begin{cases} d_{i-1} + a & x \in [(i-1)/k, (i-1/3)/k)\,, \\ d_i & x \in [(i-1)/k + 2/(3k), i/k)\,, \end{cases}$$

and

$$g'_i = \begin{cases} d_{i-1} & x \in [(i-1)/k, (i-2/3)/k)\,, \\ d_i - a = d_{i-1} + 3a & x \in [(i-1)/k + 1/(3k), i/k)\,. \end{cases}$$

It takes a moment to verify that $\int_{A_i} f'_i = \int_{A_i} g'_i$ so that f_i and g_i make equal jumps. Another routine computation shows that $\int_{A_i} f_i = \int_{A_i} g_i$ as well. Thus, the functions will suit us, provided that the total integral is 1. To do this, we observe that

$$\int_{A_i} f'_i = \frac{2}{3k}(4a(i-1) + a) + \frac{4ai}{3k} = \frac{4ai - 2a}{k}\,,$$

and similarly for $\int_{A_i} g_i'$. Hence, at i/k, the value of any f in our class is

$$A + \sum_{j=1}^{i} \frac{4aj - 2a}{k} = A + \frac{2ai^2}{k}.$$

For future reference, we note that $f_k(1) = g_k(1) = A + 2ak$. Next, we compute the integrals:

$$\begin{aligned}\int_{A_i} f_i &= \int_0^{2/3k} (f_i((i-1)/k) + (4(i-1)a + a)x)dx \\ &\quad + \int_0^{1/3k} (f_i(i/k) - 4aix)dx \\ &= \frac{A}{k} + \frac{a(18i^2 - 18i + 22)}{9k^2}.\end{aligned}$$

Thus, the integral of any f in our class is the sum over the k individual pieces, which is

$$A + \frac{a(6k^2 + 16)}{9k}.$$

As $A = 1/3$ and the integral must be 1, we must take

$$a = \frac{6k}{6k^2 + 16}.$$

We verify quickly that $a \le 1/k$, and that for $k \ge 2$, $a \ge 3/(5k)$. Finally, we insure that each f is bounded by $7/3$:

$$f_k(1) = \frac{1}{3} + \frac{12k^2}{6k^2 + 16} = \frac{42k^2 + 16}{18k^2 + 48} < \frac{7}{3}.$$

This ends the construction of our parametric family with 2^k members.

We now apply a minimax lower bound method pioneered by Assouad (1983), in the form given in Devroye (1987, p. 60): we need to compute two lower bounds α and β, where α is a uniform lower bound on $\int_{A_i} |f_i - g_i|$, and β is a uniform lower bound on $\int \sqrt{fg}$, where f and g are two of the 2^k functions that differ only on one interval. The required uniformity is with respect to i, the interval index. Clearly,

$$\int_{A_i} |f_i - g_i| = 2\left(\int_0^{1/(3k)} axdx + \int_0^{1/(6k)} 2axdx\right) = \frac{a}{6k^2}.$$

Thus, we may set

$$\alpha = \frac{a}{6k^2}.$$

Next, pick i, and let $I = A_i$, $O = [0,1] - A_i$. Let f and g be two of our functions that are equal on all intervals except A_i. Note that on A_i, $|f_i - g_i|$

is maximal at $1/3k$ from the left boundary, and the value there is $a/(3k)$. Then since

$$\int \sqrt{fg} = 1 - \frac{1}{2} \int \left(\sqrt{f} - \sqrt{g}\right)^2$$

and

$$\begin{aligned}
\int \left(\sqrt{f} - \sqrt{g}\right)^2 &= \int \frac{(f-g)^2}{\left(\sqrt{f}+\sqrt{g}\right)^2} \\
&\le \frac{3}{4} \int (f-g)^2 \\
&\qquad \text{(since } f, g \ge 1/3\text{)} \\
&= \frac{3}{4} \int_{A_i} (f-g)^2 \\
&= \frac{3}{2} \left(\int_0^{1/(3k)} (ax)^2 dx + \int_0^{1/(6k)} (2ax)^2 dx \right) \\
&= \frac{a^2}{36k^3} \,,
\end{aligned}$$

we have

$$\int \sqrt{fg} \ge 1 - \frac{a^2}{72k^3} \ge 1 - \frac{1}{72k^5} \,.$$

We set $1-\beta = 1/(72k^5)$. By Assouad's theorem (as in Devroye, 1987, p.60),

$$\begin{aligned}
\inf_{f_n} \sup_{f \in \mathcal{F}_{7/3}} \mathbf{E} \int |f_n - f| &\ge \frac{k\alpha}{2} \left(1 - \sqrt{2n(1-\beta)}\right) \\
&= \frac{ka}{12k^2} \left(1 - \sqrt{\frac{n}{36k^5}}\right) \\
&\ge \frac{1}{20k^2} \left(1 - \sqrt{\frac{1}{4}}\right) \\
&\qquad \text{(provided } k \ge 2 \text{ and } k^5 \ge n/9\text{)} \\
&= \frac{1}{40k^2} \\
&= \frac{1}{40\lceil (n/9)^{1/5} \rceil^2} \\
&\qquad \text{(upon taking } k = \lceil (n/9)^{1/5} \rceil\text{)} \\
&\ge \frac{1}{40(1+(n/9)^{1/5})^2} .
\end{aligned}$$

Note that the condition $k \ge 2$ holds if $n \ge 288$. In that case, $1 \le (n/9)^{1/5}/2$, and the lower bound is thus at least

$$\frac{1}{40((3/2)(n/9)^{1/5})^2} \ge \frac{1}{37.37192\ldots n^{2/5}}$$

which was to be shown. □

We may now combine Lemmas 1 and 2 to prove Theorem 1. Indeed,

$$\begin{aligned}
&\inf_{H:\mathbb{R}^{n+1}\to(0,\infty)} \sup_{f\in\mathcal{F}_{7/3}} \frac{\mathbf{E}\int |f_{n,H(x)}(x)-f(x)|dx}{\inf_{h:\mathbb{R}\to(0,\infty)} \mathbf{E}\int |f_{n,h(x)}(x)-f(x)|dx} \\
\geq\;& \frac{\inf_{f_n} \sup_{f\in\mathcal{F}_{7/3}} \mathbf{E}\int |f_{n,H(x)}(x)-f(x)|dx}{\sup_{f\in\mathcal{F}_{7/3}} \inf_{h:\mathbb{R}\to(0,\infty)} \mathbf{E}\int |f_{n,h(x)}(x)-f(x)|dx} \\
\geq\;& \frac{\frac{1}{37.37192\ldots n^{2/5}}}{\sqrt{\frac{28\rho}{3n}}} \\
\geq\;& \frac{n^{\frac{1}{10}}}{41\sqrt{\rho}}.
\end{aligned}$$

This concludes the proof of Theorem 1.

From looking at Lemma 1, one is tempted to try to find data-based variable bandwidths that achieve $O(1/\sqrt{n})$ error rates over the given class of convex densities, but Theorem 1 shows that in a uniform sense, no such rate is possible. It does not imply that there always exists one convex density for which the rate is worse. To prove that this is the case—and thus, that the "bad density" does not change with n, as in minimax results—, we need additional work, which is presented in the next section.

Sain and Scott (1997) give a detailed and lucid account of the zero-bias bandwidth for locally convex densities. They consider pointwise L2 errors (or MSE) and observe $O(1/n)$ pointwise rates if $h(x)$ were known. They try to estimate $h(x)$ by cross-validation. Unfortunately, they report that in practice the $O(1/n)$ rate (which corresponds to $O(1/\sqrt{n})$ L1 errors) is not achievable, which in view of our results is to be expected. We would like to make one fundamental remark however about the MISE $\mathbf{E}\int(f_n-f)^2$. If f_n is the kernel estimate with optimal $h(x)$ and $f\in\mathcal{F}_B$, then the MISE is in general infinite, while the L1 error by Lemma 1 is $O(1/\sqrt{n})$. Indeed, due to the squaring, the variance is $O(\int_0^1 1/(nh(x))dx)$, and as $h(x)$ near 0 varies about like x, we see that the integrated variance blows up.

3 Bad densities for the entire sequence

Assume that we are given an entire sequence of local data-based bandwidths, where the n-th mapping is $H_n:\mathbb{R}^{n+1}\to(0,\infty)$. We write H_n also for $H_n(x;X_1,\ldots,X_n)$. In this section, we show (Theorem 2 below) that if $a_n\downarrow 0$ arbitrarily slowly, then there exists a monotone bounded piecewise convex density on $[0,1]$ for which the ratio

$$\frac{\mathbf{E}\int |f_{n,H_n(x)}(x)-f(x)|dx}{\inf_{h:\mathbb{R}\to(0,\infty)} \mathbf{E}\int |f_{n,h(x)}(x)-f(x)|dx}$$

is infinitely often more than $a_n n^{1/10}$. Thus, the same density can be used as a counterexample no matter how large the sample size is. In our proof, we follow to some extent the lead of Birgé (1986).

Lemma 3. *Let K be a symmetric nonnegative square-integrable kernel on the real line with shape parameter ρ. Partition the line into an infinite number of intervals of length l_i and weight p_i. On the i-th interval, let f be a nondecreasing convex nonnegative function of integral p_i and taking maximal value m_i. Assume that $\sum_i p_i = 1$. Then*

$$\inf_{h:\mathbb{R}\to(0,\infty)} \mathbf{E}\int |f_{n,h(x)}(x) - f(x)|dx \le \sum_i \min\left(\sqrt{\frac{4l_i m_i \rho}{n}}, 2p_i\right) .$$

Proof. We refer to the proof of Lemma 1, and only modify a few bounds. Consider the i-th interval, and assume without loss of generality that it is $[0, l_i]$. Define $h(x)$ as in the proof of Lemma 1, and note that $h(x)$ is at least equal to the minimum distance from x to the border of its interval divided by c. As intervals can thus be treated separately, we may argue as on one interval, as in Lemma 1. For $x \in [0, l_i]$, define $\delta(x) = \min(x, l_i - x)$. Note that arguing as in Lemma 1, $h(x) \ge \delta(x)/c$, where $[-c, c]$ is the support of K.

We also note that by the choice of $h(x)$, and positivity of K, we have

$$\mathbf{E}\{|f_{n,h}(x) - f(x)|\} \le f(x) + \mathbf{E}f_{n,h}(x) = 2f(x) .$$

Thus, for all x in the i-th interval,

$$\begin{aligned}\mathbf{E}\{|f_{n,h}(x) - f(x)|\} &\le \min\left(\sqrt{\frac{m_i\int K^2}{nh(x)}}, 2f(x)\right)\\ &\le \min\left(\sqrt{\frac{m_i c\int K^2}{n\delta(x)}}, 2f(x)\right)\\ &= \min\left(\sqrt{\frac{m_i\rho}{2n\delta(x)}}, 2f(x)\right)\end{aligned}$$

so that, taking integrals then yields

$$\begin{aligned}&\mathbf{E}\int |f_{n,h}(x) - f(x)|\,dx\\ &\le \sum_i \int_{i\text{-th interval}} \min\left(\sqrt{\frac{m_i\rho}{2n\delta(x)}}, 2f(x)\right)\,dx\\ &\le \sum_i \min\left(\int_{i\text{-th interval}} \sqrt{\frac{m_i\rho}{2n\delta(x)}}\,dx, \int_{i\text{-th interval}} 2f(x)\,dx\right)\end{aligned}$$

$$= \sum_i \min\left(\sqrt{\frac{m_i\rho}{2n}} 2 \int_0^{l_i/2} 1/\sqrt{x}dx, 2p_i\right)$$

$$= \sum_i \min\left(\sqrt{\frac{4l_i m_i \rho}{n}}, 2p_i\right) . \qquad \square$$

Lemma 3 provides us with a rich enough family of densities from which to draw examples: it is applicable to basically all bounded piecewise convex densities. The freedom in the choice of the Lemma's parameters will be useful further on. The following Lemma allows us to use minimax lower bounds for densities on a fixed interval for mixtures of densities.

Lemma 4. *Let g be a fixed density supported outside $[0,1]$, and let $\mathcal{F}_B$ be as in Lemma 2. Let $p \in (0,1)$ and let $\mathcal{G}$ be the class of densities of the form $pf + (1-p)g : f \in \mathcal{F}_{7/3}$. For $n \geq 24$, we have*

$$\inf_{f_n} \sup_{f \in \mathcal{G}} \mathbf{E} \int |f_n - f| \geq \frac{p}{106.02476\ldots n^{2/5}},$$

where the infimum is over all density estimates.

Proof. We mimic the proof of Lemma 2, and make only changes where appropriate. The 2^k-member subclass construction is as in Lemma 2, except for the multiplicative factor p. This leads to the choice $1-\beta = p/6k^5$. By Assouad's theorem,

$$\begin{aligned}
\inf_{f_n} \sup_{f \in \mathcal{F}_{7/3}} \mathbf{E} \int |f_n - f| &\geq \frac{kp\alpha}{2}\left(1 - \sqrt{2n(1-\beta)}\right) \\
&= \frac{kpa}{12k^2}\left(1 - \sqrt{\frac{np}{36k^5}}\right) \\
&\geq \frac{p}{20k^2}\left(1 - \sqrt{\frac{1}{4}}\right) \\
&\qquad \text{(provided } k \geq 2 \text{ and } k^5 \geq n/9) \\
&= \frac{p}{40k^2} \\
&= \frac{p}{40\lceil (n/9)^{1/5} \rceil^2} \\
&\qquad \text{(upon taking } k = \lceil (n/9)^{1/5} \rceil) \\
&\geq \frac{p}{40(1 + (n/9)^{1/5})^2}.
\end{aligned}$$

Note that the condition $k \geq 2$ holds if $n \geq 288$. In that case, $1 \leq (n/9)^{1/5}/2$, and the lower bound is thus at least

$$\frac{p}{40((3/2)(n/9)^{1/5})^2} \geq \frac{p}{37.37192\ldots n^{2/5}}$$

which was to be shown. $\square$

Theorem 2. *Let a_n be a strictly decreasing sequence of positive numbers with zero limit.*

A. Let f_n be any density estimate. Then there exists a piecewise convex nondecreasing density f on $[-1,0]$ bounded by 5 and a subsequence n_j such that along this subsequence

$$\mathbf{E}\int |f - f_{n_j}| \geq a_{n_j} n_j^{-2/5}.$$

B. Let $f_{n,H_n(x)}$ denote any variable kernel estimate with kernel K and local bandwidth H_n. If K is symmetric, nonnegative, and has shape parameter $\rho < \infty$, then there exists a piecewise convex nondecreasing density f on $[-1,0]$ bounded by 5 and a subsequence n_j such that along this subsequence

$$\frac{\mathbf{E}\int |f_{n_j,H_{n_j}(x)} - f|}{\inf_{h:\mathbb{R}\to(0,\infty)} \mathbf{E}\int |f_{n_j,h(x)}(x) - f(x)|dx} \geq a_{n_j} n_j^{1/10}.$$

Proof. We consider an infinite subfamily of densities within the conditions of Lemma 3. Let $p_1 > p_2 > \ldots$ be a probability vector, and let $B_i = [-1+1/2^i, -1+2/2^i)$ for $i \geq 1$. On B_i, define a class of densities as in the proof of Lemma 2, parametrized by k_i, the number of partitions, and scale each density by $l_i = 1/2^i$. Each density is characterized by a bit vector b_i with k_i bits. The density for b_i on B_i is denoted by f_{i,b_i}. If $b = (b_1, b_2, \ldots)$, define the density

$$f_b = p_1 f_{1,b_1} + p_2 f_{2,b_2} + \cdots$$

and consider the class $\mathcal{F}$ of all these densities. Note that on B_i, each f_b takes values in $[p_i/3l_i, (7/3)p_i/l_i]$, is nondecreasing and piecewise convex. Formally, in Lemma 3, we have $m_i = 7p_i/3l_i$, and $l_i = 1/2^i$. Each f_b is supported on $[-1,0]$, and is nondecreasing provided that

$$\frac{7p_{i+1}}{3l_{i+1}} \leq \frac{p_i}{3l_i}$$

for all i. This translates into the condition

$$\frac{p_{i+1}}{p_i} \leq \frac{1}{14}.$$

Finally, note that each f_b is bounded by $(7/3)p_1/l_1 \leq 14/3$. Our class is thus contained in the class of bounded support bounded monotone densities. (With some extra effort, we could have made the densities continuous as well within the support.) By Lemma 3, for all b,

$$\inf_{h:\mathbb{R}\to(0,\infty)} \mathbf{E}\int |f_{n,h(x)}(x) - f_b(x)|dx \leq \frac{\sum_i \sqrt{4l_i m_i \rho}}{\sqrt{n}} = \frac{\sqrt{\frac{28\rho}{3}}\sum_i \sqrt{p_i}}{\sqrt{n}}.$$

It remains to show how to replace Lemma 2. Let $\{f_n\}$ be any sequence of density estimates. Then, let $288 < n_1 < n_2 < \cdots$ be a specially selected sequence of sample sizes. The interval B_i is targeted when $n = n_i$, and we do not care about sample sizes n outside the collection of n_i's. The observations in the sample are all determined from b and the i.i.d. pairs $(Z_1, U_1), (Z_2, U_2), \ldots$, where the Z_i's are discrete and take the value j with probability p_j, $j \geq 1$, and the U_i's are uniform $[0,1]$. Z_i picks the mixture component for X_i, and U_i is used in the probability integral transform to obtain X_i from the density f_{Z_i,b_i}. Note in particular that with this embedding, if $N_j = 0$ (where N_j is the number of X_i's in interval B_j), then the sample is unaffected by changes of b_j. The existence of a bit vector b will be established inductively. Assume that we have fixed $b_1, \ldots, b_{j-1}$. We have $n = n_j$, by assumption. Let E denote the event that $N = \sum_{k>j} N_k = 0$. Let $\mathcal{F}$ denote the family of densities f_b with the first $j-1$ components of b fixed as above and with the components b_k, $k > j$ all zero. The family has 2^{k_j} members distinguished by all possible values for b_j. Then at $n = n_j$,

$$\sup_{b_j} \mathbf{E} \int |f_n - f_b| \geq \sup_{b_j} \mathbf{E} \int_{B_j} |f_n - p_j f_{b_j}| \geq p_j n^{-2/5}/38$$

if $n_j \geq 288$, by Lemma 4. Pick b_j so that $\mathbf{E} \int |f_{n_j} - f_b| \geq p_j n_j^{-2/5}/38$. Continue in this fashion, and find $b = (b_1, b_2, \ldots)$. Then, at $n = n_j$, we have, setting $b^* = b$ in its first j components, but zero for the other components,

$$\begin{aligned}
\mathbf{E} \int |f_n - f_b| &\geq \mathbf{E} \left\{ I_E \int |f_n - f_{b^*}| \right\} \\
&= \mathbf{E} \left\{ \int |f_n - f_{b^*}| \Big| E \right\} \mathbf{P}\{E\} \\
&\geq \frac{p_j}{p_1 + \cdots + p_j} n_j^{-2/5} \mathbf{P}\{E\}/38 \\
&\geq p_j n_j^{-2/5} \mathbf{P}\{E\}/38 \,.
\end{aligned}$$

Here the ratio comes from the observation that conditional on E, the sample may be thought to be drawn from mixture densities with only the first j components. Now, $\mathbf{P}\{E\} \geq 1 - n_j \sum_{k>j} p_k \geq 1/2$ by our choice of p_j and n_j. As a_j is decreasing, we will make $p_j n_j^{-2/5}/76$ greater than or equal to $n_j^{-2/5} a_{n_j}$ for all j. From the sequence a_n, pick a subsequence a_{n_j}, with the property that $\sum_j a_{n_j} < 1/76$, $a_{n_2} \leq 1/152$, and such that $a_{n_{j+1}} < a_{n_j}/(2n_j)$. Clearly, $\sum_{k>j} a_{n_k} < a_{n_j}/n_j$ for all j. Set $p_j = 76 a_{n_j}$ so that we have the desired inequality, and the p_j's sum to less than 1. Give the remaining mass to p_1, to make a probability vector. Note that

$$n_j \sum_{k>j} p_k = n_j \sum_{k>j} 76 a_{n_j} \leq 76 a_{n_j} \leq \frac{1}{2}$$

for $j > 1$ by our choice of a subsequence. Thus, $\mathbf{P}\{E\} \geq 1/2$ as required. Also, $p_{j+1}/p_j = a_{n_{j+1}}/a_{n_j} \leq 1/2n_j \leq 1/2 \cdot 288 < 1/14$ as required for monotonicity. Therefore, for our recursively constructed b, we have at all $j > 1$,

$$\mathbf{E} \int |f_{n_j} - f_b| \geq n_j^{-2/5} a_{n_j}.$$

This concludes the proof of part A of the Theorem. For part B, we use the same subsequence, but note an earlier bound for the denominator in the ratio and obtain for all $j > 1$,

$$\begin{aligned} \frac{\mathbf{E} \int |f_{n_j, H_{n_j}(x)} - f_b|}{\inf_{h:\mathbb{R}\to(0,\infty)} \mathbf{E} \int |f_{n,h(x)}(x) - f_b(x)| dx} &\geq \frac{n_j^{-2/5} a_{n_j}}{\frac{\sqrt{\frac{28\rho}{3}} \sum_i \sqrt{p_i}}{\sqrt{n_j}}} \\ &= \frac{n_j^{1/10} a_{n_j}}{\sqrt{\frac{28\rho}{3}} \sum_i \sqrt{p_i}} \\ &\geq \frac{n_j^{1/10} a_{n_j}}{\sqrt{\rho a_{n_1}}} \times \frac{\sqrt{2}-1}{\sqrt{28 \times 152/3}} \end{aligned}$$

because

$$\begin{aligned} &\sum_i \sqrt{p_i} \\ &\quad = \sum_i \sqrt{76 a_{n_i}} \\ &\quad \leq \sqrt{76} \Big(\sqrt{a_{n_1}} + \sqrt{a_{n_1}/(2n_1)} \\ &\qquad + \sqrt{a_{n_1}/(2^2 n_1 n_2)} + \sqrt{a_{n_1}/(2^3 n_1 n_2 n_3)} + \cdots \Big) \\ &\quad \leq \sqrt{76 a_{n_1}} \sum_{i=0}^{\infty} \frac{1}{(\sqrt{2})^i} \\ &\quad = \sqrt{152 a_{n_1}} \frac{1}{\sqrt{2}-1} . \end{aligned}$$

By replacing the a_n's by $\sqrt{a_n}$ at the outset of the proof, it is trivial to see that the ratio studied in part B of the Theorem is infinitely often greater than $a_n n^{1/10}$, as required. This concludes the proof of Theorem 2. □

4 Extensions and generalizations

The proofs show that Lemma 1 and Theorems 1 and 2, with suitable changes of the constants, remain valid for piecewise convex densities, and in particular, for any piecewise linear density. However, for piecewise linear densities with a finite number of breakpoints, it may be possible to design a variable kernel density estimate with L1 error rate $O(1/\sqrt{n})$. Indeed, this is possible for the uniform density on $[0,1]$ and for trapezoidal densities. And since it is possible to estimate the breakpoints efficiently, so that the class of piecewise linear densities with $\leq k$ breakpoints is truly a parametric class.

In $\mathbb{R}^d$, if we use bounded support positive symmetric product kernels, the same methodology would work on the class of all convex densities on $[0,1]^d$ bounded by a constant B. There is no problem with the generalization of Lemma 1: the rate $O(1/\sqrt{n})$ would be achievable for the best $h(x)$ (assuming the same h is used in all dimensions). The minimax rate for the new class needs to be determined however. Clearly, it is going to be worse than in the one-dimensional case, so that even if we pick h identical in all directions (an easier problem than picking it separately for each dimension), a phenomenon similar to that described by Theorems 1 and 2 should occur.

A more subtle situation occurs when the kernel K is of order $2s$ for some positive integer s, that is, K is symmetric, $\int K = 1$, $\int x^i K(x)dx = 0$ for $1 \leq i < 2s$ and $\int x^{2s} K(x)dx = S \neq 0$. Even here, there is an analog to Theorem 1. The variance bound as described in Lemma 1 remains obviously valid uniformly over all bounded densities on $[0,1]$ that have for each $x \in (0,1)$ a bandwidth $h(x)$ that makes the bias zero. What one needs is the property that at every x, locally, $f * K_h$ increases initially when h increases from 0. By Taylor's series expansion with remainder, this is easy to establish if at every $x \in (0,1)$, f^{2s} is of the same sign as S, and if K vanishes off $[-1,1]$. The minimax lower bound over such densities on $[0,1]$ bounded by a constant B is of the order of $n^{-2s/(4s+1)}$ (we could not find a reference for this though), and therefore, the factor $n^{1/10}$ in Theorem 1 should be replaced by $n^{1/(8s+2)}$, which still tends to infinity, albeit more slowly.

5 Remarks on unbiasedness

We showed that it is futile to look for the zero-bias choice $h(x)$ even for convex densities. Of course, we knew since Rosenblatt (1956) that no universally unbiased nonnegative density estimate f_n exists:

$$\inf_{f_n : f_n \geq 0} \sup_f \int |\mathbf{E} f_n - f| > 0 .$$

On the other hand, for small classes $(\mathcal{F})$ of densities, unbiased density estimates exist:

$$\inf_{f_n} \sup_{f \in \mathcal{F}} \int |\mathbf{E} f_n - f| = 0 .$$

An example is the class of all normal densities with unknown mean and variance, for which an unbiased estimate was found by Basu (1964)—see also exercise 7.14 of Devroye (1987). A second example is the class $A_{T,s,C}$ of Devroye and Györfi (1985, page 142), which roughly speaking is a subclass of all densities with bounded support characteristic function and absolute s-th integrated derivative of the characteristic function bounded by C. Here the unbiased estimate is the ordinary kernel estimate with a superkernel and a bandwidth less than a constant depending upon T only. For more information on unbiasedness, see Lumelskii and Sapozhnikov (1969), Wertz (1975), Guttmann and Wertz (1976), and Seheult and Quesenberry (1971).

One may ask then where the boundary is for $\mathcal{F}$? Which classes are too large to find unbiased density estimates for all members in the class? We offer the following result, which relates the non-existence of unbiased estimates to the richness of the class of densities under consideration.

Theorem 3. *Let $\mathcal{F}$ be a class of uniformly bounded densities, and let $R_m(\mathcal{F})$ denote its minimax risk for sample size m:*

$$R_m(\mathcal{F}) = \inf_{f_m} \sup_{f \in \mathcal{F}} \mathbf{E} \int |f_m - f| \ .$$

Let f_n be a density estimate, where n is a fixed integer. If

$$\lim_{m \to \infty} \sqrt{m} R_m(\mathcal{F}) = \infty,$$

then either f_n is not unbiased ($\sup_{f \in \mathcal{F}} \int |\mathbf{E} f_n - f| > 0$) or

$$\sup_{f \in \mathcal{F}} \int \sqrt{\mathbf{E}\{f_n^2\}} = \infty.$$

Let us illustrate this Theorem. The fact that the minimax risk increases faster than $1/\sqrt{n}$ usually is accepted as an indication that $\mathcal{F}$ is "nonparametric". An example of a rich class is the class of all densities on $[0,1]$ with 25 continuous derivatives on the real line, each of which is bounded in absolute value by B for a sufficiently large constant B. With the knowledge that f is in this class, one would be tempted to construct density estimates bounded by B. But Theorem 3 then says that f_n cannot be unbiased for all f in the class! The price to pay for unbiasedness is unboundedness in expectation as in the last condition of Theorem 3. In particular, for such rich nonparametric classes, no bounded and compactly supported unbiased density estimate exists, even if we know a uniform bound on the densities and the support in the class.

While Theorem 3 does not supersede Rosenblatt's result, it complements it by addressing the question of the size of the classes. The two examples cited earlier of course had minimax risks of the order of $1/\sqrt{n}$. Note also that Theorem 3 does not say that $\int |f_n| = \infty$ with probability 1: indeed, we could have $\int |f_n| < \infty$ with probability 1.

Proof. Assume that f_n is unbiased for all $f \in \mathcal{F}$. We then construct the following density estimate for sample size mn:

$$f_{mn}(x) = \frac{1}{m} \sum_{j=1}^{m} f_n(x; X_{(j-1)n+1}, \ldots, X_{(j-1)n+n})$$

which is unbiased, and a sum of m independent summands. Therefore,

$$\mathrm{Var}\{f_{mn}(x)\} = \frac{1}{m} \mathrm{Var}\{f_n(x)\}.$$

But then

$$\begin{aligned} \int \mathbf{E}\{|f_{nm} - f|\} &= \int \mathbf{E}\{|f_{nm} - \mathbf{E}f_{nm}|\} \\ &\le \int \sqrt{\mathrm{Var}\{f_{nm} - \mathbf{E}f_{nm}\}} \\ &= \frac{1}{\sqrt{m}} \int \sqrt{\mathrm{Var}\{f_n\}}\ . \end{aligned}$$

Taking supremums shows that

$$\sup_{f \in \mathcal{F}} \sqrt{nm}\mathbf{E} \int |f_{nm} - f| \le \sup_{f \in \mathcal{F}} \frac{\sqrt{nm}}{\sqrt{m}} \int \sqrt{\mathrm{Var}\{f_n\}}.$$

As $m \to \infty$, the left-hand side tends to ∞ by assumption. As n is fixed, this shows that

$$\sup_{f \in \mathcal{F}} \int \sqrt{\mathrm{Var}\{f_n\}} = \infty.$$

By the uniform boundeness of f, this implies that

$$\sup_{f \in \mathcal{F}} \int \sqrt{\mathbf{E}\{f_n^2\}} = \infty.$$

□

References

1. I. Abramson, On bandwidth variation in kernel estimates—a square root law, *Annals of Statistics*, vol. 10, pp. 1217–1223, 1982.

2. H. Akaike, An approximation to the density function, *Annals of the Institute of Statistical Mathematics*, vol. 6, pp. 127–132, 1954.

3. P. Assouad, Deux remarques sur l'estimation, *Comptes Rendus de l'Académie des Sciences de Paris*, vol. 296, pp. 1021–1024, 1983.

4. A. P. Basu, Estimates of reliability for some distributions useful in life testing, *Technometrics*, vol. 6, pp. 215–219, 1964.

5. L. Breiman, W. Meisel, and E. Purcell, Variable kernel estimates of multivariate densities, *Technometrics*, vol. 19, pp. 135–144, 1977.

6. L. Birgé, Non-asymptotic minimax risk for Hellinger balls, *Probability and Mathematical Statistics*, vol. 5, pp. 21–29, 1985.

7. L. Birgé, On estimating a density using Hellinger distance and some other strange facts, *Probability Theory and Related Fields*, vol. 71, pp. 271–291, 1986.

8. L. Birgé, On the risk of histograms for estimating decreasing densities, *Annals of Statistics*, vol. 15, pp. 1013–1022, 1987a.

9. L. Birgé, Estimating a density under order restrictions: nonasymptotic minimax risk, *Annals of Statistics*, vol. 15, pp. 995–1012, 1987b.

10. L. Birgé, The Grenander estimator: a nonasymptotic approach, *Annals of Statistics*, vol. 17, pp. 1532–1549, 1989.

11. L. Devroye, A note on the L1 consistency of variable kernel estimates, *Annals of Statistics*, vol. 13, pp. 1041–1049, 1985.

12. L. Devroye, *A Course in Density Estimation*, Birkhäuser, Boston, 1987.

13. L. Devroye, Another proof of a slow convergence result of Birgé, *Statistics and Probability Letters*, vol. 23, pp. 63–67, 1995.

14. L. Devroye, Universal smoothing factor selection in density estimation: theory and practice, *Test*, vol. 6, pp. 223–320, 1997.

15. L. Devroye and L. Györfi, *Nonparametric Density Estimation: The L1 View*, John Wiley, New York, 1985.

16. L. Devroye and G. Lugosi, A universally acceptable smoothing factor for kernel density estimation, *Annals of Statistics*, vol. 24, pp. 2499–2512, 1996.

17. L. Devroye and G. Lugosi, Non-asymptotic universal smoothing factors, kernel complexity and Yatracos classes, *Annals of Statistics*, vol. 25, pp. 2626–2637, 1997.

18. L. Devroye, G. Lugosi, and F. Udina, Inequalities for a new data-based method for selecting nonparametric density estimates, Technical Report, Facultat de Ciencies Economiques, Universitat Pompeu Fabra, Barcelona, 1998.

19. L. Devroye and C. S. Penrod, The strong uniform convergence of multivariate variable kernel estimates, *Canadian Journal of Statistics*, vol. 14, pp. 211–219, 1986.

20. M. Farmen, The smoothed bootstrap for variable bandwidth selection and some results in nonparametric logistic regression, Ph.D. Dissertation, Department of Statistics, University of North Carolina, Chapel Hill, 1996.

21. H. Guttmann and W. Wertz, Note on estimating normal densities, *Sankhya, Series B*, vol. 38, pp. 231–236, 1976.

22. J. D. F. Habbema, J. Hermans, and J. Remme, Variable kernel density estimation in discriminant analysis, in: *COMPSTAT 1978: Proceedings*, ed. L. C. A. Corsten and J. Hermans, Birkhauser, Basel, 1978.

23. P. Hall, On global properties of variable bandwidth density estimators, *Annals of Statistics*, vol. 20, pp. 762–778, 1992.

24. P. Hall and J. S. Marron, Variable window width kernel estimates, *Probability Theory and related Fields*, vol. 80, pp. 37–49, 1988.

25. P. Hall and W. R. Schucany, A local cross-validation algorithm, *Statistics and Probability Letters*, vol. 8, pp. 109–117, 1989.

26. M. Hazelton, Bandwidth selection for local density estimation, *Scandinavian Journal of Statistics*, vol. 23, pp. 221–232, 1996.

27. M. C. Jones, Variable kernel density estimates and variable kernel density estimates, *Australian Journal of Statistics*, vol. 32, pp. 361–371, 1990.

28. Ya. P. Lumelskii and P. N. Sapozhnikov, Unbiased estimates of density functions, *Theory of Probability and its Applications*, vol. 14, pp. 357–364, 1969.

29. J. S. Marron, P. Hall, and T. C. Hu, Improved variable window estimators of probability densities, *Annals of Statistics*, vol. 23, pp. 1–10, 1995.

30. J. Mielniczuk, P. Sarda, and P. Vieu, Local data-driven bandwidth choice for density estimation, *Journal of Statistical Planning and Inference*, vol. 23, pp. 53–69, 1989.

31. E. Parzen, On the estimation of a probability density function and the mode, *Annals of Mathematical Statistics*, vol. 33, pp. 1065–1076, 1962.

32. J. W. Raatgever and R. P. W. Duin, On the variable kernel model for multivariate nonparametric density estimation, in: *COMPSTAT 1978: Proceedings*, ed. L. C. A. Corsten and J. Hermans, Birkhauser, Basel, 1978.

33. M. Rosenblatt, Remarks on some nonparametric estimates of a density function, *Annals of Mathematical Statistics*, vol. 27, pp. 832–837, 1956.

34. S. R. Sain, Adaptive kernel density estimation, Ph.D. Dissertation, Department of Statistics, Rice University, Houston, 1994.

35. S. R. Sain and D. W. Scott, On locally adaptive density estimation, *Journal of the American Statistical Association*, vol. 91, pp. 1525–1534, 1996.

36. S. R. Sain and D. W. Scott, Zero-bias locally adaptive density estimators, Technical Report, Rice University, Houston, 1997.

37. A. H. Seheult and C. P. Quesenberry, On unbiased estimation of density functions, *Annals of Mathematical Statistics*, vol. 42, pp. 1434–1438, 1971.

38. S. J. Sheather, A data-based algorithm for choosing the window width when estimating the density at a point, *Computational Statistics and Data Analysis*, vol. 1, pp. 229–239, 1983.

39. S. J. Sheather, An improved data-based algorithm for choosing the window width when estimating the density at a point, *Computational Statistics and Data Analysis*, vol. 4, pp. 61–65, 1986.

40. G. R. Terrell and D. W. Scott, Variable kernel density estimation, *Annals of Statistics*, vol. 20, pp. 1236–1265, 1992.

41. L. A. Thombs and S. J. Sheather, Local bandwidth selection for density estimation, in: *Proceedings of the 22nd Symposium on the Interface*, pp. 111–116, Springer-Verlag, New York, 1992.

42. W. Wertz, On unbiased density estimation, *An. Acad. Brasil. Cienc.* vol. 47, pp. 65–72, 1975.

43. Y. Yang, Minimax Optimal Density Estimation, Ph.D. Dissertation, Yale University, 1996.

Luc Devroye
School of Computer Science
McGill University
Montreal, Canada H3A 2A7
e-mail: luc@cs.mcgill.ca

Gábor Lugosi
Department of Economics
Pompeu Fabra University
Ramon Trias Fargas, 25-27
08005 Barcelona, Spain
e-mail: gabor.lugosi@econ.upf.ed

Almost Sure Asymptotic Optimality of Cross Validation for Spline Smoothing

Paul Eggermont and Vincent LaRiccia

ABSTRACT We prove the almost sure consistency and asymptotic optimality of generalized cross validation for the selection of the smoothing parameter in nonparametric regression. The consistency result has the effect of restricting the range of smoothing parameters that need to be considered for asymptotic optimality. The essential tools are exponential inequalities for weighted sums of iid normal random variables, and a simple bound for the expected value of the supremum (over a smoothing parameter) of a family of random functions.

1 Introduction

In this paper we study the almost sure convergence of generalized cross validation for the smoothing parameter selection in spline smoothing problems. The general strategy follows that of Li (1986) who proves asymptotic optimality in probability, but it seems that the present approach is easily adaptable to other situations, once consistency has been established.

The nonparametric regression problem under consideration is to estimate a smooth function $f_o(x)$ for $0 \leqslant x \leqslant 1$, according to the model

$$y_{in} = f_o(x_{in}) + d_{in},\ i = 1, 2, \cdots, n \ , \tag{1.1}$$

where $x_{1,n}, x_{2,n}, \cdots, x_{n,n}$ are known design points, and $d_{1,n}, d_{2,n}, \cdots, d_{n,n}$ are iid normal random variables with mean 0 and standard deviation σ,

$$d_n = (d_{1,n}, d_{2,n}, \cdots, d_{n,n})^T \sim N(0, \sigma^2 I) \ , \tag{1.2}$$

and $y_n = (y_{1,n}, y_{2,n}, \cdots, y_{n,n})^T$ is the data.

For simplicity we assume a periodic setting, so that boundary effects are absent. We also assume that the design points are equally spaced, say

$$x_{in} = i/n \ , \quad i = 1, 2, \cdots, n \ . \tag{1.3}$$

As an estimator of f_o we consider a smoothing spline of order m, i.e., the solution of

$$\begin{array}{ll} \text{minimize} & \frac{1}{n} \sum_{i=1}^{n} | f(x_{in}) - y_{in} |^2 + h^{2m} \| f^{(m)} \|^2_{L^2(0,1)} \\ \text{subject to} & f \in H^m(0,1) \ , \end{array} \tag{1.4}$$

where $\| \quad \|_{L^2(0,1)}$ denotes the norm on $L^2(0,1)$, and $f^{(m)}$ denotes the m-th derivative. The space $H^m(0,1)$ is a Hilbert space, and is defined as the completion of the space of m times continuously differentiable functions in the topology defined by the norm

$$\| f \|_{H^m(0,1)} = \Big\{ \sum_{p \in \mathbb{Z}} | 1 + (2\pi p)^m |^2 \, | \widehat{f}_p |^2 \Big\}^{1/2} , \tag{1.5}$$

where

$$\widehat{f}_p = \int_0^1 f(x) \, e^{2\pi i p x} \, dx \ , \quad p \in \mathbb{Z} \ , \tag{1.6}$$

are the Fourier coefficients of f. Note that (1.4) makes sense for any $m \geqslant 0$. In (1.5) it is required that $m > \frac{1}{2}$, since then the evaluation functionals $f \longmapsto f(x_{in})$ are well defined for $f \in H^m(0,1)$. Thus, we assume throughout that

$$f_o \in H^{\kappa}(0,1) \ \text{ for some } \kappa > \tfrac{1}{2} \ , \ \text{and } f_o \text{ is not constant} \ . \tag{1.7}$$

For more on the spaces H^m, see ADAMS (1975).

It is well-known, REINSCH (1967), that the problem (1.4) is equivalent to

$$\begin{array}{ll} \text{minimize} & \frac{1}{n} \| z - y_n \|^2 + h^{2m} \frac{1}{n} z^T M_n z \\ \text{subject to} & z \in \mathbb{R}^n \ , \end{array} \tag{1.8}$$

for a suitable semi-positive definite matrix $M_n \in \mathbb{R}^{n \times n}$, dependent on m and the design points $x_{1,n}, x_{2,n}, \cdots, x_{n,n}$ only. In (1.8), the norm $\| \quad \|$ is the Euclidean norm on $\mathbb{R}^n$. The solution of (1.8) is $z = S_n(h) \, y_n$ with

$$S_n(h) = (\, I + h^{2m} M_n \,)^{-1} \ , \tag{1.9}$$

but most of the time we write S_n instead of $S_n(h)$.

To have an effective estimation procedure, the smoothing parameter h must be chosen in a rational manner. It is customary to define the "optimal" choice H_{OPT} of h to be any solution of

$$\text{minimize } \ \mathfrak{L}_n(h) \stackrel{\text{def}}{=} \tfrac{1}{n} \| y_o - S_n(h) \, y_n \|^2 \ \text{ over } \ h > 0 \ . \tag{1.10}$$

Here $y_o = (f_o(x_{1,n}), f_o(x_{2,n}), \cdots, f_o(x_{n,n}))^T$ actually depends on n, but this dependence is suppressed in the notation. Of course, the loss function in (1.10) is unknown, so in order to develop rational procedures for determining an appropriate h, it must be estimated. We verify that

$$\mathfrak{M}_n(h) \stackrel{\text{def}}{=} \tfrac{1}{n} \| y_n - S_n(h) \, y_n \|^2 + \tfrac{1}{n} \sigma^2 \, \text{trace}\big(-I + 2 \, S_n(h) \big) \tag{1.11}$$

is an unbiased estimate of $\mathfrak{L}_n(h)$, so a good h ought to be obtained by

$$\text{minimizing } \ \mathfrak{M}_n(h) \ \text{ over } \ h > 0 \ . \tag{1.12}$$

This method is the C_L method of C.L. MALLOWS (1972). The drawback is that σ must be known. If σ is not known, then "generalized" cross validation may be used; a useful approach to its derivation is the zero-trace idea of LI (1985). Suppose we replace the estimate $S_n y_n$ of y_o by

$$\widetilde{S_n}(h)\, y_n = \{\, -\alpha I + (1+\alpha)\, S_n(h) \,\}\, y_n \ , \tag{1.13}$$

for a suitable α. Then

$$\widetilde{\mathfrak{M}}_n(h) \stackrel{\text{def}}{=} \tfrac{1}{n} \| \, y_n - \widetilde{S_n}(h)\, y_n \, \|^2 + \sigma^2 \tfrac{1}{n} \operatorname{trace}(\, -I + 2\, \widetilde{S_n}(h)\,) \tag{1.14}$$

is an unbiased estimator of

$$\widetilde{\mathfrak{L}}_n(h) = \tfrac{1}{n} \| \, y_o - \widetilde{S_n}(h)\, y_n \, \|^2 \ . \tag{1.15}$$

It now turns out that α may be chosen such that the trace term in (1.14) just about vanishes. For

$$1 + \alpha = \left(\tfrac{1}{n} \operatorname{trace}(\, I - S_n(h)\,)\,\right)^{-1} \ , \tag{1.16}$$

we get

$$\widetilde{\mathfrak{M}}_n(h) = \frac{\tfrac{1}{n} \| \, y_n - S_n(h)\, y_n \, \|^2}{\left(\tfrac{1}{n} \operatorname{trace}(\, I - S_n(h))\,\right)^2} - \sigma^2 \ . \tag{1.17}$$

This is the "generalized" cross validation functional of GOLUB, HEATH and WAHBA (1979), although it does not explain its original motivation. For reasons that will become clear later on, cf. (4.12)-(4.17), it is better to consider

$$\mathrm{GCV}_n(h) \stackrel{\text{def}}{=} \frac{\tfrac{1}{n} \| \, y_n - S_n(h)\, y_n \, \|^2}{\left(\tfrac{1}{n} \operatorname{trace}(\, I - S_n(h))\,\right)^2} - \tfrac{1}{n} \| \, d_n \, \|^2 \ . \tag{1.18}$$

Note that the unknown σ and $\| \, d_n \, \|$ appearing in $\widetilde{\mathfrak{M}}_n(h)$ and $\mathrm{GCV}_n(h)$ do not affect the location of their (joint) minimizers. Thus a good way to select h ought to be by minimizing $\mathrm{GCV}_n(h)$ over h, provided the resulting h is such that the corresponding α satisfies $a \approx 0$. Any minimizer of $\mathrm{GCV}_n(h)$ is denoted by H_{GCV}. The investigation of the "gcv" procedure goes hand in hand with that of the "optimal" method, and splits into the consistency and asymptotic optimality parts.

Theorem 1. (CONSISTENCY) *Under the assumption (1.7) the following three statements are equivalent. Here $h = h_n$ varies with n, deterministically or randomly.*

(a) $h \longrightarrow_{as} 0 \ , \quad nh \longrightarrow_{as} \infty \, ;$
(b) $\mathfrak{L}_n(h) \longrightarrow_{as} 0 \, ;$
(c) $\mathrm{GCV}_n(h) \longrightarrow_{as} 0 \, .$

Theorem 2. (ASYMPTOTIC OPTIMALITY) *Under the assumption (1.7)*

$$\frac{\mathfrak{L}_n(H_{GCV})}{\mathfrak{L}_n(H_{OPT})} \longrightarrow_{as} 1 .$$

Given the consistency, the asymptotic optimality is implied by the following stability result. Let $\mathcal{H}_n = [\frac{1}{n}, 1]$. Then $H_{OPT}, H_{GCV} \in \mathcal{H}_n$ almost surely for $n \longrightarrow \infty$.

Theorem 3. (STABILITY) *Under the assumption (1.7)*

(a) $\dfrac{\mathbb{E}[\,GCV_n(h)\,]}{\mathbb{E}[\,\mathfrak{L}_n(h)\,]} \longrightarrow_{as} 1$ *for* $h = H_{OPT}, H_{GCV}$;

(b) $\sup\limits_{h\in\mathcal{H}_n} \left| \dfrac{\mathfrak{L}_n(h)}{\mathbb{E}[\,\mathfrak{L}_n(h)\,]} - 1 \right| \longrightarrow_{as} 0$;

(c) $\sup\limits_{h\in\mathcal{H}_n} \left| \dfrac{GCV_n(h)}{\mathbb{E}[\,GCV_n(h)\,]} - 1 \right| \longrightarrow_{as} 0$.

The stability theorem and the inequalities

$$1 \leqslant \frac{\mathfrak{L}_n(H_{GCV})}{\mathfrak{L}_n(H_{OPT})} \leqslant \frac{\mathfrak{L}_n(H_{GCV})}{GCV_n(H_{GCV})} \cdot \frac{GCV_n(H_{OPT})}{\mathfrak{L}_n(H_{OPT})}$$

prove the asymptotic optimality. Here we used, and the stability theorem shows, that asymptotically $GCV_n(h)$ is strictly positive. (Part (a) shows that $\mathbb{E}[\,GCV_n(h)\,] > 0$.)

Essential to the proof of the stability theorem is the computation of

$$\mathbb{E}[\sup\{\,|\,a(h)\,|\; : \; h \in \mathcal{H}_n\,\}]$$

for suitable functions $a(h) = a(h; d_n)$, differentiable with respect to h. The identity $a(h) = a(1) - \int_h^1 a'(t)\,dt$ leads to the useful bound

$$\mathbb{E}[\sup_{h\in\mathcal{H}_n} |\,a(h)\,|\,] \leqslant \mathbb{E}[\,|\,a(1)\,|\,] + \int_{1/n}^{1} \mathbb{E}[\,|\,a'(h)\,|\,]\,dh . \tag{1.19}$$

2 Preliminaries

Here we discuss some well-known results and their consequences regarding spline smoothing.

The first results concern the eigenvalue-eigenvector decomposition of the system matrix M_n, see CRAVEN AND WAHBA (1979). Since M_n is a circulant matrix, it is diagonalized by the discrete Fourier transformation with *unitary* matrix $F_n \in \mathbb{C}^{n\times n}$. Thus

$$M_n = F_n^* \Lambda_n F_n \; , \tag{2.1}$$

with Λ_n a (real) diagonal matrix with diagonal elements

$$(2.2) \qquad 0 = \lambda_{1,n} < \lambda_{2,n} \leqslant \cdots \leqslant \lambda_{n,n} \,,$$

with $\lambda_{1,n}$ corresponding to the constant eigenvector $(1, 1, \cdots, 1)^T$. Part (a) of the lemma below regarding the asymptotic behavior of the $\lambda_{i,n}$ is well known, and Part (b) is an easy extension.

Lemma 1. (Craven and Wahba, 1979) *There exists a positive constant c such that for all even n,*

(a) $c^{-1}\, i^{2m} \leqslant \lambda_{i,n} \leqslant c\, i^{2m}\,,\ i = 1, 2, \cdots, n\,,$

(b) $c^{-1}\, i^{2m} \leqslant \lambda_{i,n} - \lambda_{i-n/2,n} \leqslant c\, i^{2m}\,,\ i = n/2+1, \cdots, n\,.$

We usually write statements like Lemma 1(a) as

$$(2.3) \qquad \lambda_{i,n} \asymp i^{2m}\,, \text{ for } n \to \infty\,, \text{ uniformly in } i \leqslant n\,.$$

The above lemma has consequences for the asymptotic behavior of traces and "spreads". From now on we write S_n for $S_n(h)$. We define

$$(2.4) \qquad \operatorname{spread}(h) = \tfrac{1}{n}\operatorname{trace}(\,(I - S_n\,)^2\,) - (\,\tfrac{1}{n}\operatorname{trace}(\,I - S_n\,)\,)^2\,.$$

Lemma 2. *For all integers ℓ, and for $n \to \infty$, uniformly in $h > 0$,*

(a) $\frac{1}{n}\,\mathit{trace}(\,(\,I - S_n\,)^{\ell}\,) \asymp \dfrac{(nh)^{2m\ell}}{(1+nh)^{2m\ell}}\,;$

(b) $\frac{1}{n}\,\mathit{trace}(\,(S_n)^{\ell}\,) \asymp (\,1 + nh\,)^{-1}\,;$

(c) $\mathit{spread}(h) \asymp \dfrac{(nh)^{4m}}{(\,1+nh\,)^{4m+1}}\,.$

Proof. We have $\operatorname{trace}(\,(\,I - S_n\,)^{\ell}\,) = \sum_{i=1}^{n} (w_i)^{\ell}\,,$ with

$$(2.5) \qquad w_i = h^{2m}\,\lambda_{i,n}\,(\,1 + h^{2m}\,\lambda_{i,n}\,)^{-1}$$

and so, for $n \to \infty$, uniformly in h,

$$(2.6) \qquad \operatorname{trace}(\,(\,I - S_n\,)^{\ell}\,) \asymp \sum_{i=1}^{n} \{\,(ih)^{2m}\,(\,1 + (ih)^{2m}\,)^{-1}\,\}^{\ell}\,.$$

Since $x^{2m}(\,1+x^{2m}\,)^{-1}$ is increasing, the sum behaves like the corresponding integral,

$$(2.7) \qquad \int_0^n \frac{(hx)^{2m\ell}}{(\,1+(hx)^{2x}\,)^{\ell}}\,dx = h^{-1}\int_0^{nh} \frac{x^{2m\ell}}{(\,1+x^{2x}\,)^{\ell}}\,dx\,,$$

and Part (a) follows. By a similar argument and carefully matching the behavior for $nh \longrightarrow 0$ and $nh \to \infty$, we obtain Part (b).

For the "spread" we may write

$$\text{spread}(h) = \tfrac{1}{n} \sum_{i=1}^{n} |\, w_i - \overline{w}\,|^2 , \tag{2.8}$$

with $\overline{w} = \frac{1}{n}\,\text{trace}(\, I - S_n\,) = \frac{1}{n} \sum_{i=1}^{n} w_i$. Now, for even n,

$$\tfrac{1}{n} \sum_{i=1}^{n} |\, w_i - \overline{w}\,|^2 = \tfrac{1}{n} \sum_{i=1}^{n/2} |\, w_i - \overline{w}\,|^2 + |\, w_{i+n/2} - \overline{w}\,|^2 , \tag{2.9}$$

and

$$\begin{aligned} |\, w_i - \overline{w}\,|^2 + |\, w_{i+n/2} - \overline{w}\,|^2 &\geqslant \min_{\beta} \{\, |\, w_i - \beta\,|^2 + |\, w_{i+n/2} - \beta\,|^2 \,\} \\ &= \tfrac{1}{2}\, |\, w_{i+n/2} - w_i\,|^2 . \end{aligned} \tag{2.10}$$

Now, for $n \to \infty$, uniformly in $h > 0$,

$$\begin{aligned} w_{i+n/2} - w_i &= \frac{h^{2m}(\,\lambda_{i+n/2,n} - \lambda_{i,n}\,)}{(\,1 + h^{2m}\lambda_{i+n/2,n}\,)(\,1 + h^{2m}\lambda_{i,n}\,)} \\ &\asymp \frac{\big((\,i + \frac{1}{2}\,n\,)\,h\,\big)^{2m}}{\big(\,1 + (\,i + \frac{1}{2}\,n\,)\,h\,\big)^{2m}(\,1 + ih\,)^{2m}} , \end{aligned} \tag{2.11}$$

whence

$$\begin{aligned} \tfrac{1}{n} \sum_{i=1}^{n/2} |\, w_{i+n/2} - w_i\,|^2 &\asymp (nh)^{-1} \int_o^{nh} \frac{(x + \frac{1}{2}\,nh)^{4m}}{(1 + x + \frac{1}{2}\,nh)^{4m}(1+x)^{4m}} dx \\ &\asymp \frac{(nh)^{4m}}{(1+nh)^{4m+1}} , \end{aligned} \tag{2.12}$$

the last statement, by carefully matching the behavior for $nh \longrightarrow 0$ and for $nh \to \infty$. Q.E.D.

Finally we need lower bounds on expected errors and so on.

Lemma 3. *If f_o satisfies (1.7) then*

(a) $\liminf_{n\to\infty} \inf_{h>0} h^{-4m}(1+h)^{4m}\, \frac{1}{n}\, \|\, y_o - S_n y_o\,\|^2 > 0\,;$

(b) $\liminf_{n\to\infty} n^{4m/(4m+1)} \inf_{h>0} \mathbb{E}[\,\mathfrak{L}_n(h)\,] > 0\,;$

(c) $\liminf_{n\to\infty} n^{4m/(4m+1)} \inf_{h>0} \mathbb{E}[\,\text{GCV}_n(h)\,] > 0\,.$

Proof. With (2.1)–(2.2) we have

$$\|\, y_o - S_n\, y_o\,\|^2 = \sum_{i=1}^{n} |\, w_i\, \widehat{y}_{oi}\,|^2 , \tag{2.13}$$

with w_i as in (2.5), and $\widehat{y_o} = F_n\, y_n$ is the discrete Fourier transform of y_n, see § 2. Since $0 = w_1 < w_2 \leqslant \cdots \leqslant w_n$, then

$$(2.14)\quad \| \, y_o - S_n\, y_n \, \|^2 \geqslant (w_2)^2 \sum_{i=2}^{n} | \, \widehat{y}_{oi} \, |^2 = (w_2)^2 \, \| \, y_o - \overline{y_o} \, \|^2 \ ,$$

where $\overline{y_o}$ is the average of the y_{oi}, due to the fact that the zero eigenvalue of M_n corresponds to the constant eigenvector. Thus, if f_o is not constant, then we have $\frac{1}{n} \, \| \, y_o - \overline{y_o} \, \|^2 \geqslant c$ for some positive constant depending on f_o. This shows Part (a). The asymptotic lower bound on $\mathbb{E}[\, \mathfrak{L}_n(h) \,]$ is now obtained as follows. We have, for suitable positive constants c, c_1,

$$(2.15)\quad \begin{aligned} \mathbb{E}[\, \mathfrak{L}_n(h) \,] &= \tfrac{1}{n} \, \| \, y_o - S_n\, y_o \, \|^2 + \sigma^2 \, \tfrac{1}{n} \, \mathrm{trace}(S_n^2) \\ &\geqslant c\, h^{4m} (\, 1 + h^{2m} \,)^{-2} + c_1 \, (\, 1 + nh \,)^{-2} \ , \end{aligned}$$

where we used Lemma 2(b), and Part (a) of the present lemma. Minimizing over h gives the bound of Part (b).

For Part (c) we verify that

$$\big(\tfrac{1}{n} \, \mathrm{trace}(\, I - S_n \,) \big)^2 \, \mathbb{E}[\, \mathrm{GCV}_n(h) \,] = \tfrac{1}{n} \, \| \, y_o - S_n\, y_o \, \|^2 + \sigma^2 \, \mathrm{spread}(h) \ ,$$

and so, with Lemma 2(a), (c) and Part (a) above,

$$(2.16)\qquad \mathbb{E}[\, \mathrm{GCV}_n(h) \,] \asymp \frac{h^{4m}}{(\, 1 + h \,)^{4m}} \, \frac{(\, 1 + nh \,)^{4m}}{(nh)^{4m}} + \frac{1}{1 + nh} \ .$$

We verify that for $h = n^{-1/2}$, say, we have $\mathbb{E}[\, \mathrm{GCV}_n(h) \,] \longrightarrow 0$. Since $(1 + nh)^{-1} \longrightarrow 0$ only if $nh \to \infty$, we may ignore the cumbersome factor $(nh)^{-4m} (1 + nh)^{4m}$ in the first term. Minimization over h then gives the required lower bound. Q.E.D.

Note that Part (b) is a verification of the condition $n \, \mathbb{E}[\, \mathfrak{L}_n(h) \,] \to \infty$ of Li (1986), but here a rate n^{ε} for some $\varepsilon > 0$ is needed.

3 Consistency

Here we prove Theorem 1. The following lemma is helpful.

Lemma 4. *For $n \to \infty$, uniformly in $h > 0$, and $\ell = 1, 2$,*

(a) $\langle \, (S_n)^{\ell} \, d_n \, , \, d_n \, \rangle - \mathbb{E}[\, \langle \, (S_n)^{\ell} \, d_n \, , \, d_n \, \rangle \,] =_{as} \mathcal{O}\big(\, (n \log n)^{1/2} \, \big) \, ;$

(b) $\dfrac{\| \, (\, I - S_n \,) \, d_n \, \|^2 - \mathbb{E}[\, \| \, (\, I - S_n \,) \, d_n \, \|^2 \,]}{(\, \frac{1}{n} \, \mathrm{trace}(\, I - S_n \,) \,)^2} =_{as} \mathcal{O}\big(\, (n \log n)^{1/2} \, \big) \, .$

In the proof of the lemma the following inequality, which may be traced back to ABEL (1828), is useful: if $p_1 \geqslant p_2 \geqslant \cdots p_n \geqslant 0$, then

$$(3.1) \qquad \Big| \sum_{i=1}^{n} p_i\, \delta_i \Big| \leqslant p_1 \max_{1 \leqslant j \leqslant n} \Big| \sum_{i=1}^{j} \delta_i \Big| \,.$$

The proof is based on writing $p_i = \sum_{i=1}^{n} q_j$ with *nonnegative* q_j and changing the order of summation. Its usefulness stems from the fact that the weights p_i may depend on a random parameter (h), so (3.1) "decouples" it from the δ_i.

Proof of Lemma 4. We prove Part (a) with $\ell = 2$. The case $\ell = 1$ follows similarly. With p_i defined by $p_i = (1 + h^{2m} \lambda_{i,n})^{-1}$ we have

$$(3.2) \qquad \| S_n\, d_n \|^2 - \mathbb{E}[\, \| S_n\, d_n \|^2 \,] = \sum_{i=1}^{n} (p_i)^2 \,(\, \delta_{in}^2 - \sigma^2 \,) \,,$$

with $\delta_n = F_n d_n$. Since the p_i are decreasing with i for fixed n, then, from the inequality (3.1)

$$(3.3) \qquad \Big| \| S_n\, d_n \|^2 - \mathbb{E}[\, \| S_n\, d_n \|^2 \,] \Big| \leqslant (p_1)^2 \max_{1 \leqslant j \leqslant n} \Big| \sum_{i=1}^{j} (\, \delta_{in}^2 - \sigma^2 \,) \Big| \,,$$

and, since $p_1 \leqslant 1$, this is $\mathcal{O}\big(\, (n \log n)^{1/2} \,\big)$ almost surely, cf. the proof of (4.16) below. Moreover, it holds uniformly in h.

For Part(b) we have similar to (3.3)

$$\Big| \| (\, I - S_n \,)\, d_n \|^2 - \mathbb{E}[\, \| (\, I - S_n \,)\, d_n \|^2 \,] \Big| \leqslant$$
$$(w_n)^2 \max_{1 \leqslant j \leqslant n} \Big| \sum_{i=j}^{n} (\, \delta_{in}^2 - \sigma^2 \,) \Big| =_{\text{as}} \mathcal{O}\big(\, (n \log n)^{1/2}\, (w_n)^2 \,\big) \,,$$

with w_n given by (2.5). Since the weights w_i are increasing, we get the reverse order of the partial sums. By Lemma 2(a) we have

$$(w_n)^2 \asymp \tfrac{1}{n}\, \mathrm{trace}\big(\, (\, I - S_n \,)^2 \,\big) \asymp \big(\, \tfrac{1}{n}\, \mathrm{trace}(\, I - S_n \,) \,\big)^2 \,,$$

and Part (b) follows. Q.E.D.

Proof of Theorem 1. The proofs of the statements (a) $\Longrightarrow$ (b), (c) are straightforward, and are omitted.

(b) $\Longrightarrow$ (a): Using $(a+b)^2 \leqslant 2a^2 + 2b^2$, write

$$(3.4) \qquad \begin{aligned} \tfrac{1}{n} \| y_o - S_n\, y_n \|^2 &\geqslant \tfrac{1}{2}\, \mathrm{bias}(h) - \tfrac{1}{n} \| S_n\, d_n \|^2 \\ &= \tfrac{1}{2}\, \mathrm{bias}(h) - \tfrac{1}{n}\, \sigma^2\, \mathrm{trace}(\, S_n^2 \,) \\ &\qquad - \{ \tfrac{1}{n} \| S_n\, d_n \|^2 - \mathbb{E}[\, \tfrac{1}{n} \| S_n\, d_n \|^2 \,] \} \,, \end{aligned}$$

where

$$\text{bias}(h) = \tfrac{1}{n}\,\|\, y_o - S_n\, y_o \,\|^2 \ , \tag{3.5}$$

and use Lemmas 2(b), 3(a) and 4(a). Then, for suitable positive constants c and c_1,

$$\mathfrak{L}_n(h) \geqslant_{\text{as}} c\,H - c_1\,(\,1 + nh\,)^{-1} - \mathcal{O}(\,(n^{-1}\log n)^{1/2}\,)\ . \tag{3.6}$$

Here $H = h^{4m}/(1+h^{4m})$. Now assume that $h = h_n \geqslant_{\text{as}} C > 0$ for some subsequence of $\{h_n\}_n$. Then the last two terms on the right of (3.6) tend to 0 for $n \to \infty$, and thus $\mathfrak{L}_n(h)$ does not tend to 0. This is a contradiction, and thus $h \longrightarrow_{\text{as}} 0$.

To show that $nh \longrightarrow_{\text{as}} \infty$, we use the companion inequality to (3.6),

$$\mathfrak{L}_n(h) \geqslant c\,(1+nh)^{-1} - \text{bias}(h) - \mathcal{O}(\,(n^{-1}\log n)^{1/2}\,)\ , \tag{3.7}$$

for a suitable positive constant. Now $h \longrightarrow_{\text{as}} 0$ implies $\text{bias}(h) \longrightarrow_{\text{as}} 0$, since (a) $\Longrightarrow$ (b). By (3.7), then $\mathfrak{L}_n(h) \longrightarrow_{\text{as}} 0$ implies $(\,1+nh\,)^{-1} \longrightarrow_{\text{as}} 0$. It follows that $nh \longrightarrow_{\text{as}} \infty$.

(c) $\Longrightarrow$ (a) : Let

$$T_n(h) = \big(\,\tfrac{1}{n}\,\text{trace}(\,I - S_n\,)\,\big)^2\ . \tag{3.8}$$

We now expand

$$\|\, y_n - S_n\, y_n \,\|^2 = \|\, y_o - S_n\, y_o \,\|^2 + \\ \|\,(\,I - S_n\,)\, d_n \,\|^2 + 2\,\langle\,(\,I - S_n\,)^2\, y_o\,,\, d_n\,\rangle\ ,$$

so that

$$\text{GCV}_n(h) = \frac{\text{bias}(h) + \text{stuff}(h) + \sigma^2\,\text{spread}(h) + \tfrac{2}{n}\langle (I - S_n)^2\, y_o, d_n\rangle}{T_n(h)}\ , \tag{3.9}$$

with bias(h) as in (3.5), spread (h) as in (2.4), and

$$\text{stuff}(h) = \tfrac{1}{n}\,\|\,(\,I - S_n\,)\, d_n \,\|^2 - \mathbb{E}[\,\tfrac{1}{n}\,\|\,(\,I - S_n\,)\, d_n\,\|^2\,] \\ - \big(\,\tfrac{1}{n}\,\|\, d_n \,\|^2 - \sigma^2\,\big)\ T_n(h)\ . \tag{3.10}$$

Now, the last term is $\mathcal{O}(\,(n\log n)^{1/2}\,T_n(h)\,)$ a.s. (uniformly in h), cf. the proof of (4.16) below, and then Lemma 4(b) shows that also uniformly in $h > 0$

$$\text{stuff}(h) =_{\text{as}} \mathcal{O}(\,(n^{-1}\log n)^{1/2}\,T_n(h)\,)\ ,$$

and we need to worry about the inner product. Again, using the eigenvalue-eigenvector decomposition (2.1)–(2.2), with

$$v_i = (h^{2m}\lambda_{i,n})^2\,(\,1 + h^{2m}\lambda_{i,n}\,)^{-2}\ , \tag{3.11}$$

we have by the equality (3.1)

$$(3.12)\qquad \bigl|\,\langle\,(\,I - S_n\,)^2\, y_o\,,\, d_n\,\rangle\,\bigr| = \Bigl|\sum_{i=1}^{n} v_i\, \widehat{y}_{oi}\, \delta_{in}\Bigr| \leqslant v_n \max_{1\leqslant j\leqslant n} Z_{j,n}\ ,$$

where $Z_{j,n} = |\sum_{i=j}^{n} \widehat{y}_{oi}\,\delta_{in}|$. The "order" is reversed since the v_i are increasing with i. Note that by Lemma 2(a)

$$(3.13)\qquad v_n \asymp (nh)^{4m}(1+nh)^{-4m} \asymp \bigl(\tfrac{1}{n}\,\mathrm{trace}(I - S_n)\bigr)^2 = T_n(h)\ .$$

Also note that since $d_n \sim N(0, \sigma^2 I)$, then $\delta_n \sim N(0, \sigma^2 I)$, so that the $\delta_{in}, i = 1, 2, \cdots, n$, are independent random variables with mean 0. Thus

$$(3.14)\qquad \mathbb{E}[\,Z_{j,n}\,|\,\delta_{j+1,n}, \cdots, \delta_{n,n}\,] \geq Z_{j+1,n}\ ,$$

and $Z_{j,n}, j = 1, 2, \cdots, n$, is a supermartingale. Then for every θ, so is $\exp(\,\theta\, Z_{j,n}\,), j = 1, 2, \cdots, n$. By DOOB's submartingale inequality,

$$(3.15)\qquad \mathbb{P}[\max_{1\leqslant j\leqslant n} Z_{j,n} > t\,] \leqslant \exp(\,-\,\theta\, t\,)\ \mathbb{E}[\,\exp(\,\theta\,|\,s_n\,|\,)\,]\ ,$$

where $s_n = \sum_{i=1}^{n} \widehat{y}_{oi}\,\delta_{i,n}$. We verify that

$$\mathbb{E}[\,\exp(\,\theta\,|\,s_n\,|\,)\,] \leqslant \mathbb{E}[\,\exp(\,\theta\, s_n\,)\,] + \mathbb{E}[\,\exp(\,-\,\theta\, s_n\,)\,] = 2\,\mathbb{E}[\,\exp(\,\theta\, s_n\,)\,]\ ,$$

and

$$\begin{aligned}\mathbb{E}[\,\exp(\,\theta\, s_n\,)\,] &= \exp\bigl(\,\tfrac{1}{2}\,\sigma^2\,\theta^2 \sum_{i=1}^{n} |\,\widehat{y}_{oi}\,|^2\,\bigr)\\ &= \exp\bigl(\,\tfrac{1}{2}\,\sigma^2\,\theta^2\,\|\,y_o\,\|^2\,\bigr) \leqslant \exp(\,c\,n\,\theta^2\,)\ ,\end{aligned}$$

for a suitable constant c. The choice $\theta = t\,/(\,2\,c\,n\,)$ then gives the bound, with c replaced by $1/(2\,c)$,

$$(3.16)\qquad \mathbb{P}[\max_{1\leqslant j\leqslant n} Z_{j,n} > t\,] \leqslant 2\,\exp(-c\,t^2/n)\ .$$

It follows by the BOREL-CANTELLI lemma that

$$(3.17)\qquad \max_{1\leqslant j\leqslant n} Z_{j,n} =_{\text{as}} \mathcal{O}\bigl(\,(\,n\log n\,)^{1/2}\,\bigr)\ .$$

Putting all this together shows that uniformly in $h > 0$,

$$(3.18)\qquad \langle (I - S_n)^2 y_o, d_n\rangle =_{\text{as}} \mathcal{O}\bigl(\,(\,n\log n\,)^{1/2}\ T_n(h)\,\bigr)\ .$$

From (3.9), and Lemmas 2(a) and 4(b), we obtain for suitable positive constants c and c_1,

$$\mathrm{GCV}_n(h) \geqslant_{\text{as}} \frac{c\,h^{4m}}{(\,1+h\,)^{4m}}\ \frac{(\,1+nh\,)^{4m}}{(nh)^{4m}} + \frac{c_1}{1+nh} - \mathcal{O}\bigl(\,(n^{-1}\log n)^{1/2}\,\bigr)\ .$$

Thus, if $\mathrm{GCV}_n(h) \longrightarrow_{\text{as}} 0$ then the second term must tend to 0, so that $nh \longrightarrow_{\text{as}} \infty$. Then the first term behaves as $h^{4m}(1+h)^{-4m}$, and this tends to 0 only if $h \longrightarrow_{\text{as}} 0$. Q.E.D.

4 Stability

Here we prove the stability theorem. Strictly speaking Part (a) is not a stability result, but we let that pass. Also, the proofs of Parts (b) and (c) are similar, so only Part (c) is proved.

Proof of Theorem 3(a). From (3.9) we obtain

$$T_n(h)\,\mathbb{E}[\,\mathrm{GCV}_n(h)\,] = \mathrm{bias}(h) + \sigma^2\,\mathrm{spread}(h)\ ,$$

with spread(h) as in (2.4), so that

$$\begin{aligned} T_n(h)\left|\frac{\mathbb{E}[\,\mathrm{GCV}_n(h)\,]}{\mathbb{E}[\,\mathfrak{L}_n(h)\,]} - 1\right| &= \frac{\mathrm{bias}(h)\,(\,1 - T_n(h)\,) + \sigma^2\,V_n(h)}{\mathrm{bias}(h) + \sigma^2\,\frac{1}{n}\,\mathrm{trace}(S_n^2)} \\ &\leqslant 1 - T_n(h) + \frac{V_n(h)}{\frac{1}{n}\,\mathrm{trace}(S_n^2)}\ , \end{aligned} \tag{4.1}$$

with

$$V_n(h) = \tfrac{1}{n}\,\mathrm{trace}(\,(\,I - S_n\,)^2\,) - T_n(h) - \tfrac{1}{n}\,\mathrm{trace}(\,S_n^2\,)\,T_n(h)\ .$$

With p_i as in (3.2), and $\overline{p} = \frac{1}{n}\sum_{i=1}^n p_i$, then $T_n(h) = (\,1 - \overline{p}\,)^2$, and

$$V_n(h) = -\overline{p}^{\,2} + (\,\tfrac{1}{n}\sum_{i=1}^{n}(p_i)^2\,)\,(\,1 - (\overline{p} - 1)^2\,)\ .$$

With $h = H_{\mathrm{OPT}}, H_{\mathrm{GCV}}$, then by Lemma 2(b), since $nh \longrightarrow_{\mathrm{as}} \infty$,

$$\overline{p} = \tfrac{1}{n}\,\mathrm{trace}(\,S_n\,) \asymp_{\mathrm{as}} (nh)^{-1}\ ,\quad \tfrac{1}{n}\sum_{i=1}^{n}(p_i)^2 = \tfrac{1}{n}\,\mathrm{trace}(\,S_n^2\,) \asymp_{\mathrm{as}} (nh)^{-1}\ .$$

Thus, $V_n(h) =_{\mathrm{as}} \mathcal{O}(\,(nh)^{-2}\,)$, and so

$$\frac{V_n(h)}{\frac{1}{n}\,\mathrm{trace}(S_n^2)} =_{\mathrm{as}} \mathcal{O}(\,(nh)^{-1}\,) \longrightarrow_{\mathrm{as}} 0\ .$$

Since $T_n(h) \longrightarrow_{\mathrm{as}} 1$, then (4.1) shows Part (a). Q.E.D.

We now get to the actual stability result.

Proof of Theorem 3(c). The starting point is (3.9), which gives

$$a(h) \stackrel{\mathrm{def}}{=} \frac{\mathrm{GCV}_n(h)}{\mathbb{E}[\,\mathrm{GCV}_n(h)\,]} - 1 = \frac{\mathrm{stuff}(h) + \frac{2}{n}\,\langle\,(\,I - S_n\,)^2\,y_o\,,\,d_n\,\rangle}{\mathcal{G}_n(h)}\ , \tag{4.2}$$

with "stuff" as in (3.10), and

$$\mathcal{G}_n(h) = \mathrm{bias}(h) + \sigma^2\,\mathrm{spread}(h)\ . \tag{4.3}$$

It follows that $|\,a(h)\,| \leqslant |\,b(h)\,| + 2\,|\,c(h)\,|$ with

$$b(h) = \{\, \mathcal{G}_n(h) \,\}^{-1} \operatorname{stuff}(h) \;, \tag{4.4}$$

and

$$c(h) = \{\, n\mathcal{G}_n(h) \,\}^{-1} \langle\, (\, I - S_n\,)^2\, y_o \,,\, d_n \,\rangle \;. \tag{4.5}$$

We begin by bounding $c(h)$. We first bound $c(h)$ as

$$\sup_{h\in\mathcal{H}_n} |\,c(h)\,| \leqslant \{\, \inf_{h>0}\, n\,\mathcal{G}_n \,\}^{-1/2} \sup_{h\in\mathcal{H}_n} |\,e(h)\,| \;, \tag{4.6}$$

with

$$e(h) = \{\, n \operatorname{bias}(h) \,\}^{-1/2} \langle\, (\, I - S_n\,)^2\, y_o \,,\, d_n \,\rangle \;,. \tag{4.7}$$

By Lemma 3(c) we have

$$\{\, \inf_{h>0}\, n\,\mathcal{G}_n(h) \,\}^{-1/2} = \mathcal{O}\big(\, n^{-1/(8m+2)} \,\big) \;,$$

and thus we must show that $e(h)$ is not too large.

By the standard arguments we have that

$$\mathbb{P}[\, \sup_{h\in\mathcal{H}_n} |\,e(h)\,| \geqslant t \,] \leq \exp(-\,t\,)\, \mathbb{E}[\, \sup_{h\in\mathcal{H}_n} \exp(\,|\,e(h)\,|\,) \,] \;. \tag{4.8}$$

Now $\exp(\,|\,t\,|\,) \leq \exp(\,t\,) + \exp(-\,t\,)$, so that the expected value is dominated by

$$\mathbb{E}[\, \sup_{h\in\mathcal{H}_n} \exp(\,e(h)\,) \,] + \mathbb{E}[\, \sup_{h\in\mathcal{H}_n} \exp(\,-\,e(h)\,) \,] = 2\,\mathbb{E}[\, \sup_{h\in\mathcal{H}_n} \exp(\,e(h)\,) \,] \;,$$

the last equality by the symmetry of $e(h)$. Below we show that

$$\mathbb{E}[\, \sup_{h\in\mathcal{H}_n} \exp(\,e(h)\,) \,] = \mathcal{O}\big(\, \log n \,\big) \;, \tag{4.9}$$

so then, for a suitable constant c

$$\mathbb{P}[\, \sup_{h\in\mathcal{H}_n} |\,e(h)\,| > t \,] \leq c\, \exp(-\,t\,)\, \log n \;. \tag{4.10}$$

By the BOREL-CANTELLI lemma it follows that

$$\sup_{h\in\mathcal{H}_n} |\,e(h)\,| =_{\text{as}} \mathcal{O}\big(\, \log n \,\big) \;,$$

and so

$$\sup_{h\in\mathcal{H}_n} |\,c(h)\,| =_{\text{as}} \mathcal{O}\big(\, n^{-1/(8m+2)}\, \log n \,\big) \;, \tag{4.11}$$

The term $b(h)$ of (4.4) can be treated similarly. First we note that "stuff" defined in (3.10) may be written as

$$\begin{aligned} \text{stuff}(h) = {} & \tfrac{1}{n}\left(\| S_n\, d_n \|^2 - \mathbb{E}[\, \| S_n\, d_n \|^2 \,] \right) \\ & - \tfrac{2}{n}\left(\langle S_n\, d_n\,,\, d_n \rangle - \mathbb{E}[\, \langle S_n\, d_n\,,\, d_n \rangle \,] \right) \\ & - \left(\tfrac{1}{n} \| d_n \|^2 - \sigma^2 \right) \left(1 - T_n(h) \right) . \end{aligned} \tag{4.12}$$

Now note that

$$1 - T_n(h) = \tfrac{2}{n}\,\text{trace}(\, S_n\,) - (\, \tfrac{2}{n}\,\text{trace}(\, S_n\,)\,)^2 \leqslant \tfrac{1}{n}\,\text{trace}(\, S_n\,) \asymp \text{spread}(h)\ ,$$

where the asymptotic equivalence in the last step follows from Lemma 2. Consequently, the last term in (4.12) is $\mathcal{O}\big((n^{-1}\log n)^{1/2}\, \mathcal{G}_n(h) \big)$, so that its contribution to $b(h)$ in (4.4) may be ignored. As far as the remaining terms are concerned, we have

$$|\, b(h) \,| \leqslant K \,\{\, \inf_{h>0}\, n\, \mathcal{G}_n(h) \,\}^{-1/2} \,\{\, |\, \beta_2(h) \,| + 2\,|\, \beta_1(h) \,| \,\}\ , \tag{4.13}$$

with $K^2 = \frac{1}{n}\,\text{trace}(\, S_n^{\,2}) \,/\, \text{spread}(h) = \mathcal{O}(\,1\,)$, (by Lemma 2), and

$$\beta_\ell(h) = \{\, \text{trace}(\, S_n^{2\,\ell}\,) \,\}^{-1/2} \,\{\, \langle\, S_n^{\,\ell}\, d_n\,,\, d_n \,\rangle - \mathbb{E}[\, \langle\, S_n^{\,\ell}\, d_n\,,\, d_n \,\rangle \,] \,\}\ . \tag{4.14}$$

As before, the standard arguments give, for θ yet to be chosen,

$$\mathbb{P}[\, \sup_{h\in\mathcal{H}_n}\, |\, \beta_\ell(h) \,| > t \,] \leqslant \exp(-\theta\, t\,)\, \mathbb{E}[\, \sup_{h\in\mathcal{H}_n}\, \exp(\, \theta\, |\, \beta_\ell(h) \,|\,) \,]\ , \tag{4.15}$$

and surely

$$\exp(\, \theta\, |\, \beta_\ell(h) \,|\,) \leqslant \exp(\, \theta\, \beta_\ell(h)\,) + \exp(-\theta\, \beta_\ell(h)\,)\ .$$

Below we show, similar to (4.9), that for $\theta = \{\, 8\,\sigma^2 \,\}^{-1}$

$$\mathbb{E}[\, \sup_{h\in\mathcal{H}_n}\, \exp(\pm\, \theta\, \beta_\ell(h)\,) \,] = \mathcal{O}(\, \log n\,)\ , \tag{4.16}$$

and then, similar to the above

$$\sup_{h\in\mathcal{H}_n}\, |\, b(h) \,| =_{\text{as}} \mathcal{O}\big(\, n^{-1/(8m+2)}\, \log n \,\big)\ , \tag{4.17}$$

and Theorem 3(c) is proved. Part (b) is similar, and is omitted. Q.E.D.

Proof of (4.9). It turns out that (1.19) is good enough for our purposes. Thus,

$$\mathbb{E}[\, \sup_{h\in\mathcal{H}_n}\, \exp(\, e(h)\,) \,] \leqslant \tag{4.18}$$

$$\mathbb{E}[\, \exp(\, e(1)\,) \,] + \int_{\mathcal{H}_n} \mathbb{E}[\, |\, e'(h) \,|\, \exp(\, e(h)\,) \,]\, dh\ .$$

To compute $e'(h)$ we use the eigenvalue-eigenvector decomposition and write, with v_i as in (3.11),

$$e(h) = \frac{\sum_{i=1}^{n} v_i\, \widehat{y}_{oi}\, \delta_{i,n}}{\sqrt{n\, \mathrm{bias}(h)}} \; . \tag{4.19}$$

Then, observing that

$$v_i{}' = \frac{d}{dh}\, v_i = 4\, mh^{-1}(h^{2m}\lambda_{i,n})^2\, (\, 1 + h^{2m}\lambda_{i,n}\,)^{-3} \; ,$$

we obtain

$$|\, v_i{}' \,| \leqslant 4\, mh^{-1}\, v_i \; , \quad |\, v_i{}' \,|^2 \leqslant (\, 4\, mh^{-1}\,)^2\, v_i \; . \tag{4.20}$$

For $e'(h)$ we have

$$e'(h) = \frac{\sum_{i=1}^{n} v_i'\, \widehat{y}_{oi}\, \delta_{in}}{\sqrt{n\, \mathrm{bias}(h)}} - \tfrac{1}{2}\, e(h)\, \frac{\mathrm{bias}'(h)}{\mathrm{bias}(h)} \; . \tag{4.21}$$

Now observe that

$$\mathbb{E}[\, |\, \sum_{i=1}^{n} v_i'\, \widehat{y}_{oi}\, \delta_{in}\, |^2\,] = \sigma^2 \sum_{i=1}^{n} |\, v_i'\, \widehat{y}_{oi}\, |^2 \leqslant (\, 4\, m\sigma h^{-1})^2\, n\, \mathrm{bias}(h) \; . \tag{4.22}$$

where we used (4.20).

For the second term on the right of (4.21) we note that

$$\mathrm{bias}'(h) = \tfrac{1}{n} \sum_{i=1}^{n} v_i'\, |\, \widehat{y}_{oi}\, |^2 \leqslant (\, 4\, m\sigma h^{-1})^2\, n\, \mathrm{bias}(h) \; ,$$

where (4.20) was used again. Since $|\, v_i \,| \leqslant 1$ we also have, similar to (4.21),

$$\mathbb{E}[\, |\, e(h)\, |^2\,] = \frac{\sigma^2 \sum_{i=1}^{n} |\, v_i\, \widehat{y}_{oi}\, |^2}{n\, \mathrm{bias}(h)} \leqslant \sigma^2 \; . \tag{4.23}$$

Consequently, we have shown that

$$|\, e'(h)\, | \leq h^{-1}\, |\, \varepsilon(h)\, | \; , \tag{4.24}$$

where $\varepsilon(h)$ is a weighted sum of independent Normal$(0, \sigma^2)$ random variables, and

$$\mathbb{E}[\, |\, \varepsilon(h)\, |^2\,] = \mathcal{O}(\, 1\,) \; . \tag{4.25}$$

Now, with Cauchy-Schwarz,

$$\mathbb{E}[\, |\, e'(h)\, |\, \exp(\, e(h)\,)\,] \leqslant h^{-1}\, \big(\, \mathbb{E}[\, |\, \varepsilon(h)\, |^2\,]\,\big)^{1/2}\, \big(\, \mathbb{E}[\exp(\, 2\, e(h)\,)\,]\,\big)^{1/2} \; .$$

Since $e(h)$ is a weighted sums of iid normals with mean 0, then

$$\mathbb{E}[\exp(e(h))] = \exp(\tfrac{1}{2}\mathbb{E}[|e(h)|^2]) .$$

Of course, $\mathbb{E}[e(h)^2] = \mathcal{O}(1)$, so that the above shows

$$\mathbb{E}[|e'(h)| \exp(e(h))] = \mathcal{O}(h^{-1}) . \tag{4.26}$$

Since $\mathbb{E}[\exp(c(1))] = \mathcal{O}(1)$, then (4.18) shows (4.9). Q.E.D.

Proof of (4.16). We only show the bound on $\mathbb{E}[\sup_{h\in\mathcal{H}_n} \exp(+\beta_\ell(h))]$. Again we use (1.19), and we show that

$$\mathbb{E}[|\beta_\ell'(h)| \{\exp(\theta\,\beta_\ell(h)) + \exp(-\theta\,\beta_\ell(h))\}] \leqslant c\,h^{-1} , \tag{4.27}$$

for a suitable constant c, independent of n and h.

Using the eigenvalue-eigenvector decomposition we write

$$\beta_\ell(h) = \frac{\sum_{i=1}^n p_i^{\ \ell}\,\eta_{i,n}}{\{\sum_{i=1}^n p_i^{\ 2\ell}\}^{1/2}} ,$$

with $p_i = (1 + h^{2m}\lambda_{i,n})^{-1}$, and $\eta_{i,n} = \delta_{i,n}^2 - \sigma^2$. Then

$$(p_i^{\ \ell})' = -2\,m\,\ell\,h^{2m-1}\lambda_{in}\,(1 + h^{2m}\lambda_{in})^{-\ell-1} ,$$

so that

$$|(p_i^{\ \ell})'| \leq 2\,m\,\ell\,h^{-1}\,p_i^{\ \ell} .$$

Later on we also use this with ℓ replaced by $2\,\ell$.

Then, from

$$\beta_\ell'(h) = \frac{\sum_{i=1}^n (p_i^{\ \ell})'\,\eta_{i,n}}{\{\sum_{i=1}^n p_i^{\ 2\ell}\}^{1/2}} - \tfrac{1}{2}\,\beta_\ell(h)\,\frac{\sum_{i=1}^n (p_i^{\ 2\ell})'}{\sum_{i=1}^n p_i^{\ 2\ell}} ,$$

it follows that

$$\mathbb{E}[|\beta_\ell'(h)|^2] \leq c\,h^{-2} .$$

Finally then, with Cauchy-Schwarz,

$$\mathbb{E}[|\beta_\ell'(h)| \exp(\pm\,\theta\,\beta_\ell(h))] \leqslant \\ (\mathbb{E}[|\beta_\ell'(h)|^2])^{1/2}\,(\mathbb{E}[|\exp(\pm\,2\,\theta\,\beta_\ell(h))])^{1/2} . \tag{4.28}$$

The first factor on the right is $\mathcal{O}(h^{-1})$. The second factor is equal to the product of the square roots of the expectations

$$\mathbb{E}[\exp(\pm\,2\,\theta\,w_i\,\eta_{i,n})] , \quad i = 1, 2, \cdots, n , \tag{4.29}$$

with $w_i = p_i^{\,\ell} \left\{ \sum_{i=1}^n p_i^{\,2\ell} \right\}^{-1/2}$. Note that $0 < w_i < 1$ for all i. We verify that if $X \sim N(0, \sigma^2)$, then

$$\mathbb{E}[\exp(a X^2)] = (1 - 2 a \sigma^2)^{-1/2} , \quad a < (2\sigma^2)^{-1} ,$$

and so, provided $0 < \theta < 1/(\max_i 4 w_i \sigma^2)$,

$$\mathbb{E}[\exp(\pm 2\theta \beta_\ell(h))] = \left\{ \prod_{i=1}^n \frac{\exp(\pm A_i)}{1 \pm A_i} \right\}^{1/2} , \tag{4.30}$$

where $A_i = 4\theta w_i \sigma^2$. Now, since $0 < w_i < 1$ for all i, if we take $\theta = (8\sigma^2)^{-1}$, then $A_i \leqslant \frac{1}{2}$ for all i, and there exists a constant c such that

$$\begin{aligned} \exp(\pm A_i) &\leqslant 1 \pm A_i + c A_i^{\,2} , \\ (1 \pm A_i)^{-1} &\leqslant 1 \mp A_i + c A_i^{\,2} , \end{aligned}$$

so that for another constant $\widetilde{c}$

$$(1 \pm A_i)^{-1} \exp(\pm A_i) \leqslant 1 + \widetilde{c} A_i^{\,2} \leqslant \exp(\widetilde{c} A_i^{\,2}) .$$

Consequently, the right-hand side of (4.30) is bounded by

$$\exp\Bigl(\widetilde{c} \sum_{i=1}^n A_i^{\,2} \Bigr) = \exp(\tfrac{1}{4} \widetilde{c}) ,$$

and so (4.28)–(4.30) prove (4.16). Q.E.D.

Acknowledgement. The authors thank the referee for patiently pointing out various blunders in earlier versions of the paper.

References

N.H. Abel (1828), Recherches sur la série $1 + \frac{m}{1} x + \frac{m(m-1)}{1 \cdot 2} x^2 + \frac{m(m-1)(m-2)}{1 \cdot 2 \cdot 3} x^3 + \cdots$. *J. Reine Angew. Math.* **1**. Reprinted in: *Ouevres complètes de Niels Henrik Abel.* L. Sylow, S. Lie (Eds.). Høgdahl & Søn, Christiania, 1881, pp. 219–250.

R.A. Adams (1975), *Sobolev Spaces*, Academic Press, New York.

P. Craven, G. Wahba (1979), Smoothing noisy data with spline functions, *Numer. Math.* **31**, 377–403.

G.H. Golub, M. Heath, G. Wahba (1979), Generalized cross-validation as a method for choosing a good ridge parameter, *Technometrics* **21**, 215–223.

K.-C. Li (1985), From Stein's unbiased risk estimates to the method of generalized cross validation, *Ann. Statist.* **13**, 1352–1377.

K.-C. Li (1986), Asymptotic optimality of C_L and generalized cross validation in ridge regression with application to spline smoothing, *Ann. Statist.* **14**, 1101–1112.

C.L. Mallows (1972), Some comments on C_p, *Technometrics* **15**, 661–675.

Ch. Reinsch (1967), Smoothing by spline functions, *Numer. Math.* **10**, 177–183.

Paul Eggermont and Vincent LaRiccia
Food and Resource Economics
University of Delaware
Newark, DE 19716
e-mail: eggermon@udel.edu
e-mail: lariccia@udel.edu

Rademacher Processes and Bounding the Risk of Function Learning

Vladimir Koltchinskii* and
Dmitriy Panchenko**

ABSTRACT We construct data dependent upper bounds on the risk in function learning problems. The bounds are based on local norms of the Rademacher process indexed by the underlying function class, and they do not require prior knowledge about the distribution of training examples or any specific properties of the function class. Using Talagrand's type concentration inequalities for empirical and Rademacher processes, we show that the bounds hold with high probability that decreases exponentially fast when the sample size grows. In typical situations that are frequently encountered in the theory of function learning, the bounds give nearly optimal rate of convergence of the risk to zero.

1 Local Rademacher norms and bounds on the risk: main results

Let $(S, \mathcal{A})$ be a measurable space and let $\mathcal{F}$ be a class of $\mathcal{A}$-measurable functions from S into $[0, 1]$. Denote $\mathcal{P}(S)$ the set of all probability measures on $(S, \mathcal{A})$. Let $f_0 \in \mathcal{F}$ be an unknown *target function.* Given a probability measure $P \in \mathcal{P}(S)$ (also unknown), let $(X_1, \dots, X_n)$ be an i.i.d. sample in $(S, \mathcal{A})$ with common distribution P (defined on a probability space $(\Omega, \Sigma, \mathbb{P})$). In computer learning theory, the problem of estimating f_0, based on the labeled sample $(X_1, Y_1), \dots, (X_n, Y_n)$, where $Y_j := f_0(X_j)$, $j = 1, \dots, n$, is referred to as A *function learning problem.* The so called *concept learning* is a special case of function learning. In this case, $\mathcal{F} := \{I_C : C \in \mathcal{C}\}$, where $\mathcal{C} \subset \mathcal{A}$ is called a class of concepts (see Vapnik (1998), Vidyasagar (1997), Devroye, Györfi and Lugosi (1996) for the account on statistical learning theory). The goal of function learning is to find an estimate $\hat{f}_n := \hat{f}_n((X_1, Y_1), \dots, (X_n, Y_n))$ of the unknown target function such that

---

*The research of V. Koltchinskii is partially supported by NSA Grant MDA904-99-1-0031.

**The research of D. Panchenko was partially supported by Boeing Computer Services Grant 3-48181.

the L_1-distance between $\hat{f}_n$ and f_0 becomes small with high probability as soon as the sample size becomes large enough. The L_1-distance $P|\hat{f}_n - f_0|$ is often called *the risk* (also the generalization, or prediction error) of the estimate $\hat{f}_n$. A class $\mathcal{F}$ is called *probably approximately correctly (PAC) learnable* iff for all $\varepsilon > 0$

$$\pi_n(\mathcal{F};\varepsilon) := \sup_{P\in\mathcal{P}(S)} \sup_{f_0\in\mathcal{F}} \mathbb{P}\{P|\hat{f}_n - f_0| \geq \varepsilon\} \to 0 \text{ as } n \to \infty.$$

The bounds on the probability $\pi_n(\mathcal{F};\varepsilon)$ are of importance in the theory. Such bounds allow one to determine the quantity

$$N_{\mathcal{F}}(\varepsilon;\delta) := \inf\{n : \pi_n(\mathcal{F};\varepsilon) \leq \delta\},$$

which is called *the sample complexity of learning.* Unfortunately, a bound that is uniform in the class of all distributions $\mathcal{P}(S)$ is not necessarily tight for a particular distribution P and often such a bound does not provide a reasonable estimate of the minimal sample size needed to achieve certain accuracy of learning in the case of a particular P.

A natural approach to the function learning problem (in the case when $f_0 \in \mathcal{F}$) is to find $\hat{f}_n \in \mathcal{F}$ such that $\hat{f}_n(X_j) = f_0(X_j) = Y_j$ for all $j = 1,\dots,n$. In learning theory, such an estimate $\hat{f}_n$ is called *consistent* (this notion should not be confused with consistency in a statistical sense).

We construct below a data dependent bound on the risk of a consistent estimate $\hat{f}_n$. More precisely, given $\delta > 0$, we define a quantity

$$\hat{\beta}_n(\mathcal{F};\delta) = \hat{\beta}_n(\mathcal{F};\delta;(X_1,Y_1),\dots,(X_n,Y_n))$$

such that for any consistent estimate $\hat{f}_n$

$$\sup_{P\in\mathcal{P}(S)} \sup_{f_0\in\mathcal{F}} \mathbb{P}\{P|\hat{f}_n - f_0| \geq \hat{\beta}_n(\mathcal{F};\delta)\} \leq \delta. \tag{1.1}$$

We will consider a couple of important examples in which the bound we suggest gives nearly optimal rate of convergence of the risk to 0 as the sample size tends to infinity.

Given a class $\mathcal{G}$ of $\mathcal{A}$-measurable functions from S into $[0,1]$ with $0 \in \mathcal{G}$, let $\hat{\mathcal{G}}_n$ denote the restriction of the class $\mathcal{G}$ on the sample $(X_1,\dots,X_n)$. Consider a quantity

$$\hat{\gamma}_n(\mathcal{G};\delta) = \hat{\gamma}_n(\hat{\mathcal{G}}_n;\delta;X_1,\dots,X_n)$$

such that the bound

$$\sup_{P\in\mathcal{P}(S)} \mathbb{P}\{P\hat{g}_n \geq \hat{\gamma}_n(\mathcal{G};\delta)\} \leq \delta$$

holds for any class $\mathcal{G}$ and for any function $\hat{g}_n \in \mathcal{G}$ satisfying the conditions $\hat{g}_n(X_j) = 0$ for all $j = 1,\dots,n$.

Define

$$\mathcal{F}(f_0) := \{|f - f_0| : f \in \mathcal{F}\}$$

(note that the values of the functions from this class are known on the sample $(X_1, \dots, X_n)$) and

$$\begin{aligned}\hat{\mathcal{F}}_n(f_0) &:= \{(|f - f_0|(X_j) : 1 \le j \le n) : f \in \mathcal{F}\} \\ &= \{(|f(X_j) - Y_j| : 1 \le j \le n) : f \in \mathcal{F}\}.\end{aligned}$$

If $\hat{f}_n$ is a consistent estimate, then the function $\hat{g}_n := |\hat{f}_n - f_0| \in \mathcal{F}(f_0)$ satisfies the condition $\hat{g}_n(X_j) = 0$ for all $j = 1, \dots, n$. Then, clearly, for any consistent estimate $\hat{f}_n$,

$$\sup_{P \in \mathcal{P}(S)} \sup_{f_0 \in \mathcal{F}} \mathbb{P}\{P|\hat{f}_n - f_0| \ge \hat{\gamma}_n(\mathcal{F}(f_0); \delta)\} \le \delta.$$

Therefore if one defines (for $Y_j = f_0(X_j)$)

$$\hat{\beta}_n(\mathcal{F}; \delta; (X_1, Y_1), \dots, (X_n, Y_n)) := \hat{\gamma}_n(\hat{\mathcal{F}}_n(f_0); \delta; X_1, \dots, X_n),$$

then (1.1) holds.

These considerations show that the problem can always be reduced to the case $f_0 \equiv 0$. To simplify the notations, we make this assumption in what follows.

We also assume for simplicity that $\mathcal{F}$ is a countable class of functions. This condition can be easily replaced by standard measurability assumptions known in the theory of empirical processes (see, e.g., [4] or [13]; we do not make a countability assumption in some of the examples below). Estimates $\hat{f}_n$ are supposed to be $\Sigma \times \mathcal{A}$-measurable. We denote by P_n the empirical measure based on the sample $(X_1, \dots, X_n)$:

$$P_n := n^{-1} \sum_{j=1}^{n} \delta_{X_j},$$

where δ_x is the probability measure concentrated at the point $x \in S$. We also use the notation $\|\cdot\|_{\mathcal{F}}$ for the sup-norm of functions from the class $\mathcal{F}$ into $\mathbb{R}$:

$$\|Y\|_{\mathcal{F}} := \sup_{f \in \mathcal{F}} |Y(f)|.$$

Our approach is based on the following simple idea. Denote $B(r) := \{f : P|f| \le r\}$ and set $r_0^n = 1$. It is clear that for any consistent estimate $\hat{f}_n$, $P_n \hat{f}_n = 0$ and, hence,

$$P\hat{f}_n \le P_n\hat{f}_n + \|P_n - P\|_{\mathcal{F}} = \|P_n - P\|_{\mathcal{F}} = \|P_n - P\|_{\mathcal{F} \cap B(r_0^n)} =: r_1^n.$$

Therefore, $\hat{f}_n \in \mathcal{F} \bigcap B(r_1^n)$. It means that actually

$$P\hat{f}_n \le P_n\hat{f}_n + \|P_n - P\|_{\mathcal{F} \cap B_{r_1^n}} = \|P_n - P\|_{\mathcal{F} \cap B_{r_1^n}}.$$

We can repeat this recursive procedure infinitely many times. Namely, if $r_{k+1}^n := \|P_n - P\|_{\mathcal{F}\cap B(r_k^n)}$, then, by induction, $P\hat{f}_n \le r_k^n$ for any natural k. It is also clear that the sequence $\{r_k^n\}$ is nonincreasing Indeed, by a simple induction argument, we have that $r_k^n \le r_{k-1}^n$ implies that

$$r_{k+1}^n = \|P_n - P\|_{\mathcal{F}\cap B(r_k^n)} \le \|P_n - P\|_{\mathcal{F}\cap B(r_{k-1}^n)} = r_k^n.$$

Thus, the following proposition holds.

Proposition 1. *The sequence $\{r_k^n\}_{k\ge 1}$ is nonincreasing and for any consistent estimate $\hat{f}_n$ $P\hat{f}_n \le \inf_{k\ge 0} r_k^n$.*

The sequence $\{r_k^n\}_{k\ge 1}$ depends not only on the data; it also depends explicitly on the unknown distribution P, so it can not be used for the purposes of bounding the risk. However, there is a simple bootstrap type approach that allows one to get around this difficulty.

The Rademacher process indexed by the function class $\mathcal{F}$ is defined as

$$R_n = \frac{1}{n}\sum_{i=1}^n \varepsilon_i \delta_{X_i},$$

where $\{\varepsilon_i\}$ is a Rademacher sequence (an i.i.d. sequence of random variables taking the values $+1$ and -1 with probability $1/2$ each) independent of $\{X_i\}$. It has been used for a long time to obtain the bounds on the sup-norm of the empirical process indexed by functions (in the so-called symmetrization inequalities, see [13]). Recently, Koltchinskii (1999) (see also Koltchinskii et al. (1999)) suggested using $\|R_n\|_{\mathcal{F}}$ as a data-based measure of the accuracy of empirical approximation $\|P_n - P\|_{\mathcal{F}}$ in learning problems and developed a version of structural risk minimization in which the norms of the Rademacher process play the role of data-dependent penalties. Lozano (1999) compared this method of penalization with the method based on VC-dimensions and the cross-validation method and found that in the so-called problem of the "intervals model selection" the Rademacher penalization performs better than other methods. Hush and Scovel (1999) used Rademacher norms to obtain posterior performance bounds for machine learning. However, the "global" norm of the Rademacher process does not allow one to recover the rate of convergence of the risk to 0 in the case when $f_0 \in \mathcal{F}$ (the so-called zero error case). To address this problem, we define below a sequence of localized norms of the Rademacher process that majorizes the sequence $\{r_k^n\}$ defined above.

Given $\varepsilon > 0$, let $\bar{\varphi}$ be a (random) function defined by

$$\bar{\varphi}(r) := \bar{K}_1 \|R_n\|_{\mathcal{F}\cap B_{2r}^e} + \bar{K}_2\sqrt{r\varepsilon} + \bar{K}_3\varepsilon,$$

where $B_r^e = \{f \in \mathcal{F} : P_n f \le r\}$ and $\bar{K}_1, \bar{K}_2, \bar{K}_3 > 0$ are numerical constants.

We introduce the following data-dependent sequence

$$\{\bar r_k^n\}_{k\geq 0} = \{\bar r_k^n(X_1,\dots,X_n;\varepsilon_1,\dots,\varepsilon_n)\}_{k\geq 0},$$

$$\bar r_0^n = 1,\ \bar r_{k+1}^n = \bar\varphi(\bar r_k^n)\wedge 1, \qquad k=0,1,2,\dots \ . \tag{1.2}$$

Since the function $\bar\varphi$ is nondecreasing, a simple induction shows that the sequence $\{\bar r_k^n\}$ is nonincreasing.

Theorem 2. *There is a choice of numerical constants $\bar K_1, \bar K_2, \bar K_3 > 0$ such that for all $P\in\mathcal{P}(S)$, for all $N\geq 1$ and for any consistent estimate $\hat f_n$,*

$$\mathbb{P}\{P\hat f_n \geq \bar r_N^n\} \leq 2Ne^{-\frac{n\varepsilon}{2}}.$$

Thus, if one chooses $N\geq 1$ and, for a given $\delta>0$, $\varepsilon > (\log 2N\delta)/n$, then one can define $\beta_n(\mathcal{F};\delta) := \bar r_N^n$ to get the bound (1.1). The question to be answered is how large should be the number of iterations N to achieve a reasonably good upper bound on the risk in such a way (if it is possible at all). Surprisingly, under rather general conditions the upper bound becomes sharp after very few iterations (roughly, the number of iterations N is of the order $\log_2\log_2(\frac{1}{\varepsilon})$).

In what follows, given a (pseudo)metric space $(M;d)$, we denote $N_d(M;\varepsilon)$ the minimal number of balls of radius ε, covering M, and $H_d(M;\varepsilon) := \log N_d(M;\varepsilon)$. Also, for a probability measure Q on $(S,\mathcal{A})$, $d_{Q,2}$ denotes the metric of the space $L_2(S;dQ)$.

Given a class of functions $\mathcal{F}$, assume that

$$\mathbb{E}_\varepsilon\|n^{-1/2}\sum_{i=1}^n \varepsilon_i\delta_{X_i}\|_{B^{\varepsilon}(r)\cap\mathcal{F}} \leq \hat\psi_n(\sqrt{r})$$

for some concave nondecreasing (random) function $\hat\psi_n$. Usually the role of $\hat\psi_n$ will be played by the random entropy integral

$$\hat\psi_n(r) = K\int_0^r H^{1/2}_{d_{P_n,2}}(\mathcal{F},u)du$$

or by some further upper bound on the random entropy integral. Let us denote by $\hat\delta_n := \hat\delta_n(X_1,\dots,X_n)$ the solution of the equation

$$\hat\delta_n = n^{-1/2}\hat\psi_n\big(\sqrt{\hat\delta_n}\big).$$

The following theorem gives the upper bound on the quantity $\bar r_N^n$.

Theorem 3. *If the number of iterations is equal to $N = [\log_2\log_2\varepsilon^{-1}]+1$, then for some numerical constant $c>0$ and for all $P\in\mathcal{P}(S)$*

$$\mathbb{P}\left(\bar r_N^n \geq c(\hat\delta_n\vee\varepsilon)\right) \leq ([\log_2\log_2\varepsilon^{-1}]+1)e^{-\frac{n\varepsilon}{2}}.$$

Example 1. Learning a concept from a VC-class. Consider the case of the concept learning, when $\mathcal{F} := \{I_C : C \in \mathcal{C}\}$. Given a sample $(X_1, \dots, X_n)$ with unknown common distribution $P \in \mathcal{P}(S)$, we observe the labels $\{Y_j := I_{C_0}(X_j) : 1 \le j \le n\}$ for an unknown target concept $C_0 \in \mathcal{C}$. An estimate $\hat{C}_n = \hat{C}_n((X_1, Y_1), \dots, (X_n, Y_n))$ of the target concept C_0 is called consistent iff $I_{\hat{C}_n}(X_j) = Y_j$ for all $j = 1, \dots, n$. Let

$$\Delta^{\mathcal{C}}(X_1, \dots, X_n) := \mathrm{card}\big(\{C \cap \{X_1, \dots, X_n\} : C \in \mathcal{C}\}\big).$$

Then

$$\hat{\psi}_n(r) := K(\log \Delta^{\mathcal{C}}(X_1, \dots, X_n))^{1/2} r$$

is an upper bound on the random entropy integral, which yields the value of $\hat{\hat{\delta}}_n$

$$\hat{\hat{\delta}}_n = K^2 \frac{\log \Delta^{\mathcal{C}}(X_1, \dots, X_n)}{n}.$$

Thus, with the same choice of N we get for some numerical constant $c > 0$ the bound

$$\mathbb{P}\left(\bar{r}_N^n \ge c\Big(\frac{\log \Delta^{\mathcal{C}}(X_1, \dots, X_n)}{n} \vee \varepsilon\Big)\right) \le ([\log_2 \log_2 \varepsilon^{-1}] + 1) e^{-\frac{n\varepsilon}{2}}.$$

Theorem 2 implies at the same time that for any consistent estimate $\hat{C}_n$ we have $P(\hat{C}_n \triangle C_0) \le \bar{r}_N^n$ with probability at least $1 - 2Ne^{-n\varepsilon/2}$. This shows that for a VC-class of concepts $\mathcal{C}$ with VC-dimension $V(\mathcal{C})$, the local Rademacher norm $\bar{r}_N^n$ (which, according to Theorem 2, is an upper bound on the risk of consistent concepts $\hat{C}_n$) is bounded from above by the quantity $O(V(\mathcal{C}) \log n / n)$. Up to a logarithmic factor, this is the optimal (in a minimax sense) convergence rate of the generalization error to 0 (see, e.g., [3]).

Next we consider the conditions in terms of entropy with bracketing $H_{[\,]}(\mathcal{F}, \varepsilon) := \log N_{[\,]}(\mathcal{F}, \varepsilon)$. Here $N_{[\,]}(\mathcal{F}, \varepsilon)$ denotes the minimal number of "brackets" $[f^-, f^+] := \{f : f^- \le f \le f^+\}$ with $d_{P,2}(f^-, f^+) \le \varepsilon$ (f^-, f^+ being two measurable functions from S into $[0, 1]$, such that $f^- \le f^+$). Let

$$\psi_{[\,]}(r) = \int_0^r \left(H_{[\,]}(\mathcal{F}, u) + 1\right)^{1/2} du,$$

and let $\delta_{[n]} = \delta_{[n]}(P)$ be the solution of the equation

$$\delta_{[n]} = n^{-1/2} \psi_{[\,]}(\sqrt{\delta_{[n]}}).$$

Again, for some $\varepsilon > 0$, we set $N := [\log_2 \log_2 \varepsilon^{-1}] + 1$. Then the following theorem holds.

Theorem 4. *There exists a constant $c > 0$ such that for all $P \in \mathcal{P}(S)$*

$$\mathbb{P}\left(\bar{r}_N^n \ge c(\delta_{[n]}(P) \vee \varepsilon)\right) \le ([\log_2 \log_2 \varepsilon^{-1}] + 1) e^{-\frac{n\varepsilon}{2}}.$$

In particular, if $H_{[\,]}(\mathcal{F};u) = O(u^{-\gamma})$, where $\gamma < 2$, then $\psi_{[\,]}(r) \asymp r^{1-\gamma/2}$ and $\delta_{[n]} \asymp n^{-\frac{2}{2+\gamma}}$.

Example 2. Learning a concept from a d-dimensional cube. Let $S = [0,1]^d$. We consider a problem of estimation of a set (a concept) $C_0 \subset [0,1]^d$, based on the observations (X_j, Y_j), $j = 1, \dots, n$, where X_j, $j = 1, \dots, n$ are i.i.d. points in $[0,1]^d$ with common distribution P and $Y_j := I_{C_0}(X_j)$, $j = 1, \dots, n$. Such a model frequently occurs in the problems of edge estimation in image analysis (see Mammen and Tsybakov (1995)). Assume that the distribution P has a density p such that for some $B > 0$

$$B^{-1} \le p(x) \le B, \; x \in [0,1]^d.$$

Let $\mathcal{C}$ be a class of Borel subsets in $[0,1]^d$ such that $\mathcal{C} \ni C_0$. Let λ be the Lebesgue measure on $[0,1]^d$. Denote $N_I(\mathcal{C};\varepsilon)$ the minimal number of brackets $[C^-, C^+] := \{C : C^- \subset C \subset C^+\}$ with $\lambda(C^+ \setminus C^-) \le \varepsilon$ (C^-, C^+ being two measurable subsets in $[0,1]^d$ such that $C^- \subset C^+$). Let $H_I(\mathcal{C};\varepsilon) := \log N_I(\mathcal{C};\varepsilon)$. This version of entropy with bracketing is often called "entropy with inclusion". We define

$$\psi_I(r) = \int_0^r (H_I(\mathcal{C},u) + 1)^{1/2}\,du,$$

and let $\delta_n^I = \delta_n^I(P)$ be the solution of the equation

$$\delta_n^I = n^{-1/2}\psi_I(\sqrt{\delta_n^I}).$$

If we have

$$H_I(\mathcal{C};u) = O(u^{-\gamma}),$$

then Theorem 4 easily implies that with some constant $c > 0$

$$\mathbb{P}\left(\bar{r}_N^n \ge c(\delta_n^I \vee \varepsilon)\right) \le ([\log_2 \log_2 \varepsilon^{-1}] + 1)e^{-\frac{n\varepsilon}{2}},$$

where $\delta_n^I \asymp n^{-\frac{1}{1+\gamma}}$. By Theorem 2, for any consistent estimate $\hat{C}_n$ of the set C_0 (i.e. such that $I_{\hat{C}_n}(X_j) = Y_j$, $j = 1, \dots, n$), the quantity $\bar{r}_N^n$ is an upper bound (up to a constant) on $\lambda(\hat{C}_n \triangle C_0)$.

In particular, if $\mathcal{C}$ is the class of sets with α-smooth boundary in $[0,1]^d$, then well-known bounds on the bracketing entropy due to Dudley (see e.g. Dudley (1999)) imply that $\gamma = \frac{d-1}{\alpha}$ and $\delta_n^I = n^{-\frac{\alpha}{d-1+\alpha}}$. Similarly, if $\mathcal{C}$ is the class of closed convex subsets of $[0,1]^d$, the rate becomes $\delta_n^I = n^{-\frac{2}{d+1}}$. It was shown by Mammen and Tsybakov (1995) that both rates are optimal in a minimax sense.

The examples above show that the local Rademacher penalties (based on the only data and using neither prior information about the underlying distribution, nor the specific properties of the function class) can recover the optimal convergence rates of the estimates in function learning problems.

2 Proofs of the main results

The proofs of the results are based on a version of Talagrand's concentration inequalities for empirical processes, see [11], [12]. The version of the inequalities we are using, with explicit numerical values of the constants involved (that determine the values of the constants in our procedures, such as $\bar{K}_1, \bar{K}_2, \bar{K}_3$ above) are due to Massart (1999). These inequalities are also very convenient for applications since the quantity σ^2 (the sup-norm of the variances, see below) they involve is very easy to bound. It should also be mentioned that the idea of using Talagrand's concentration inequalities to bound the risk in nonparametric estimation and, especially, in model selection problems goes back to Birgé and Massart (see [2], [1] and references therein).

We now formulate Massart's inequality in a form convenient for our purposes.

Theorem 5. *Let $\mathcal{F}$ be some countable family of real valued measurable functions, such that $\|f\|_\infty \le b < \infty$ for every $f \in \mathcal{F}$. Let Z denote either $\|P_n - P\|_{\mathcal{F}}$ or $\|R_n\|_{\mathcal{F}}$. Let $\sigma^2 = n \sup \mathrm{Var}(f(X_1))$. Then for any positive real number x and $0 < \gamma < 1$,*

$$\mathbb{P}(Z \ge (1+\gamma)\mathbb{E}Z + [\sigma\sqrt{2kx} + k(\gamma)bx]/n) \le e^{-x}, \tag{2.1}$$

where k and $k(\gamma)$ can be taken equal to $k = 4$ and $k(\gamma) = 3.5 + 32\gamma^{-1}$. Moreover,

$$\mathbb{P}(Z \le (1-\gamma)\mathbb{E}Z - [\sigma\sqrt{2k'x} - k'(\gamma)bx]/n) \le e^{-x}, \tag{2.2}$$

where $k' = 5.4$ and $k'(\gamma) = 3.5 + 43.2\gamma^{-1}$.

Proof of Theorem 2. For any fixed real positive number r, let

$$\varphi_1(r) = \|P_n - P\|_{\mathcal{F}\cap B(r)},$$

$$\varphi_2(r) = (1+\gamma)\mathbb{E}\|P_n - P\|_{\mathcal{F}\cap B(r)} + 2\sqrt{r\varepsilon} + (1.75 + 16\gamma^{-1})\varepsilon,$$

$$\begin{aligned}\varphi_3(r) = \frac{2(1+\gamma)}{1-\gamma'}\Big[\|R_n\|_{\mathcal{F}\cap B(r)} &+ \sqrt{5.4r\varepsilon} + (1.75 + 21.6\gamma'^{-1})\varepsilon\Big] \\ &+ 2\sqrt{r\varepsilon} + (1.75 + 16\gamma^{-1})\varepsilon.\end{aligned}$$

Then, for any $r > 0$,

$$\mathbb{P}\Big(\varphi_1(r) \le \varphi_2(r) \le \varphi_3(r)\Big) \ge 1 - 2e^{-\frac{n\varepsilon}{2}}. \tag{2.3}$$

Indeed, in order to apply inequalities (2.1) and (2.2), we notice that for every $f \in \mathcal{F}\bigcap B(r)$ the sup-norm $\|f\|_\infty \le b = 1$ and

$$\sigma^2 = \sup_{\mathcal{F}\cap B_r} n\mathrm{Var}(f(X)) \le \sup_{\mathcal{F}\cap B(r)} nPf^2 \le \sup_{\mathcal{F}\cap B(r)} nPf \le nr.$$

Moreover, if we set $x = n\varepsilon/2$, then (2.1) implies

$$\mathbb{P}\Big(\|P_n - P\|_{\mathcal{F}\cap B(r)} \geq (1+\gamma)\mathbb{E}\|P_n - P\|_{\mathcal{F}\cap B(r)} + 2\sqrt{r\varepsilon}$$
$$+(1.75 + 16\gamma^{-1})\varepsilon\Big) \leq e^{-\frac{n\varepsilon}{2}},$$

and (2.2) implies

$$\mathbb{P}\Big(\mathbb{E}\|R_n\|_{\mathcal{F}\cap B(r)} \geq (1-\gamma')^{-1}[\|R_n\|_{\mathcal{F}\cap B(r)} + \sqrt{5.4r\varepsilon}$$
$$+(1.75 + 21.6\gamma'^{-1})\varepsilon]\Big) \leq e^{-\frac{n\varepsilon}{2}}.$$

Taking into account the symmetrization inequality

$$\mathbb{E}\|P_n - P\|_{\mathcal{F}\cap B(r)} \leq 2\mathbb{E}\|R_n\|_{\mathcal{F}\cap B(r)},$$

we get (2.3).

We set

$$\bar{K}_1 := \frac{2(1+\gamma)}{1-\gamma'},\ \bar{K}_2 := \frac{2\sqrt{5.4}(1+\gamma)}{1-\gamma'} + 2,$$
$$\bar{K}_3 := \frac{2(1+\gamma)}{1-\gamma'}(1.75 + 21.6\gamma'^{-1}) + (1.75 + 16\gamma^{-1}).$$

Let us introduce the following sequence: $\hat{r}_0^n := 1$ and $\hat{r}_{k+1}^n = \varphi_2(\hat{r}_k^n) \wedge 1$ for $k = 0, 1, 2, \ldots$. Since φ_2 is nondecreasing, it's easy to prove by induction that the sequence $\{\hat{r}_k^n\}$ is nonincreasing.

We will also prove by induction that for all $k \geq 0$,

$$\mathbb{P}\Big\{r_i^n \leq \hat{r}_i^n \leq \bar{r}_i^n,\ i \leq k\Big\} \geq 1 - 2ke^{-\frac{n\varepsilon}{2}}. \tag{2.4}$$

For $k = 0$ (2.4) is trivial since $r_0^n = \hat{r}_0^n = \bar{r}_0^n = 1$. We proceed by the induction argument. Let us introduce the events

$$\mathcal{A}_k = \{r_i^n \leq \hat{r}_i^n \leq \bar{r}_i^n,\ i \leq k\} \text{ and } \mathcal{B}_k = \{\varphi_1(\hat{r}_k^n) \leq \varphi_2(\hat{r}_k^n) \leq \varphi_3(\hat{r}_k^n)\}.$$

To make the induction step, let us assume that we have already proven that

$$\mathbb{P}(\mathcal{A}_k) \geq 1 - 2ke^{-\frac{n\varepsilon}{2}}.$$

Then (2.3) implies

$$\mathbb{P}(\mathcal{B}_k) \geq 1 - 2e^{-\frac{n\varepsilon}{2}}.$$

On the event $\mathcal{A}_k \bigcap \mathcal{B}_k$,

$$\mathcal{F} \cap B(\hat{r}_k^n) \subseteq \mathcal{F} \cap B^e(2\hat{r}_k^n),$$

since for $f \in \mathcal{F} \bigcap B(\hat{r}_k^n)$

$$\begin{aligned} P_n f &\leq Pf + \|P_n - P\|_{\mathcal{F}\cap B(\hat{r}_k^n)} \leq \hat{r}_k^n + \|P_n - P\|_{\mathcal{F}\cap B(\hat{r}_k^n)} \\ &= \hat{r}_k^n + \varphi_1(\hat{r}_k^n) \leq \hat{r}_k^n + \varphi_2(\hat{r}_k^n) = \hat{r}_k^n + \hat{r}_{k+1}^n \leq 2\hat{r}_k^n, \end{aligned}$$

which implies that the inequalities $\varphi_3(\hat{r}_k^n) \le \bar{\varphi}(\hat{r}_k^n) \le \bar{\varphi}(\bar{r}_k^n) = \bar{r}_{k+1}^n$ hold. Therefore, on the event $\mathcal{A}_k \bigcap \mathcal{B}_k$,

$$r_{k+1}^n = \varphi_1(r_k^n) \le \varphi_1(\hat{r}_k^n) \le \varphi_2(\hat{r}_k^n) = \hat{r}_{k+1}^n \le \varphi_3(\hat{r}_k^n) \le \bar{r}_{k+1}^n.$$

So, $\mathcal{A}_k \bigcap \mathcal{B}_k \subseteq \mathcal{A}_{k+1}$, which completes the proof of the induction step

$$\mathbb{P}\left(\mathcal{A}_{k+1}\right) \ge 1 - 2(k+1)e^{-\frac{n\varepsilon}{2}}.$$

It follows that

$$\mathbb{P}(r_N^n > \bar{r}_N^n) \le 2Ne^{-\frac{n\varepsilon}{2}},$$

and since, by Proposition 1, $P\hat{f}_n \le r_N^n$, we conclude that

$$\mathbb{P}\{P\hat{f}_n > \bar{r}_N^n\} \le 2Ne^{-\frac{n\varepsilon}{2}}.$$

Proof of Theorem 3. Let $(\Omega_\varepsilon, \Sigma_\varepsilon, \mathbb{P}_\varepsilon)$ denote the probability space on which the Rademacher sequence $\varepsilon_1, \dots, \varepsilon_n, \dots$ is defined, $\mathbb{E}_\varepsilon$ being the expectation with respect to $\mathbb{P}_\varepsilon$. We introduce the function

$$\varphi_4(r) = \frac{2(1+\gamma)}{1-\gamma'}\Big[(1+\gamma''^{-1})\mathbb{E}_\varepsilon\|R_n\|_{\mathcal{F}\cap B^e(2r)} + 2\sqrt{r\varepsilon}$$
$$+(1.75+16\gamma''^{-1})\varepsilon + \sqrt{5.4r\varepsilon} + (1.75+21.6\gamma'^{-1})\varepsilon\Big]$$
$$+2\sqrt{r\varepsilon} + (1.75+16\gamma^{-1})\varepsilon, \tag{2.5}$$

where $\gamma'' > 0$. The inequalities (2.1) and (2.2) also hold for the conditional probability $\mathbb{P}_\varepsilon$ and the process $Z = R_n$ with fixed $X_1, \dots, X_n$. Therefore, for any $r > 0$,

$$\mathbb{P}_\varepsilon(\bar{\varphi}(r) \le \varphi_4(r)) \ge 1 - e^{-\frac{n\varepsilon}{2}}.$$

Define a sequence

$$\check{r}_0^n = \varphi_4(1), \ \ \check{r}_{k+1}^n = \varphi_4(\check{r}_k^n) \wedge 1, \ \ k = 0, 1, 2, \dots\ .$$

By an induction argument, similar to the one we used in the proof of Theorem 2, we get

$$\mathbb{P}_\varepsilon\Big(\bigcap_{i=1}^N \{\bar{r}_i^n \le \check{r}_i^n\}\Big) \ge 1 - Ne^{\frac{n\varepsilon}{2}}.$$

If we prove that $\check{r}_k^n \le a_k$ for a sequence a_k, independent of $\varepsilon_1, \dots, \varepsilon_n$, then the unconditional probability

$$\mathbb{P}\Big(\bigcap_{i=1}^N \{\bar{r}_i^n \le a_i\}\Big) \ge 1 - Ne^{\frac{n\varepsilon}{2}}.$$

By the assumption we have

$$\mathbb{E}_\varepsilon \|n^{-1} \sum_{i=1}^{n} \varepsilon_i \delta_{X_i}\|_{B^e(r)\cap\mathcal{F}} \le \hat{\psi}_n(\sqrt{r}). \tag{2.6}$$

Hence, we can choose $c \ge 1$, depending on the parameters $\gamma, \gamma', \gamma''$ in the definition (2.5) of the function φ_4, in such a way that

$$\breve{r}_{k+1}^n = \varphi_4(\breve{r}_k^n) \le c\left(\varepsilon + (\breve{r}_k^n\varepsilon)^{1/2} + n^{-1/2}\hat{\psi}_n\left(\sqrt{\breve{r}_k^n}\right)\right).$$

The above inequality implies by induction that the sequence

$$r_0 = 1,\ r_{k+1} = c\left(\varepsilon + (r_k\varepsilon)^{1/2} + n^{-1/2}\hat{\psi}_n\left(\sqrt{r_k}\right)\right) \wedge 1,$$

majorizes the sequence $\breve{r}_k^n$.

It is clear that in the case when $r_1 < 1$, the sequence r_k is decreasing and it converges to the solution δ of the equation

$$\delta = c\left(\varepsilon + (\delta\varepsilon)^{1/2} + n^{-1/2}\psi\left(\sqrt{\delta}\right)\right).$$

Let us study the behaviour of the difference $d_k := r_k - \delta$. Since the function $\hat{\psi}_n$ is concave, we have

$$\hat{\psi}_n'(\sqrt{\delta}) \le \hat{\psi}_n(\sqrt{\delta})/\sqrt{\delta}.$$

The definition of δ implies that

$$c\left(n^{-1/2}\hat{\psi}_n(\sqrt{\delta}) + \sqrt{\delta\varepsilon}\right) \le \delta.$$

Therefore

$$\begin{aligned} d_{k+1} &= r_{k+1} - \delta = c\left(n^{-1/2}\hat{\psi}_n(\sqrt{r_k}) - n^{-1/2}\hat{\psi}_n(\sqrt{\delta}) + \sqrt{r_k\varepsilon} - \sqrt{\delta\varepsilon}\right) \\ &\le c\left(n^{-1/2}\hat{\psi}_n'(\delta) + \sqrt{\varepsilon}\right)\sqrt{r_k - \delta} \le c\left(n^{-1/2}\hat{\psi}_n(\sqrt{\delta}) + \sqrt{\delta\varepsilon}\right)/\sqrt{\delta}\sqrt{d_k} \\ &\le \sqrt{\delta d_k}. \end{aligned}$$

We have proven that the sequence d_k satisfies the following inequality,

$$d_{k+1} \le \sqrt{\delta d_k},\ k \ge 0.$$

Now it is easy to show by induction that

$$d_N \le \delta^{2^{-1}+\ldots+2^{-N}} = \delta^{1-2^{-N}}.$$

Going back to the sequence r_k, we get that

$$r_N = \delta + d_N \le \delta\left(1 + \delta^{-2^{-N}}\right).$$

Since the definition of δ implies that $\delta^{-1} < \varepsilon^{-1}$, the choice of

$$N = \left[\log_2 \log_2 \varepsilon^{-1}\right] + 1$$

guarantees that $\delta^{-2^{-N}} \leq 2$ and, hence, $r_N \leq (1+2)\delta = 3\delta$. What remains to do in order to finish the proof of the theorem is to bound δ by the maximum of ε and the solution $\hat{\delta}_n$ of the equation $\hat{\delta}_n = n^{-1/2}\hat{\psi}_n(\sqrt{\hat{\delta}_n})$. Actually, we will prove that δ is bounded dy $\delta'' := (3c)^2\delta'$, where $\delta' = \left(\hat{\delta}_n \vee \varepsilon.\right)$ First of all let us notice that the fact that $\hat{\psi}_n$ is concave and $\hat{\psi}_n(0) = 0$ implies that for $c \geq 1$ $\hat{\psi}_n(cx) \leq c\hat{\psi}_n(x)$. Also note that, since $\delta' \geq \hat{\delta}_n$, the concavity of $\hat{\psi}_n$ and the definition of $\hat{\delta}_n$ imply

$$n^{-1/2}\hat{\psi}_n\left(\sqrt{\delta'}\right) \leq \frac{n^{-1/2}\hat{\psi}_n(\sqrt{\hat{\delta}_n})}{\sqrt{\hat{\delta}_n}}\sqrt{\delta'} = \sqrt{\hat{\delta}_n}\sqrt{\delta'} \leq \delta'.$$

Combining these properties, we get

$$\begin{aligned} &c\left(\varepsilon + (9c^2\delta'\varepsilon)^{1/2} + n^{-1/2}\hat{\psi}_n\left(3c\sqrt{\delta'}\right)\right) \\ &\leq c\left(2\sqrt{(3c)^2\delta'} + \delta'\right) \leq 9c^2\delta' = \delta''. \end{aligned}$$

With necessity it means that $\delta \leq \delta'' = 9c^2(\hat{\delta}_n \vee \varepsilon)$. And, hence, $\bar{r}_N^n \leq \delta'' \leq 27c^2(\hat{\delta}_n \vee \varepsilon)$.

The theorem is proved.

Proof of Theorem 4. In order to bound $\bar{r}_k$, we first construct the bound on $\|R_n\|_{\mathcal{F}\cap B^e(2\bar{r}_k)}$ in terms of $\mathbb{E}\|P_n - P\|_{\mathcal{F}\cap B(\check{r}_k)}$ for properly defined sequence $\check{r}_k$. Afterwards, the expectation can be majorized by the bracketing entropy integral. We will show that the sequence $\check{r}_k$ can be chosen as follows,

$$\check{r}_0 = 1, \;\; \check{r}_{k+1} = \left(\tilde{c}_1\mathbb{E}\|P_n - P\|_{\mathcal{F}\cap B(3\check{r}_k)} + \tilde{c}_2\sqrt{\varepsilon\check{r}_k} + \tilde{c}_3\right) \wedge 1,$$

for some large enough constants $\tilde{c}_1, \tilde{c}_2, \tilde{c}_3 > 0$. One can argue similarly to the proof of Theorem 3 to show that the following bound holds:

$$\mathbb{P}\Big(\bigcap_{k\leq i}\{\bar{r}_k \leq \check{r}_k\}\Big) \geq 1 - 2ie^{-\frac{n\varepsilon}{2}}. \tag{2.7}$$

We will prove an even stronger assertion that for the event

$$\mathcal{A}_i = \bigcap_{k\leq i}\Big(\{\bar{r}_k \leq \check{r}_k\} \cap \{\mathcal{F}\cap B^e(2\bar{r}_k) \subseteq \mathcal{F}\cap B(3\check{r}_k)\}\Big)$$

we have

$$\mathbb{P}(\mathcal{A}_i) \geq 1 - 2ie^{-\frac{n\varepsilon}{2}}. \tag{2.8}$$

Let us choose the constants $c'_1, c'_2, c'_3 > 0$ and $\tilde{c}_1, \tilde{c}_2, \tilde{c}_3 > 0$ in such a way that for the functions

$$\varphi_5(r) = \left(c'_1 \|P_n - P\|_{\mathcal{F}\cap B(r)} + c'_2\sqrt{\varepsilon r} + c'_3\varepsilon\right)$$

and

$$\varphi_6(r) = \left(\tilde{c}_1 \mathbb{E}\|P_n - P\|_{\mathcal{F}\cap B(r)} + \tilde{c}_2\sqrt{\varepsilon r} + \tilde{c}_3\varepsilon\right),$$

the inequalities of Massart (see Theorem 5) would imply that for any fixed $r > 0$

$$\varphi_3(r) \le \varphi_5(r) \le \varphi_6(r)$$

with probability at least $1 - 2e^{-\frac{n\varepsilon}{2}}$ (the function φ_3 was defined in the proof of Theorem 2). Clearly, we have $\check{r}_{k+1} = \varphi_6(\check{r}_k) \wedge 1$.

First observe that (2.8) holds for $i = 0$ (since $\bar{r}_0 = \check{r}_0 = 1$). Define

$$\mathcal{B}_i := \{\varphi_3(3\check{r}_i) \le \varphi_5(3\check{r}_i) \le \varphi_6(\check{r}_i)\}.$$

Then

$$\mathbb{P}(\mathcal{B}_i) \ge 1 - 2e^{-\frac{n\varepsilon}{2}}.$$

To make an induction step, we first of all notice that on the event $\mathcal{A}_i \cap \mathcal{B}_i$, we have

$$\bar{r}_{i+1} = \bar{\varphi}(\bar{r}_i) \wedge 1 \le \varphi_3(3\check{r}_i) \wedge 1 \le \varphi_5(3\check{r}_i) \wedge 1 \le \varphi_6(3\check{r}_i) \wedge 1 = \check{r}_{i+1}.$$

Also, on the event $\mathcal{A}_i \cap \mathcal{B}_i$, we have $\mathcal{F} \cap B^e(2\bar{r}_{i+1}) \subseteq \mathcal{F} \cap B(3\check{r}_{i+1})$. Indeed, if $f \in \mathcal{F} \cap B^e(2\bar{r}_{i+1})$, then

$$\begin{aligned} Pf &\le 2\bar{r}_{i+1} + \|P_n - P\|_{\mathcal{F}\cap B^e(2\bar{r}_{i+1})} \le 2\bar{r}_{i+1} + \|P_n - P\|_{\mathcal{F}\cap B^e(2\bar{r}_i)} \\ &\le 2\bar{r}_{i+1} + \|P_n - P\|_{\mathcal{F}\cap B(3\check{r}_i)} \le 2\bar{r}_{i+1} + \varphi_5(3\check{r}_i) \wedge 1 \\ &\le 2\bar{r}_{i+1} + \varphi_6(3\check{r}_i) \wedge 1 = 2\bar{r}_{i+1} + \check{r}_{i+1} \le 3\check{r}_{i+1} \end{aligned}$$

(to show that $\|P_n - P\|_{\mathcal{F}\cap B(3\check{r}_i)} \le \varphi_5(3\check{r}_i) \wedge 1$ we used the fact that the constant c'_1 in the definition of φ_5 is larger than 1). Thus, $\mathcal{A}_i \cap \mathcal{B}_i \subset \mathcal{A}_{i+1}$ and

$$\mathbb{P}(\mathcal{A}_{i+1}) \ge 1 - 2(i+1)e^{-\frac{n\varepsilon}{2}}.$$

The proof of the induction step and of the bounds (2.8) and (2.7) is complete.

To finish the proof of the theorem one has to bound $\mathbb{E}\|P_n - P\|_{\mathcal{F}\cap B(r)}$. Since for all $g \in \mathcal{F} \bigcap B(r)$ we have $\|g\|_{P,2} \le (Pg)^{1/2} \le \sqrt{r}$ and $|g| \le 1$, by Theorem 2.14.2 in [13]

$$\mathbb{E}\|P_n - P\|_{\mathcal{F}\cap B(r)} \le c\left(n^{-1/2}\psi_{[\,]}\left(\sqrt{r}\right) + I\{1 > \sqrt{n}a(\sqrt{r})\}\right),$$

where

$$a(\sqrt{r}) = \sqrt{r}/\sqrt{1 + H_{[\,]}(\mathcal{F}, \sqrt{r})}.$$

We can assume that $\check{r}_N \geq \delta_{[n]}$, otherwise, bound (2.7) immediately implies the assertion of the theorem. Therefore, $\check{r}_k \geq \delta_{[n]}$ for all $k \leq N$, which implies that $1 \leq \sqrt{n}a(\sqrt{3\check{r}_k})$. Indeed, using concavity of $\psi_{[\,]}$ and the definition of $\delta_{[n]}$, we have

$$\frac{\psi_{[\,]}(\sqrt{3\check{r}_k})}{\sqrt{3\check{r}_k}} \leq \frac{\psi_{[\,]}(\sqrt{\delta_{[n]}})}{\sqrt{\delta_{[n]}}} = \sqrt{n}\sqrt{\delta_{[n]}} \leq \sqrt{n}\sqrt{3\check{r}_k},$$

which implies

$$3\check{r}_k \geq n^{-1/2}\psi_{[\,]}\left(\sqrt{3\check{r}_k}\right) \geq n^{-1/2}\left(3\check{r}_k\right)^{1/2}\left(1 + H_{[\,]}\left(\mathcal{F}, \sqrt{3\check{r}_k}\right)\right)^{1/2}.$$

Hence, $1 \leq \sqrt{n}a(\sqrt{3\check{r}_k})$ and

$$\mathbb{E}\|P_n - P\|_{\mathcal{F}\cap B(3\check{r}_k)} \leq cn^{-1/2}\psi_{[\,]}\left(\sqrt{3\check{r}_k}\right).$$

Finally, with some constant $c > 0$

$$\check{r}_{k+1} \leq c\left(n^{-1/2}\psi_{[\,]}(\sqrt{3\check{r}_k}) + \varepsilon + \sqrt{\varepsilon\check{r}_k}\right).$$

The proof can be completed by the argument we used in Theorem 3.

Acknowledgement. The authors are thankful to Jon Wellner for pointing out the recent paper of Massart (1998).

References

[1] Barron, A., Birgé, L. and Massart, P. (1999) Risk Bounds for Model Selection via Penalization. *Probability Theory and Related Fields* **113**, 301–413.

[2] Birgé, L. and Massart, P. (1997) From Model Selection to Adaptive Estimation. In: *Festschrift for L. Le Cam. Research Papers in Probability and Statistics*, D. Pollard, E. Torgersen and G. Yang (Eds.), pp. 55–87. Springer-Verlag, New York.

[3] Devroye, L., Györfi, L. and Lugosi, G. (1996) *A probabilistic theory of pattern recognition*, Springer-Verlag, New York.

[4] Dudley, R.M. (1999) *Uniform Central Limit Theorems*, Cambridge University Press, Cambridge, UK.

[5] Hush, D. and Scovel, C. (1999) Posterior performance bounds for machine learning, Los Alamos National Laboratory, preprint.

[6] Koltchinskii, V. (1999) Rademacher penalties and structural risk minimization, preprint.

[7] Koltchinskii, V., Abdallah, C.T., Ariola, M., Dorato, P., Panchenko, D. (1999) Statistical Learning Control of Uncertain Systems: It is better than it seems. Preprint, UNM.

[8] Lozano, F. (1999) Model selection using Rademacher penalization, Proceedings of the ICSC Symposia on Neural Computation (NC'2000), 2000, Berlin, Germany.

[9] Massart, P. (1998) About the constants in Talagrand's concentration inequalities for empirical processes, Université Paris-Sud, preprint.

[10] Mammen, E. and Tsybakov, A. (1995) Asymptotical minimax recovery of sets with smooth boundaries, *Ann. Statist.* **23**, 502–524.

[11] Talagrand, M. (1996) A new look at independence. *Ann. Probab.* 24, 1-34.

[12] Talagrand, M. (1996) New concentration inequalities in product spaces, *Invent. Math.* **126**, 505–563.

[13] van der Vaart, A. and Wellner, J. (1996) *Weak convergence and Empirical Processes with Applications to Statistics*, Springer-Verlag, New York.

[14] Vapnik, V. (1998) *Statistical Learning Theory*, John Wiley & Sons, New York.

[15] Vidyasagar, M. (1997) *A theory of learning and generalization*, Springer-Verlag, New York.

Vladimir Koltchinskii and Dmitriy Panchenko
Department of Mathematics and Statistics
The University of New Mexico
Albuquerque NM 87131-1141
e-mail: `vlad@math.unm.edu`
e-mail: `panchenk@math.unm.edu`

PART 10

Statistics in a Multidimensional Setting

Bootstrapping Empirical Distributions under Auxiliary Information

Erich Haeusler and Christiane Plies

ABSTRACT Being the nonparametric maximum likelihood estimator, the classical empirical distribution function is the estimator of choice for a completely unknown distribution function. As shown by Qin and Lawless (1994), in the presence of some auxiliary information the nonparametric maximum likelihood estimator is a modified empirical distribution function. It puts random masses on the observations in order to take the available information into account. Zhang (1997) has proved a functional central limit theorem for this modified empirical distribution function. The centered Gaussian limit process in this fclt has a complicated covariance structure so that the result is not directly applicable in statistical problems, e.g. for the construction of confidence bands. It is shown here that the bootstrap is one possible remedy.

1 Introduction and main results

Let $(X_i)_{i\geq 1}$ be a sequence of independent and identically distributed random variables with common distribution function F. The nonparametric maximum likelihood estimator for estimating F on the basis of the first n of these random variables is the classical empirical distribution function F_n which has been studied extensively.

Suppose now that auxiliary information about F is available in the sense that for known functions $g_1(x), \ldots, g_r(x)$ we have

$$E(g(X_1)) = 0, \tag{1.1}$$

where $g(x) = (g_1(x), \ldots, g_r(x))$. Examples are a known mean μ of X_1, where $g(x) = x - \mu$, a known median m, where $g(x) = 1_{(-\infty,m]}(x) - \frac{1}{2}$ provided that F is continuous, or a known mean μ and known variance σ^2, where $g(x) = (x - \mu, x^2 - \sigma^2 - \mu^2)$.

For model (1.1) (extended by an additional parameter θ which for simplicity we will not consider here), using ideas from the concept of empirical likelihood as developed by Owen (1988, 1990, 1991), Qin and Lawless (1994) have derived the nonparametric maximum likelihood estimator $\hat{F}_n$ for F. It turns out that $\hat{F}_n$ is a distribution function which puts random masses on the observed data $x_1, \ldots, x_n$ from $X_1, \ldots, X_n$. These masses are obtained by maximizing the likelihood function

$$L(F) = \prod_{i=1}^{n} dF(x_i) = \prod_{i=1}^{n} p_i,$$

where $p_i = dF(x_i) = P(X_i = x_i)$, subject to the restrictions

$$p_i \geq 0, \qquad \sum_{i=1}^{n} p_i = 1, \qquad \sum_{i=1}^{n} p_i g(x_i) = 0.$$

Note that the information provided by the model (1.1) is taken into account through the condition $\sum_{i=1}^{n} p_i g(x_i) = 0$.

If 0 is an interior point of the convex hull of the points $g(x_1), \ldots, g(x_n)$, then a unique maximizer $(\hat{p}_{n,1}, \ldots, \hat{p}_{n,n})$ of $L(F)$ with $0 < \hat{p}_{n,i} < 1$ for $i = 1, \ldots, n$ exists and can be found by the method of Lagrange multipliers; see Owen (1990), p. 100, and Qin and Lawless (1994), p. 304/305. We have

$$\hat{p}_{n,i} = \frac{1}{n(1 + \eta^t g(x_i))}, \quad 1 \leq i \leq n, \tag{1.2}$$

where the superscript t denotes transposition of the r-dimensional Lagrange multiplier η. This Lagrange multiplier η itself is a solution of the equation

$$\sum_{i=1}^{n} \frac{g(x_i)}{1 + \eta^t g(x_i)} = 0 \tag{1.3}$$

and because of $\hat{p}_{n,i} < 1$ belongs to the open convex set

$$M_n = \{\eta \in \mathbb{R}^r : 1 + \eta^t g(x_i) > \frac{1}{n} \quad \text{for} \quad i = 1, \ldots, n\}.$$

Moreover, equation (1.3) has exactly one solution in M_n provided that the $r \times r$–matrix $\sum_{i=1}^{n} g(x_i) g(x_i)^t$ is positive definite; see Owen (1990), p. 105, or use the mean value theorem for a direct proof. Notice that $\sum_{i=1}^{n} g(x_i) g(x_i)^t$ is always non-negative definite by construction.

Consequently, under model (1.1) the appropriate empirical distribution function for the observed data $x_1, \ldots, x_n$ is defined by

$$\hat{F}_n(x) = \sum_{i=1}^{n} \hat{p}_{i,n} 1_{\{x_i \leq x\}}, \quad -\infty < x < \infty, \tag{1.4}$$

where $\hat{p}_{n,i}$ is given by (1.2) with η being the unique solution of (1.3) in M_n.

For $\hat{F}_n$ to be a reasonable estimator it has to be defined by (1.2)–(1.4) at least with probability converging to one as n goes to infinity. For this, the regularity condition

$$\sigma = E(g(X_1) g(X_1)^t) \quad \text{is positive definite} \tag{1.5}$$

is sufficient. Under (1.1) and (1.5) the proof of Lemma 2 in Owen (1990) adapted to the present situation gives as $n \to \infty$,

$$P(0 \text{ is an interior point of the convex hull of } \{g(X_1), \dots, g(X_n)\}) \to 1,$$

whereas the law of large numbers implies

$$P(\sum_{i=1}^{n} g(X_i)g(X_i)^t \text{ is positive definite}) \to 1.$$

Hence with probability converging to one we observe samples $x_1, \dots, x_n$ for which $\hat{F}_n$ is well-defined through (1.2)–(1.4). From now on we will drop the distinction between the observed data $x_1, \dots, x_n$ and the random variables $X_1, \dots, X_n$ and will always assume that $\hat{F}_n$ is well-defined for $X_1, \dots, X_n$.

According to Theorem 3 of Qin and Lawless (1994) the modified empirical distribution function $\hat{F}_n$ is an efficient estimator for F under the model (1.1). In his Theorem 3.1, Zhang (1997) established consistency in the extended model including the parameter θ. In our model his result gives that under (1.5) and

$$E(\|g(X_1)\|^3) < \infty, \tag{1.6}$$

where $\| \ \|$ denotes the Euclidean norm, we have in probability as $n \to \infty$,

$$\sup_{-\infty < x < \infty} |\hat{F}_n(x) - F(x)| \to 0. \tag{1.7}$$

Also, the following functional central limit theorem for $\hat{F}_n$ is contained in Theorem 3.3 of Zhang (1997): Under (1.5) and (1.6) we have as $n \to \infty$

$$n^{1/2}(\hat{F}_n - F) \xrightarrow{\mathcal{L}} W \quad \text{in} \quad D[-\infty, \infty], \tag{1.8}$$

where $D[-\infty, \infty]$ is the Skorohod space over the compact time interval $[-\infty, \infty]$, where $\xrightarrow{\mathcal{L}}$ denotes convergence in distribution and W is a centered Gaussian process with covariance function

$$E(W(x)W(y)) = F(\min(x, y)) - F(x)F(y) - U(x)^t \sigma^{-1} U(y),$$

where $U(x) = E(g(X_1)1_{\{X_1 \le x\}})$. This result is not immediately applicable in statistical problems, e.g. for the construction of confidence bands for F, since the limit process depends heavily on the unknown F through the quantities σ and U. The dependence on F does not disappear if the Kolmogorov-Smirnov statistics are considered, in contrast to the classical case, where these statistics are distribution-free even for finite sample sizes as long as the underlying F is continuous. It is the aim of the present paper to show that the bootstrap is one possible remedy here, provided that F is continuous, which will always be assumed from now on.

The bootstrap will, of course, be based on resampling from $\hat{F}_n$. For this, let $X^*_{n,1}, \ldots, X^*_{n,n}$ be independent and identically distributed random variables with common distribution function $\hat{F}_n$, given $X_1, \ldots, X_n$. Then by definition of $\hat{F}_n$,

$$E(g(X^*_{n,1})|X_1, \ldots, X_n) = \sum_{i=1}^{n} \hat{p}_{n,i} g(X_i) = 0$$

so that the bootstrap random variables $X^*_{n,1}, \ldots, X^*_{n,n}$ also satisfy the basic equation (1.1). Hence we can define the bootstrap version $\hat{F}^*_n$ of $\hat{F}_n$ by

$$\hat{F}^*_n(x) = \sum_{i=1}^{n} \frac{1}{n(1 + \eta^{*t} g(X^*_{n,i}))} 1_{\{X^*_{n,i} \leq x\}}, \quad -\infty < x < \infty,$$

where η^* is the solution of the equation

$$\sum_{i=1}^{n} \frac{g(X^*_{n,i})}{1 + \eta^{*t} g(X^*_{n,i})} = 0$$

in the set

$$M^*_n = \{\eta^* \in \mathbb{R}^r : 1 + \eta^{*t} g(X^*_{n,i}) > \frac{1}{n} \quad \text{for} \quad i = 1, \ldots, n\}.$$

Our main result is the following "in probability" functional central limit theorem for this bootstrap procedure.

Theorem 1. *For continuous F, under model (1.1) and the assumptions (1.5) and (1.6), as $n \to \infty$ in probability,*

$$n^{1/2}(\hat{F}^*_n - \hat{F}_n) \xrightarrow{\mathcal{L}} W \quad \text{in} \quad D[-\infty, \infty]$$

under the conditional probability $P(\cdot|X_1, \ldots, X_n)$, where the limit process W is the same as in (1.8).

As usual, let $\| \ \|_\infty$ denote the supremum norm. Since W has continuous sample paths, it follows from Ylvisaker (1968) that $\|W\|_\infty$ has a continuous distribution function. Therefore, Theorem 1 in conjunction with (1.8) implies by a well-known argument the following consistency result for the Kolmogorov-Smirnov statistics.

Corollary 1. *Under the conditions of Theorem 1,*

$$P(n^{1/2}\|\hat{F}^*_n - \hat{F}_n\|_\infty \leq x|X_1, \ldots, X_n) - P(n^{1/2}\|\hat{F}_n - F\|_\infty \leq x) \to 0$$

as $n \to \infty$ in probability, uniformly in $-\infty < x < \infty$.

For the construction of asymptotic confidence bands for the distribution function F, let $H_n^*(u)$, $0 < u < 1$, denote the random quantile function of the random distribution function

$$P(n^{1/2}\|\hat{F}_n^* - \hat{F}_n\|_\infty \leq x|X_1, \dots, X_n), \quad -\infty < x < \infty.$$

From (1.8) and Theorem 1 we obtain by usual arguments

Corollary 2. *Under the conditions of Theorem 1, for all* $0 < u < 1$ *as* $n \to \infty$

$$P(n^{1/2}\|\hat{F}_n - F\|_\infty \leq H_n^*(u)) \to u.$$

According to this corollary the intervals

$$\text{(1.9)} \qquad [\hat{F}_n(x) - \frac{H_n^*(1-\alpha)}{n^{1/2}}, \hat{F}_n(x) + \frac{H_n^*(1-\alpha)}{n^{1/2}}], \quad -\infty < x < \infty,$$

form a simultaneous confidence band for F of asymptotic coverage level $1 - \alpha$.

2 Proof of Theorem 1

The proof of Theorem 1 is a "bootstrap version" of the proof of (1.8) in Zhang (1997). Setting

$$\sigma_n = E(g(X_{n,1}^*)g(X_{n,1}^*)^t|X_1 \dots, X_n)$$

and noting that the matrix σ_n is symmetric and nonnegative definite, we write on the event $\{\det(\sigma_n) > 0\}$, motivated by Theorem 3.2 in Zhang (1997), for $-\infty < x < \infty$,

$$\hat{F}_n^*(x) - \hat{F}_n(x) = \frac{1}{n}\sum_{i=1}^n Y_{n,i}^*(x) + R_n^*(x),$$

where

$$Y_{n,i}^*(x) = 1_{\{X_{n,i}^* \leq x\}} - \hat{F}_n(x) - U_n(x)^t\sigma_n^{-1}g(X_{n,i}^*)$$

and

$$U_n(x) = E(g(X_{n,1}^*)1_{\{X_{n,1}^* \leq x\}}|X_1, \dots, X_n).$$

Then the proof consists in showing that as $n \to \infty$ in probability,

$$\text{(2.1)} \qquad n^{-1/2}\sum_{i=1}^n Y_{n,i}^* \xrightarrow{\mathcal{L}} W \quad \text{in } D([-\infty, \infty]) \quad \text{under} \quad P(\cdot|X_1, \dots, X_n)$$

and

$$\text{(2.2)} \qquad \sup_{-\infty < x < \infty} |R_n^*(x)| = o_P(n^{-1/2}) \quad \text{under} \quad P(\cdot|X_1, \dots, X_n).$$

The convergence in law in (2.1) is shown by proving convergence of the finite dimensional distributions and tightness. We will consider convergence of the finite dimensional distributions first. For all $-\infty < x_1 < \cdots < x_k < \infty$ and $z_1, \ldots, z_k \in \mathbb{R}$ with $z_{(m)} = \max_{1 \le j \le k} |z_j| > 0$ we have to show, by the Cramér-Wold device, that in probability as $n \to \infty$,

$$\sum_{j=1}^{k} z_j n^{-1/2} \sum_{i=1}^{n} Y^*_{n,i}(x_j) \xrightarrow{\mathcal{L}} \sum_{j=1}^{k} z_j W(x_j). \tag{2.3}$$

Setting, for $1 \le i \le n$ and $n \ge 1$,

$$Z^*_{n,i} = n^{-1/2} \sum_{j=1}^{k} z_j Y^*_{n,i}(x_j)$$

and noting that $\sum_{j=1}^{k} z_j W(x_j)$ is centered normal with variance

$$\sum_{j,l=1}^{k} z_j z_l \mathrm{cov}(W(x_j), W(x_l)) = \sum_{j,l=1}^{k} z_j z_l E(W(x_j)W(x_l)) = V,$$

say, (2.3) will follow from the central limit theorem, provided that for the triangular array $Z^*_{n,i}, 1 \le i \le n,\ n \ge 1$, we can verify Lindeberg's condition and

$$\sum_{i=1}^{n} \mathrm{Var}(Z^*_{n,i} | X_1, \ldots, X_n) \to V \tag{2.4}$$

in probability as $n \to \infty$.

For the proof of the variance condition (2.4) note that

$$\sum_{i=1}^{n} \mathrm{Var}(Z^*_{n,i} | X_1, \ldots, X_n) = \sum_{j,l=1}^{k} z_j z_l E(Y^*_{n,1}(x_j) Y^*_{n,1}(x_l) | X_1, \ldots, X_n)$$

because under $P(\cdot | X_1, \ldots, X_n)$ the random variables $Y^*_{n,1}(x_j), \ldots, Y^*_{n,n}(x_j)$ are centered and the variables $Y^*_{n,1}(x_j)Y^*_{n,1}(x_l), \ldots, Y^*_{n,n}(x_j)Y^*_{n,n}(x_l)$ are independent and identically distributed, for fixed $1 \le j, l \le k$. Clearly,

$$\begin{aligned} & E(Y^*_{n,1}(x_j) Y^*_{n,1}(x_l) | X_1, \ldots, X_n) \\ = \ & \hat{F}_n(\min(x_j, x_l)) - \hat{F}_n(x_j)\hat{F}_n(x_l) - U_n(x_j)^t \sigma_n^{-1} U_n(x_l) = \Gamma_n(j, l), \end{aligned}$$

say, and it remains to show $\Gamma_n(j,l) \to E(W(x_j)W(x_l))$ as $n \to \infty$ in probability. Now

$$\begin{aligned} & |\Gamma_n(j,l) - E(W(x_j)W(x_l))| \\ \le \ & |\hat{F}_n(\min(x_j, x_l)) - F(\min(x_j, x_l))| + |\hat{F}_n(x_j)\hat{F}_n(x_l) - F(x_j)F(x_l)| \\ & + |U_n(x_j)^t \sigma_n^{-1} U_n(x_l) - U(x_j)^t \sigma^{-1} U(x_l)| = V_{n,1} + V_{n,2} + V_{n,3}, \end{aligned}$$

say, where the first two summands on the right-hand side converge to zero by consistency (1.7) of $\hat{F}_n$. For the third summand we obtain

$$\begin{aligned} V_{n,3} \leq & \|U_n(x_j) - U(x_j)\| \, \|\sigma_n^{-1}\| \, \|U_n(x_l)\| \\ & + \|U(x_j)\| \, \|\sigma_n^{-1}\| \, \|U_n(x_l) - U(x_l)\| \\ & + \|U(x_j)\| \, \|\sigma_n^{-1} - \sigma^{-1}\| \, \|U(x_l)\| \end{aligned}$$

so that it is enough to verify in probability as $n \to \infty$,

$$\sigma_n \to \sigma, \tag{2.5}$$

which gives $\sigma_n^{-1} \to \sigma^{-1}$ because σ is regular by assumption (1.5), and for all $-\infty < x < \infty$,

$$U_n(x) \to U(x). \tag{2.6}$$

We consider (2.5) first. Note that

$$E(g(X_{n,1}^*)g(X_{n,1}^*)^t | X_1, \ldots, X_n) = \sum_{i=1}^n \hat{p}_{n,i} g(X_i) g(X_i)^t$$

so that

$$\sigma_n - \sigma = \sum_{i=1}^n (\hat{p}_{n,i} - \frac{1}{n}) g(X_i) g(X_i)^t + \frac{1}{n} \sum_{i=1}^n g(X_i) g(X_i)^t - E(g(X_1) g(X_1)^t).$$

Clearly, the second and third summands on the right-hand side combined converge to zero by the law of large numbers, in view of (1.6). The norm of the first one equals

$$\begin{aligned} & \|\frac{1}{n} \sum_{i=1}^n \frac{\eta^t g(X_i)}{1 + \eta^t g(X_i)} g(X_i) g(X_i)^t \| \\ \leq & \max_{1 \leq i \leq n} |\eta^t g(X_i)| \max_{1 \leq i \leq n} \frac{1}{|1 + \eta^t g(X_i)|} \frac{1}{n} \sum_{i=1}^n \|g(X_i) g(X_i)^t\|. \end{aligned} \tag{2.7}$$

Here the first factor on the right-hand side converges to zero in probability, see formula (2.15) in Owen (1990) and Lemma 2.1 in Zhang (1997). Consequently,

$$\max_{1 \leq i \leq n} \frac{1}{|1 + \eta^t g(X_i)|} = O_P(1). \tag{2.8}$$

Clearly, by the law of large numbers, the third factor on the right-hand side of (2.7) is also bounded in probability. Thus, the whole product converges to zero in probability, which proves (2.5). For the proof of (2.6) we write

$$\begin{aligned}\|U_n(x) - U(x)\| &\le \|\sum_{i=1}^n (\hat{p}_{n,i} - \frac{1}{n}) g(X_i) 1_{\{X_i \le x\}}\| \\ &+ \|\frac{1}{n}\sum_{i=1}^n g(X_i) 1_{\{X_i \le x\}} - E(g(X_1)1_{\{X_1 \le x\}})\|.\end{aligned}$$

Here the second summand on the right-hand side converges to zero in probability by the law of large numbers, whereas the first one is bounded by

$$\max_{1\le i\le n} |\eta^t g(X_i)| \max_{1\le i\le n} \frac{1}{|1+\eta^t g(X_i)|} \frac{1}{n} \sum_{i=1}^n \|g(X_i)\|$$

which converges to zero in probability for the same reasons which applied to the right-hand side in (2.7). This completes the proof of (2.6) and hence of (2.4).

We now turn to the proof of Lindeberg's condition for the array $Z^*_{n,i}$, $1 \le i \le n$, $n \ge 1$. Since the random variables $Z^*_{n,i}$ are centered under $P(\cdot|X_1, \ldots, X_n)$, we have to show that for every $\varepsilon > 0$,

$$L_n(\varepsilon) = \sum_{i=1}^n E(Z^{*2}_{n,i} I(|Z^*_{n,i}| \ge \varepsilon)|X_1, \ldots, X_n) \to 0$$

in probability as $n \to \infty$. Using the crude bounds

$$Z^{*2}_{n,i} \le \frac{k^2}{n} \sum_{j=1}^k z_j^2 Y^*_{n,i}(x_j)^2$$

and

$$I(|Z^*_{n,i}| \ge \varepsilon) \le \sum_{l=1}^k I(|Y^*_{n,i}(x_l)| \ge \varepsilon_1 n^{1/2}),$$

where $\varepsilon_1 = \varepsilon k^{-1} z^{-1}_{(m)}$, we obtain the inequality

$$L_n(\varepsilon) \le k^2 \sum_{j,l=1}^k z_j^2 \frac{1}{n} \sum_{i=1}^n E(Y^*_{n,i}(x_j)^2 I(|Y^*_{n,i}(x_l)| \ge \varepsilon_1 n^{1/2})|X_1, \ldots, X_n).$$

Hence it is enough to show that, for fixed $1 \le j, l \le k$,

$$\frac{1}{n} \sum_{i=1}^n E(Y^*_{n,i}(x_j)^2 I(|Y^*_{n,i}(x_l)| \ge \varepsilon_1 n^{1/2})|X_1, \ldots, X_n) \to 0$$

in probability as $n \to \infty$. For all $1 \le j, l \le k$ the bivariate random variables $(Y^*_{n,1}(x_j), Y^*_{n,1}(x_l)), \ldots, (Y^*_{n,n}(x_j), Y^*_{n,n}(x_l))$ are independent and identically distributed under $P(\cdot|X_1,\ldots,X_n)$. Hence

$$\begin{aligned}
&\frac{1}{n}\sum_{i=1}^{n} E(Y_{n,i}^*(x_j)^2 I(|Y_{n,i}^*(x_l)| \geq \varepsilon_1 n^{1/2})|X_1,\ldots,X_n) \\
=\;& E(Y_{n,1}^*(x_j)^2 I(|Y_{n,1}^*(x_l)| \geq \varepsilon_1 n^{1/2})|X_1,\ldots,X_n) \\
\leq\;& 2(I_n + II_n + III_n + IV_n),
\end{aligned}$$

where

$$\begin{aligned}
I_n &= E(|1_{\{X_{n,1}^*\leq x_j\}} - \hat{F}_n(x_j)|^2 \\
&\quad \times I(|1_{\{X_{n,1}^*\leq x_l\}} - \hat{F}_n(x_l)| \geq \varepsilon_1 n^{1/2}/2)|X_1,\ldots,X_n), \\
II_n &= E(|1_{\{X_{n,1}^*\leq x_j\}} - \hat{F}_n(x_j)|^2 \\
&\quad \times I(|U_n(x_l)^t\sigma_n^{-1} g(X_{n,1}^*)| \geq \varepsilon_1 n^{1/2}/2)|X_1,\ldots,X_n), \\
III_n &= E(|U_n(x_j)^t\sigma_n^{-1} g(X_{n,1}^*)|^2 \\
&\quad \times I(|1_{\{X_{n,1}^*\leq x_l\}} - \hat{F}_n(x_l)| \geq \varepsilon_1 n^{1/2}/2)|X_1,\ldots,X_n), \\
IV_n &= E(|U_n(x_j)^t\sigma_n^{-1} g(X_{n,1}^*)|^2 \\
&\quad \times I(|U_n(x_l)^t\sigma_n^{-1} g(X_{n,1}^*)| \geq \varepsilon_1 n^{1/2}/2)|X_1,\ldots,X_n).
\end{aligned}$$

Clearly, $I_n = III_n = 0$ for all sufficiently large n. Using the trivial bound

$$|1\{X_{n,1}^* \leq x_j\} - \hat{F}_n(x_j)|^2 \leq 1$$

and Markov's inequality, we get

$$\begin{aligned}
II_n &\leq \frac{4}{\varepsilon_1^2 n} E(|U_n(x_l)^t\sigma_n^{-1} g(X_{n,1}^*)|^2|X_1,\ldots,X_n) \\
&\leq \frac{4}{\varepsilon_1^2 n}\|U_n(x_l)^t\sigma_n^{-1}\|^2 E(\|g(X_{n,1}^*)\|^2|X_1,\ldots,X_n),
\end{aligned}$$

since $U_n(x_l)^t\sigma_n^{-1}$ is measurable w.r.t. the σ-field $\sigma(X_1,\ldots,X_n)$ generated by $X_1,\ldots,X_n$. Now, $\|U_n(x_l)^t\sigma_n^{-1}\|^2 = O_P(1)$ by (2.5) and (2.6), and

$$E(\|g(X_{n,1}^*)\|^2|X_1,\ldots,X_n) = \sum_{i=1}^{n}\hat{p}_{n,i}\|g(X_i)\|^2 = O_P(1), \tag{2.9}$$

because by definition of $\hat{p}_{n,i}$ and (2.8),

$$\max_{1\leq i\leq n}\hat{p}_{n,i} = O_P(\frac{1}{n})$$

whereas assumption (1.6) implies

$$\frac{1}{n}\sum_{i=1}^{n}\|g(X_i)\|^2 = O_P(1).$$

Thus we see that II_n converges to zero in probability. Finally, by Hölder's and Markov's inequality we obtain

$$\begin{aligned} IV_n &\le E(|U_n(x_j)^t\sigma_n^{-1}g(X_{n,1}^*)|^3|X_1,\dots,X_n)^{2/3} \\ &\quad \times P(|U_n(x_l)^t\sigma_n^{-1}g(X_{n,1}^*)| \ge \varepsilon_1 n^{1/2}/2|X_1,\dots,X_n)^{1/3} \\ &\le 2\varepsilon_1^{-1}n^{-1/2}E(|U_n(x_j)^t\sigma_n^{-1}g(X_{n,1}^*)|^3|X_1,\dots,X_n)^{2/3} \\ &\quad \times E(|U_n(x_l)^t\sigma_n^{-1}g(X_{n,1}^*)|^3|X_1,\dots,X_n)^{1/3} \\ &\le 2\varepsilon_1^{-1}n^{-1/2}\|U_n(x_j)^t\sigma_n^{-1}\|^2\|U_n(x_l)^t\sigma_n^{-1}\| \\ &\quad \times E(\|g(X_{n,1}^*)\|^3|X_1,\dots,X_n). \end{aligned}$$

As above, $\|U_n(x_j)^t\sigma_n^{-1}\|^2$ and $\|U_n(x_l)^t\sigma_n^{-1}\|$ are bounded in probability. Moreover, using assumption (1.6) and the argument employed above to establish (2.9), we obtain

$$E(\|g(X_{n,1}^*)\|^3|X_1,\dots,X_n) = \sum_{i=1}^n \hat{p}_{n,i}\|g(X_i)\|^3 = O_P(1).$$

Thus we see that IV_n converges to zero in probability, too. This concludes the proof of Lindeberg's condition.

Finally, we give the tightness proof for the processes $n^{-1/2}\sum_{i=1}^n Y_{n,i}^*$. Since we want to show our bootstrap fclt in probability, we have to show for all $\varepsilon_1, \varepsilon_2 > 0$ that

$$\lim_{\delta\downarrow 0}\limsup_{n\to\infty} P(P(\omega_\infty(n^{-1/2}\sum_{i=1}^n Y_{n,i}^*,\delta) \ge \varepsilon_1|X_1,\dots,X_n) \ge \varepsilon_2) = 0, \tag{2.10}$$

where the appropriate modulus of continuity $\omega_\infty(f,\delta)$ for functions f defined on $[-\infty,\infty]$ and $\delta > 0$ is given by

$$\begin{aligned} \omega_\infty(f,\delta) &= \sup\{|f(x)-f(y)| : -\infty \le x,y \le \infty,\ \rho(x,y) \le \delta\} \\ &= \sup_{\rho(x,y)\le\delta} |f(x)-f(y)|, \end{aligned}$$

for short, with ρ being a metric for the compact real line $[-\infty,\infty]$. For the proof of (2.10) we write

$$\begin{aligned} n^{-1/2}\sum_{i=1}^n Y_{n,1}^*(x) &= n^{-1/2}\sum_{i=1}^n (1_{\{X_{n,i}^*\le x\}} - \hat{F}_n(x)) \\ &\quad - U_n(x)^t\sigma_n^{-1}n^{-1/2}\sum_{i=1}^n g(X_{n,i}^*) \\ &= Q_n(x) - T_n(x), \end{aligned}$$

say, and note that

$$\omega_\infty(n^{-1/2}\sum_{i=1}^n Y^*_{n,i}, \delta) \le \omega_\infty(Q_n, \delta) + \omega_\infty(T_n, \delta),$$

so that we can treat Q_n and T_n separately. Here Q_n is the classical empirical process for the distribution function $\hat{F}_n$. Therefore, we introduce the uniform empirical distribution function G_n based on random variables which are independent of the sequence $(X_i)_{i\ge 1}$. Then under $P(\cdot|X_1,\ldots,X_n)$,

$$Q_n \stackrel{\mathcal{L}}{=} n^{-1/2}(G_n(\hat{F}_n(x)) - \hat{F}_n(x))_{-\infty\le x\le\infty},$$

where $\stackrel{\mathcal{L}}{=}$ denotes equality in law. Writing $\alpha_n(t) = n^{1/2}(G_n(t) - t)$ for the uniform empirical process pertaining to G_n, under $P(\cdot|X_1,\ldots,X_n)$ we obtain

$$\begin{aligned} \omega_\infty(Q_n, \delta) &\stackrel{\mathcal{L}}{=} \omega_\infty(\alpha_n(\hat{F}_n), \delta) \\ &\le \omega_\infty(\alpha_n(F), \delta) + 2 \sup_{-\infty\le x\le\infty} |\alpha_n(\hat{F}_n(x)) - \alpha_n(F(x))|. \end{aligned}$$

By tightness of α_n we get for all $\varepsilon_1 > 0$,

$$\lim_{\delta\downarrow 0}\limsup_{n\to\infty} P(\omega_\infty(\alpha_n(F), \delta) \ge \varepsilon_1) = 0\,, \tag{2.11}$$

which by independence of α_n and $X_1,\ldots,X_n$ gives

$$\lim_{\delta\downarrow 0}\limsup_{n\to\infty} P(P(\omega_\infty(\alpha_n(F), \delta) \ge \varepsilon_1|X_1,\ldots,X_n) \ge \varepsilon_2) = 0.$$

Moreover, with

$$\omega(f,\delta) = \sup\{|f(x) - f(y)| : 0 \le x, y \le 1,\ |x-y| \le \delta\}$$

denoting the usual modulus of continuity for functions f defined on the unit interval and $\delta > 0$, for all $\varepsilon_1, \varepsilon_2, \theta > 0$ by independence of α_n and $X_1,\ldots,X_n$ and (1.7),

$$\begin{aligned} &P(P(\sup_{-\infty\le x\le\infty} |\alpha_n(\hat{F}_n(x)) - \alpha_n(F(x))| \ge \varepsilon_1|X_1,\ldots,X_n) \ge \varepsilon_2) \\ \le\ & P(1_{\{\omega(\alpha_n,\theta)\ge\varepsilon_1\}} \ge \varepsilon_2) + P(\sup_{-\infty\le x\le\infty} |\hat{F}_n(x) - F(x)| > \theta) \\ \le\ & P(\omega(\alpha_n,\theta) \ge \varepsilon_1) + o(1). \end{aligned}$$

Tightness of α_n now gives, as $n\to\infty$, for all $\varepsilon_1, \varepsilon_2 > 0$,

$$P(P(\sup_{-\infty\le x\le\infty} |\alpha_n(\hat{F}_n(x)) - \alpha_n(F(x))| \ge \varepsilon_1|X_1,\ldots,X_n) \ge \varepsilon_2) \to 0,$$

which when combined with (2.11) yields

$$\lim_{\delta \downarrow 0} \limsup_{n \to \infty} P(P(\omega_\infty(Q_n, \delta) \geq \varepsilon_1 | X_1, \ldots, X_n) \geq \varepsilon_2) = 0.$$

This proves tightness of Q_n. Therefore, it remains to show that for all $\varepsilon_1, \varepsilon_2 > 0$,

$$\lim_{\delta \downarrow 0} \limsup_{n \to \infty} P(P(\omega_\infty(T_n, \delta) \geq \varepsilon_1 | X_1, \ldots, X_n) \geq \varepsilon_2) = 0.$$

Now

$$\omega_\infty(T_n, \delta) \leq \omega_\infty(U_n, \delta) \; \|\sigma_n^{-1}\| \; \|n^{-1/2} \sum_{i=1}^n g(X_{n,i}^*)\|,$$

where the second and third factors on the right-hand side are clearly bounded in probability. Applying the Cauchy-Schwarz inequality we get

$$\begin{aligned} \omega_\infty(U_n, \delta) \quad &\leq \quad E(\|g(X_{n,1}^*)\|^2 | X_1, \ldots, X_n)^{1/2} \\ &\quad \times \sup_{\rho(x,y) \leq \delta} E((1_{\{X_{n,1}^* \leq x\}} - 1_{\{X_{n,1}^* \leq y\}})^2 | X_1, \ldots, X_n)^{1/2}, \end{aligned}$$

where the first factor on the right-hand side is bounded in probability by (2.9). Moreover, we have

$$\begin{aligned} &\sup_{\rho(x,y) \leq \delta} E((1_{\{X_{n,1}^* \leq x\}} - 1_{\{X_{n,1}^* \leq y\}})^2 | X_1, \ldots, X_n)^{1/2} \\ \leq \quad &\sup_{\rho(x,y) \leq \delta} |\hat{F}_n(x) - \hat{F}_n(y)|^{1/2} \\ \leq \quad &\sup_{\rho(x,y) \leq \delta} |F(x) - F(y)|^{1/2} + 2 \sup_{-\infty \leq x \leq \infty} |\hat{F}_n(x) - F(x)|^{1/2} \\ \leq \quad &\sup_{\rho(x,y) \leq \delta} |F(x) - F(y)|^{1/2} + o_P(1) \end{aligned}$$

by (1.7). Since

$$\lim_{\delta \downarrow 0} \sup_{\rho(x,y) \leq \delta} |F(x) - F(y)| = 0$$

by continuity of F, we get

$$\lim_{\delta \downarrow 0} \limsup_{n \to \infty} P(P(\omega_\infty(T_n, \delta) \geq \varepsilon_1 | X_1, \ldots, X_n) \geq \varepsilon_2) = 0,$$

which proves tightness of T_n and concludes the proof of (2.10) and hence of (2.1).

For the proof of (2.2) we have to produce bounds on

$$\sup_{-\infty < x < \infty} |R_n^*(x)|$$

which are analogous to the bounds derived by Zhang (1997) in the proof of his Theorem 3.2 on the remainder

$$\sup_{-\infty<x<\infty} |R_n(x)|,$$

appearing in his representation (3.1) for $\hat{F}_n(x) - F(x)$. This can be achieved by carefully carrying over all the arguments in Zhang (1997) from the random variables $X_1, \ldots, X_n$ to the bootstrap variables $X^*_{n,1}, \ldots, X^*_{n,n}$ which, as above, always have to be considered conditionally under $P(\cdot|X_1, \ldots, X_n)$. The technical details are of the same type as in the proof of (2.1) presented above and will not be given here.

3 Some simulations

Here we present the results of a small simulation study concerning the finite sample coverage levels of the asymptotic confidence bands (1.9) and the decrease of width of these bands when compared to the confidence bands for F obtained from the classical empirical process $n^{1/2}(F_n - F)$ which apply if no auxiliary information about F is available. We consider the model (1.1) for the following four cases:

Case 1: Known mean, standard normal distribution,

Case 2: Known median, standard normal distribution,

Case 3: Known mean, exponential distribution with mean one,

Case 4: Known median, exponential distribution with mean one.

In each case we generated 1000 samples of the sizes $n = 20, 40, 60, 80, 100$ and for each sample used 1000 independent bootstrap samples to obtain Monte Carlo estimates for the bootstrap quantiles $H^*_n(1-\alpha)$. The nominal confidence levels are the standard levels $1 - \alpha = 0.9, 0.95, 0.99$. The finite sample coverage levels observed in our simulations (i.e. the fraction of simulated confidence bands which covered the true underlying distribution function) are given in Tables 1-4 below. The results show that the finite sample coverage levels are fairly close to the nominal ones already for small and medium sample sizes.

Table 1. Simulated finite sample coverage levels for the confidence band (1.9) in Case 1: Known mean, standard normal distribution

n	20	40	60	80	100
$1-\alpha=$0.90	0.908	0.887	0.892	0.894	0.875
0.95	0.948	0.944	0.951	0.953	0.931
0.99	0.994	0.992	0.991	0.996	0.987

Table 2. Simulated finite sample coverage levels for the confidence band (1.9) in Case 2: Known median, standard normal distribution

n	20	40	60	80	100
$1-\alpha=$0.90	0.906	0.896	0.904	0.900	0.899
0.95	0.952	0.949	0.944	0.952	0.952
0.99	0.991	0.991	0.984	0.990	0.987

Table 3. Simulated finite sample coverage levels for the confidence band (1.9) in Case 3: Known mean, exponential distribution with mean one

n	20	40	60	80	100
$1-\alpha=$0.90	0.895	0.891	0.899	0.901	0.898
0.95	0.947	0.944	0.950	0.950	0.952
0.99	0.982	0.986	0.988	0.989	0.991

Table 4. Simulated finite sample coverage levels for the confidence band (1.9) in Case 4: Known median, exponential distribution with mean one

n	20	40	60	80	100
$1-\alpha=0.90$	0.901	0.902	0.903	0.902	0.902
0.95	0.942	0.951	0.949	0.954	0.958
0.99	0.987	0.994	0.987	0.989	0.989

We now compare the widths of the confidence bands (1.9) with the widths of the bands obtained from the classical empirical process $n^{1/2}(F_n - F)$, i.e. with the the widths of the bands

$$(3.1) \qquad [F_n(x) - \frac{Q_n^c(1-\alpha)}{n^{1/2}}, F_n(x) + \frac{Q_n^c(1-\alpha)}{n^{1/2}}], \quad -\infty < x < \infty,$$

where Q_n^c denotes the (nonrandom) quantile function of $\|n^{1/2}(F_n - F)\|_\infty$. For the quantile function Q^c of $\|B^0\|_\infty$ with B^0 being the Brownian bridge we have for all $0 < \alpha < 1$

$$Q_n^c(1-\alpha) \to Q^c(1-\alpha) \quad \text{as } n \to \infty$$

by convergence in distribution of the classical Kolmogorov-Smirnov statistics. For the random quantile function H_n^* we get from (1.8) as $n \to \infty$ in probability,

$$H_n^*(1-\alpha) \to Q^W(1-\alpha).$$

Here Q^W denotes the quantile function of $\|W\|_\infty$ with W being the limiting process in the fclt (1.8). Thus for the relative decrease of width of the confidence bands (1.9) when compared with the classical bands (3.1) we obtain as $n \to \infty$ in probability,

$$\frac{Q_n^c(1-\alpha) - H_n^*(1-\alpha)}{Q_n^c(1-\alpha)} \to \frac{Q^c(1-\alpha) - Q^W(1-\alpha)}{Q^c(1-\alpha)}.$$

Note that $Q_n^c(1-\alpha)$ does not depend on the underlying continuous F and that the confidence band (3.1) constructed with this quantile has exact coverage probability for each finite sample size n. Under model (1.1) even the asymptotic quantile $Q^W(1-\alpha)$ depends on F and, of course, on the function g defining the model. For a numerical comparison we used the averages $\overline{H}_n^*(1-\alpha)$ of the simulated $H_n^*(1-\alpha)$-values and the known values of $Q_n^c(1-\alpha)$ to obtain the estimates

$$\frac{Q_n^c(1-\alpha) - \overline{H}_n^*(1-\alpha)}{Q_n^c(1-\alpha)}$$

for the relative decrease of width of the bands (1.9) when compared to the classical bands (3.1) for finite n. Our results were fairly stable over the considered range of sample sizes $n = 20, 40, 60, 80, 100$ so that we report them only for the sample size $n = 100$.

Table 5. Relative decrease of width of the confidence bands (1.9) when compared with the classical bands (3.1) for n = 100 in %

	Case 1	Case 2	Case 3	Case 4
α =0.90	33	27	24	28
0.95	33	27	24	27
0.99	35	28	25	27

Thus, in the cases considered in our simulations, the bands (1.9) under the model (1.1) are considerably smaller than the classical bands (3.1) and might be preferred in practice.

References

A.B. Owen (1988), Empirical likelihood ratio confidence intervals for a single functional, *Biometrika* **75**, 237–249.

A.B. Owen (1990), Empirical likelihood confidence regions, *Ann. Statist.* **18**, 90–120.

A.B. Owen (1991), Empirical likelihood for linear models *Ann. Statist.* **19**, 1725–1747.

J. Qin and J. Lawless (1994), Empirical likelihood and general estimating equations, *Ann. Statist.* **22**, 300–325.

D. Ylvisaker (1968), A note on the absence of tangencies in Gaussian sample paths, *Ann. Math. Statist.* **39**, 261–262.

B. Zhang (1997), Estimating a distribution function in the presence of auxiliary information, *Metrika* **46**, 221–244.

Mathematical Institute
University of Giessen
D-35392 Giessen
Germany
e-mail: erich.haeusler@math.uni-giessen.de

On the Characteristic Function of the Matrix von Mises–Fisher Distribution with Application to $SO(N)$–Deconvolution

Peter T. Kim*

ABSTRACT Characteristic functions provide a useful method for analyzing probability distributions. In most instances, the domain is taken to be Euclidean space and although the integral transforms may not have a simple expression, some qualitative features about the underlying probability distribution can often be extracted. Although there is some research efforts in spaces more complicated then the Euclidean case, very little is available on characteristic functions for probability distributions on these general domains. In this paper we will calculate the characteristic function of a probability distribution on a more general domain, in particular, the $SO(N)$–version of the matrix von Mises-Fisher distribution. It is found that when Fourier transforms of the latter are taken with respect to the irreducible representations of $SO(N)$, a concrete expression for the characteristic function is obtained. In addition, some qualitative properties are found which in turn are used to obtain L^2–rates of convergence for the $SO(N)$–deconvolution problem.

1 Introduction

Fourier analysis on groups, through the theory of group representations, have appeared in the probability and statistics literature; see for example Hannan (1965), Marcus and Pisier (1981), Diaconis (1988), Lo and Ng (1988), Rosenthal (1994), and Porod (1996) to name just a few. Nevertheless there has been little in the way of using them for characterizing certain probability distributions on non-Abelian continuous groups in the same way that they are used to characterize distributions in Euclidean space. This undoubtedly stems from the fact that for non-Abelian continuous groups, explicit probability densities are somewhat rare. For certain spaces of matrices however, Downs (1972) put forward the notion of a matrix von Mises-Fisher distribution. Although the functional form appears very

*This research was supported in part by a grant from the Natural Sciences and Engineering Research Council of Canada, OGP46204.

simple, beyond that, little is known about its Fourier properties. Consequently, it will be the subject of this paper to understand the Fourier properties of the matrix von Mises-Fisher distribution and to use these properties for a statistical inverse problem.

Let $SO(N)$ be the space of $N \times N$ real orthogonal matrices of determinant 1. Consider the $SO(N)$–version of the matrix von Mises-Fisher (Langevin) distribution (introduced in Downs (1972) and further studied in Khatri and Mardia (1977) and Chikuse (1998)) having mean direction $g \in SO(N)$ and concentration parameter $\kappa > 0$,

$$f_{\kappa,g}(x) = c(\kappa) \exp\left\{\kappa \operatorname{tr} x g^{-1}\right\}, \quad x \in SO(N) \tag{1.1}$$

where $c(\kappa)$ normalizes the distribution, "tr" denotes trace and the density is absolutely continuous with respect to the probability Haar measure on $SO(N)$. In this paper we derive the characteristic function of (1.1) where by the latter we mean the Fourier transform with respect to some complete orthonormal bases of $L^2(SO(N))$. The particular basis that will be used are the irreducible representations of $SO(N)$ which among other features, are the eigenfunctions of the Laplace-Beltrami operator on $SO(N)$. This usage conforms with that forwarded by Hendriks (1990) and coincides with the Euclidean version. Note that other non-Euclidean integral transforms have been called by such a name, see Peddada and Richards (1991) and Richards (1995), however, for the purposes of this paper and in light of the Peter-Weyl theorem, we will adopt the Hendriks interpretation.

We now provide a summary of what is to follow.

In Section 2, a sketch of the necessary representation theory tools will be outlined. In particular, an understanding of how to take apart tensor products of irreducible representations of $SO(N)$ and re-express them as direct sums will be needed. In fact, we will apply the generalization of the Clebsch-Gordan formula to $SO(N)$. The nice feature about this is that computations become explicit and we can actually calculate tail behavior up to multiplicative constants. This result can then be applied to re-expressing powers of the standard irreducible character as a weighted sum of other irreducible characters.

In Section 3, we apply the findings of Section 2 to calculate the Fourier coefficients of (1.1). The Fourier series for (1.1) can then be written. Although the exact quantities of the Fourier coefficients are difficult to compute, the variational range of the operator norm is still tractable, which in turn allows us to estimate the tail behavior of the characteristic function of (1.1). This then allows us to statistically recover the true underlying density for the $SO(N)$–deconvolution problem when noise is according to (1.1), which is the topic of Section 4. As one might suspect, the tails of the characteristic function of (1.1) decrease exponentially, consequently, recovery of the true underlying density can only take place logarithmically. This is in line with the Euclidean case; see Fan (1991). All proofs will be deferred to the last section.

2 Irreducible representations of $SO(N)$

The necessary background for reading this section concerns the representation theory of compact Lie groups in general and $SO(N)$ in particular. Some of the properties are spelled out in Fegan (1991), Gong (1991), Rosenthal (1994), Porod (1996) and Kim (1998), although more general results will be needed for this paper. Representation theory is a very rich area of mathematics having deep historical roots. There are a vast number of writings on the subject; however, this author has a particular prejudice toward Bröcker and tom Dieck (1985) although among many to choose from, Hewitt and Ross (1970), Humphreys (1972), Helgason (1984), and Fulton and Harris (1991) are excellent choices as well. Since what is to be covered involves a major part of the subject, and having to explain every detail about the algebraic construction of the subject could turn this paper into a book, we will take a somewhat sketchy approach, leaving details to be checked against standard texts such as Bröcker and tom Dieck (1985). Indeed, since definitions from the latter will be used here we suggest that the reader have that book available for reference if need be. For representation theory workers, or readers who know representation theory, what will appear is quite standard. The author would however like to point out that the main purpose of using representation theory is to calculate the characteristic function of (1.1), followed by applying these findings to the $SO(N)$–deconvolution problem.

The dimension of $SO(N)$ as a manifold is

$$\text{(2.1)} \qquad \dim SO(N) = \sum_{l=1}^{N-1} l = \frac{N(N-1)}{2}.$$

This arises from the fact that the Lie algebra of $SO(N)$ (the tangent space at the unit element) is $\mathfrak{so}(N)$, the space of $N \times N$ skew symmetric matrices with $0's$ on the main diagonal. What will follow is a brief sketch of the structure of $SO(N)$.

Let I^* denote a k–fold integer lattice and let $I^*_m = \{z \in \mathbb{Z}^k : |z_l| \le m, l = 1, \dots, k\}$ where $\mathbb{Z}$ denotes the set of all integers. For $N = 2k+1$ odd, let

$$\text{(2.2)} \qquad \bar{K} \cap I^*_m = \{j \in \mathbb{Z}^k : m \ge j_1 \ge j_2 \ge \cdots \ge j_k \ge 0\}.$$

For $N = 2k$ even, let

$$\text{(2.3)} \qquad \bar{K} \cap I^*_m = \{j \in \mathbb{Z}^k : m \ge j_1 \ge j_2 \ge \cdots \ge |j_k| \ge 0\}.$$

One notices that in the even case, an extra set of indices come from the relation $|j_k|$. To get all of the irreducible representations, let $m \to \infty$ for both the even and odd cases and define

$$\text{(2.4)} \qquad \bar{K} \cap I^* = \bigcup_{m \ge 0} \bar{K} \cap I^*_m.$$

In the above notation, $\bar{K} \cap I^*_m$ is known as the *Weyl polyhedra*, while $\bar{K} \cap I^*$ is the integer portion of the *fundamental Weyl chamber* whose elements uniquely index all of the inequivalent irreducible representations of $SO(N)$. In fact, this is true for all connected simple compact Lie groups where the $\bar{K}$ is determined by the root structure of the group classified according to the Dynkin diagrams. The particular form of $\bar{K} \cap I^*$ for $SO(N)$ when $N = 2k+1$ reflects the B_k root structure $k \geq 2$, while $N = 2k$ reflects the D_k root structure $k \geq 3$. It is worth pointing out that the Dynkin diagrams are attributable to E.B. Dynkin whose work in Markov processes may be more familiar to the current reader.

Let $\mathfrak{W}$ be the Weyl group of $SO(N)$ and let wj denote the action of $w \in \mathfrak{W}$ on $j \in \bar{K} \cap I^*$ which permutes the indices with possible sign changes. Define $\mathfrak{W}j = \{wj : w \in \mathfrak{W}, j \in \bar{K} \cap I^*\}$ and for each $j \in \bar{K} \cap I^*$, let $\mathrm{Con}(\mathfrak{W}j)$ be the convex hull of $\mathfrak{W}j$ intersected with $\bar{K} \cap I^*$. For $N = 2k+1$ the Weyl group is the semi-direct product of $\mathbb{Z}/2$ with $\mathbb{S}(k)$, the symmetric group on $k-$letters, while for $N = 2k$ the Weyl group is the semi-direct product of $\mathbb{Z}/2$ with the alternating group $\mathbb{A}(k)$, the even permutations of $\mathbb{S}(k)$.

A partial ordering on $\bar{K} \cap I^*$ is available. Indeed define

$$r \preceq s \quad \Leftrightarrow \quad r \in \mathrm{Con}(\mathfrak{W}s) \tag{2.5}$$

for $r, s \in \bar{K} \cap I^*$.

Since the collection of inequivalent irreducible representations of $SO(N)$ can be uniquely indexed by $j \in \bar{K} \cap I^*$, for each irreducible representation U of $SO(N)$ write U_j, along with it's *irreducible character* $\chi_j = \chi_{U_j} = \mathrm{tr}\, U_j$ and it's *dimension* $d_j = d_{U_j} = \chi_j(e)$, where $e \in SO(N)$ denotes the unit element.

We shall need to decompose products of irreducible characters. Indeed, for $SO(3)$, we have the Clebsch-Gordan formula,

$$\chi_l \chi_p = \chi_{|l-p|} + \chi_{|l-p|+1} + \cdots + \chi_{l+p} \tag{2.6}$$

for $l, p = 0, 1, \ldots$. Notice that in this case we have to deal with only a single integer since $k = 1$ for $SO(3)$. In the higher order cases, we have to deal with multiple integers. The generalization can be found in Steinberg (1961) or Proposition 3.4 and exercise 6 in Bröcker and tom Dieck (1985, 259-260). It goes as follows.

Let Φ_+ denote the set of positive roots of $SO(N)$. In particular, denote by e_j, $j = 1, \ldots, k$, the standard basis for $\mathbb{R}^k$. Then for $N = 2k+1$, $\Phi_+ = \{e_i \pm e_j : 1 \leq i < j \leq k\} \cup \{e_i : 1 \leq i \leq k\}$ and for $N = 2k$, $\Phi_+ = \{e_i \pm e_j : 1 \leq i < j \leq k\}$. Define $\rho = \frac{1}{2}\sum_{\alpha \in \Phi_+} \alpha$ so that for $N = 2k+1$, $\rho = ((2k-1)/2, \ldots, 3/2, 1/2)$ and for $N = 2k$, $\rho = (k-1, \ldots, 2, 1, 0)$. For $\zeta \in I^*$, a family of nonnegative integers $\{n_\alpha : \alpha \in \Phi_+\}$ such that $\zeta = \sum_{\alpha \in \Phi_+} n_\alpha \cdot \alpha$ is called a *decomposition into positive roots*. Let $\mathfrak{p}(\zeta)$ be the cardinality of the set of all decompositions of ζ into positive roots.

We have the following generalization of the Clebsch-Gordan formula.

Lemma 1. (Steinberg) *Suppose χ_r and χ_s are two irreducible characters. Then*

$$\chi_r \chi_s = \sum_{t \in \bar{K} \cap I^*} m(r, s, t) \chi_t,$$

where

$$m(r, s, t) = \sum_{v, w \in \mathfrak{W}} \det(vw) \cdot \mathfrak{p}(v(r + \rho) + w(s + \rho) - (t + 2\rho)).$$

In particular, $m(r, s, r + s) = 1$ and $m(r, s, t) \neq 0$ implies $t \preceq r + s$. □

This result is true for all compact semisimple connected Lie groups which of course includes $SO(N)$. Although this is an explicit formula for counting the multiplicities, it is quite cumbersome when in fact the Weyl group is large. In particular, $\mathfrak{W}$ is a finite group whose order is $2^k k!$ for $SO(2k+1)$ and $2^{k-1} k!$ for $SO(2k)$. As one can see, counting the multiplicities involves a double summation over the Weyl group.

We will need a variant of Lemma 1 for powers of the irreducible character of the *standard representation*, $1 = (1, 0, \ldots, 0)$. Consequently, as motivation for the next result, consider $SO(3)$ so that by the Clebsch-Gordan formula (2.6),

$$\chi_1^2 = \chi_0 + \chi_1 + \chi_2. \tag{2.7}$$

By successively applying (2.6) we get

$$\begin{aligned} \chi_1^\nu &= (\chi_0 + \chi_1 + \chi_2) \chi_1^{\nu - 1} \\ &= \sum_{j=0}^{\nu} \mu(\nu, j) \chi_j. \end{aligned} \tag{2.8}$$

We note that $1 \leq \mu(\nu, j) \leq 3^\nu$ where the upper bound uses the fact that $\{\chi_j : j = 0, 1, \ldots\}$ is an orthonormal set along with $|\chi_1| \leq \chi_1(e) = 3$ and the Cauchy-Schwarz inequality.

The generalization of (2.7) for $SO(N)$ is

$$\chi_1^2 = \chi_0 + \chi_{(1,1,0,\ldots,0)} + \chi_2, \tag{2.9}$$

where $0 = (0, \ldots, 0)$ and $2 = (2, 0, \ldots, 0)$, and by successively applying (2.9) a generalization of (2.8) can be made for other root structures which include types B_k, $k \geq 2$ and D_k, $k \geq 3$.

Define $\mathfrak{B}(0) = \{(0, \ldots, 0)\}$, the latter indexing the trivial representation and for $\nu \geq 1$ and $j \in \bar{K} \cap I^*$ define

$$\mathfrak{B}(\nu) = \mathrm{Con}(\mathfrak{W}(\nu, 0, \ldots, 0)), \tag{2.10}$$

where $|j| = j_1 + j_2 + \cdots + |j_k|$. We have the following.

Lemma 2. *For* $\chi_1 = \chi_{(1,0,\ldots,0)}$ *and* $\nu \geq 0$,

$$\chi_1^\nu = \sum_{j \in \mathfrak{B}(\nu)} \mu(\nu, j)\chi_j$$

where $0 \leq \mu(\nu, j) \leq N^\nu$ *when* $j \in \mathfrak{B}(\nu)$ *and zero otherwise.*

We note that by Lemma 1, $\mathfrak{B}(\nu)$ defines the set over which $\mu(\nu, j)$ is a nonnegative integer when $j \in \mathfrak{B}(\nu)$. A refinement therefore is to define

$$\mathfrak{A}(\nu) = \mathfrak{B}(\nu) \cap \{\mu(\nu, j) \geq 1\}, \tag{2.11}$$

where $j \in \bar{K} \cap I^*$ and $\nu \geq 0$. Then

$$1 \leq \mu(\nu, j) \leq N^\nu, \quad \text{for} \quad j \in \mathfrak{A}(\nu) \tag{2.12}$$

and zero otherwise.

We note that $\mathfrak{A}(\nu) \neq \emptyset$. This follows from Lemma 1 in that for every $\nu \geq 0$, $(\nu, 0, \ldots, 0) \in \mathfrak{A}(\nu)$.

The exact calculation of $\mu(\nu, \ell)$ can be obtained from Lemma 1; however, one can use the Weyl character formula and the Weyl integral formula to express it in terms of trigonometric functions. In particular, let $\mathbb{T} \simeq \oplus_k SO(2)$ be the maximal torus in $SO(N)$, where $\oplus_k$ means k-fold direct sum and $\simeq$ means isomorphism. For $\gamma \in \bar{K} \cap I^*$ and $\rho = \frac{1}{2}\sum_{\alpha \in \Phi_+} \alpha$, let $c(\gamma + \rho)$ denote the evaluation of χ_γ by the Weyl character formula, which for $SO(N)$ is displayed in Gong (1991, 122-123), Rosenthal (1994, 401-402) and in Porod (1996); and for the general case in Bröcker and tom Dieck (1985, 244).

By the Weyl integral formula,

$$\mu(\nu, \ell) = \frac{1}{|\mathfrak{W}|} \int_{\mathbb{T}} (c(1 + \rho))^\nu \overline{c(\ell + \rho)} \eta(t) dt \tag{2.13}$$

for $\nu \geq 1$, where overbar denotes complex conjugation, dt is Haar measure on $\mathbb{T}$ and $\eta(t)$ is proportional to

$$\prod_{j \leq l} (\cos\theta_j - \cos\theta_l)^2 \quad \text{for} \quad N = 2k$$

and

$$\prod_i (1 - \cos\theta_i) \prod_{j \leq l} (\cos\theta_j - \cos\theta_l)^2 \quad \text{for} \quad N = 2k + 1,$$

see Gong (1991, 125). Again however, the exact calculation of (2.13) is somewhat difficult to compute.

In our situation (2.9) can also be used to define an alternative distribution on $SO(N)$ motivated by the Watson distribution on a unit sphere. This distribution serves as a model for the real projective space $\mathbb{RP}^{k-1}$; see Fisher Lewis and Embleton (1993, 89-92).

Indeed, define

$$d(\kappa)\exp\left\{\kappa\left(\operatorname{tr} xg^{-1}\right)^2\right\}, \quad x \in SO(N) \tag{2.14}$$

for $\kappa > 0$ and fixed $g \in SO(N)$. Since

$$\left(\operatorname{tr} xg^{-1}\right)^2 = \chi_1^2(xg^{-1}),$$

for $\kappa > 0$ and $x, g \in SO(N)$, we will call (2.14), the *matrix Watson distribution.*

It turns out that the tail behavior of the characteristic function of the matrix Watson distribution behaves similarly to the characteristic function of the matrix von Mises-Fisher distribution, so that any follow-up on the former will not be pursued in this paper.

2.1 Fourier Analysis on $SO(N)$

The collection

$$\{d_j^{1/2} U_j : j \in \bar{K} \cap I^*\} \tag{2.15}$$

is a complete orthonormal basis for $L^2(SO(N))$. This follows from the fact that (2.15) are the eigenfunctions of the Laplace-Beltrami operator on $SO(N)$. This means that for $f \in L^2(SO(N))$, the *Fourier transform* is defined by

$$\hat{f}(j) = \int_{SO(N)} U_j(g^{-1}) f(g) dg, \tag{2.16}$$

where dg is the probability Haar measure on $SO(N)$ and $j \in \bar{K} \cap I^*$. *Fourier inversion* can then be expressed as

$$f(g) = \sum_{j \in \bar{K} \cap I^*} d_j \operatorname{tr} U_j(g) \hat{f}(j), \tag{2.17}$$

for $g \in SO(N)$ where (2.17) should be interpreted in the L^2–sense, although with sufficient smoothness it can hold pointwise almost everywhere in the Haar measure of $SO(N)$.

If $f \in L^2(SO(N))$ is a probability density function, the *characteristic function* of f will be (2.16). This coincides with the usual notion of characteristic function for distributions defined on $\mathbb{R}$ in that among other things, (2.15) is a complete orthonormal basis for $L^2(SO(N))$. Furthermore, one can show that the mapping $f \mapsto \hat{f}$ is one-to-one as also in the Euclidean case. This is a consequence of the Peter-Weyl theorem.

Some notation that will be used in this paper is as follows. For two positive sequences $\{a_n\}$ and $\{b_n\}$, $a_n \ll b_n$ will denote $a_n = O(b_n)$ as $n \to \infty$. If $a_n \ll b_n$ and $b_n \ll a_n$ then we will use the symbol $a_n \asymp b_n$. If $a_n/b_n \to 1$ as $n \to \infty$ this will be denoted by $a_n \sim b_n$. Furthermore, $\|\cdot\|_2$ will denote the usual L^2–norm while $\|\cdot\|_{\mathrm{op}}$ will denote the usual operator norm.

3 Main results

We now state the main results for the characteristic function of (1.1).

Theorem 1. *For the von Mises-Fisher distribution on $SO(N)$,*

$$\hat{f}_{\kappa,g}(\ell) = \frac{c(\kappa)}{d_\ell} \sum_{\{\nu \geq 0 | \ell \in \mathfrak{A}(\nu)\}} \frac{\kappa^\nu \mu(\nu,\ell)}{\nu!} U_\ell(g^{-1})$$

where $\ell \in \bar{K} \cap I^*$, $\nu \geq 0$, $\kappa > 0$, and $g \in SO(N)$. In addition, for $\nu \geq 0$

$$1 \leq \mu(\nu,\ell) \leq N^\nu,$$

where $\ell \in \mathfrak{A}(\nu)$ and $\mu(\nu,\ell) = 0$ if $\ell \notin \mathfrak{A}(\nu)$.

Theorem 2. *The von Mises-Fisher distribution has Fourier series*

$$f_{\kappa,g}(x) = c(\kappa) \sum_{\ell \in \bar{K} \cap I^*} \sum_{\ell \in \mathfrak{A}(\nu)} \frac{\kappa^\nu \mu(\nu,\ell)}{\nu!} \chi_\ell(xg^{-1}),$$

$\kappa > 0$, $x, g \in SO(N)$.

As a consequence of Theorem 2,

$$c(\kappa) = \int_{SO(N)} \exp\{\kappa \chi_1(x)\}\, dx = \left[\sum_{\nu \geq 0} \frac{\kappa^\nu \mu(\nu,0)}{\nu!}\right]^{-1},$$

for some $\kappa > 0$.

One can therefore see that the integral involves those studied by Selberg and others. In particular, the evaluation of the integral can be expressed in terms of zonal polynomials, see Richards (1989) and Bingham, Chang and Richards (1992).

As mentioned previously, although $\mu(\nu,\ell)$ can be explicitly written as in (2.13), the exact calculation is somewhat difficult to perform in general. Nevertheless, an understanding of the variational range of the operator norm is manageable and can be written in the following form.

Theorem 3. *For the von Mises-Fisher distribution, if $\ell \in \mathfrak{A}(\nu)$ then*

$$\frac{c(\kappa)}{d_\ell}\,\mathrm{e}^{-0.5\log|\ell|+(1+\log\kappa)|\ell|-|\ell|\log|\ell|} \ll \|\hat{f}_{\kappa,g}(\ell)\|_{\mathrm{op}}$$

$$\ll \frac{c(\kappa)}{d_\ell}\,\mathrm{e}^{-0.5\log|\ell|+(1+\log(\kappa N))|\ell|-|\ell|\log|\ell|},$$

as $|\ell| \to \infty$ for some $\kappa > 0$.

We can now see how the condition $\ell \in \mathfrak{A}(\nu)$ is required to get the lower bound in Theorem 3; however, the author is unable to resolve whether a more general condition is enough as $|\ell| \to \infty$. The latter is certainly true for $SO(3)$ and $SO(4)$, however, it would be of interest to know whether

$$\mu(\nu,\ell) \geq 1 \quad \text{for all } \ell \preceq (|\nu|,0,\ldots,0) \quad \text{as } |\ell| \to \infty.$$

The consequence, if true, would be that the von Mises-Fisher distribution has non-trivial Fourier coefficients for all $\ell \in \bar{K} \cap I^*$; see Hewitt and Ross (1970 II 153), and Richards (1989) for a discussion of this problem which has been addressed by Hermann Weyl.

4 Application to $SO(N)$–deconvolution

In statistical deconvolution, the interest is in reconstruction of the true density, given observed samples from some density convolved with a noise density. The rate at which the true density can be recovered is determined by the tail behavior of the characteristic function of the noise density. This situation for the Euclidean case is quantified in Fan(1991), where in fact the smoother the characteristic function tails, the more difficult the reconstruction of the true density becomes. Non-Euclidean generalizations of this appear in Rooij and Ruymgaart (1991), Healy and Kim (1996), Healy, Hendriks and Kim (1998), and Kim (1998).

As discussed in Kim (1998), suppose Y is a random quantity over $SO(N)$ and is made up of the true measurement X composed with noise ϵ. The true measurement can then be viewed as some random quantity on $SO(N)$ along with the error being some random quantity also on $SO(N)$. Consequently, the observations consist of

$$Y = X\epsilon,$$

where the multiplication is with respect to the group action $SO(N) \times SO(N) \to SO(N)$.

The density of Y is then the convolution of the densities of ϵ and X,

$$f_Y(u) = f_X * f_\epsilon(u) = \int_{SO(N)} f_X(v^{-1}u) f_\epsilon(v)\,dv.$$

Therefore, we can write

$$\hat{f}_X(j) = \hat{f}_Y(j)\hat{f}_{\epsilon^{-1}}(j),$$

provided that $\hat{f}_\epsilon(j)$ is invertible, where for ease of notation we will henceforth define $[\hat{f}_\epsilon(j)]^{-1} = \hat{f}_{\epsilon^{-1}}(j)$.

In general f_Y is assumed unknown hence $\hat{f}_Y(j)$ is unknown. We assume however that a random sample $Y_1, \ldots, Y_n$ is available so that we can form the empirical characteristic function

$$\hat{f}_Y^n(j) = \frac{1}{n}\sum_{l=1}^{n} U_j(Y_l^{-1}), \tag{4.1}$$

in direct analogy to the Euclidean case. Following this by using (4.1) in the Fourier inversion formula, we can obtain a nonparametric deconvolution density estimator for f_X by

$$f_X^n(g) = \sum_{j\in\bar{K}\cap I_m^*} d_j \mathrm{tr}\left\{U_j(g)\hat{f}_Y^n(j)\hat{f}_{\epsilon^{-1}}(j)\right\}, \tag{4.2}$$

where $m = m(n) \to \infty$ as $n \to \infty$.

Alternatively, define

$$K_n^\epsilon(g) = \sum_{j\in\bar{K}\cap I_m^*} d_j \mathrm{tr}\left\{U_j(g)\hat{f}_{\epsilon^{-1}}(j)\right\}.$$

Then (4.2) can be written in the more familiar kernel form,

$$f_X^n(g) = \frac{1}{n}\sum_{l=1}^{n} K_n^\epsilon(Y_l^{-1}g), \tag{4.3}$$

for $g \in SO(N)$.

The concern is to see how fast f_X can be recovered. This of course depends on the tail of the characteristic function of f_ϵ and in Kim (1998), the condition

$$\|\hat{f}_{\epsilon^{-1}}(j)\|_{\mathrm{op}} \ll d_j^u \tag{4.4}$$

for $u \geq 0$ as $d_j \to \infty$ is invoked. This provides recovery of f_X at the following rate,

$$E\|f_X^n - f_X\|_2^2 \ll n^{-2s/[2s+(\dim SO(N)/\mathbb{T})u+\dim SO(N)]}$$

as $n \to \infty$.

By Theorem 3, condition (4.4) does not apply when the error distribution is (1.1). Indeed, a separate result is needed.

Theorem 4. *Suppose f_ϵ is distributed according to the von Mises-Fisher distribution, f_Y is bounded and f_X is $s-$times continuously differentiable with square integrable derivatives, $s \geq 1$. Then*

$$E\|f_X^n - f_X\|_2^2 \ll (\dim \mathbb{T})^{2s} \left(\frac{\log n}{\log\log n}\right)^{-2s}$$

as $n \to \infty$ where $m \asymp \log n/(k \log\log n)$.

Consequently, we have the following.

Corollary 5. *Suppose f_ϵ is distributed according to the von Mises-Fisher distribution, f_Y is bounded and f_X is $s-$times continuously differentiable with square integrable derivatives, $s \geq 1$. Then*

$$E\|f_X^n - f_X\|_2^2 \ll (\dim \mathbb{T})^{2s} (\log n)^{-2s/(1+\beta)}$$

as $n \to \infty$ for any $\beta > 0$.

We note that in Theorem 3.2 of Kim (1998) the error structure having the operator bound (4.4) can be said to be ordinary smooth. Thus a polynomial L^2-rate of convergence is obtained in that result. For the von Mises-Fisher case however, the noise distribution is exponentially smooth and in fact through Corollary 5, bears resemblance to what is called super-smooth error distribution in the Euclidean case, see Fan (1991). The consequence of super-smooth errors results in logarithmic L^2-rate of convergence in Euclidean deconvolution, Fan (1991), and the same phenomena is observed for $SO(N)-$deconvolution.

5 Proofs

We now present the proofs of all of our results.

Proof of Lemma 2. According to the Dynkin diagrams $SO(2k+1)$ has root structure B_k, $k \geq 2$. Thus the Weyl group is $\mathfrak{W} \simeq (\mathbb{Z}/2) \wr \mathbb{S}(k)$ where $\wr$ denotes the wreath product. For $SO(2k)$ the root structure is D_k, $k \geq 3$. Thus the Weyl group is $\mathfrak{W} \simeq (\mathbb{Z}/2) \wr \mathbb{A}(k)$.

Consequently for any $j = (j_1, \dots, j_k)$ with $j_1 \geq \cdots \geq |j_k| \geq 0$, $\mathfrak{W}j$ consists of the elements derived from $(\pm j_{\sigma(1)}, \dots, \pm j_{\sigma(k)})$ for $\sigma \in \mathbb{S}(k)$ for $SO(2k+1)$ and $\sigma \in \mathbb{A}(k)$ for $SO(2k)$. In the case where some of the indices are equal, there are multiciplities, while if $j_1 > \cdots > j_k > 0$, $2^k k!$ distinct points are obtained for $SO(2k+1)$ and $2^{k-1}k!$ distinct points are obtained for $SO(2k)$.

Now for any $\ell \in \bar{K} \cap I^*$ such that $\ell_1 \geq \cdots \geq \ell_k \geq 0$, take the distinct points of $\mathfrak{W}\ell$, join them by straight lines and observe the bounding convex set. Take this convex set and intersect it with $\bar{K} \cap I^* = \{j_1 \geq \cdots \geq j_k \geq 0\}$ and that intersection consists of all points that are $\preceq \ell$.

By successively applying Lemma 1,

$$\chi_1^\nu = \sum_{(j_1,\dots,j_k)\in\mathfrak{B}(\nu)} \mu(\nu,j)\chi_j$$

where $0 \le \mu(\nu,j) \le N^\nu$ when $(j_1,\dots,j_k)\in\mathfrak{B}(\nu)$ and zero otherwise.

For the upper bound, note that $\{\chi_j : j \in \bar{K}\cap I^*\}$ is an orthonormal set. Thus

$$\begin{aligned}\int \chi_1^\nu \bar{\chi}_\ell &= \int \sum_h \mu(\nu,h)\chi_h\bar{\chi}_\ell \\ &= \mu(\nu,\ell).\end{aligned}$$

Now apply the Cauchy-Schwarz inequality bearing in mind that $|\chi_1| \le \chi_1(e) = N$. We therefore have

$$0 \le \mu(\nu,\ell) \le N^\nu,$$

for $\nu \ge 0$. □

Proof of Theorem 1. We have

$$\begin{aligned}\int_{SO(N)} \exp(\kappa\chi_1(x))\bar{\chi}_\ell(x)dx &= \int \sum_\nu \frac{(\kappa\chi_1)^\nu}{\nu!}\bar{\chi}_\ell \\ &= \sum_\nu \frac{\kappa^\nu}{\nu!}\int \chi_1^\nu\bar{\chi}_\ell \\ &= \sum_\nu \frac{\kappa^\nu}{\nu!}\sum_h \mu(\nu,h)\int \chi_h\bar{\chi}_\ell \\ &= \sum_{\ell\in\mathfrak{A}(\nu)} \frac{\kappa^\nu}{\nu!}\mu(\nu,\ell).\end{aligned}$$

The second equality is justified by the analyticity of $\exp x$ and the third equality follows from applying Lemma 2. The last equality follows from the orthonormality of the irreducible characters. □

Proof of Theorem 2. We have

$$\begin{aligned}f_{\kappa,g}(x) &= \sum_{\ell\in\bar{K}\cap I^*} d_\ell \mathrm{tr}\left\{U_\ell(x)\hat{f}_{\kappa,g}(\ell)\right\} \\ &= \sum_{\ell\in\bar{K}\cap I^*} d_\ell \mathrm{tr}\left\{\frac{c(\kappa)}{d_\ell}\sum_{\ell\in\mathfrak{A}(\nu)}\frac{\kappa^\nu\mu(\nu,\ell)}{\nu!}U_\ell(x)U_\ell(g^{-1})\right\} \\ &= c(\kappa)\sum_{\ell\in\bar{K}\cap I^*}\sum_{\ell\in\mathfrak{A}(\nu)}\frac{\kappa^\nu\mu(\nu,\ell)}{\nu!}\mathrm{tr}\left\{U_\ell(xg^{-1})\right\} \\ &= c(\kappa)\sum_{\ell\in\bar{K}\cap I^*}\sum_{\ell\in\mathfrak{A}(\nu)}\frac{\kappa^\nu\mu(\nu,\ell)}{\nu!}\chi_\ell(xg^{-1}),\end{aligned}$$

for some $\kappa > 0$. □

Proof of Theorem 3. Arguing as in the proof of Lemma 2, fix an $\ell \in \mathfrak{A}(\nu)$. One can now see how the variational range of $\|\hat{f}_{\kappa,g}(\ell)\|_{\mathrm{op}}$ can be controlled. In particular, since the components of the sum are always positive,

$$\sum_{\ell \in \mathfrak{A}(\nu)} \frac{\kappa^{\nu} \mu(\nu, \ell)}{\nu!} \geq \frac{\kappa^{|\ell|}}{|\ell|!}.$$

On the other hand for any $\ell \in \bar{K} \cap I^*$, note that

$$\mathfrak{A}(\nu) \subset \mathrm{Con}(\mathfrak{W}(\nu, 0, \ldots, 0)),$$

for $\nu \geq 0$.

Therefore,

$$\sum_{\ell \in \mathfrak{A}(\nu)} \frac{\kappa^{\nu} \mu(\nu, \ell)}{\nu!} \leq \sum_{\nu \geq |\ell|} \frac{(N\kappa)^{\nu}}{\nu!} \leq \frac{2(N\kappa)^{|\ell|}}{|\ell|!},$$

as $|\ell| \to \infty$. Here the inequality

$$\exp x - \sum_{s=0}^{n-1} \frac{x^s}{s!} \leq \frac{2x^n}{n!}, \tag{5.1}$$

for $0 < x < n/2$ is used.

Finally, apply Stirling's approximation,

$$\Gamma(u+1) \sim \sqrt{2\pi u}(u/\mathrm{e})^u,$$

as $u \to \infty$.

We are able to get

$$\frac{c(\kappa)}{d_\ell} \mathrm{e}^{-0.5 \log |\ell| + (1+\log \kappa)|\ell| - |\ell| \log |\ell|} \ll \|\hat{f}_{\kappa,g}(\ell)\|_{\mathrm{op}}$$

$$\ll \frac{c(\kappa)}{d_\ell} \mathrm{e}^{-0.5 \log |\ell| + (1+\log(N\kappa))|\ell| - |\ell| \log |\ell|}$$

as $|\ell| \to \infty$ for some $\kappa > 0$. □

Proof of Theorem 4. The proof can be reconstructed as in Kim (1998). Indeed, first take the integrated variance bias decomposition,

$$E\|f_X^n - f_X\|_2^2 = E\|f_X^n - Ef_X^n\|_2^2 + \|Ef_X^n - f_X\|_2^2. \tag{5.2}$$

As in Kim (1998),

$$E\|f_X^n - Ef_X^n\|_2^2 \ll \frac{1}{n} \sum_{j \in \bar{K} \cap I_m^*} d_j^2 \|\hat{f}_{\epsilon^{-1}}(j)\|_{\mathrm{op}}^2. \tag{5.3}$$

Now according to Theorem 3,

$$\|\hat{f}_{\epsilon^{-1}}(j)\|_{\text{op}}^2 \ll d_j^2 \exp\{\log|j| - 2(1+\log\kappa)|j| + 2|j|\log|j|\},$$

as $|j| \to \infty$.

Using this bound in (5.3), and noting that $d_j \ll m^{k^2}$ when $N = 2k+1$ and $d_j \ll m^{k(k-1)}$ when $N = 2k$, for $j \in \bar{K} \cap I_m^*$, we have

$$E\|f_X^n - Ef_X^n\|_2^2 \ll n^{-1} m^{\dim(SO(N)/\mathbb{T})+\dim SO(N)} \times e^{\log km - 2km(1+\log\kappa) + 2km\log km} \tag{5.4}$$

as $n \to \infty$. Also from Kim (1998), it is established that

$$\|Ef_X^n - f_X\|_2^2 \ll m^{-2s}, \tag{5.5}$$

as $n \to \infty$. Putting (5.4) and (5.5) into (5.2), we get

$$E\|f_X^n - f_X\|^2 \ll n^{-1} m^{\dim(SO(N)/\mathbb{T})+\dim SO(N)} \times e^{\log km - 2km(1+\log\kappa) + 2km\log km} + m^{-2s}, \tag{5.6}$$

as $n \to \infty$. This rate is optimized when

$$m \asymp \frac{\log n}{k \log\log n},$$

as $n \to \infty$. □

References

Bingham, C., Chang, T., Richards, D. (1992). Approximating the matrix Fisher and Bingham distributions: Applications to spherical regression and Procrustes analysis. *J. Multivariate Anal.* **41**, 314–337.

Bröcker, T., tom Dieck, T. (1985). *Representations of Compact Lie Groups.* New York: Springer-Verlag.

Chikuse, Y. (1993). High dimensional asymptotic expansions for the matrix Langevin distributions on the Stiefel manifold. *J. Multivariate Anal.* **44**, 82–101.

Chikuse, Y. (1998). Density estimation on the Stiefel Manifold. *J. Multivariate Anal.* **66**, 188–206.

Diaconis, P. (1988). *Group Representations in Probability and Statistics.* Hayward: Institute of Mathematical Statistics Lecture Notes- Monograph Series.

Downs, T.D. (1972). Orientation statistics. *Biometrika* **59**, 665–676.

Dynkin, E.B. (1950). The structure of semisimple Lie algebras. *Uspekhi Mat. Nauk. (NS)* **2**, 59–127. Also appears in *Amer. Math. Soc. Translations* **17**(1970).

Fan, J. (1991). Global behavior of deconvolution kernel estimates. *Statist. Sinica* **1**, 541–551.

Fegan, H.D. (1991). *Introduction to Compact Lie Groups.* Singapore: World Scientific.

Fisher, N.I., Lewis, T., Embleton, B.J. (1993). *Statistical analysis of spherical data.* Cambridge: Cambridge University Press.

Fulton, W., Harris, J. (1991). *Representation Theory: A First Course.* New York: Springer-Verlag.

Gong, S. (1991). *Harmonic Analysis on Classical Groups.* Berlin: Springer-Verlag.

Gross, K.I., Richards, D.S. (1995). Total positivity, finite reflection groups and a formula of Harish-Chandra. *J. Approx. Theory* **82**, 60–87.

Hannan, E.J. (1965). Group representations and applied probability. *J. Appl. Prob.* **2**, 1–68.

Healy, D.M., Kim, P.T. (1996). An empirical Bayes approach to directional data and efficient computation on the sphere. *Ann. Statist.* **24**, 232–254.

Healy, D.M., Hendriks, H., Kim, P.T. (1998). Spherical deconvolution. *J. Multivariate Anal.* **67**, 1–22.

Helgason, S. (1984). *Groups and Geometric Analysis.* Orlando: Academic Press.

Hendriks, H. (1990). Nonparametric estimation of a probability density on a Riemannian manifold using Fourier expansions. *Ann. Statist.* **18**, 832–849.

Hewitt, E., Ross, K.A. (1970). *Abstract Harmonic Analysis*, Vol. I and Vol. II. New York: Springer-Verlag.

Humphreys, J.E. (1972) *Introduction to Lie Algebras and Representation Theory.* New York: Springer-Verlag.

Khatri, C.G., and Mardia, K.V. (1977). The von Mises-Fisher distribution in orientation statistics. *J. Roy. Statist. Soc. Ser. B* **39**, 95–106.

Kim, P.T. (1998). Deconvolution density estimation on $SO(N)$. *Ann. Statist.* **26**, 1083–1102.

Lo, J.T., Ng, S. (1988). Characterizing Fourier series representations of probability distributions on compact Lie groups. *SIAM J. Appl. Math.* **48**, 222–228.

Marcus, M.B., Pisier, G. (1981). *Random Fourier series with application to Harmonic Analysis*. Princeton: Princeton University Press.

Peddada, S.D., Richards, D.S. (1991). Proof of a conjecture of M.L. Eaton on the characteristic function of the Wishart distribution. *Ann. Probab.* **19**, 868–874.

Porod, U. (1996). The cut-off phenomena for random reflections. *Ann. Probab.* **24**, 74–96.

Richards, D. S. (1989). Analogs and extensions of Selberg's integral. In *q-Series and Combinatorics* (D. Stanton, ed.), IMA Volumes in Mathematics and Applications **18**, 109–137. New York: Springer-Verlag.

Richards, D.S. (1995). Exact asymptotics for some probability distributions on compact manifolds. *Ann Statist* **23**, 1582–1586.

Rooij, A.C.M. van and Ruymgaart, F.H. (1991). Regularized deconvolution on the circle and the sphere. In *Nonparametric Functional Estimation and Related Topics* (G. Roussas, ed.), 679–690. Netherlands: Kluwer Academic Publishers.

Rosenthal, J.S. (1994). Random rotations: Characters and random walks on $SO(N)$. *Ann. Probab.* **22**, 398–423.

Steinberg, R. (1961). A general Clebsch-Gordan theorem. *Bull. Amer. Math. Soc.* **67**, 406–407.

Department of Mathematics and Statistics
University of Guelph
Guelph, Ontario N1G 2W1 Canada
e-mail: `pkim@msnet.mathstat.uoguelph.ca`

Testing for Ellipsoidal Symmetry of a Multivariate Distribution

Vladimir Koltchinskii* and
Lyudmila Sakhanenko**

ABSTRACT An approach to testing a multivariate distribution for ellipsoidal symmetry is suggested. The test statistics are defined as functionals of empirical processes indexed by special classes of functions. The parameters of the underlying distribution (the center of symmetry, the scaling transformation, etc.) are not supposed to be known and are estimated based on the data. General limit theorems for empirical processes that imply consistency of the tests against any asymmetric alternative (subject to some smoothness of the underlying density) and justify the use of a version of the bootstrap to evaluate the critical values of the test are proved.

1 Main results

In this paper, we develop an approach to the problem of testing a multivariate distribution for ellipsoidal symmetry. The test statistics we suggest are based on functionals of empirical processes indexed by special classes of functions. The functionals provide measures of (ellipsoidal) asymmetry of the empirical distribution. The parameters of the underlying elliptically contoured (under the null hypothesis) distribution are not supposed to be known and are estimated based on the data. The limit distributions of the test statistics depend on the unknown parameters, and we develop a version of the bootstrap to determine the critical regions of the tests. The problem of testing for symmetry in the univariate case has a long history (see Shorack and Wellner (1986) for some references). The results of the paper of Arcones and Giné (1991), who developed bootstrap tests for symmetry in the univariate case, are very close to what we are trying to accomplish for ellipsoidal symmetry. We refer to the papers of Beran (1979), Romano

*The research of V. Koltchinskii is partially supported by NSA Grant MDA904-99-1-0031 and by a Humboldt Research Fellowship.

**The research of L. Sakhanenko was partially supported by Boeing Computer Services Grant 3-48181.

(1989), Baringhaus (1991), Heathcote, Rachev and Cheng (1995), Koltchinskii and Li (1998), Quiroz, Manzotti and Pérez (1999) for other results on multivariate symmetry testing.

A Borel probability measure P on $\mathbf{R}^d$ is called *ellipsoidally symmetric* iff there exist an affine nonsingular transformation A from $\mathbf{R}^d$ onto $\mathbf{R}^d$ and a spherically symmetric about 0 Borel probability measure Π such that $P = \Pi \circ A^{-1}$. In other words, if X is a random vector with the distribution P, then there exist a vector $\theta_0 \in \mathbf{R}^d$ (the center of symmetry) and a nonsingular linear transformation $V_0 : \mathbf{R}^d \to \mathbf{R}^d$ (the scaling transformation) such that the random vector $V_0^{-1}(X - \theta_0)$ is spherically symmetric. Clearly, the linear transformation V_0 is not unique (one could take instead, for example, $V_0 O$ for an arbitrary orthogonal transformation O). But, on the other hand, one could always fix V_0 assuming that it is symmetric, positive definite and has operator norm equal to 1 (or normalized in any other proper way). For a Borel probability measure Π, spherically symmetric about 0, define

$$\pi_0(\rho) := \Pi(\{x : |x| \leq \rho\}), \ \rho > 0.$$

The triple (θ_0, V_0, π_0) will be called specifiers of the ellipsoidally symmetric probability measure P. We denote by $\mathcal{E}(\mathbf{R}^d)$ the set of all ellipsoidally symmetric distributions on $\mathbf{R}^d$.

To simplify the matter, we suppose in what follows that $\mathbb{E}|X|^2 < +\infty$, which allows one to define $\theta_0 = \theta_0(P) := \mathbb{E}X = \int_{\mathbf{R}^d} xP(dx)$ and to set $V_0 = V_0(P)$ to be the square root of the covariance operator of X. Obviously, these parameters are defined for any P such that the corresponding X is square-integrable (not necessarily ellipsoidally symmetric). We also can define in such generality $\pi_0 = \pi_{0,P}$ as the distribution of the random variable $|V_0^{-1}(X - \theta_0)|$. Given P, let P^s denote the ellipsoidally symmetric distribution with specifiers (θ_0, V_0, π_0). It will be called *the ellipsoidal symmetrization* of P.

Let now $(X_1, ..., X_n)$ be an i.i.d. sample from the distribution P, defined on a probability space $(\Omega, \Sigma, \mathbb{P})$, and let P_n be the empirical distribution based on the sample $(X_1, ..., X_n)$. Our goal is to suggest and to study several tests for ellipsoidal symmetry of the distribution P, based on the observations $(X_1, \ldots, X_n)$.

Let m denote the uniform distribution (the normalized Haar measure) on the unit sphere S^{d-1}. Given a bounded Borel function f on $\mathbf{R}^d$, we define

$$m_f(\rho) := \int_{S^{d-1}} f(\rho v)\ m(dv), \ \rho > 0. \tag{1.1}$$

It is easy to check that for an ellipsoidally symmetric P (with specifiers (θ_0, V_0, π_0)),

$$\int_{\mathbf{R}^d} f(V_0^{-1}(x - \theta_0))P(dx) = \int_0^{+\infty} m_f(\rho)\pi_0(d\rho). \tag{1.2}$$

If the class $\mathcal{F}$ characterizes the distribution, i.e.

$$\int_{\mathbf{R}^d} f dQ_1 = \int_{\mathbf{R}^d} f dQ_2 \text{ for all } f \in \mathcal{F}$$

implies that $Q_1 = Q_2$, then P is ellipsoidally symmetric iff (1.2) holds for all $f \in \mathcal{F}$. In general, we call P $\mathcal{F}$-asymmetric iff there exists an $f \in \mathcal{F}$ such that (1.2) does not hold.

Next we define

$$\theta_n := \theta_n(X_1, \ldots, X_n) := \bar{X}_n := n^{-1} \sum_{i=1}^{n} X_i$$

and

$$V_n := V_n(X_1, \ldots, X_n) := \Big(n^{-1} \sum_{i=1}^{n} (X_i - \bar{X}_n) \otimes (X_i - \bar{X}_n)\Big)^{1/2}.$$

[Here $u \otimes v$ denotes the linear transformation $x \mapsto \langle v, x \rangle u$ from $\mathbf{R}^d$ into $\mathbf{R}^d$, $u, v \in \mathbf{R}^d$]. Under the condition $\mathbb{E}|X|^2 < +\infty$, the law of large numbers implies that $V_n \to V_0$ as $n \to \infty$ a.s. Since V_0 is nonsingular, V_n becomes nonsingular with probability 1 for large enough n. *The scaled residuals* of the observations $(X_1, ..., X_n)$ are defined as

$$Z_j := Z_{j,n} := V_n^{-1}(X_j - \theta_n), \ j = 1, \ldots n.$$

Let π_n denote the empirical distribution based on the sample $(|Z_1|, ..., |Z_n|)$.

Consider a class $\mathcal{F}$ of Borel functions from $\mathbf{R}^d$ into $\mathbf{R}$. Our approach to the problem of testing for ellipsoidal symmetry will be to use functionals (such as the sup-norm) of the stochastic process

$$\xi_n(f) := n^{1/2}\Big(\int_{\mathbf{R}^d} f(V_n^{-1}(x - \theta_n)) P_n(dx) - \int_0^{+\infty} m_f(\rho) d\pi_n(\rho)\Big)$$

$$= n^{-1/2} \sum_{j=1}^{n} \big[f(Z_j) - m_f(|Z_j|)\big], \ f \in \mathcal{F}$$

as test statistics. In view of the relationship (1.2), such functionals are "measures of asymmetry" of the empirical distribution. Note that a nonsingular affine transformation of the data $(X_1, \ldots, X_n)$ results in an orthogonal transformation of the scaled residuals. If the class $\mathcal{F}$ is invariant with respect to all orthogonal transformations (i.e. for all $f \in \mathcal{F}$ and any orthogonal transformation O we have $f \circ O \in \mathcal{F}$), then the test statistic defined as the sup-norm of the process ξ_n is affine invariant. This is the case in the following examples.

Example 1. Let

$$\mathcal{C} := \{\{v \in S^{d-1} : \langle v, l \rangle \geq c\} : l \in S^{d-1}, c \in \mathbf{R}_+\}$$

be the class of "caps" on the unit sphere S^{d-1}. Consider the class

$$\mathcal{F} := \{I_C(\frac{x}{|x|})I_{\{0<|x|\le t\}} : C \in \mathcal{C}, t \in \mathbf{R}\}.$$

For $f(\cdot) = I_C(\frac{\cdot}{|\cdot|})I_{\{0<|\cdot|\le t\}}$, we have $m_f(\rho) = m(C)I_{(0,t]}(\rho),\ \rho > 0$. The process ξ_n becomes now

$$\xi_n(C,t) = n^{-1/2}\sum_1^n (I_C(\frac{Z_j}{|Z_j|}) - m(C))I_{\{0<|Z_j|\le t\}},\ C \in \mathcal{C}, t > 0.$$

The test statistic $T_n := \sup_{t>0, C\in\mathcal{C}} |\xi_n(C,t)|$ can be also represented as

$$T_n = n^{-1/2} \max_{1\le j\le n} \sup_{C\in\mathcal{C}} \Big|\sum_{k=1}^{j} (I_C(\frac{Z_{[k]}}{|Z_{[k]}|}) - m(C))\Big|,$$

where $Z_{[j]},\ j = 1,\dots,n$ is the rearrangement of $Z_1,\dots,Z_n$, such that $|Z_{[1]}| \le \dots \le |Z_{[n]}|$.

Example 2. Let $\mathcal{G}_l$ denote the linear space of spherical harmonics of degree less than or equal to l in $\mathbf{R}^d$, and let $\mathcal{B}_l$ be the unit ball in $\mathcal{G}_l \cap L_2(S^{d-1}, dm)$. Denote

$$\mathcal{F} := \{I_{\{0<|x|\le t\}}\psi(\frac{x}{|x|}) : \psi \in \mathcal{B}_l, t > 0\}.$$

Then for $f \in \mathcal{F}$, $f(x) = I_{\{0<|x|\le t\}}\psi(\frac{x}{|x|})$, we have $m_f(\rho) = m(\psi)I_{(0,t]}(\rho)$, where $m(\psi)$ is the average of ψ on S^{d-1}. In this case, the process ξ_n becomes

$$\xi_n(t,\psi) = n^{-1/2}\sum_1^n \big(I_{\{0<|Z_j|\le t\}}(\psi(\frac{Z_j}{|Z_j|}) - m(\psi))\big),\ t > 0, \psi \in \mathcal{B}_l.$$

The statistic $T_n := \sup_{t>0,\psi\in\mathcal{B}_l} |\xi_n(t;\psi)|$ becomes

$$T_n = n^{-1/2} \max_{1\le j\le n} \Big(\sum_{s=1}^{\dim(\mathcal{G}_l)} \sum_{k=1}^{j} (\psi_s(\frac{Z_{[k]}}{|Z_{[k]}|}) - \delta_{s1}))^2\Big)^{1/2},$$

where $\{\psi_s,\ s = 1,\dots,\dim(\mathcal{G}_l)\}$ denotes an orthonormal basis of the space $\mathcal{G}_l$, $\psi_1 \equiv 1$, $\delta_{ij} = 1$ for $i = j$ and $= 0$ otherwise. Quiroz and Dudley (1991) used a similar approach to testing for multivariate normality. Quiroz, Manzotti and Pérez (1999) developed tests for ellipsoidal symmetry based on spherical harmonics.

Example 3. Let

$$\mathcal{H} := \Big\{\{x \in \mathbf{R}^d : \langle x, u\rangle \le c\} : u \in S^{d-1}, c \in \mathbf{R}\Big\}$$

be the class of all half-spaces in $\mathbf{R}^d$. For $f := I_H$, $H := \{x \in \mathbf{R}^d : \langle x, u\rangle \leq c\}$, we have $m_f(\rho) := \gamma(\frac{c}{\rho})$, where $\gamma(c) := m\{v : \langle v, u\rangle \leq c\}$. The process ξ_n in this case is

$$\xi_n(u,c) := n^{-1/2}\sum_{j=1}^{n}\big(I_{\{\langle Z_j,u\rangle\leq c\}} - \gamma(\frac{c}{|Z_j|})\big),\ u \in S^{d-1}, c \in \mathbf{R}$$

and the test statistic can be defined as $T_n := \sup_{c\in\mathbf{R}}\sup_{u\in S^{d-1}}|\xi_n(u,c)|$.

Example 4. Let

$$\mathcal{F} := \Big\{\frac{|\cdot - t| - |\cdot|}{|t|} : t \in \mathbb{R}^d,\ t \neq 0\Big\}.$$

For $f(\cdot) = \frac{|\cdot-t|-|\cdot|}{|t|}$, we have $m_f(\rho) = \rho\frac{\beta(t)-1}{|t|}$, where $\beta(t) := \int_{S^{d-1}}|v - t|m(dv)$. Thus, the process ξ_n becomes

$$\xi_n(t) := n^{-1/2}|t|^{-1}\sum_{i=1}^{n}\big(|Z_i - t| - |Z_i|\beta(t)\big),\ t \in \mathbf{R}^d$$

and the test statistic can be choosen as $T_n := \sup_{t\in\mathbf{R}^d, t\neq 0}|\xi_n(t)|$.

For any polynomial p on $\mathbf{R}^m$ of degree less than or equal to r we will call $\{x \in \mathbf{R}^m : p(x) \geq 0\}$ *a polynomial set of degree less than or equal to* r in $\mathbf{R}^m$. Let $\mathcal{P}_{r,m}$ denote the class of all polynomial sets in $\mathbf{R}^m$ of degree less than or equal to r. Then any set from the union $\bigcup\{\mathcal{A}(B_1,\ldots,B_l) : B_1,\ldots,B_l \in \mathcal{P}_{r,m}\}$ is called a *semi-algebraic set of degree less than or equal to r and order less than or equal to l*, $\mathcal{A}(B_1,\ldots,B_l)$ being the minimal set algebra generated by $B_1,\ldots,B_l$. Let $\mathcal{SA}_{r,m,l}$ denote the class of all semialgebraic sets of degree less than or equal to r and order less than or equal to l in $\mathbf{R}^m$. A class $\mathcal{G}$ of functions on $\mathbf{R}^d$ is *a semi-algebraic subgraph class* iff for some r, l for all functions g from $\mathcal{G}$ the set $\{(x,t) : g(x) \geq t \geq 0 \ \text{ or } \ g(x) \leq t \leq 0\}$ belongs to $\mathcal{SA}_{r,d+1,l}$ and for all functions g_1, g_2 from $\mathcal{G}$ the set $\{(x,y,t) : g_1(x)g_2(y) \geq t \geq 0 \ \text{ or } \ g_1(x)g_2(y) \leq t \leq 0\}$ belongs to $\mathcal{SA}_{r,2d+1,l}$.

It is not hard to see that the function classes defined in Examples 1–4 are semi-algebraic subgraphs.

Let

$$\mathbf{E}(f;\theta,V) := \int_{\mathbf{R}^d}\big[f(V^{-1}(x-\theta)) - m_f(|V^{-1}(x-\theta)|)\big]P(dx).$$

It follows from (1.2) that, for an ellipsoidally symmetric P and for all f,

$$\mathbf{E}(f;\theta_0,V_0) = 0.$$

Let W_P° denote the P-Brownian bridge, i.e. a centered Gaussian process indexed by functions in $L_2(\mathbf{R}^d; dP)$ with the covariance

$$\mathbb{E}W_P^\circ(f)W_P^\circ(g) = P(fg) - P(f)P(g).$$

We will frequently use integral notation for $W_P^\circ(f)$:

$$W_P^\circ(f) = \int_{\mathbf{R}^d} f(x) W_P^\circ(dx).$$

As always, $\ell^\infty(\mathcal{F})$ denotes the space of all uniformly bounded functions on $\mathcal{F}$ with the sup-norm $\|Y\|_{\mathcal{F}} := \sup_{f\in\mathcal{F}}|Y(f)|$, $Y \in \ell^\infty(\mathcal{F})$.

A sequence of stochastic processes $\zeta_n : \mathcal{F} \mapsto \mathbf{R}$ is said to converge weakly in $\ell^\infty(\mathcal{F})$ (in the sense of Hoffmann-Jørgensen) to the stochastic process $\zeta : \mathcal{F} \mapsto \mathbf{R}$ iff there exists a Radon probability measure γ on $\ell^\infty(\mathcal{F})$ such that γ is the distribution of ζ and, for all bounded and $\|\cdot\|_{\mathcal{F}}$-continuous functionals $\Phi : \ell^\infty(\mathcal{F}) \mapsto \mathbf{R}$, we have $\mathbb{E}^*\Phi(\zeta_n) \to \int \Phi(x)\gamma(dx)$.

We assume in what follows that the class $\mathcal{F}$ satisfies standard measurability assumptions used in the theory of empirical processes (see Dudley (1999), or van der Vaart and Wellner (1996)). We also introduce the following smoothness conditions (S) on the distribution P and the class $\mathcal{F}$:

$(S1)$ P is absolutely continuous with a uniformly bounded and continuously differentiable density p such that for some $A > d+1$

$$\sup_{x\in\mathbf{R}^d} (1+|x|)^A |p'(x)| < +\infty,$$

where p' denotes the derivative of the density p.

$(S2)$ The class $\mathcal{F}$ is uniformly bounded and for all $\varepsilon > 0$ and $R > 0$

$$\sup_{f\in\mathcal{F}} \text{mes}\Big\{x \in \mathbf{R}^d : |x| \le R \text{ and } \omega_f(x;\delta) \ge \varepsilon\Big\} \to 0 \text{ as } \delta \to 0.$$

Here mes denotes Lebesgue measure in $\mathbf{R}^d$ and

$$\omega_f(x;\delta) := \sup\{|f(x_1) - f(x_2)| : |x_1 - x| \le \delta, |x_2 - x| \le \delta\}.$$

Theorem 1. *Suppose that $\mathcal{F}$ is a semialgebraic subgraph class, the smoothness conditions (S) hold and $\int_{\mathbf{R}^d} |x|^4 dP < +\infty$. Then the sequence of stochastic processes*

$$\{\xi_n(f) - n^{1/2}\mathbf{E}(f;\theta_0,V_0) : f \in \mathcal{F}\}$$

converges weakly in the space $\ell^\infty(\mathcal{F})$ to a Gaussian stochastic process ξ_P. In particular, if P is ellipsoidally symmetric with specifiers (θ_0, V_0, π_0), then the sequence ξ_n converges weakly in the space $\ell^\infty(\mathcal{F})$ to the process ξ_P.

Let $T_n := \|\xi_n\|_{\mathcal{F}}$. Given $\alpha > 0$, let

$$t_\alpha := \inf\Big\{t : \mathbb{P}\{\|\xi_P\|_{\mathcal{F}} \ge t\} \le \alpha\Big\}.$$

Let H_0 be the hypothesis that $P \in \mathcal{E}(\mathbf{R}^d)$ and let H_a be the alternative that $P \notin \mathcal{E}(\mathbf{R}^d)$. Also, denote by $H_a(\mathcal{F})$ the alternative that P is $\mathcal{F}$-asymmetric. Then Theorem 1 implies that under the hypothesis H_0,

$$\mathbb{P}\{T_n \ge t_\alpha\} \to \alpha$$

and under the alternative $H_a(\mathcal{F})$,

$$\mathbb{P}\{T_n \geq t_\alpha\} \to 1 \text{ as } n \to \infty.$$

In particular, if $\mathcal{F}$ characterizes the distribution, then under the alternative H_a (i.e. for a fixed ellipsoidally asymmetric distribution P)

$$\mathbb{P}\{T_n \geq t_\alpha\} \to 1 \text{ as } n \to \infty.$$

[The proof of these facts uses the well-known theorem of Cirel'son on continuity of the distribution of the sup-norm of Gaussian process, see Cirel'son (1975)]. The class $\mathcal{F}$ characterizes the distribution in the case of Examples 1,3,4 above. This is a well-known property of the class used in Example 3. As to Example 4, we refer, e.g., to the paper of Koldobskii (1990) for similar statements. To prove that this is the case in Example 1, consider the map $\mathbf{R}^d \setminus \{0\} \ni x \mapsto (|x|, \frac{x}{|x|}) \in \mathbf{R}_+ \times S^{d-1}$. Since this map is a Borel isomorphism (even a homeomorphism), it suffices to show that for any two finite measures P, Q in $\mathbf{R}_+ \times S^{d-1}$ the condition

$$P((0,t] \times C) = Q((0,t] \times C) \text{ for all } t > 0,\ C \in \mathcal{C}$$

implies $P = Q$. We will prove that, in fact, for any two finite measures P, Q on $\mathbf{R}_+ \times \mathbf{R}^d$ the condition

$$P((0,t] \times H) = Q((0,t] \times H) \text{ for all } t > 0,\ H \in \mathcal{H},$$

where $\mathcal{H}$ is the class of all halfspaces in $\mathbf{R}^d$, implies that $P = Q$ (the previous statement then follows, since one can consider two measures in $\mathbf{R}_+ \times \mathbf{R}^d$, both supported in $\mathbf{R}_+ \times S^{d-1}$). The condition

$$P((0,t] \times H) = Q((0,t] \times H) \text{ for all } t > 0,\ H \in \mathcal{H}$$

is equivalent to the following one,

$$\int_{\mathbf{R}_+} \int_{\mathbf{R}^d} I_{(0,t]}(u) I_{(-\infty,c]}(\langle l, x\rangle) P(du, dx)$$

$$= \int_{\mathbf{R}_+} \int_{\mathbf{R}^d} I_{(0,t]}(u) I_{(-\infty,c]}(\langle l, x\rangle) Q(du, dx)$$

for all $l \in S^{d-1}, t > 0, c \in \mathbf{R}$. Using a standard approximation of Borel functions by simple functions, we extend this to the equality

$$\int_{\mathbf{R}_+} \int_{\mathbf{R}^d} \varphi(u)\psi(\langle l, x\rangle) P(du, dx) = \int_{\mathbf{R}_+} \int_{\mathbf{R}^d} \varphi(u)\psi(\langle l, x\rangle) Q(du, dx)$$

that holds for all bounded Borel functions φ, ψ. If we set $\varphi(u) := e^{isu}$ and $\psi(u) := e^{iu}$, we obtain that the characteristic functions of P and Q are equal, which implies that $P = Q$.

In most cases, however, the limit distributions of such statistics as T_n depend on the unknown parameters of the distribution P. Thus, to implement the test one has to evaluate the distribution of the test statistic using, for instance, the bootstrap. We describe below a version of the bootstrap for ellipsoidal symmetry testing.

Denote by P_n^s the ellipsoidally symmetric distribution with specifiers (θ_n, V_n, π_n). Let $(X_1^s, ..., X_n^s)$ be an i.i.d. sample from the distribution P_n^s, defined on a probability space $(\hat{\Omega}, \hat{\Sigma}, \hat{P})$. One can construct such a sample using the following procedure. Take an i.i.d. sample $(U_1, ..., U_n)$ uniformly distributed on S^{d-1} and an i.i.d. sample $(\hat{R}_1, ..., \hat{R}_n)$ from π_n independent of $(U_1, ..., U_n)$. Define

$$X_j^s := \theta_n + V_n U_j \hat{R}_j, \; j = 1, ..., n.$$

Then $(X_1^s, ..., X_n^s)$ is an i.i.d. sample from P_n^s.

Let $\hat{P}_n$ denote the empirical measure based on the sample $(X_1^s, ..., X_n^s)$, and let

$$\hat{\theta}_n := \theta_n(X_1^s, ..., X_n^s), \quad \hat{V}_n := V_n(X_1^s, ..., X_n^s).$$

Define *the bootstrapped scaled residuals* as

$$\hat{Z}_j := \hat{Z}_{j,n} := (\hat{V}_n)^{-1}(X_j^s - \hat{\theta}_n), \; j = 1, \dots, n.$$

Let $\hat{\pi}_n$ denote the empirical distribution based on the sample $(|\hat{Z}_1|, \dots, |\hat{Z}_n|)$.

The bootstrap version of ξ_n is the process

$$\hat{\xi}_n(f) := n^{1/2}\Big(\int_{\mathbf{R}^d} f((\hat{V}_n)^{-1}(x - \hat{\theta}_n))\hat{P}_n(dx) - \int_0^{+\infty} m_f(\rho) d\hat{\pi}_n(\rho)\Big)$$

$$= n^{-1/2} \sum_{j=1}^n \big[f(\hat{Z}_j) - m_f(|\hat{Z}_j|)\big], \; f \in \mathcal{F}.$$

Now we are going to consider a bootstrap version of Theorem 1. Let $BL_1(\ell^\infty(\mathcal{F}))$ denote the set of all functionals $\Phi : \ell^\infty(\mathcal{F}) \mapsto \mathbf{R}$ such that for all $Y \in \ell^\infty(\mathcal{F})$ $|\Phi(Y)| \le 1$ and for all $Y_1, Y_2 \in \ell^\infty(\mathcal{F})$ $|\Phi(Y_1) - \Phi(Y_2)| \le \|Y_1 - Y_2\|_{\mathcal{F}}$. Given two stochastic processes $\zeta_1, \zeta_2 : \Omega \times \hat{\Omega} \times \mathcal{F} \mapsto \mathbf{R}$, we define the following bounded Lipschitz distance:

$$d_{BL}(\zeta_1, \zeta_2) := \sup_{\Phi \in BL_1(\ell^\infty(\mathcal{F}))} |\hat{\mathbf{E}}^* \Phi(\zeta_1) - \hat{\mathbf{E}}^* \Phi(\zeta_2)|.$$

Theorem 2. *Suppose that $\mathcal{F}$ is a semi-algebraic subgraph class, the smoothness conditions (S) hold and $\int_{\mathbf{R}^d} |x|^4 dP < +\infty$. Then the sequence of stochastic processes $\{\hat{\xi}_n\}$ converges weakly in the space $\ell^\infty(\mathcal{F})$ to a version $\hat{\xi}_{P^s}$ of the process ξ_{P^s} (defined on the probability space $(\hat{\Omega}, \hat{\Sigma}, \hat{\mathbb{P}})$) in probability $\mathbb{P}$. More precisely,*

$$d_{BL}(\hat{\xi}_n; \hat{\xi}_{P^s}) \to 0 \text{ as } n \to \infty \text{ in probability } \mathbb{P}.$$

In particular, if P is ellipsoidally symmetric, $\hat{\xi}_n$ converges weakly to a version of the process ξ_P.

Let $\hat{T}_n := \|\hat{\xi}_n\|_{\mathcal{F}}$. Given $\alpha > 0$, let

$$\hat{t}_{n,\alpha} := \inf\Big\{t : \hat{\mathbb{P}}\{\hat{T}_n \geq t\} \leq \alpha\Big\}.$$

Then Theorems 1 and 2 imply that under the hypothesis H_0,

$$\mathbb{P}\{T_n \geq \hat{t}_{n,\alpha}\} \to \alpha$$

and under the alternative $H_a(\mathcal{F})$,

$$\mathbb{P}\{T_n \geq \hat{t}_{n,\alpha}\} \to 1 \text{ as } n \to \infty.$$

In particular, if $\mathcal{F}$ characterizes the distribution, the bootstrap test is consistent against any asymmetric alternative (subject to the smoothness conditions (S)): under the alternative H_a,

$$\mathbb{P}\{T_n \geq \hat{t}_{n,\alpha}\} \to 1 \text{ as } n \to \infty.$$

2 Proofs

We start with introducing some notation. Let $\mathcal{SP}$ denote the set of all symmetric positive definite linear transformations in $\mathbf{R}^d$. Given θ, V, denote $\tau_{\theta,V} f(\cdot) := f(V^{-1}(\cdot - \theta))$. For a function f on $\mathbf{R}^d$, let

$$\tilde{f}(x) := f(x) - m_f(|x|),\ x \in \mathbf{R}^d,$$

$$f_{\theta,V}(x) := f(V^{-1}(x-\theta)) = \tau_{\theta,V} f(x),\ x \in \mathbf{R}^d, V \in \mathcal{SP}.$$

Given a class $\mathcal{G}$ of functions, define $\tilde{\mathcal{G}} := \{\tilde{g} :\ g \in \mathcal{G}\}$,

$$\mathcal{G}_{\text{aff}} := \{g_{\theta,V} : \theta \in \mathbf{R}^d, V \in \mathcal{SP}\} = \{\tau_{\theta,V} g : \theta \in \mathbf{R}^d, V \in \mathcal{SP}\}.$$

Now the process ξ_n is represented as

$$\xi_n(f) = n^{1/2} \int_{\mathbf{R}^d} (\tilde{f})_{\theta_n,V_n} dP_n,\ f \in \mathcal{F}$$

and the process $\hat{\xi}_n$ as

$$\hat{\xi}_n(f) = n^{1/2} \int_{\mathbf{R}^d} (\tilde{f})_{\hat{\theta}_n,\hat{V}_n} d\hat{P}_n,\ f \in \mathcal{F}.$$

It is also clear that

$$\mathbf{E}(f;\theta;V) = \int_{\mathbf{R}^d} (\tilde{f})_{\theta,V} dP.$$

Define

$$\mathbf{E}^s(f;\theta;V) = \int_{\mathbf{R}^d} (\tilde{f})_{\theta,V} dP^s.$$

Given a function g on $\mathbf{R}^d$, we can write

$$\int_{\mathbf{R}^d} g(x) P_n^s(dx) = \int_0^{+\infty} \int_{S^{d-1}} g(\theta_n + V_n u\rho) m(du) \pi_n(d\rho)$$

$$(2.1) \qquad = \int_{\mathbf{R}^d} M_g(\theta_n; V_n; x) P_n(dx),$$

where

$$M_g(\theta; V; x) := \int_{S^{d-1}} g(\theta + Vu|V^{-1}(x-\theta)|) m(du).$$

A similar computation shows that

(2.2)

$$\int_{\mathbf{R}^d} g(x) P^s(dx) = \int_{\mathbf{R}^d} M_g(\theta_0; V_0; x) P(dx) = \int_{\mathbf{R}^d} M_g(\theta_0; V_0; x) P^s(dx).$$

Let

$$\Gamma(g;\theta;V) := \int_{\mathbf{R}^d} M_g(\theta; V; x) P(dx).$$

Given a class $\mathcal{G}$ of functions, define

$$\mathcal{G}^2 := \{gh : g \in \mathcal{G}, h \in \mathcal{G}\},$$

$$\mathcal{M}(\mathcal{G}) := \{M_g(\theta, V, \cdot) : g \in \mathcal{G}, \theta \in \mathbf{R}^d, V \in \mathcal{S}P\}.$$

We will use the following lemma that describes smoothness properties of the functions introduced above. We skip its rather lengthy and technical proof. The assumption (S1) above is used mostly in the proof of differentiability properties below; the assumption (S2) is needed to prove continuity statements (such as (C1) and (C6)).

Lemma 1. *Suppose that P and $\mathcal{F}$ satisfy the smoothness conditions (S). Then the following statements hold:*
(C1) If $\theta \to \theta_0$ and $V \to V_0$, then

$$\sup_{f\in\mathcal{F}} \int_{\mathbf{R}^d} |\tilde{f}_{\theta,V} - \tilde{f}_{\theta_0,V_0}|^2 P(dx) \to 0$$

and

$$\sup_{f\in\mathcal{F}} \int_{\mathbf{R}^d} |\tilde{f}_{\theta,V} - \tilde{f}_{\theta_0,V_0}|^2 P^s(dx) \to 0.$$

(C2) The function $\mathbf{E}(f;\theta,V)$ is differentiable at the point (θ_0, V_0) for any $f \in \mathcal{F}$, and the Taylor expansion of the first order

$$\mathbf{E}(f;\theta,V) = \mathbf{E}(f;\theta_0,V_0) + \mathbf{E}'_\theta(f;\theta_0,V_0)(\theta-\theta_0) + \mathbf{E}'_V(f;\theta_0,V_0)(V-V_0)$$

$$+o(|\theta - \theta_0| + |V - V_0|)$$

holds uniformly in $f \in \mathcal{F}$.
(C3) Similarly, the function $\mathbf{E}^{\mathbf{s}}(f;\theta, V)$ *is differentiable at the point* (θ_0, V_0) *for any* $f \in \mathcal{F}$, *and the Taylor expansion of the first order*

$$\mathbf{E}^{\mathbf{s}}(f;\theta, V) = \mathbf{E}^{\mathbf{s}}(f;\theta_0, V_0) + (\mathbf{E}^{\mathbf{s}})'_\theta(f;\theta_0, V_0)(\theta - \theta_0)$$

$$+(\mathbf{E}^{\mathbf{s}})'_V(f;\theta_0, V_0)(V - V_0) + o(|\theta - \theta_0| + |V - V_0|)$$

holds uniformly in $f \in \mathcal{F}$.
(C4) The function $\Gamma(g;\theta;V)$ *is continuous with respect to* (θ, V) *at* (θ_0, V_0) *uniformly in* $g \in ((\tilde{\mathcal{F}})_{\mathrm{aff}})^2$.
(C5) The function $\Gamma(g;\theta, V)$ *is differentiable at the point* (θ_0, V_0) *for any* $g \in (\tilde{\mathcal{F}})_{\mathrm{aff}}$, *and, moreover, the Taylor expansion of the first order*

$$\Gamma(g;\theta, V) = \Gamma(g;\theta_0, V_0) + \Gamma'_\theta(g;\theta_0, V_0)(\theta - \theta_0) + \Gamma'_V(g;\theta_0, V_0)(V - V_0)$$

$$+o(|\theta - \theta_0| + |V - V_0|)$$

holds uniformly in $g \in (\tilde{\mathcal{F}})_{\mathrm{aff}}$.
The matrix-valued functions $(\theta, V) \mapsto \Gamma'_\theta((\tilde{f})_{\theta,V};\theta_0, V_0)$ *and* $(\theta, V) \mapsto \Gamma'_V((\tilde{f})_{\theta,V};\theta_0, V_0)$ *are continuous at* (θ_0, V_0) *uniformly in* $f \in \mathcal{F}$.
(C6) if $\theta \to \theta_0$ *and* $V \to V_0$, *then for all* $\delta > 0$,

$$\sup\{\int_{\mathbf{R}^d} |M_{\tilde{f}_{t,W}}(\theta;V;x) - M_{\tilde{f}_{t,W}}(\theta_0;V_0;x)|^2 P(dx) : t \in \mathbf{R}^d, |t - \theta_0| \le \delta,$$

$$W \in \mathcal{SP}, \quad |W - V_0| \le \delta\} \to 0$$

and

$$\sup_{f\in\mathcal{F}} \int_{\mathbf{R}^d} |M_{(\tilde{f})_{\theta,V}}(\theta_0;V_0;x) - M_{(\tilde{f})_{\theta_0,V_0}}(\theta_0;V_0;x)|^2 P(dx) \to 0.$$

We also need the following lemma to prove Theorems 1 and 2.

Lemma 2. *For a uniformly bounded semialgebraic subgraph class* $\mathcal{F}$, *the classes* $(\tilde{\mathcal{F}})_{\mathrm{aff}}$, $\mathcal{M}((\tilde{\mathcal{F}})_{\mathrm{aff}})$ *and* $\mathcal{M}((\tilde{\mathcal{F}})^2_{\mathrm{aff}})$ *are uniformly Donsker and uniformly Glivenko-Cantelli.*

Proof. We will give here only the proof for the class $(\tilde{\mathcal{F}})_{\mathrm{aff}}$. Since $\mathcal{F}$ is a semi-algebraic subgraph class, so is the class

$$\mathcal{F}_{\mathrm{aff}} = \{f(V^{-1}(\cdot - \theta)) : f \in \mathcal{F},\ \theta \in \mathbf{R}^d, V \in \mathcal{SP}\}.$$

It follows that the last class is VC-subgraph. Consider now the class

$$\mathcal{H} := \{m_f(|V^{-1}(\cdot - \theta)|) : f \in \mathcal{F},\ \theta \in \mathbf{R}^d, V \in \mathcal{SP}\}.$$

Let

$$g_{f,\theta,V}(u,x) := f(|V^{-1}(x-\theta)|u) \quad \text{for} \quad f \in (\tilde{\mathcal{F}})_{\text{aff}}, \theta \in \mathbf{R}^d, V \in \mathcal{SP},$$

$$\mathcal{G} = \{g_{f,\theta,V} : f \in \mathcal{F}, \theta \in \mathbf{R}^d, V \in \mathcal{SP}(\mathbf{R}^d)\},$$

and

$$\mathcal{G}^\bullet = \{g(u,\cdot) : g \in \mathcal{G}, u \in S^{d-1}\},$$

$$\mathcal{G}_\bullet = \{g(\cdot,x) : g \in \mathcal{G}, x \in \mathbf{R}^d\}.$$

We will prove that
(a) $\mathcal{G}^\bullet$ is a VC-subgraph class
and that
(b) $\mathcal{H}$ belongs to the closed convex hull of $\mathcal{G}^\bullet$.

Both (a) and (b) follow from the fact that the classes $\mathcal{G}_\bullet$ and $\mathcal{G}^\bullet$ are semi-algebraic subgraphs. Let us establish this fact, for instance, for the class $\mathcal{G}^\bullet$. To this end, consider the set

$$\{(x,t) : f(|V^{-1}(x-\theta)|u) \geq t \geq 0\} \cup \{(x,t) : f(|V^{-1}(x-\theta)|u) \leq t \leq 0\}$$

for a function f from the class $\mathcal{F}$. By definition of the semi-algebraic subgraph property, for fixed finite l, r, this set belongs to the set algebra generated by at most l sets of the form

$$\{(x,t) : p(|V^{-1}(x-\theta)|u, t) \geq 0\},$$

where p is a polynomial of degree less than or equal to r. Note that $p(|V^{-1}(x-\theta)|u,t)$ can be written as $b(x,t)|V^{-1}(x-\theta)| - a(x,t)$ for some polynomials $a(x,t), b(x,t)$ of degree $\leq 2r$. Thus, we can consider the set

$$\{(x,t) : b(x,t)|V^{-1}(x-\theta)| - a(x,t) \geq 0\},$$

which is the union of the following three sets:

$$\{(x,t) : b(x,t) > 0, a(x,t) > 0, |V^{-1}(x-\theta)|^2 b^2(x,t) - a^2(x,t) \geq 0\},$$

$$\{(x,t) : b(x,t) > 0, a(x,t) \leq 0\},$$

$$\{(x,t) : b(x,t) \leq 0, a(x,t) \leq 0, |V^{-1}(x-\theta)|^2 b^2(x,t) - a^2(x,t) \leq 0\}.$$

Hence, the class of sets

$$\Big\{\{(x,t) : f(|V^{-1}(x-\theta)|u) \geq t \geq 0\} \cup \{(x,t) : f(|V^{-1}(x-\theta)|u) \leq t \leq 0\} :$$

$$f \in \mathcal{F}, \theta \in \mathbf{R}^d, V \in \mathcal{SP}\Big\}$$

is semi-algebraic subgraph. Therefore, (a) holds. To prove (b), note that $\mathcal{G}_\bullet$ is an m-Glivenko-Cantelli class of functions on S^{d-1} (since it is a semi-algebraic subgraph and, hence, a VC–subgraph). This implies that for an i.i.d. sequence $\{U_k\}$ of uniformly distributed vectors on S^{d-1}, we have

$$\sup_{x\in\mathbf{R}^d}\left|n^{-1}\sum_{k=1}^{n} g_{f,\theta,V}(U_k;x) - m_f(|V^{-1}(x-\theta)|)\right| \to 0 \text{ as } n\to\infty \text{ a.s..}$$

Since

$$n^{-1}\sum_{k=1}^{n} g_{f,\theta,V}(U_k;\cdot) \in \operatorname{conv}(\mathcal{G}^\bullet),$$

(b) follows.

Now (a) implies that for some $\alpha > 0$,

$$\sup_{Q\in\mathcal{P}(\mathbf{R}^d)} N_{d_{Q,2}}(\mathcal{G}^\bullet;\varepsilon) = O(\varepsilon^{-\alpha})$$

and, by a well-known property of the uniform entropy of convex hulls (due, essentially, to Dudley, see, e.g., van der Vaart and Wellner (1996), p.142), (b) implies that

$$\sup_{Q\in\mathcal{P}(\mathbf{R}^d)} H_{d_{Q,2}}(\mathcal{H};\varepsilon) \le \sup_{Q\in\mathcal{P}(\mathbf{R}^d)} H_{d_{Q,2}}(\operatorname{conv}(\mathcal{G}^\bullet);\varepsilon) = O(\varepsilon^{-\frac{2\alpha}{\alpha+2}}).$$

Here $d_{Q,2}$ denotes the metric of the space $L_2(\mathbf{R}^d;Q)$, $N_{d_{Q,2}}(\mathcal{F};\varepsilon)$ is the minimal number of balls of radius ε with respect to the metric $d_{Q,2}$, covering the class $\mathcal{F}$, $H_{d_{Q,2}}(\mathcal{F};\varepsilon) = \log N_{d_{Q,2}}(\mathcal{F};\varepsilon)$, and $\mathcal{P}(\mathbf{R}^d)$ denotes the set of all Borel probability measures on $\mathbf{R}^d$. Therefore, we get

$$\int_0^\infty \sup_{Q\in\mathcal{P}(\mathbf{R}^d)} H^{1/2}_{d_{Q,2}}(\mathcal{H};\varepsilon)d\varepsilon < +\infty.$$

Since $\mathcal{F}_{\text{aff}}$ is a VC-subgraph, we also have

$$\int_0^\infty \sup_{Q\in\mathcal{P}(\mathbf{R}^d)} H^{1/2}_{d_{Q,2}}(\mathcal{F}_{\text{aff}};\varepsilon)d\varepsilon < +\infty,$$

and the representation

$$\tilde{f}_{\theta,V}(x) = f(V^{-1}(x-\theta)) - m_f(|V^{-1}(x-\theta)|)$$

implies that

$$\int_0^\infty \sup_{Q\in\mathcal{P}(\mathbf{R}^d)} H^{1/2}_{d_{Q,2}}((\tilde{\mathcal{F}})_{\text{aff}};\varepsilon)d\varepsilon < +\infty.$$

It follows that $(\tilde{\mathcal{F}})_{\text{aff}}$ is uniformly Donsker. This completes the proof for the class $(\tilde{\mathcal{F}})_{\text{aff}}$. The proof for the two remaining classes follows the same lines, but is a bit more involved.

Proof of Theorem 1. Define a process

$$\eta_n(f;\theta;V) := n^{1/2}(P_n - P)((\tilde{f})_{\theta,V}), f \in \mathcal{F}, \theta \in \mathbf{R}^d, V \in \mathcal{S}P.$$

Since, by Lemma 2, $(\tilde{\mathcal{F}})_{\text{aff}}$ is a P-Donsker class and the statement (C1) of Lemma 1 holds, we can use asymptotic equicontinuity to get for all $\varepsilon > 0$

$$\lim_{\delta\to 0}\limsup_{n\to\infty} \mathbb{P}^*(\{\sup_{|\theta-\theta_0|\le\delta}\sup_{|V-V_0|\le\delta}\sup_{f\in\mathcal{F}}|\eta_n(f;\theta;V)-\eta_n(f;\theta_0;V_0)| \ge \varepsilon\}) = 0.$$

Clearly,

$$\begin{aligned}&\xi_n(f) - n^{1/2}\mathbf{E}(f;\theta_0;V_0)\\ &= \eta_n(f;\theta_n;V_n) + n^{1/2}\big(\mathbf{E}(f;\theta_n;V_n) - \mathbf{E}(f;\theta_0;V_0)\big) = \eta_n(f;\theta_0,V_0)\\ &+n^{1/2}(\mathbf{E}(f;\theta_n,V_n) - \mathbf{E}(f;\theta_0,V_0)) + (\eta_n(f;\theta_n,V_n) - \eta_n(f;\theta_0,V_0)).\end{aligned}$$

If $|\theta_n - \theta_0| \le \delta$ and $|V_n - V_0| \le \delta$, then

$$\begin{aligned}&\sup_{f\in\mathcal{F}}|\xi_n(f) - n^{1/2}\mathbf{E}(f;\theta_0;V_0) - \eta_n(f;\theta_0,V_0)\\ &-n^{1/2}(\mathbf{E}(f;\theta_n,V_n) - \mathbf{E}(f;\theta_0,V_0))|\\ &\le \sup_{|\theta-\theta_0|\le\delta}\sup_{|V-V_0|\le\delta}\sup_{f\in\mathcal{F}}|\eta_n(f;\theta;V) - \eta_n(f;\theta_0;V_0)|. \qquad (2.3)\end{aligned}$$

Note that

$$\eta_n(f;\theta_0,V_0) = n^{1/2}\int_{\mathbf{R}^d}(\tilde{f})_{\theta_0,V_0}d(P_n - P).$$

Using (2.3) and consistency of (θ_n, V_n) we get (uniformly in $f \in \mathcal{F}$)

$$\begin{aligned}\xi_n(f) - n^{1/2}\mathbf{E}(f;\theta_0;V_0) &= n^{1/2}\int_{\mathbf{R}^d}(\tilde{f})_{\theta_0,V_0}d(P_n - P)\\ &+n^{1/2}(\mathbf{E}(f;\theta_n,V_n) - \mathbf{E}(f;\theta_0,V_0)) + o_p(1). \qquad (2.4)\end{aligned}$$

It follows from the statement (C1) of Lemma 1 and $n^{1/2}$-consistency of (θ_n, V_n) that uniformly in $f \in \mathcal{F}$,

$$\begin{aligned}n^{1/2}(\mathbf{E}(f;\theta_n,V_n) - \mathbf{E}(f;\theta_0,V_0)) &= n^{1/2}\mathbf{E}'_\theta(f,\theta_0,V_0)(\theta_n - \theta_0)\\ &+n^{1/2}\mathbf{E}'_V(f,\theta_0,V_0)(V_n - V_0) + o_p(1),\ n\to\infty. \qquad (2.5)\end{aligned}$$

Representations (2.4) and (2.5), the fact that $(\tilde{\mathcal{F}})_{\text{aff}}$ is a uniform Donsker class (see Lemma 2) and the statement (C1) of Lemma 1 imply that the sequence $\xi_n(f) - n^{1/2}\mathbf{E}(f;\theta_0;V_0)$ converges weakly in the space $\ell^\infty(\mathcal{F})$ to the Gaussian stochastic process ξ_P, defined by

$$\xi_P(f) := W_P^\circ((\tilde{f})_{\theta_0,V_0}) + \mathbf{E}'_\theta(f;\theta_0,V_0)(\int_{\mathbf{R}^d} xW_P^\circ(dx))$$

$$+\mathbf{E}'_V(f;\theta_0,V_0)\Big(\int_{\mathbf{R}^d}(x-\theta_0)\otimes(x-\theta_0)W_P^\circ(dx)\Big), f\in\mathcal{F},$$

whose distribution is a Radon measure in $\ell^\infty(\mathcal{F})$ (as can easily be shown). This implies the first statement of the theorem. If P is ellipsoidally symmetric, then $\mathbf{E}(f;\theta_0;V_0)=0$ and we obtain the second statement.

Proof of Theorem 2. Define a process

$$\hat{\eta}_n(f;\theta;V) := n^{1/2}(\hat{P}_n - P_n^s)((\tilde{f})_{\theta,V}), f\in\mathcal{F}, \theta\in\mathbf{R}^d, V\in\mathcal{SP}.$$

Since, by Lemma 2, the class $\mathcal{M}(((\tilde{\mathcal{F}})_{\text{aff}})^2)$ is uniformly Glivenko-Cantelli, the statement (C4) of Lemma 1 and representations (2.1), (2.2) imply that uniformly in $g,h\in(\tilde{\mathcal{F}})_{\text{aff}}$

$$\int_{\mathbf{R}^d} ghdP_n^s \to \int_{\mathbf{R}^d} ghdP^s \text{ as } n\to\infty \text{ a.s..}$$

Similarly, since the class $\mathcal{M}((\tilde{\mathcal{F}})_{\text{aff}})$ is uniformly Glivenko-Cantelli, by the statement (C5) of Lemma 1 and representations (2.1), (2.2), we get uniformly in $g\in(\tilde{\mathcal{F}})_{\text{aff}}$,

$$\int_{\mathbf{R}^d} gdP_n^s \to \int_{\mathbf{R}^d} gdP^s \text{ as } n\to\infty \text{ a.s..}$$

Since $(\tilde{\mathcal{F}})_{\text{aff}}$ is a uniform Donsker class, we can use Corollary 2.7 in Giné and Zinn (1991) to prove that a.s. $n^{1/2}(\hat{P}_n - P_n^s)$ converges weakly in the space $\ell^\infty((\tilde{\mathcal{F}})_{\text{aff}})$ to the same limit as $n^{1/2}(\tilde{P}_n - P^s)$ ($\tilde{P}_n$ being the empirical measure based on a sample from P^s), i.e. to the P^s-Brownian bridge $W_{P^s}^\circ$. The statement (C1) of Lemma 1 and asymptotic equicontinuity allow us to establish that for all $\varepsilon>0$, $\mathbb{P}$ a.s.,

(2.6)
$$\lim_{\delta\to 0}\limsup_{n\to\infty}\hat{\mathbb{P}}^*\Big(\Big\{\sup_{|\theta-\theta_n|\le\delta}\sup_{|V-V_n|\le\delta}\sup_{f\in\mathcal{F}}|\hat{\eta}_n(f;\theta;V)-\hat{\eta}_n(f;\theta_n;V_n)|\ge\varepsilon\Big\}\Big)=0.$$

Define

$$\hat{\mathbf{E}}_n(f;\theta,V) := \int_{\mathbf{R}^d}(\tilde{f})_{\theta,V}dP_n^s.$$

Since

$$\hat{\mathbf{E}}_n(f;\theta_n,V_n)=0,$$

we can write

$$\hat{\xi}_n(f)=\hat{\eta}_n(f;\hat{\theta}_n;\hat{V}_n)+n^{1/2}\hat{\mathbf{E}}_n(f;\hat{\theta}_n;\hat{V}_n)=\hat{\eta}_n(f;\theta_n,V_n)$$

$$+n^{1/2}(\hat{\mathbf{E}}_n(f;\hat{\theta}_n,\hat{V}_n)-\hat{\mathbf{E}}_n(f;\theta_n,V_n))+(\hat{\eta}_n(f;\hat{\theta}_n,\hat{V}_n)-\hat{\eta}_n(f;\theta_n,V_n)).$$

If $|\hat{\theta}_n - \theta_n| \le \delta$ and $|\hat{V}_n - V_n| \le \delta$, then

$$\sup_{f\in\mathcal{F}} |\hat{\xi}_n(f) - \hat{\eta}_n(f;\theta_n, V_n) - n^{1/2}(\hat{\mathbf{E}}_n(f;\hat{\theta}_n,\hat{V}_n) - \hat{\mathbf{E}}_n(f;\theta_n,V_n))|$$

$$\le \sup_{|\theta-\theta_n|\le\delta} \sup_{|V-V_n|\le\delta} \sup_{f\in\mathcal{F}} |\hat{\eta}_n(f;\theta;V) - \hat{\eta}_n(f;\theta_n;V_n)|. \tag{2.7}$$

Note that

$$\hat{\eta}_n(f;\theta_n,V_n) = n^{1/2}\int_{\mathbf{R}^d} (\tilde{f})_{\theta_n,V_n} d(\hat{P}_n - P_n^s).$$

Using (2.6), (2.7) and standard asymptotic properties of the estimators (θ_n, V_n), we get (uniformly in $f \in \mathcal{F}$)

$$\hat{\xi}_n(f) = n^{1/2}\int_{\mathbf{R}^d} (\tilde{f})_{\theta_n,V_n} d(\hat{P}_n - P_n^s)$$

$$+n^{1/2}(\hat{\mathbf{E}}_n(f;\hat{\theta}_n,\hat{V}_n) - \hat{\mathbf{E}}_n(f;\theta_n,V_n)) + o_p(1). \tag{2.8}$$

[Here and in what follows the remainder $o_p(1)$ converges to 0 uniformly in $f \in \mathcal{F}$ in probability $\mathbb{P}\times\hat{\mathbb{P}}$].

Applying the asymptotic equicontinuity condition to the process $n^{1/2}(\hat{P}_n - P_n^s)$ and using the statement (C1) of Lemma 1, we get

$$n^{1/2}\int_{\mathbf{R}^d} (\tilde{f})_{\theta_n,V_n} d(\hat{P}_n - P_n^s) = n^{1/2}\int_{\mathbf{R}^d} (\tilde{f})_{\theta_0,V_0} d(\hat{P}_n - P_n^s) + o_p(1) \tag{2.9}$$

as $n \to \infty$. Now we can write

$$n^{1/2}(\hat{\mathbf{E}}_n(f;\hat{\theta}_n,\hat{V}_n) - \hat{\mathbf{E}}_n(f;\theta_n,V_n))$$

$$= n^{1/2}(\hat{\mathbf{E}}_n(f;\hat{\theta}_n,\hat{V}_n) - \mathbf{E}^{\mathrm{s}}(f;\hat{\theta}_n,\hat{V}_n))$$

$$-n^{1/2}(\hat{\mathbf{E}}_n(f;\theta_n,V_n) - \mathbf{E}^{\mathrm{s}}(f;\theta_n,V_n))$$

$$+n^{1/2}(\mathbf{E}^{\mathrm{s}}(f;\hat{\theta}_n,\hat{V}_n) - \mathbf{E}^{\mathrm{s}}(f;\theta_n,V_n)). \tag{2.10}$$

Note that by (2.1) and (2.2)

$$n^{1/2}(\hat{\mathbf{E}}_n(f;\theta,V) - \mathbf{E}^{\mathrm{s}}(f;\theta,V)) = n^{1/2}(P_n - P)(M_{(\tilde{f})_{\theta,V}}(\theta_n;V_n;\cdot))$$

$$+n^{1/2}(\Gamma((\tilde{f})_{\theta,V};\theta_n;V_n) - \Gamma((\tilde{f})_{\theta,V};\theta_0;V_0))$$

$$= n^{1/2}(P_n - P)(M_{(\tilde{f})_{\theta,V}}(\theta_0;V_0;\cdot))$$

$$+\big[n^{1/2}(P_n - P)(M_{(\tilde{f})_{\theta,V}}(\theta_n;V_n;\cdot)) - n^{1/2}(P_n - P)(M_{(\tilde{f})_{\theta,V}}(\theta_0;V_0;\cdot))\big]$$

$$+n^{1/2}(\Gamma((\tilde{f})_{\theta,V};\theta_n;V_n) - \Gamma((\tilde{f})_{\theta,V};\theta_0;V_0)).$$

Since, by Lemma 2, $\mathcal{M}((\tilde{\mathcal{F}})_{\mathrm{aff}})$ is uniformly Donsker and the statements (C5), (C6) of Lemma 1 hold, it is easy to prove the weak convergence of the processes

$$\big\{n^{1/2}(\hat{\mathbf{E}}_n(f;\theta,V) - \mathbf{E}^{\mathrm{s}}(f;\theta,V)),\ f\in\mathcal{F},\ \theta\in B(\theta_0),\ V\in B(V_0)\big\}$$

in the space $\ell^\infty(\mathcal{F} \times B(\theta_0) \times B(V_0))$, where $B(\theta_0)$ is a ball in $\mathbf{R}^d$ with the center θ_0 and $B(V_0)$ is a ball in $\mathcal{SP}$ with the center V_0. Using the asymptotic equicontinuity and the statement (C6) of Lemma 1, we get

$$\begin{aligned} & n^{1/2}(\hat{\mathbf{E}}_n(f;\hat{\theta}_n,\hat{V}_n) - \mathbf{E}^s(f;\hat{\theta}_n,\hat{V}_n)) \\ & -n^{1/2}(\hat{\mathbf{E}}_n(f;\theta_n,V_n) - \mathbf{E}^s(f;\theta_n,V_n)) = o_p(1) \text{ as } n \to \infty. \end{aligned} \tag{2.11}$$

It follows from the statement (C3) of Lemma 1 and standard asymptotic properties of the estimators (θ_n, V_n), $(\hat{\theta}_n, \hat{V}_n)$ that uniformly in $f \in \mathcal{F}$

$$\begin{aligned} & n^{1/2}(\mathbf{E}^s(f;\hat{\theta}_n,\hat{V}_n) - \mathbf{E}^s(f;\theta_n,V_n)) = n^{1/2}(\mathbf{E}^s)'_\theta(f,\theta_0,V_0)(\hat{\theta}_n - \theta_n) \\ & + n^{1/2}(\mathbf{E}^s)'_V(f,\theta_0,V_0)(\hat{V}_n - V_n) + o_p(1) \\ & = (\mathbf{E}^s)'_\theta(f,\theta_0,V_0)\big(n^{1/2}\int_{\mathbf{R}^d} x d(\hat{P}_n - P_n^s)\big) \\ & + (\mathbf{E}^s)'_V(f,\theta_0,V_0)n^{1/2}\Big(\int_{\mathbf{R}^d}(x \otimes x)d(\hat{P}_n - P_n^s) \\ & - \int_{\mathbf{R}^d} x d(\hat{P}_n - P_n^s) \otimes \int_{\mathbf{R}^d} x dP_n^s \\ & - \int_{\mathbf{R}^d} x dP_n^s \otimes \int_{\mathbf{R}^d} x d(\hat{P}_n - P_n^s)\Big) + o_p(1), \ n \to \infty. \end{aligned} \tag{2.12}$$

Relationships (2.8)-(2.12) along with, again, Corollary 2.7 in Giné and Zinn (1991), imply the statement of the theorem.

Acknowledgement. We would like to thank the referee for a number of comments and suggestions that improved the paper.

References

[1] Arcones, M. and Giné, E. (1991) Some bootstrap tests for symmetry for univariate distributions. *Ann. Statist.* **19**, 1496–1511.

[2] Baringhaus, L. (1991) Testing for spherical symmetry of a multivariate distribution. *Ann. Statist.* **19**, 899–917.

[3] Beran, R. (1979) Testing for ellipsoidal symmetry of a multivariate density. *Ann. Statist.* **7**, 150–162.

[4] Cirel'son, B. (1975) The density of the distribution of the maximum of a Gaussian process. *Theory of Probability and Appl.* **20**, 847–855.

[5] Dudley, R.M. (1999) Uniform Central Limit Theorem. Cambridge University Press, Cambridge.

[6] Giné, E. and Zinn, J. (1990) Bootstrapping general empirical measures. *Ann. Probab.* **18**, 851–869.

[7] Giné, E. and Zinn, J. (1991) Gaussian characterization of uniform Donsker classes. *Ann. Probab.* **19**, 758–782.

[8] Heathcote, C.R., Rachev, S.T. and Cheng, B. (1995) Testing Multivariate Symmetry. *J. Multivariate Analysis* **54**, 91–112.

[9] Koldobskii, A.L. (1990) Inverse problem for potentials of measures in Banach spaces. In: *Probab. Theory and Math. Stat.*, B. Grigelionis, Yu. Prohorov, V. Sazonov et al (Eds), Vol. 1, pp. 627–637, VSP-Mokslas, Vilnius-Utrecht.

[10] Koltchinskii, V. and Li, L. (1998) Testing for spherical symmetry of a multivariate distribution. *J. Multivariate Analysis* **65**, 228–244.

[11] Quiroz, A.J. and Dudley, R. (1991) Some New Tests for Multivariate Normality. *Probab. Theory and Related Fields.* **87**, 521–546.

[12] Quiroz, A.J., Manzotti, A. and Pérez, F.J. (1999) A bootstrap procedure for testing the null hypothesis of elliptical symmetry. Preprint.

[13] Romano, J. (1989) Bootstrap and randomization tests of some nonparametric hypotheses. *Ann. Statist.* **17**, 141–159.

[14] Shorack, G. and Wellner, J. (1986) Empirical Processes with Applications to Statistics. Wiley, New York.

[15] van der Vaart, A.W. and Wellner, J. (1996) Weak Convergence and Empirical Processes with Applications to Statistics. Springer-Verlag, New York.

Vladimir Koltchinskii and Lyudmila Sakhanenko
Department of Mathematics and Statistics
The University of New Mexico
Albuquerque, NM 87131-1141
e-mail: `vlad@math.unm.edu`
e-mail: `sakhlyu@math.unm.edu`

Progress in Probability

Editors

Professor Thomas M. Liggett
Department of Mathematics
University of California
Los Angeles, CA 90024-1555

Professor Charles Newman
Courant Institute of
Mathematical Sciences
251 Mercer Street
New York, NY 10012

Professor Loren Pitt
Department of Mathematics
University of Virginia
Charlottesville, VA 22903-3199

Progress in Probability is designed for the publication of workshops, seminars and conference proceedings on all aspects of probability theory and stochastic processes, as well as their connections with and applications to other areas such as mathematical statistics and statistical physics. It acts as a companion series to *Probability and Its Applications,* a context for research level monographs and advanced graduate texts.

We encourage preparation of manuscripts in some form of TeX for delivery in camera-ready copy, which leads to rapid publications, or in electronic form for interfacing with laser printers or typesetters.

Proposals should be sent directly to the editors or to:
Birkhäuser Boston, 675 Massachusetts Avenue, Cambridge, MA 02139, U.S.A.

12 ÇINLAR/CHUNG/GETOOR. Seminar on Stochastic Processes, 1985
13 ÇINLAR/CHUNG/GETOOR/GLOVER. Seminar on Stochastic Processes, 1986
14 DEVROYE. A Course in Density Estimation
15 ÇINLAR/CHUNG/GETOOR/GLOVER. Seminar on Stochastic Processes, 1987
16 KIPER. Random Perturbations of Dynamical Systems
17 ÇINLAR/CHUNG/GETOOR/GLOVER. Seminar on Stochastic Processes, 1988
18 ÇINLAR/CHUNG/GETOOR/FITZSIMMONS/WILLIAMS. Seminar on Stochastic Processes, 1989
19 ALEXANDER/WATKINS. Spatial Stochastic Processes: A Festschrift in Honor of Ted Harris on his 70th Birthday
20 HAAGERUP/HOFFMANN-JØRGENSEN/NIELSEN. Probability in Banach Spaces 6: Proceedings of the Sixth International Conference, Sandbjerg, Denmark 1986
21 EBERLEIN/KUELBS/MARCUS. Probability in Banach Spaces 7: Proceedings of the Seventh International Congress
22 PINSKY. Diffusion Processes and Related Problems in Analysis, Vol. I: Diffusions in Analysis and Geometry
23 HAHN/MASON/WEINER. Sums, Trimmed Sums and Extremes
24 ÇINLAR/CHUNG. Seminar on Stochastic Processes, 1990
25 CAMBANIS/SAMORODNITSKY/TAQQU. Stable Processes and Related Topics
26 CRUZEIRO/ZAMBRINI. Stochastic Analysis and Applications: Proceedings of the 1989 Lisbon Conference
27 PINSKY/WIHSTUTZ. Diffusion Processes and Related Problems in Analysis, Vol. II. Stochastic Flows
28 DURRETT/KESTEN. Random Walks, Brownian Motion and Interacting Particle Systems

29 ÇINLAR/CHUNG/SHARPE. Seminar on Stochastic Processes, 1991
30 DUDLEY/HAHN/KUELBS. Probability in Banach Spaces 8, 1992
31 KÖREZLIOĞLU/ÜSTÜNEL. Stochastic Analysis and Related Topics, 1992
32 NUALART/SOLE. Barcelona Seminar on Stochastic Analysis: St. Feliu de Guixols, 1991
33 ÇINLAR/CHUNG/SHARPE. Seminar on Stochastic Processes, 1992
34 FRIEDLIN. The Dynkin Festschrift: Markov Processes and their Applications, 1994
35 HOFFMAN-JØRGENSEN/KUELBS/MARCUS. Probability in Banach Spaces 9, 1993
36 BOLTHAUSEN/DOZZI/RUSSO. Seminar on Stochastic Analysis, Random Fields and Applications, 1993
37 BANDT/GRAF/ZÄHLE. Fractal Geometry and Stochastics, 1995
38 KÖREZLIOĞLU/ØKSENDAL/ÜSTÜNEL. Stochastic Analysis and Related Topics V: The Silivri Workshop, 1994
39 ADLER/MÜLLER/ROZOVSKII. Stochastic Modelling in Physical Oceanography
40 CHAUVIN/COHEN/ROUAULT. Trees: Workshop in Versailles, June 14–16, 1995
41 BOVIER/PICCO. Mathematical Aspects of Spin Glasses and Neural Networks.
42 DECREUSEFOND/GJERDE/ØKSENDAL/ÜSTÜNEL. Stochastic Analysis and Related Topics VI: The Geilo Workshop, 1996
43 EBERLEIN/HAHN/TALAGERAND. High Dimensional Probability
44 BRAMSON/DURRETT. Perplexing Problems in Probability: Festschrift in Honor of Harry Kesten
45 DALANG/DOZZI/RUSSO. Seminar on Stochastic Analysis, Random Fields and Applications, 1996
46 BANDT/GRAF/ZÄHLE. Fractal Geometry and Stochastics II
47 GINÉ/MASON/WELLNER. High Dimensional Probability II